MOLECULAR MECHANISMS OF DEMENTIA

MOLECULAR MECHANISMS OF DEMENTIA

Biomarkers, Neurochemistry, and Therapy

AKHLAQ A. FAROOQUI
Department of Molecular and Cellular Biochemistry, The Ohio State University, Columbus, OH, United States

ACADEMIC PRESS
An imprint of Elsevier
ELSEVIER

Academic Press is an imprint of Elsevier
125 London Wall, London EC2Y 5AS, United Kingdom
525 B Street, Suite 1650, San Diego, CA 92101, United States
50 Hampshire Street, 5th Floor, Cambridge, MA 02139, United States
The Boulevard, Langford Lane, Kidlington, Oxford OX5 1GB, United Kingdom

British Library Cataloguing-in-Publication Data
A catalogue record for this book is available from the British Library

Library of Congress Cataloging-in-Publication Data
A catalog record for this book is available from the Library of Congress

ISBN: 978-0-12-816347-4

For Information on all Academic Press publications
visit our website at https://www.elsevier.com/books-and-journals

Publisher: Nikki Levy
Acquisition Editor: Melanie Tucker
Editorial Project Manager: Kristi Anderson
Production Project Manager: Sujatha Thirugnana Sambandam
Cover Designer: Miles Hitchen

Typeset by MPS Limited, Chennai, India

Dedication

This monograph is dedicated to my beloved father, the "late Sharafyab Ahmed Sahab," whose guidance and influence continue to inspire and support me.

Akhlaq A. Farooqui

Contents

About the Author

Dr. Akhlaq A. Farooqui is a leader in the field of signal transduction, brain phospholipases A_2, bioactive ether lipid metabolism, polyunsaturated fatty acid metabolism, glycerophospholipid-, sphingolipid-, and cholesterol-derived lipid mediators, glutamate-induced neurotoxicity, and modulation of signal transduction by phytochemicals. Dr. Farooqui has discovered the stimulation of plasmalogen-selective phospholipase A_2 (PlsEtn-PLA_2) and diacyl- and monoacylglycerol lipases in the brains of patients with Alzheimer's disease. Stimulation of PlsEtn-PLA_2 produces plasmalogen deficiency and increases levels of eicosanoids that may be related to the loss of synapses in brains of patients with Alzheimer's disease. Dr. Farooqui has published cutting edge research on the generation and identification of glycerophospholipid-, sphingolipid-, and cholesterol-derived lipid mediators in kainic acid-mediated neurotoxicity by lipidomics. Dr. Farooqui has authored numerous monographs: *Glycerophospholipids in Brain: Phospholipase A_2 in Neurological Disorders* (2007); *Neurochemical Aspects of Excitotoxicity* (2008); *Metabolism and Functions of Bioactive Ether Lipids in Brain* (2008); and *Hot Topics in Neural Membrane Lipidology* (2009); *Beneficial Effects of Fish Oil in Human Brain* (2009); *Neurochemical Aspects of Neurotraumatic and Neurodegenerative Diseases* (2010); *Lipid Mediators and Their Metabolism in the Brain* (2011); *Phytochemicals, Signal Transduction, and Neurological Disorders* (2012); *Metabolic Syndrome: An Important Risk Factor for Stroke, Alzheimer Disease, and Depression* (2013); *Inflammation and Oxidative Stress in Neurological Disorders* (2014); *High Calorie Diet and the Human Brain* (2015); and *Therapeutic Potentials of Curcumin for Alzheimer Disease* (2016). All of the 12 monographs above were published by Springer, New York and Springer International Publishing Switzerland. Monographs on *Neurochemical Aspects of Alzheimer's Disease* (2017) and *Ischemic, Traumatic Brain, and Spinal Cord Injuries: Mechanisms and Potential Therapies* (2018) were published by Academic Press: an imprint of Elsevier, San Diego, CA.

In addition, Dr. Akhlaq A. Farooqui has edited 11 books: *Biogenic Amines: Pharmacological, Neurochemical and Molecular Aspects in the CNS* (2010), Nova Science Publisher, Hauppauge, NY; *Molecular Aspects of Neurodegeneration and Neuroprotection* (2011), Bentham Science Publishers Ltd; *Phytochemicals and Human Health: Molecular and Pharmacological*

Aspects (2011), Nova Science Publisher, Hauppauge, NY; *Molecular Aspects of Oxidative Stress on Cell Signaling in Vertebrates and Invertebrates* (2012), Wiley Blackwell Publishing Company, New York; *Beneficial Effects of Propolis on Human Health in Chronic Diseases*, volume 1 (2012), Nova Science Publishers, Hauppaauge, NY; *Beneficial Effects of Propolis on Human Health in Chronic Diseases*, volume 2 (2012), Nova Science Publishers, Hauppaauge, NY; *Metabolic Syndrome and Neurological Disorders* (2013), Wiley Blackwell Publishing Company, New York; *Diet and Exercise in Cognitive Function and Neurological Diseases* (2015), Wiley Blackwell Publishing Company, New York; *Trace Amines and Neurological Disorders: Potential Mechanisms and Risk Factors* (2016), Elsevier, New York; *Neuroprotective Effects of Phytochemicals in Neurological Disorders* (2017), Wiley-Blackwell, John Wiley and Sons, Inc., Hoboken, NJ; and *Role of the Mediterranean Diet in the Brain and Neurodegenerative Diseases* (2018), Elsevier, New York.

Preface

Dementia is a progressive neurodegenerative syndrome, which is characterized by the deterioration in cognitive function (i.e., the ability to process thought) beyond what might be expected from normal aging. Dementia affects memory, thinking, orientation, comprehension, calculation, learning capacity, language, and judgment. Dementia is accompanied by the decrease in blood flow, imbalances in neurotransmitters and neurotrophins, changes in brain volume, and neurogenesis dysfunction. The increase in life expectancy (aging) is a major risk factor for dementia. Dementia is linked with many neurological disorders, such as traumatic brain injury, stroke, Alzheimer's disease (AD), and Parkinson's disease. In 2010, there were 35.6 million people suffering from dementia worldwide and this number is expected to increase twofold by 2030 and threefold by 2050. Although younger-onset dementia is being increasingly recognized, dementia is most commonly a disease that affects the elderly (age 65–85). Dementia causes great physical, emotional, and social stress because the caregiving process is long in duration, unfamiliar, unpredictable, and ambiguous. In the later stages of dementia, many family caregivers relocate their relative to a residential aged care facility, most often when the burden of care outweighs the means of the caregiver. The total estimated worldwide cost for dementia in 2010 was USD 604 billion, representing about 1% of the global gross domestic product. By 2030, the total cost for the social care of dementias is projected to increase by 85%, which amounts to approximately USD 1.117 trillion. Several types of dementia have been reported to occur in human patients including Alzheimer's type of dementia, Lewy body dementia, vascular dementia, frontotemporal dementia, and dementia as a result of neurological disorders such as stroke, AIDS, or multiple sclerosis. Among the above mentioned dementia subtypes, AD type of dementia is the most common cause of dementia (70%) and is rare before 60 years of age. AD type of dementia is characterized clinically by progressive memory and orientation loss and other cognitive deficits, including impaired judgment and decision-making, apraxia, and language disturbances. The second most frequent subtype is vascular dementia (17% of all dementia cases), and the third one is dementia with Lewy bodies (10%–25% of all dementia cases). These are then followed by frontotemporal dementia/degeneration (FTD). Among the above mentioned dementias, vascular dementia is defined as the loss of cognitive function resulting from ischemic or hemorrhagic brain lesions due to cerebrovascular disease or cardiovascular pathology. In addition to advancing age, other risk factors for dementia are long-term consumption of a high calorie diet, physical and cognitive inactivity, and epigenetic and environmental factors. In addition, dementia is also linked to (1) cardiovascular (e.g., hypertension, diabetes, obesity) and cerebrovascular problems (e.g., stroke); (2) excessive alcohol consumption; (3) social isolation; and (4) having one or two copies of the APOEϵ4 genetic variant. Information on underlying molecular mechanisms contributing to various types of dementia still remains speculative and controversial. However, the current pathogenic scenarios of different types of dementia are based on a number of common mechanisms of neurodegeneration, such as accumulation of misfolded proteins (within or outside cells), mitochondrial dysfunction, increase in oxidative stress, dysregulation in calcium homeostasis, and induction of neuroinflammation along with early synaptic disconnection and late apoptotic cell death. Clinical and preclinical studies have indicated that symptoms of dementia may be linked with alterations in neuroplasticity in corticolimbic brain regions. In particular, divergent responses have been reported not only in neuronal atrophy and loss of synapses in the prefrontal cortex and hippocampus, but also in synaptic density in the amygdala and nucleus accumbens. Based on these considerations, there is a growing consensus in recognizing dementia as a public health priority.

In the light of the above information, I have decided to provide readers with a comprehensive and cutting-edge description of risk factors, pathogenesis, biomarkers, and potential treatment strategies for dementia. This monograph has 10 chapters. The first chapter describes information on neurochemical aspects of various types dementias. Chapter 2, Neurochemical Aspects of Poststroke Dementia, describes information on neurochemical aspects of poststroke dementia. Chapter 3, Neurochemical Aspects of Alzheimer's Type of Dementia, describes information on neurochemical aspects of AD type of dementia. Chapter 4, Neurochemical Aspects of Lewy Body Dementia, describes cutting-edge information on neurochemical aspects of Lewy body dementia. Chapter 5, Neurochemical Aspects of Vascular Dementia, provides information on the neurochemical aspects of vascular dementia. Chapter 6, Neurochemical Aspects of Frontotemporal Dementia, describes the neurochemical aspects of FTD. Chapter 7, Potential Treatment Strategies for Dementia With Pharmacological and Nonpharmacological Interventions, details cutting-edge information on potential treatment strategies for dementia with pharmacological and nonpharmacological interventions. Chapter 8, Potential Treatment

Strategies for the Treatment of Dementia With Chinese Medicinal Plants, is devoted to cutting-edge information on potential strategies for the treatment of dementia with traditional Chinese medicinal plants. Chapter 9, Potential Treatment Strategies of Dementia With Ayurvedic Medicines, describes cutting-edge information on the treatment of dementia with Ayurvedic medicines. Finally, Chapter 10, Summary and Perspective for Future Research on Dementia, provides readers with a perspective that will be important for future research work on dementia. My presentation and demonstrated ability to present complicated information on signal transduction processes in dementia makes this book particularly accessible to neuroscience graduate students, teachers, and fellow researchers. It can be used as a supplemental text for a range of neuroscience courses. Clinicians, neuroscientists, neurologists, and pharmacologists will find this book useful for understanding the molecular aspects of dementias and their treatment. To the best of my knowledge no one has written a monograph on risk factors, pathogenesis, biomarkers, and treatment for dementias. This monograph is the first to provide a comprehensive description of signal transduction processes associated with the pathogenesis and treatment of dementias.

The choices of topics presented in this monograph are personal. They are based on my interest in the pathogenesis of various types of dementias. The key objective of this monograph is to critically evaluate the information on risk factors, pathogenesis, biomarkers, and potential treatment strategies for various types of dementia. Each chapter of this monograph contains an extensive list of references, which are arranged alphabetically, to works that are cited in the text. I have tried to ensure uniformity and mode of presentation as well as a logical progression of subjects from one topic to another and have provided an extensive bibliography. For the sake of simplicity and uniformity a large number of figures with chemical structures of dietary components along with line diagrams of colored signal transduction pathways are also included. I hope that my attempt to integrate and consolidate the knowledge on risk factors, pathogenesis, biomarkers, and potential treatment strategies for various types of dementias will initiate more studies on molecular mechanisms and the treatment of various types of dementias in human. This knowledge will be useful for the optimal health of young, boomer, and preboomer American generations.

Akhlaq A. Farooqui
Columbus, OH, United States

Acknowledgments

I thank my wife, Tahira, for critical reading of this monograph, offering valuable advice, useful discussion, and evaluation of subject matter. Without her help and participation, this monograph neither could nor would have been completed. I would also like to express my gratitude to Melanie Tucker and Kristi Anderson of Elsevier/Academic Press for their quick responses to my queries and professional manuscript handling.

Akhlaq A. Farooqui

List of Abbreviations

AD	Alzheimer's disease
ADDLs	Aβ-derived diffusible ligands
AGE	advanced glycation endproducts
APP	amyloid precursor protein
ARA	arachidonic acid
Aβ	β-amyloid
BBB	blood–brain barrier
CAT	catalase
COX	cyclooxygenase
CSF	cerebrospinal fluid
DHA	docosahexaenoic acid
EPA	eicosapentaenoic acid
EPOX	epoxygenase
GPx	glutathione peroxidase
GSK-3	glycogen synthase-3
IGF	insulin growth factor
IL	interleukin
Ins-1,4,5-P_3	inositol-1,4,5-trisphosphate
LOX	lipoxygenase
LTP	long-term potentiation
MetS	metabolic syndrome
MetS	metabolic syndrome
NFTs	neurofibrillary tangles
NO	nitric oxide
PET	positron-emission tomography
PKC	protein kinase C
PLA_2	phospholipase A_2
PtdCho	phosphatidylcholine
PtdEtn	phosphatidylethanolamine
PtdIns	phosphatidylinositol
PtdIns(4,5)P_2	phosphatidylinositol 4,5-bisphosphate
PtdIns4P	phoshatidylinositol 4-phosphate
RAGE	receptor for advanced glycation endproducts
RNS	reactive nitrogen species
ROS	reactive oxygen species
SOD	superoxide dismutase
SPECT	single-photon emission computed tomography
TNF-α	tumor necrosis factor-alpha

CHAPTER

1

Neurochemical Aspects of Dementia

INTRODUCTION

Due to the increase in life expectancy, the number of elderly across the globe is increasing at a constant rate and it is estimated that the number of seniors will increase to approximate 2.1 billion by the year 2050 worldwide (United Nations, 2015; WHO, 2016). In the United States, the number of seniors is predicted to increase from approximately 45 million currently to 70 million by the year 2030 (Ortman et al., 2014); similarly, in the European Union the number of seniors over the age of 80 is expected to grow from 5% to 12% of the population (The 2015 Ageing Report, 2015). In 2016, Canada had more persons over the age of 65 (16.9%) than under the age of 15 (16.6%) (Government of Canada, Statistics Canada, 2016). This global trend is driven largely by the baby boom generation, born between 1946 and 1964, which began entering their senior years in 2011. Increase in life expectancy is a major risk factor for cognitive deterioration. Dementia is a Latin word which means madness or mindlessness (De means without and ment means mind). In a medical context, dementia is not a name for a particular disease, but a progressive neurodegenerative syndrome, which is characterized by impairment in memory and activities of daily living, altered behavior, personality, and other cognitive dysfunctions (Sosa-Ortiz et al., 2012). Dementia mainly affects older people: only 2% of cases start before the age of 65 years. After this, the prevalence doubles with every 5-year increment in age. Dementia is one of the major causes of disability in later life. In 2010, there were 35.6 million people suffering from dementia worldwide and this number is projected to double over the subsequent 20 years (Prince et al., 2013a). Dementia not only results in deterioration in cognitive function (i.e., the ability to process thought) beyond what might be expected from normal aging, but also affects

DOI: https://doi.org/10.1016/B978-0-12-816347-4.00001-5

memory, thinking, orientation, comprehension, calculation, learning capacity, language, and judgment. The prevalence rate for dementia increases with advancing age (Sosa-Ortiz et al., 2012). Thus, persons above 60 years of age show 0.43% prevalence whereas persons aged above 65 years show 2.44% prevalence. The prevalence rate rises to 54.8% in individuals above 95 years of age (Vas et al., 2001). The impairment in cognitive function is commonly accompanied, and occasionally preceded, by deterioration in emotional control, social behavior, or motivation. Importantly, 75% of people with dementia manifest some, but not all, symptoms of dementia at a given time (Lyketsos et al., 2002). According to World Health Organization estimates 35.6 million people live with dementia, a number that is anticipated to triple by 2050 (World Health Organization, 2012).

Mild cognitive impairment (MCI) is form of predementia, which is characterized by objective impairment in cognition that is not severe enough to require help with usual activities of daily living. MCI leads to general forgetfulness in many people as they age. However, only a few MCI patients develop dementia. The mild dementia involves memory loss, confusion, personality changes, getting lost, and difficulty in planning and carrying out tasks. In moderate dementia daily life becomes more challenging, and the patient may require help in performing daily life activities. Symptoms are similar to mild dementia along with significant changes in personality (Sosa-Ortiz et al., 2012; Rizzi et al., 2014). Severe dementia involves all the symptoms of moderate dementia and the loss of the ability to communicate. The patient may need full-time care. Simple tasks, such as sitting and holding one's head up become impossible along with the loss of bladder control. According to the World Health Organization (2012) most of the increase in dementia patients will occur in low- and middle-income countries. Thus, currently 62% of all people with dementia live in such regions. This proportion may increase to 66% in 2030 and 71% in 2050. The fastest growth in the elderly population is taking place in China, India, and their south Asian and western Pacific neighbors (Dong et al., 2007; Llibre Rodriguez et al., 2008; Plassman et al., 2007). Although recent studies suggest a decline in prevalence (Matthews et al., 2013), dementia remains a devastating and costly disease. In the United States the cost of dementia has already surpassed that of cancer and heart diseases (Hurd et al., 2013). The realization of its paramount public health impact has led nations, including the United States, to develop national plans to cope with dementia and attempt to reduce its devastating effects (National Alzheimer's Project Act; Public Law 111-375).

As stated above, dementia syndrome not only results in deterioration in cognitive function (i.e., the ability to process thought) beyond what may be expected from normal aging, but also affects memory, thinking, orientation, comprehension, calculation, learning capacity, language, and judgment. A gradual age-related cognitive dysfunction, particularly in executive function and mental speed, is evident even in nondemented oldest-old. Hearing and vision losses, which are also prevalent in the oldest-old and found in some cases to precede/predict cognitive decline, may mechanically interfere in neuropsychological evaluations. As stated above, the prevalence rate for dementia increases essentially with advancing age (Savva et al., 2009). Thus, persons above 60 years of age show 0.43% prevalence, whereas persons aged above 65 years show 2.44% prevalence. The prevalence rate rises to 54.8% in individuals above 95 years of age (Vas et al., 2001; Cerejeira et al., 2012). It is stated that impairment in cognitive function in dementia is commonly accompanied, and occasionally preceded, by deterioration in emotional control, social behavior, or motivation. These processes may result in memory loss, and cognitive impairment. Among humans, the impact of dementia can be felt at three interrelated levels: the individual (patients with dementia), their family and friends, and wider society. While dementia does shorten the lives of those affected, its greatest impact is upon quality of life, both for individuals living with dementia, and for their family and carers. The total estimated worldwide costs of managing dementia were US$ 604 billion in 2010, equivalent to 1% of the world's gross domestic product (Prince et al., 2013a,b). Low-income countries accounted for just less than 1% of total worldwide costs (but 14% of the prevalence of dementia), middle-income countries for 10% of the costs (but 40% of the prevalence of dementia) and high-income countries for 89% of the costs (but 46% of the prevalence of dementia). About 70% of the global costs occurred in just two regions: Western Europe and North America. These discrepancies are accounted for by the much lower costs per person in lower income countries – US$ 868 in low-income countries, US$ 3109 in lower-middle-income, US$ 6827 in upper-middle-income, and US$ 32,865 in high-income countries (Wimo et al., 2017).

CLASSIFICATION OF DEMENTIAS

Several types of dementia have been reported to occur in the human population including Alzheimer's type of dementia (30%), vascular dementia (VaD; 26%), mixed dementia (21%), Lewy body dementia (LBD; 11%), frontotemporal dementia (FTD)/degeneration (7%), and infective dementia (5%) (Fig. 1.1). Secondary causes of dementia include vascular, CNS infections, trauma, metabolic derangements, and other reversible/treatable causes such as type 2 diabetes, stroke, AIDS, or

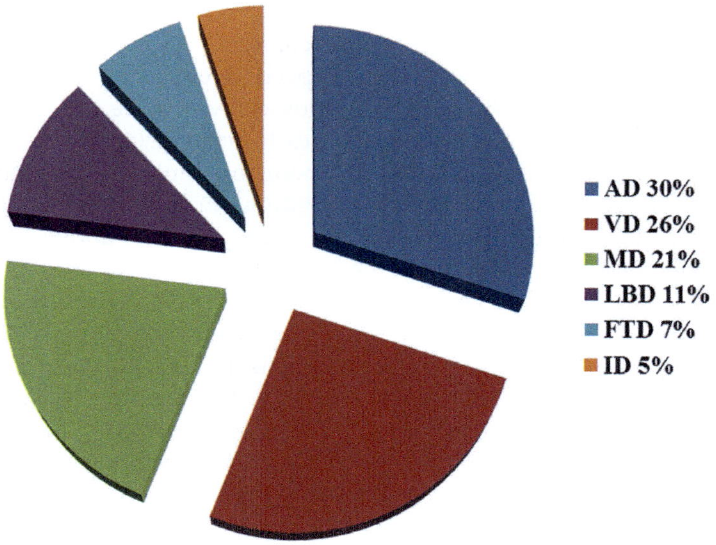

FIGURE 1.1 Proportions of various types of dementia found in the human population. *AD*, Alzheimer's type of dementia; *FTD*, frontotemporal dementia; *ID*, infective dementia; *LBD*, Lewy body dementia; *MD*, mixed dementia; *VD*, vascular dementia.

multiple sclerosis (Kabasakalian and Finney, 2009; Ironside and Bell, 2007). Sometimes dementia-mediated changes become reversible. This is called pseudodementia. Pseudodementia is caused by depression, malnourishment (vitamin deficiency), dehydration, medications, sleep deprivation, metabolic problems, excessive drinking, smoking, and infections. Symptoms of dementia are underrecognized. The understanding of behavioral and psychological symptoms of dementia (BPSD) would be helpful for an early diagnosis and better management so as to improve the patients' quality of life (Levy and Chelune, 2007; Cerejeira et al., 2012; Kales et al., 2015).

Alzheimer's disease (AD) type of dementia is the most common cause of dementia (30%) and is rare before 60 years of age. AD type of dementia is characterized clinically by progressive memory and orientation loss and other cognitive deficits, including impaired judgment and decision-making, apraxia, and language disturbances. AD type of dementia is accompanied by the accumulation of Aβ peptide in the form of senile plaques (Farooqui, 2017). These Aβ plaques are thought to be one of the major contributors to dementia caused by AD (Villemagne et al., 2013). In addition to senile plaques, AD also appears to be related to tangles in the brain, which are structural abnormalities due to defective or deficient tau proteins; tau proteins support microtubules, which help provide cell structure and movement. AD is typically diagnosed by biomarkers, such as Aβ in cerebrospinal fluid (CSF), tau proteins, and regional brain volumes; these measurable substances can be used to predict AD progression in patients with MCI. In addition, ischemic, cerebral small vessel disease, and neurodegenerative diseases have a profound impact on the onset and expression of the dementia, supporting the view that there may be reciprocal interactions between ischemia and neurodegeneration (Shi and Wardlaw, 2016). Ischemic injury results in a number of small, focal cerebral infarcts (small strokes) that may go unnoticed individually but have an additive detrimental effect as more and more small areas of the brain are destroyed by ischemic events; however, there are also a number of other causal subtypes of cerebrovascular disease (Nagata et al., 2007). The second most frequent type is VaD and the third is LBD. These are then followed by FTD.

VaD is the second most common type of dementia following AD-type dementia. The onset of VaD occurs when the blood supply to the brain is reduced by various cerebrovascular pathologies, such as hypoperfusions or hemorrhages causing disruption of the blood–brain barrier (BBB) and neurovascular units (NVUs), usually in hemispheric white matter (Román et al., 2002; Iadecola, 2013), leading to a progressive decline in memory and cognitive function. Pathological features of VaD are diffuse myelin pallor, astrocytic gliosis, and the loss of oligodendrocytes leading to rarefaction, vacuolization, and the loss of myelin and axons without definite necrosis, ultimately culminating in white matter lesions and lacunes (Román et al., 2002). The hardening of cerebral arteries in VaD leads to a reduction in blood flow, a major contributor of cognitive decline.

Changes in dementia involve the decrease in blood flow causing low perfusion pressure and hypoperfusion in several brain regions including hippocampus (American Psychiatric Association, 2013). Sustained cerebral hypoperfusion in dementia syndrome may cause white matter attenuation, a key feature common to both AD and dementia associated with cerebral small vessel disease. Under conditions of chronic hypoperfusion, white matter rarefaction, glial activation, and axon damage can promote diffused ischemic-neuronal loss (Hachinski et al., 1974; Libon et al., 2006). It is proposed that cerebral hypoperfusion is the common pathophysiological mechanism which contributes to cognitive decline and degenerative processes leading to VaD and cerebral small vessel disease. White matter changes are also closely associated with increased risk for stroke, dementia, and disability (Duncombe et al., 2017). Based on the above description, converging evidence suggests that pathological features of VaD are diffuse myelin pallor, astrocytic gliosis, and the loss of oligodendrocytes leading to rarefaction, vacuolization, and the loss of myelin and axons without definite necrosis, ultimately culminating in white matter lesions and lacunes (Román et al., 2002). These findings are supported by structural neuroimaging studies (Risacher and Saykin, 2013; Valkanova and Ebmeier, 2014). The onset of dementia syndrome may occur in several stages including MCI, mild dementia, moderate dementia, and severe dementia (Sosa-Ortiz et al., 2012; Rizzi et al., 2014).

Progressive accumulation of α-synuclein (α-syn) in selected regions of the brain contributes to the pathogenesis of Parkinson's disease (PD) and LBD (Hashimoto and Masliah, 1999). In these disorders, the abnormal accumulation of α-syn is not limited to the striatonigral system but also affects the limbic areas, the insula, frontal cortex, and subcortical nuclei (Hurtig et al., 2000; Marui et al., 2002). The FTDs are a group of heterogeneous neurodegenerative disorders characterized by progressive deterioration of behavior or language and associated pathology in the frontal or temporal lobes. Six clinical subtypes of FTD have been described in the literature. They are: (1) behavioral variant of FTD; (2) semantic variant primary progressive aphasia; (3) nonfluent agrammatic variant primary progressive aphasia; (4) corticobasal syndrome; (5) progressive supranuclear palsy; and (6) FTD associated with motor neuron disease (Finger, 2016; Olney et al., 2017). Some FTD related disorders include FTD with motor neuron disease, progressive supranuclear palsy syndrome, and corticobasal syndrome.

The abovementioned information on dementia subtypes are highly variable among countries with more variation being observed in developing countries as compared to developed countries due to factors such as cultural and socioeconomic variability and a lack of methodological uniformity (Rizzi et al., 2014). Many studies have indicated that type 2 diabetes increases risk of AD type of dementia in the elderly. Multiple possible mechanisms have been proposed to explain this association. These mechanisms include direct effects of hyperglycemia, insulin resistance, and insulin-induced Aβ amyloidosis in brain, as well as indirect ischemic effects of type 2 diabetes-promoted cerebrovascular disease (Niures et al., 2015). Among these mechanisms, induction of insulin resistance contributes to increased risk of cognitive decline and increased rates of brain atrophy with dementia (Rasgon et al., 2010; Yaffe et al., 2004).

It is well known that normal brain function is dependent on receiving 20% of the cardiac output of oxygenated blood, and both higher and lower blood pressure may reduce this cerebral blood flow (Stukas et al., 2014). Another factor, which contributes to a reduction in regional cerebral blood flow is the decrease of endothelial nitric oxide (NO) synthesis in elderly subjects. NO is synthesized by the endothelial nitric oxide synthase (eNOS). NO is not only important for cardiovascular homeostasis, but also acts as a vasodilator controlling vasomotor function and local blood flow (Katusic and Austin, 2014). Thus, vascular risk factors play an important role in the pathogenesis of dementia and AD.

This suggestion is supported by recent findings on the involvement of NVU, an entity, which is composed of astrocytes, mural vascular smooth muscle cells and pericytes, and endothelia. NVU regulates blood flow, controls the exchange across the BBB, contributes to immune surveillance in the brain, and provides trophic support to brain cells in the pathogenesis of poststroke dementia (Nelson et al., 2016). The BBB plays a critical role in maintaining CNS homeostasis and its dysfunction contributes to multiple neurological disorders. The BBB dysfunctions include (1) BBB disruption, which results in leakage of circulating substances, which may be neurotoxic, into the brain; (2) transporter dysfunction, which has consequences such as inadequate nutrient supply, build-up of toxic substances in the brain, and increased entry of compounds that are normally extruded; and (3) alterations in protein expression and secretions by endothelial cells and other cell types of the NVU that can promote inflammatory reactions and oxidative stress leading to neuronal damage. Aging is an important factor, which modulates the integrity of the NVU. The age-related physiological or pathological changes in the cellular components of the NVU increase the vulnerability of the NVU to ischemia/reperfusion injury leading to brain damage (Cai et al., 2017). Aging impairs cerebral blood flow triggered by alterations in NVU, a critical entity, which modulates and matches oxygen and nutrient delivery to the increased demands in active brain regions (Tarantini et al., 2017). These findings support the view that aging, consumption of a Western diet (meat, sweets, and high-fat dairy products), physical inactivity, and environmental factors may contribute to the pathogenesis of dementia. These factors decrease cerebral blood flow, induce brain hypofunction, and induce an onset of neuroglial crisis leading to MCI, which all ultimately contribute to dementia (Fig. 1.2).

Diagnosing VaD is not a simple matter. Currently, there is a lack of validated criteria for establishing a diagnosis of VaD, and many of the various pathologies that reduce the brain's blood supply are complex (Jellinger, 2008). Although cerebrovascular lesions can be seen using brain imaging techniques, the diagnosis of VaD remains difficult, since such lesions may or may not be contributing to dementia symptoms, and this can lead to overdiagnosis of VaD as the cause of dementia (Niemantsverdriet et al., 2015). Collective evidence suggests that age-related cerebromicrovascular dysfunction and microcirculatory damage play critical roles in the pathogenesis of many types of dementia in the elderly, including AD. Understanding and targeting of the age-related molecular mechanisms, which underlie vascular dysfunction-mediated alterations in dementia, play a major role in preserving brain health in older subjects. Maintenance of normal cerebral perfusion, protecting the microcirculation from hypertension-induced damage, and moment-to-moment adjustment of regional oxygen and nutrient supply to changes in demand are prerequisites for the prevention of cerebral ischemia and

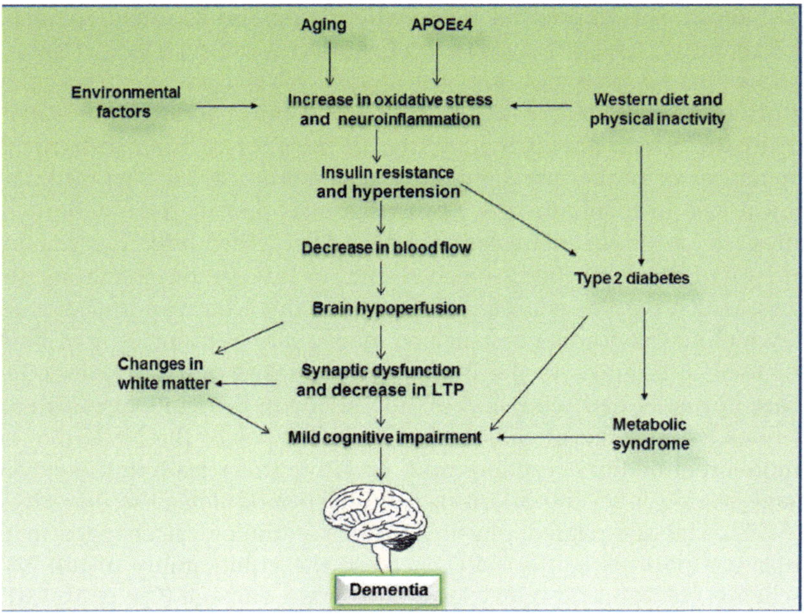

FIGURE 1.2 Factors contributing to the onset of dementia.

neuronal dysfunction. Aging is also accompanied by a marked deficiency in circulating insulin-like growth factor-1 (IGF-1), which has been shown to contribute to age-related cognitive decline. Impairment of moment-to-moment adjustment of cerebral blood flow via neurovascular coupling is thought to play a critical role in the genesis of age-related cognitive impairment (Toth et al., 2015, 2017). In contrast to AD type and LBD dementias, VaD is defined as loss of cognitive function resulting from ischemic or hemorrhagic brain lesions due to cerebrovascular disease or cardiovascular pathology. Diagnosis of VaD requires cognitive loss, often predominantly subcortical; vascular brain lesions identified by neuroimaging; a temporal link between stroke and dementia; and exclusion of other causes of dementia. Dementia syndrome is not only accompanied by the deterioration in cognitive function beyond what may be expected from normal aging, but also affects memory, thinking, orientation, comprehension, calculation, learning capacity, language, and judgment.

RISK FACTORS FOR DEMENTIAS

Major risk factors for dementia include advancing age, low body mass index (BMI), long-term consumption of a Western diet, physical and cognitive inactivity, and epigenetic and environmental factors (Fig. 1.2) (Farooqui, 2017). Other risk factors for dementia include (1) cardiovascular (e.g., hypertension, atrial fibrillation, diabetes, sleep apnea, insulin resistance, and obesity) and cerebrovascular problems (e.g., stroke); (2) excessive alcohol consumption; (3) social isolation; (4) traumatic brain injury (TBI); (5) hearing loss; and (6) having one or two copies of the APOEϵ4 genetic variant (Farooqui, 2017). A combination of all these factors is known to contribute to the pathogenesis and development of the dementia syndrome, but information on underlying molecular mechanisms contributing to dementia remains speculative and controversial. In addition, other modifiable risk factors include smoking, hypertension, lower literacy rate, nutritional status, and metabolic and cardiovascular factors (Baumgart et al., 2015). Secondary causes of dementia are associated with vascular, CNS infections, trauma, hearing loss, metabolic derangements, and other reversible/treatable causes (Kabasakalian and Finney, 2009; Lin et al., 2013; Gurgel et al., 2014).

Among these causes, hearing loss is independently associated with accelerated cognitive decline and incident cognitive impairment in community-dwelling older adults (Lin et al., 2013). A number of mechanisms have been implicated in explaining the involvement of hearing loss and cognition dysfunction in dementia. Poor verbal communication associated with hearing loss may confound cognitive testing, or vice versa may indicate that there may be an overdiagnosis of hearing loss in individuals with subclinical cognitive impairment (Gordon-Salant, 2005).

The mechanistic basis of association between hearing loss and cognitive decline in dementia is not known. However, it is proposed that hearing loss may contribute to an overall cycle of multimorbidity and frailty or synergistically interact with other known risk factors for dementia (Daviglus et al., 2011; Plassman et al., 2010), both of which may be associated with cognitive decline in older adults. However, the underlying pathway contributing to hearing loss and cognition may not mutually exclusive, and hence, multiple pathways (e.g., shared neuropathology, cognitive load, increased loneliness) may coexist and synergistically contribute to accelerated cognitive decline in demented subjects with hearing loss (Daviglus et al., 2011; Plassman et al., 2010). Another limitation of our study is that hearing loss was only measured at baseline, and information was not available on the trajectory or the possible etiology of the hearing loss.

In vitro studies have shown that inhaled anesthetic agents can promote dementia (Chen et al., 2013; Jiang et al., 2017). Very little is known on the biochemical consequences of anesthetic agent injections on pathogenesis of dementia. It is reported that regardless of the anesthetic agent used, anesthesia induces rapid and massive hyperphosphorylation of tau, rapid and prolonged hypothermia, inhibition of Ser/Thr PP2A (protein phosphatase 2A), but no changes in APP metabolism or Aβ (beta-amyloid peptide) accumulation (Chen et al., 2013; Jiang et al., 2017). Reestablishment of normothermia during anesthesia completely prevents tau phosphorylation. Detailed investigations have indicated that changes in tau phosphorylation are not a result of anesthesia per se, but a consequence of anesthesia-induced hypothermia, which led to inhibition of phosphatase activity and subsequent hyperphosphorylation of tau (Planel et al., 2007). In contrast, other animal studies have provided evidence that exposure to anesthetic agents can impair not only memory and induce caspase-3 activation, but also can increase levels of Aβ (Planel et al., 2007; Xie et al., 2008). The prevalence of dementia is consistently higher among women. This may be due to longer life expectancy in women. Lower educational levels have been found associated with higher prevalence of dementia. Within the United States, prevalence has been reported as elevated in African American and Latino populations; some authors have attributed these findings to lower education and higher cardiovascular morbidity in those populations (Graham, 2014).

SYMPTOMS OF DEMENTIAS

Neurochemical mechanisms of dementia start years before it is diagnosed clinically—the preclinical period of the disease, when neuronal degeneration has begun, but cellular and biochemical damage is not yet sufficient for symptoms to manifest (Mosconi, 2005). The symptoms of dementia in patients are heterogeneous and largely unpredictable. These symptoms include agitation, depression, apathy, repetitive questioning, psychosis, aggression, sleep disturbances, wandering, and a variety of inappropriate behaviors. Importantly, 75% of people with dementia manifest some, but not all symptoms of dementia at a given time (Fig. 1.3) (Lyketsos et al., 2002). These symptoms are among the most complex, stressful, and costly aspects of dementia care. They are accompanied by a 40% or greater loss of neocortical synapses as compared with normal adults. These symptoms may lead not only to poor neurotransmission in dementia patients, but may also contribute to healthcare problems, and income loss for family care givers (Van Den Wijngaart et al., 2007). The prevalence rate for dementia increases essentially with advancing age. Persons above 60 years of age show 0.43% prevalence whereas persons aged above 65 years show 2.44% prevalence. The prevalence rate rises to 54.8% in individuals above 95 years of age. The impairment in cognitive function is commonly accompanied,

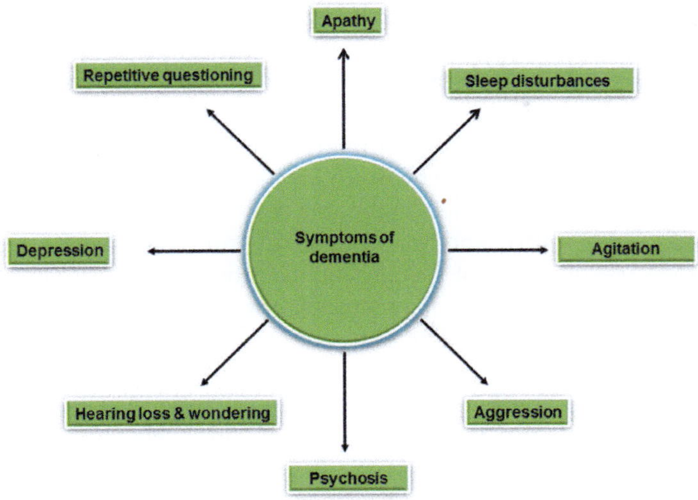

FIGURE 1.3 Symptoms of dementia.

and occasionally preceded, by deterioration in emotional control, social behavior, or motivation. The first step in recognizing the clinical manifestations of dementia is to appropriately understand the psychopathology and molecular mechanisms contributing to the symptoms of various types of dementias. This can be a very challenging issue because there is considerable overlap in pathogenic mechanisms among various types of dementias. Secondly, it is useful to evaluate whether specific symptoms occur in association with various types of neurodegenerative diseases (Lyketsos et al., 2002; Kabasakalian and Finney, 2009; Ostling et al., 2009). It is proposed that symptoms of dementia can be delayed by enhancing and encouraging cognitive and physical activity, social engagement, smoking cessation, and healthy diet, including alcohol reduction. Comorbid depression is common in older people with dementia and treating this can improve cognition. Dementia is typically diagnosed by a doctor in a clinical setting, when acquired cognitive impairment has become severe enough to compromise social and/or occupational functioning (Buntinx et al., 2011). Positron-emission tomography (PET) and single-photon emission computed tomography (SPECT) are both brain imaging methods that are most commonly used to make a dementia diagnosis by examining the physical condition of the brain (Bamford et al., 2016). However, even with these tools, the difficulties of recognizing and diagnosing dementia are apparent, and approximately half of dementia cases are currently undiagnosed (Savva and Arthur, 2015).

ETIOLOGY OF DEMENTIAS

The etiology and pathogenesis of dementia is controversial. Several genetic and environmental factors are known to increase the risk of dementia in the aged individuals. Importantly, risk/protective factors for dementia in younger elderly subjects may not pertain to the oldest-old. Moreover, postmortem studies suggest that neuropathology is abundant in the oldest-old brains, and not necessarily correlated with dementia, making the determination of the etiology a difficult mission. Information on underlying molecular mechanisms contributing to various types of dementia still remains speculative and controversial. However, the current pathogenic scenarios of different types of dementia are based on a number of common mechanisms of neurodegeneration, such as accumulation of abnormal proteins (within or outside cells), mitochondrial dysfunction, increase in oxidative stress, dysregulation in calcium homeostasis, activation of microglia and astrocytes, induction of neuroinflammation along with early synaptic disconnection, and late apoptotic cell death (Farooqui, 2010, 2017). At the molecular level, these processes are accompanied by the accumulation of abnormal proteins (Aβ, tau protein, α-synuclein), which activate microglia and astrocytes. The involvement of microglia and astrocytes in the onset and progress of the neurodegenerative process in AD and Lewy body disease suggests that these cells play an important role in the pathogenesis of dementias. It is becoming increasingly recognized that neuroinflammation and oxidative stress can have both detrimental and beneficial effects in the brain. However, little is known about the interplay of microglia, astrocytes, and neurons in the early phases of dementia. The abnormal production and secretion of proinflammatory cytokines, chemokines, and the complement system proteins, as well as reactive oxygen and nitrogen species, can disrupt nerve terminals activity inducing neural dysfunction and loss of synapses, which not only correlates with memory decline; but also contributes to impaired judgment and decision-making, apraxia, and language disturbances (Farooqui, 2017).

BIOMARKERS FOR DEMENTIAS

A biomarker is a parameter that is used as an indicator of normal biological processes, pathogenic processes, or pharmacological responses to therapeutic drugs (Biomarkers Definitions Working Group, 2001). In various types of dementia, potential biomarker information comes from multiple sources, including clinical tests for memory impairment, bodily fluid or tissues, neuroimaging, and smell tests among others. Biomarkers for dementia can be identified from two sources: (1) biofluid analytes, for example, CSF, peripheral blood samples such as urine; and (2) neuroimaging measures, for example, magnetic resonance imaging (MRI), magnetic resonance spectroscopy, or PET (Henriksen et al., 2014; Jack, 2012). The diagnosis of dementia is mainly based primarily on the clinical presence of potential biomarkers in the CSF and cognitive and functional impairment. The use of neurochemical, neurobiological, and neuroimaging studies has not only provided partial understanding of causes of dementia, but also have helped in identification of potential biomarkers (serum C-reactive protein, serum interleukin 6, plasma alpha-1-antichymotrypsin, and hyperhomocysteinemia) (Farooqui, 2010, 2017). In addition, the presence of a positive 14-3-3 protein, elevated S100B, and elevated total-tau to phospho-tau ratio is also a positive predictor of onset of dementia (Fig. 1.4) (McKhann et al., 2011; Seshadri et al., 2002; Mattsson et al., 2009). CSF analysis using a variety of immunochemical techniques allows a range of neuronal-specific or neuronal-enriched proteins to be measured. The use of neuronal-enriched CSF markers β-amyloid and tau in the routine evaluation of patients with dementia varies considerably between countries and between clinicians (Mattsson et al., 2009; McKhann et al., 2011; Seppala et al., 2012; Stomrud et al., 2007). These biomarkers have not only been used for distinguishing different aspects of the underlying pathology and detecting presymptomatic pathological

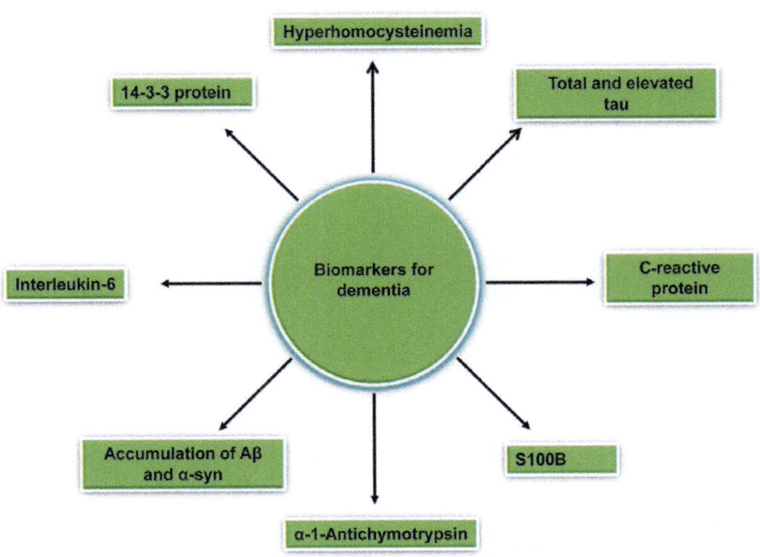

FIGURE 1.4 Biomarkers of dementia.

changes in dementia, but also in predicting decline or conversion between clinical disease states, and/or monitoring disease progression and response to treatment. MRI, PET, and SPECT imaging can be used to measure downstream neuronal injury including brain atrophy, and hypometabolism or hypoperfusion (Jagust, 2006; Koikkalainen et al., 2016). Another important feature of age-related dementia is synaptic protein loss, which may distinguish old patients with dementia from old individuals without dementia. The protein and messenger RNA (mRNA) levels of seven synaptic markers (complexin-1, complexin-2, synaptophysin, synaptobrevin, syntaxin, synaptosomal-associated protein 25 (SNAP-25), and septin-5) have been compared with the brains of nondemented and demented individuals ranging from 70 to 103 years of age. The brains of persons with dementia show significantly lower levels of gene and protein expression of synaptic markers regardless of age. Importantly, dementia is associated with reductions in all measured synaptic markers irrespective of their role(s) in synaptic function (Head et al., 2009). It is proposed that these protein levels may protect neuronal function in oldest-old individuals and reflect compensatory responses that may be involved with maintaining cognition. This suggestion is supported by the gene and protein expression levels of synaptic markers being decreased in persons with dementia, regardless of age (Schnaider Beeri et al., 2012). In addition, the establishment of specialized memory clinics is very important for the diagnosis of dementia. These clinics can provide the ability to identify and treat different types of dementia in the larger population and can also act as a resource from which information about the management of dementias can be shared with other practitioners and throughout the community.

ANIMAL MODELS FOR DEMENTIAS

As stated above, dementia is a progressive and irreversible syndrome characterized by progressive loss of memory and cognitive function. Suitable animal models have been developed to replicate symptoms of several types of dementia including pathology and neurochemical and genomic changes, and cognitive impairment in animal models. An ideal animal model for dementia should not only exhibit progressive dementia-like neuropathology and cognitive deficits, but like humans it should manifest some memory loss and cognitive deficits with advancing age.

Mice, rats, and monkeys have been used to develop AD type of dementia. Extensive information is available on mice models of AD type of dementia. Thus, studies on transgenic (Tg) mice have indicated that double-Tg mice, which overexpress human mutant APP and tau (Tg line APP^{sw}-tau^{vlw}) show several characteristics of the AD phenotype, such as accumulation of Aβ, hyperphosphorylation of tau, formation of neurofibrillary tangles, glial cell proliferation, and significant neurodegeneration in the entorhinal cortex (EC) and CA1 subfield of the hippocampus (Perez et al., 2005; Ribe et al., 2005). All the above phenotypic traits of AD type of dementia develop in these mice in an age-dependent manner and are accompanied by progressive hippocampus-dependent memory impairment. However, neurodegeneration in these mice predates overt deposition of Aβ, supporting the view that extracellular fibrillar amyloid may not be causing neuronal death. Furthermore, the extent of neurodegeneration in these mice does not correlate well with total immunostained amyloid plaque burden (Ribe et al., 2005). Detailed studies on mice models of AD type of dementia have demonstrated that only very few transgenic animal models show neuronal death. This is in contrast to human AD type of dementia, which displays massive neurodegeneration, which can be seen in postmortem human brains (Elder et al., 2010). In addition, many mice models do not show cognitive dysfunction despite overexpression of APP (Masliah et al., 2001). The development of neurofibrillary tangles (NFT) is not observed in most of the APP overexpressing models (Ribeiro et al., 2013). Converging evidence thus suggests that at the present time an ideal animal model for AD type of dementia is not available (Cuadrado-Tejedor and García-Osta, 2014). Furthermore, almost all transgenic models are only related to the familial early onset form of AD, which represents a mere 5% of AD cases. The remaining 95% are sporadic late-onset forms—the causes and pathogenesis of this form remain elusive. Converging evidence thus suggests that at present mouse models display some neurochemical, neuropathological, and behavioral alterations of AD type of dementia. However, these animal models do not recapitulate all aspects of human AD. Furthermore, failure of AD type of dementia immunotherapy in mouse models indicates that there is a need for developing superior models of the AD type of dementia pathology with cognitive dysfunction.

Gradual degeneration and loss of dopaminergic neurons in the substantia nigra, pars compacta, and subsequent reduction of dopamine levels in striatum contribute to motor deficits in PD. This neurodegenerative disease is characterized by tremor, bradykinesia, and muscle rigidity along with impaired gait and posture (Jankovic, 2008; Bronstein et al., 2009). In addition, 50% of PD patients also show frontostriatal-mediated executive dysfunction, including deficits in attention, short-term working memory, speed of mental processing, and impulsivity promoting dementia. Lewy bodies are neuronal inclusions (abnormal filamentous assemblies), which contain α-syn. In DLB, they are found in brainstem nuclei and the neocortex. The two clinical subgroups of LBD, namely dementia with Lewy bodies (DLB) and Parkinson's disease dementia (PDD), have been described in the literature (Aarsland et al., 2004). These subgroups can be distinguished by the temporal relationship between the onset of dementia and motor symptoms. Therefore, in DLB the onset of dementia occurs within 1 year of motor symptoms, while in PDD dementia occurs 1 year or more after motor symptoms (Aarsland et al., 2004; McKeith, 2009). Clinically, DLB is characterized by induction of cognitive dysfunction along with the pronounced variation in attention and alertness, recurrent visual hallucinations which are typically well formed and detailed, and spontaneous motor features of parkinsonism: repeated falls, syncope, transient loss of consciousness, systematized delusions, and hallucinations (Mayo and Bordelon, 2014). In PDD, the abovementioned symptoms appear after 1 year of onset of PDD. Converging evidence suggests that the diagnosis of DLB and PDD is based on an arbitrary distinction between the time of onset of motor and cognitive symptoms supporting the view that these syndromes share many neurobiological similarities, but there are also differences (Goldman et al., 2014). Deposition of Aβ and tau is more marked and more closely related to cognitive impairment in DLB than PDD, possibly contributing to dementia at onset. The relatively more severe executive impairment in DLB than PDD may relate to the loss of frontohippocampal projections in DLB. Visual hallucinations and delusions associate with more abundant Lewy body pathology in temporal cortex in DLB. The differential involvement of pathology in the striatum may account for the differences in parkinsonism (Goldman et al., 2014). Collective evidence suggests that DLB and PDD are common dementia syndromes, which occur in the human population with overlap in their clinical features, neuropathology, and management. They are believed to exist on a spectrum of Lewy body disease.

The causes and molecular mechanisms of PD are not known, although complex interactions between genetic and environmental factors contribute to the pathogenesis of PD. Various risk factors have been found for sporadic PD, including exposure to pesticides and other toxins, positive family history, and oophorectomy, but age remains the most important one documented so far (Bronstein et al., 2009). Chemically-induced animal models of PD are produced by administering chemical toxins to dopaminergic neurons. These toxic chemicals include reserpine, methamphetamine, 6-hydroxydopamine, and 1-methyl-4-phenyl-1,2,3,6-tetrahydropyridine. Recently, systemic administration of agricultural chemicals, such as rotenone and paraquat, has been used to produce specific features of PD in rodents, apparently via oxidative damage (Betarbet et al., 2002). Transgenic animals that overexpress α-syn are used to study the role of this protein in dopaminergic degeneration.

To produce the neurological and behavioral changes, animal models of VaD have been developed in rats (Farkas et al., 2007). These animal models are based on chronic or transient global hypoperfusion, focal hypoperfusion, embolic occlusion, or hypertension (Jiwa et al., 2010). Among above models, permanent bilateral occlusion of common carotid arteries in rodents is a reliable model for investigation of the cognitive and histopathologic consequences of chronic cerebral hypoperfusion (Venkat et al., 2015; Jiwa et al., 2010; Farkas et al., 2007; Kim et al., 2008, 2009; Du et al., 2017). The model is generated by the occlusion of the permanent bilateral common carotid artery in rats. This model shows changes in learning and memory, cerebral blood flow, energy metabolism, and neuropathology, which are initiated by ischemic injury. In this model, hypoperfusion not only produces disruption of homeostatic interactions among oxidative stress, neuroinflammation, and neurotransmitter system dysfunction, but also mitochondrial dysfunction, disturbance of lipid metabolism, and alterations of growth factors.

AUTOPHAGY AND DEMENTIAS

Autophagy is a lysosome-dependent intracellular degradation process that allows recycling of cytoplasmic constituents into bioenergetic and biosynthetic materials for the maintenance of neural homeostasis. Since the function of autophagy is particularly important in various stress conditions, perturbation of autophagy can lead to cellular dysfunction and diseases. Accumulation of abnormal protein aggregates, a common cause of neurodegenerative diseases, can be reduced through autophagic degradation. Autophagy isolates cytosolic materials within a double membrane vesicle called an "autophagosome" which then fuses with lysosome to degrade isolated substrates (Mizushima et al., 2011). Three types of autophagy—macroautophagy, microautophagy, and chaperone-mediated autophagy—have been described in mammalian systems (Boya et al., 2013; Nakatogawa et al., 2009). Defective regulation of the autophagy machinery and/or dysfunction of the lysosomal process can disrupt cellular homeostasis and lead to various disorders (Shintani and Klionsky, 2004). Three major autophagy deregulations have been described in neurodegenerative diseases: (1) insufficient autophagy activation; (2) autophagy dysfunction due to reduced lysosomal function; and (3) autophagic stress related to pathologic activation of autophagy (Cherra and Chu, 2008). These observations support the view that autophagy generally plays a neuroprotective role in the brain. Autophagy is dysregulated in neurodegenerative diseases such as AD, PD, ALS, and HIV-associated neurodegenerative disorder due to the accumulation of misfolded proteins. As stated above, AD is the most common type of a progressive dementia caused by the accumulation of beta-amyloid (Aβ). Alterations in autophagy have been implicated in the pathogenesis of AD by failing to clear aggregated Aβ and by playing a role in APP metabolism. Recent studies indicate that the autophagocytic pathology observed in AD most likely arises from impaired clearance of autophagic vacuoles rather than strong autophagy induction alone, indicating selective alterations in molecular components of the autophagy pathway (Nixon, 2007). PD and LBD are caused by the accumulation and impaired clearance of α-Syn aggregates. These processes promote abnormal signal transduction processes closely associated with neurodegeneration in PD and LBD (Bendiske and Bahr, 2003). It is reported in PD and LBD that α-syn aggregates may interfere with the autophagy related mechanisms and lead to neurodegeneration (Bendiske and Bahr, 2003; Cuervo et al., 2004). Similarly, FTD is produced by the deposition of p-tau. In cell culture models of FTD, dysregulation of autophagy results in the accumulation of autophagosomes, which is detrimental for the survival of neurons (Lee and Gao, 2009). Converging evidence suggests that the pathogenesis of the abovementioned neurodegenerative diseases is linked to the abnormal accumulation of proteins in the brain, which in turn is related to deficits in protein clearance. Autophagy is a key cellular protein clearance pathway with proteolytic cleavage and degradation via the ubiquitin–proteasome pathway representing another important clearance mechanism. Alterations in the levels of autophagy and the proteins associated with the autophagocytic pathway have been reported in various types of dementias (Kragh et al., 2012).

LINK BETWEEN DEMENTIAS AND COGNITIVE DYSFUNCTION

Cognitive dysfunction is defined as the loss of intellectual functions such as thinking, remembering, and reasoning that interfere with daily activities. Patients with cognitive dysfunction lose the ability to learn, recall, concentrate, and problem solve. Cognitive function is regulated not only by neurochemical and intricate synaptic changes, but also by neuronal and glial interactions (Morrison and Baxter, 2012). Cognitive dysfunction is one of the primary disabilities of the aging process. It predisposes individuals to dementia, neurological diseases, and psychiatric disorders, eventually affecting the quality of life. The intensity of cognitive decline is markedly increased in patients with stroke, dementia, and neurodegenerative diseases (Schuh et al., 2011). Although some progress has been made on molecular aspects of cognitive decline in aging, significant work is still needed to understand molecular mechanism of cognitive decline. Investigators have developed various cognitive tests to measure cognitive dysfunction. These tests involve response to questionnaires, blood tests, brain scans, personal history, and a specific (or more than one) cognitive test. Cognitive testing can range from a few minutes to more than two hours (Alzheimer's Australia Tests, 2016). As stated above, many risk factors exist that can elevate the odds of developing dementia. The majority of these risk factors are reversible; however, others, such as age and family history, are irreversible. Reversible risk factors include: cardiovascular diseases (CVDs), diabetes, depression, excessive alcohol intake, smoking, physical inactivity, and poor dietary habits (Alzheimer's Association, 2016). The aging process is known to substantially decrease hippocampal neurogenesis (Kuhn et al., 1996; Rao et al., 2005, 2006). It is well known that neural stem cells (NSCs) that proliferate at a given time in the subgranular zone decrease considerably between young adult age and old age resulting in a substantial reduction in the production of new cells (Kuhn et al., 1996; Rao et al., 2005, 2006; Drapeau and Abrous, 2008). In addition, with aging, cerebral blood flow is also decreased by 15%–20% (Wongrakpanich et al., 2016). Furthermore, aging is also accompanied by the age-related decrease in the synthesis of many neurotransmitters and their receptors. These include the catecholamines (adrenaline and noradrenalin), dopamine, and serotonin. These reductions can slow reaction time, impair information processing, and, sometimes, increase the risk of depression (Knight and Nigam, 2008). The onset of dementia-related neurodegenerative diseases occurs when neural cells fail to neutralize increases in oxidative stress, and consequently fail to remove the accumulated misfolded proteins, deoxyribonucleic acid (DNA), and oxidized cell membranes (Mattson and Magnus, 2006).

Long-term consumption of the Mediterranean diet not only delays the cognitive decline, but also promotes the maintenance of NSCs. In a study of 2258 community-dwelling, nondemented New Yorkers (Scarmeas et al., 2006a), adherence to the Mediterranean diet decreased the likelihood of developing AD over an approximate 4-year follow-up. Compared to individuals in the highest quartile of dietary adherence, individuals with poor Mediterranean dietary practices had an approximately 40% greater risk of developing AD (Scarmeas et al., 2006b). Although the molecular mechanisms linking the Mediterranean diet to reduced cognitive impairment are still controversial and are still under investigation, at least two studies have demonstrated that greater Mediterranean diet adherence is associated with preserved cortical thickness (Mosconi et al., 2014) and a lower incidence of MRI infarcts. A recent meta-analysis combining results across five studies has indicated that higher adherence to the Mediterranean diet is associated with a 27% reduced risk of MCI and a 36% reduced risk of AD among cognitively normal adults (Singh et al., 2014). In contrast, long-term consumption of a Western diet not only promotes insulin resistance, hypertension, and ultimately diabetes and metabolic syndrome (Farooqui, 2013), but also alters the composition of gut microbiota and promotes obesity-related inflammation and cognitive impairment. Hypertension leads to atherosclerosis and hardening of the large arteries and can lead to blockage of small blood vessels in the brain. High blood pressure can also lead to hemorrhagic stroke, since high blood pressure can weaken the blood vessels in the brain, causing them to balloon and burst. This decreases cerebral blood flow and produces brain hypofunction inducing local microenvironmental changes, which may promote and support microvascular rarefaction (Toth et al., 2013a), endothelial dysfunction (Girouard et al., 2007), oxidative stress, and neurovascular uncoupling (Kazama et al., 2004). These processes impair delivery of oxygen and glucose to the activated brain regions causing MCI and ultimately dementia (Fig. 1.5) (Iadecola et al., 2009; Iadecola, 2014). Induction of vascular oxidative stress is a key pathogenic factor in neurovascular dysfunction (Simpson et al., 2010). Experimental studies indicate that free radicals generated by the enzyme nicotinamide adenine dinucleotide phosphate oxidase are responsible for the cerebrovascular alterations induced by vascular cognitive impairment risk factors (Park et al., 2007, 2008). In old animals, the generation of free radicals can promote

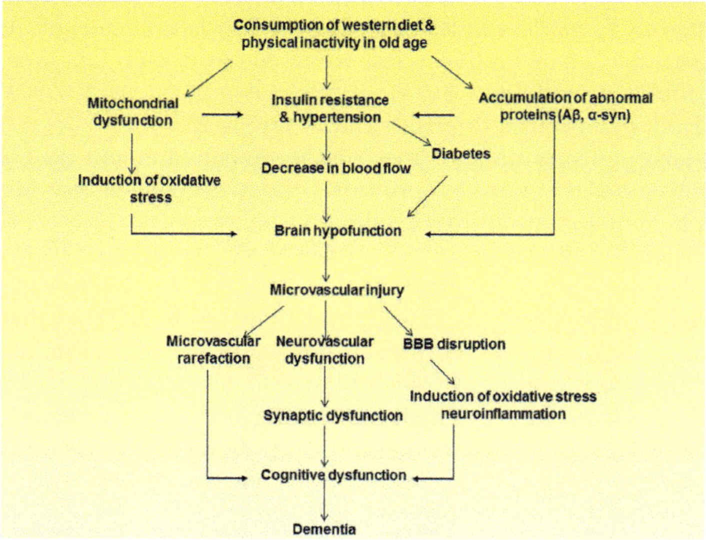

FIGURE 1.5 Neurochemical changes contributing to microvascular injury, blood–brain barrier (BBB) disruption, synaptic dysfunction in dementia.

the induction of neuroinflammation not only by activating redox-sensitive proinflammatory transcription factors, and induction of endothelial dysfunction, but also by the generation of proinflammatory eicosanoids, which promote vascular leakage, protein extravasation, and cytokine production (Marchesi et al., 2008). Inflammation, in turn, enhances oxidative stress by upregulating the expression of reactive oxygen species (ROS)-producing enzymes and downregulating antioxidant defenses (Gill et al., 2010). With aging, blood vessels of demented and nondemented subjects undergo profound changes. Thus, demented and nondemented subjects show tortuosity of the vessels and thickening of the vessel walls, which decreases vascular reactivity in the regions of the white matter with hyperintensities on MRI (Marstrand et al., 2002). Long-term consumption of the Western diet, age, and physical inactivity are the major factors that cause changes in the blood vessels responsible for the impairments in cerebral blood flow and oxygenation along with an increase in permeability (Wardlaw et al., 2009). These changes contribute to brain hypofunction. Neuroimaging studies have also indicated that patients with dementia also show white matter hyperintensities, cerebrovascular lesions, and cerebral amyloid angiopathy (Petrovitch et al., 2005). Other ultrastructural abnormalities include changes in microvasculature of demented subjects. These abnormalities include capillary wall deterioration and the accumulation of erythrocytes (Schreiber et al., 2013), basement membrane thickening, and pericyte degeneration (Farkas et al., 2000), resulting in BBB permeability (Yang and Rosenberg, 2011) and vascular cognitive impairment. In dementia, neurogenesis is decreased due to reduction in brain-derived neurotrophic factor (BDNF). Because neurogenesis in the dentate gyrus declines dramatically with aging and dementia, this may represent another possible cause of age-related impairment in hippocampal-dependent memory and cognitive decline. In dementia, induction of hypertension affects the structural and functional integrity of the cerebral microcirculation. This may contribute to microvascular damage (capillary rarefaction, BBB disruption), neurovascular uncoupling, microglia activation, and neuroinflammation (Toth et al., 2013a,b) and the genesis of cerebral microhemorrhages. Persistent disturbances of BBB due to poststroke dementia allow substantial cerebral extravasation of blood-borne potentially neurotoxic molecules, which thereby promote the synthesis and release of proinflammatory cytokines resulting in the activation of microglial phagocytes and production of ROS (Kalaria, 2010). Chronic oxidative stress and neuroinflammation in the brain results in the production of toxic mediators that compromise cell function, alter cellular phenotypes, damage DNA, and eventually lead to more neuroinflammation and neurodegeneration. These processes are supported by the decrease in cerebral blood flow, which causes hypertension, brain hypofunction and alterations in local microenvironment in the brain promoting microvascular rarefaction (Toth et al., 2013a), endothelial dysfunction (Girouard et al., 2007), and neurovascular uncoupling (Kazama et al., 2004). These processes not only impair delivery of oxygen and glucose to the activated brain regions, but decrease synaptic plasticity and long-term potentiation (LTP) producing MCI and ultimately dementia (Iadecola et al., 2009; Iadecola, 2014). In addition, during aging and in the onset of dementia, the induction of endothelial dysfunction is caused by augmented production from the intracellular enzymes NADPH oxidase and uncoupled eNOS, as well as from mitochondrial dysfunction in the absence of appropriate increases in antioxidant defenses as regulated by relevant transcription factors, such as FOXO. Interestingly, it appears that NF-κB, a critical inflammatory transcription factor, is sensitive to this age-related endothelial redox change and its activation induces transcription of proinflammatory cytokines that can further suppress endothelial function, thus creating a vicious feed-forward cycle. Dementia also promotes hypertension-mediated disruption of the BBB, microglia activation, and neuroinflammation in the hippocampus (Toth et al., 2013b). The mechanisms of hypertension-related BBB disruption are multifaceted and may involve increased endothelial oxidative stress and endothelial injury, pericyte damage, and changes in tight junctions, which form an essential structural component of the BBB (Toth et al., 2013a; Takemori et al., 2013). Pericytes are important cellular constituents of the BBB (Zlokovic, 2008), and recent studies demonstrate that pericyte deficiency in $Pdgfr\beta^{+/-}$ mice leads to significant impairment of BBB function (Bell et al., 2010) and that pericyte loss exacerbates AD-like neurodegeneration in mice (Sagare et al., 2013).

LIFESTYLE, COGNITIVE FUNCTION, AND DEMENTIAS

It is well known that age-related cognitive decline in human subjects is a complex process, which is accompanied by the erosion in learning and memory function. Several mechanisms have been reported to contribute to cognitive loss with aging. They not only include induction of oxidative stress and neuroinflammation (Craft et al., 2012) and alterations in brain neurochemistry/plasticity and connectivity (DeCarli et al., 2012), but also influence epigenetic (Kosik et al., 2012) and other environmental/psychosocial factors (Kremen et al., 2012). Reelin is a large matrix glycoprotein, which is widely expressed in interneurons throughout the brain.

Reelin interacts with lipoprotein and integrin membrane receptors in target cells. Reelin signaling regulates migration and dendritic growth in developing neurons, while it can modulate synaptic plasticity in adult neurons. Reelin is expressed not only in excitatory neurons, but also in neurons of entorhinal layer II (Ramos-Moreno et al., 2006). At the molecular level, reelin signaling negatively regulates tau phosphorylation (Hiesberger et al., 1999); conversely, reelin depletion creates a permissive environment for cellular events that are tied to the pathophysiology of AD (Hiesberger et al., 1999; Hoe et al., 2009). A marked decrease in expression of reelin has been observed in the EC of mouse models of AD and in the human AD brain (Chin et al., 2007). The functional consequences of reduced reelin expression with age-related cognitive decline have yet to be elucidated. However, it is shown that reelinergic signaling contributes to LTP at Schaffer collateral synapses in CA1 (Herz and Chen, 2006) suggesting that reelin may be associated with synaptic plasticity. In contrast, lifestyle (diet, exercise, and sleep) protect against cognitive decline and dementia (Hugo and Ganguli, 2014). In the elderly population, changes in lifestyle may prevent cognitive decline in later life (Smith and Blumenthal, 2016). Thus, long-term consumption of the Mediterranean diet and regular exercise can delay the progression of cognitive decline and the Alzheimer's society recommends the consumption of the Mediterranean diet as an approach to improve memory and cognitive function (Table 1.1) (Alzheimer's Society Mediterranean Diet, 2016; Aridi et al., 2017).

TABLE 1.1 Effects of Long-Term Consumption of Mediterranean and Western Diets on Human Brain

Mediterranean Diet + Exercise	**Effect**
Cognitive function	Maintained in elderly
Life span	Increased
Insulin resistance	Decreased
Risk of cardio and cerebrovascular diseases	Decreased
Risk of neurological diseases	Decreased
Quality of life	Increased
Western diet + exercise	Effect
Cognitive function	Decreased
Life span	Decreased
Insulin resistance	Increased
Risk of cardio and cerebrovascular diseases	Increased
Risk of neurological diseases	Increased
Quality of life	Decreased

In contrast, the consumption of the Western diet, which is rich in high lipid consumption (mainly saturated fatty acids, or saturated fatty acids) and refined carbohydrate consumption with a low ingestion of vegetables, results in the development of obesity, cardiovascular disorders, cancer, and type 2 diabetes (Farooqui, 2014). Type 2 diabetes and metabolic syndrome are important risk factors for stroke, AD, and depression (Farooqui, 2013, 2015). The early control of diets and lifestyle have been shown to lower the rate of MCI (Eshkoor et al., 2015; Lara et al., 2016). Collective evidence suggests that the consumption of the Mediterranean diet and regular exercise not only delays the onset of dementia and protects against cognitive decline (Daviglus et al., 2010), but also enhances the quality of life in older adults by increasing their ability to accomplish everyday tasks with independence (Chou et al., 2012).

DEMENTIA-LINKED CHRONIC VISCERAL AND NEUROLOGICAL DISORDERS

As stated above, the prevalence of dementia is increasing steadily throughout the world's population in most countries and in parallel the prevalences of metabolic diseases (obesity, CVDs, type 2 diabetes, and CVD) and neurological disorders (stroke, AD, PD, and dementia) are also rising (Farooqui, 2013). Obesity increases the risk of dementia in late life by contributing to the accumulation of brain lesions, through vascular and dysmetabolic pathways (Gustafson, 2006; Luchsinger and Gustafson, 2009). However, due to the decline of body weight after midlife and the subtle onset and progress of neuropathology during the long preclinical phase of dementia (Jack et al., 2010), issues of directionality may arise, since age and high BMI in late life may appear to be protective (Gustafson et al., 2009). Any excess risk is plausibly related to adiposity in midlife, when weight gain is more pronounced (Wills et al., 2010), and associations with dementia are least likely disease- and age-confounded. However, whether midlife underweight relates to dementia risk remains to be observed. Some human clinical studies have indicated that consumption of the Western diet may be related to the risk of dementia and MCI, in the form of short-term memory and executive function deficits. The molecular mechanisms linking dementia with the risk of cognitive impairment are still largely unknown but may involve insulin resistance, the gut–brain axis, and systemic and brain proinflammatory mediators leading to central neuroinflammation (Farooqui, 2013, 2015). A common feature of chronic metabolic diseases (diabetes and metabolic syndrome) is not only the induction of insulin resistance and alterations in microbiota composition, but also the onset of obesity and induction of chronic and low-grade inflammation in the brain and gut. This chronic and low-grade inflammation may eventually spread from peripheral tissue to the brain, and recent reports suggest that neuroinflammation is an important causal mechanism in cognitive decline. This inflammatory status may be triggered not only by changes in levels of inflammatory mediators (prostaglandins, leukotrienes, and thromboxanes), but also by alterations in gut microbiota composition in demented subjects with obesity. At the molecular level, metabolic syndrome is accompanied not only by dysregulation in the expression of adipocytokines, cytokines, and chemokines, but also by alterations in levels of inflammatory lipid mediators (Farooqui, 2013). As stated above, metabolic syndrome is a risk factor for neurological disorders such as stroke, depression, and AD type of dementia (Fig. 1.6). Having the metabolic syndrome approximately doubles the risk of having a stroke, AD, and depression compared with age-matched healthy humans (Farooqui, 2013). The molecular mechanism underlying the mirror relationship between metabolic syndrome and neurological disorders is not fully understood. However, all cellular and biochemical alterations observed in metabolic syndrome like onset of diabetes, impairment in endothelial cell function, abnormality in essential fatty acid metabolism,

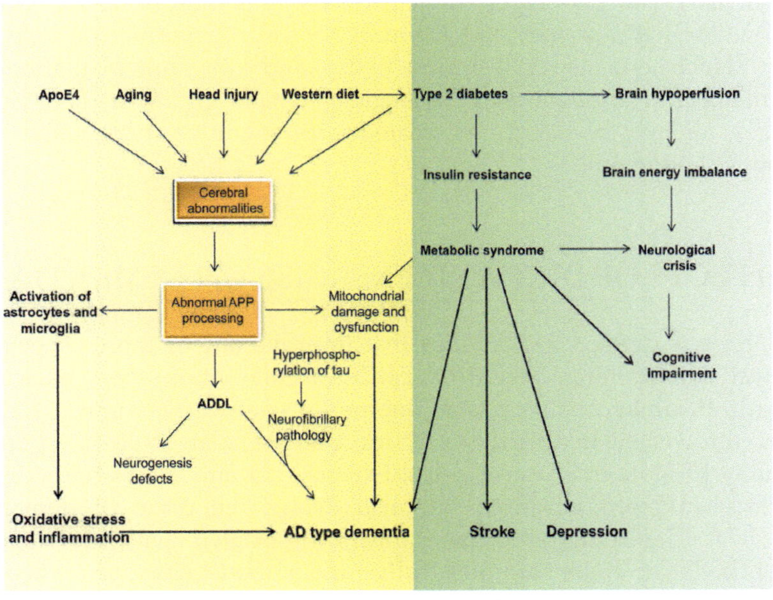

FIGURE 1.6 Interactions between risk factors for Alzheimer's disease and metabolic syndrome. *AD*, Alzheimer's disease; *ADD*, Aβ-derived diffusible ligand; *ApoE4*, apolipoprotein E4; *APP*, amyloid precursor protein.

alterations in lipid mediators, and increase in adipocytokines, cytokines, and chemokines along with abnormal insulin/leptin signaling, and insulin resistance may represent a pathological bridge between metabolic syndrome and neurological disorders such as stroke, depression, and AD type of dementia. In the elder population the simultaneous presence of cognitive dysfunction and reduced gait speed indicate early signs of dementia (presenting decades before actual presentation of cognitive impairment) (Verghese et al., 2013; Borges et al., 2018). This condition is called motoric cognitive risk (MCR) syndrome. It is not known for MCR, how interactions between the cognitive dysfunction and physical domain (e.g., walking) occur with the aging process. However, it is well known that walking requires a complex interplay of sensory, cognitive, and motor functions, and these systems may be altered early in dementias (Verghese et al., 2007). Slow gait results from neurological and non-neurological diseases (Verghese et al., 2006). The predictive validity of MCR for dementia and VaD supports the use of a gait-based phenotype (irrespective of etiology) to identify high-risk individuals (Verghese et al., 2007). White matter lesions and subcortical infarcts are associated with slow gait in aging (Rosano et al., 2006). However, one can speculate the validity of MCR for dementia solely on the basis of vascular dysfunction and pathology, given the occurrence of slow gait early in other neurodegenerative diseases such as AD, LBD, and non-Alzheimer's dementias. It can be suggested that: (1) rehabilitation exercises, which improve strength/power, and postural control; (2) individual walking trails, which enhance physical activity; and (3) implementation of patient-specific motivational strategies, which promote behavioral changes, all may be closely associated with cognitive dysfunction and the physical domain.

EFFECTS OF DIET AND EXERCISE ON DEMENTIA

Exercise is defined as activities that improve one or more aspects of physical conditioning, including cardiovascular endurance, muscular strength, flexibility, balance, and fine motor control (Farooqui, 2014). It is well established that regular physical exercise is an important part of a healthy lifestyle in all age groups. Exercise not only slows brain aging (decrease in brain atrophy), but also delays the onset of dementia (Fig. 1.7). The underlying mechanisms associated with slowing of dementia by exercise are not fully understood. However, exercise produces changes in the brain's structural integrity by enhancing neurogenesis and angiogenesis and by secretions of growth factors promoting synaptic plasticity (Gomez-Pinilla et al., 2008; Farooqui, 2014).

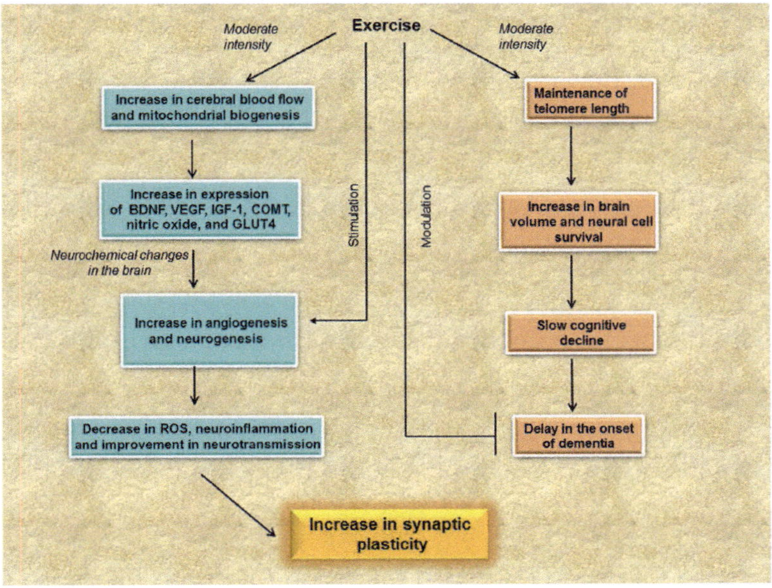

FIGURE 1.7 Effect of exercise on the onset of dementia.
BDNF, brain-derived neurotrophic factor; *COMT*, catechol-*O*-methyltransferase; *GLUT4*, glucose transporter 4; *IGF-1*, insulin-like growth factor-1; *ROS*, reactive oxygen species; *VEGF*, vascular endothelial growth factor.

These processes improve cognitive function by increasing the gray matter volume (Hillman et al., 2008) and by inducing neurogenesis in the dentate gyrus. Neurogenesis is coupled with angiogenesis, which in turn is related to cerebral blood volume (van Praag et al., 1999). It is hypothesized that the measurement of cerebral blood volume may provide an in vivo correlation between neurogenesis and increased cerebral blood flow due to exercise. At the molecular level, many molecules contribute to exercise-induced neurogenesis, angiogenesis, and cerebral blood flow. These molecules include vascular endothelial growth factor, BDNF, IGF-1, catechol-*O*-methyltransferase, endorphins, and NO (Neeper et al., 1996; Stroth et al., 2010; Camargo et al., 2013). In addition, exercise also modulates genes involved in insulin-like signaling, energy metabolism, neurogenesis, and synaptic plasticity, along with learning and memory (Reagan, 2007; van Praag et al., 2005). In addition, in mouse and rat models of AD, exercise decreases Aβ neuropathology (Kim et al., 2014). Also, long-term exercise treatment reduces oxidative stress in the hippocampus of aging rats (Marosi et al., 2012). In a large study of healthy elderly subjects, lower plasma and brain Aβ levels were observed in those reporting higher levels of physical activity (Brown et al., 2013), and similar findings have been found in preclinical

AD subjects (Okonkwo et al., 2014). A number of studies have also described that a single bout of exercise leads to immediate changes in the methylation pattern of certain genes in DNA and affects the proteins that these genes express (Hamer et al., 2013). Exercise-related methylation change appears to be stronger among older people, with age accounting for 30% of the methylation variation (Brown, 2015). In a study of 90-year-old male physicians in the United States, it was reported that maintaining active exercise is an important contributor to good quality aging for physicians who reached 90 years and beyond in good health (Yates et al., 2008). Collectively, these studies strongly support the view that exercise improves brain health by contributing to disease prevention and helping recovery from illness. It influences our cognition and psychological health, and reduces our risk of developing illnesses such as diabetes, heart disease, and stroke (Barnes and Yaffe, 2011).

WEBSITES FOR MORE INFORMATION ON DEMENTIA

- https://www.alz.org/documents_custom/141209-Cognitive AssessmentToo-kit-final.pdf
- See Dementia Module of mhGAP Intervention Guide of WHO www.who.int/mental_health/mhgap/mhGAP_intervention_guide_02/en/
- https://www.mja.com.au/journal/2016/204/5/clinical-practice-guidelines-dementia-australiawww.ipa-online.org/-international psychogeriatric association
- https://www.ipa-online.org/publications/guides-to-bpsd-IPA Complete Guides to BPSD
- https://www.oxfordscholarship.com/view/10.1093/acprof:oso/9780199561636.001.0001/acprof-9780199561636

CONCLUSION

Dementia syndrome is a disorder of cognitive impairment that interferes with everyday life, such as memory, thinking, orientation, comprehension, calculation, learning capacity, language, and judgment. The prevalence of dementia is increasing exponentially with increasing age and the chances of developing dementia doubles every 5 years of age after the age of 65. Age-related cognitive changes in the early stages of dementia are accompanied by alterations in neuronal structure without neuronal death, loss of synapses, and dysfunction of neuronal networks. Several types of dementia are known to occur in the human population; for example, Alzheimer's type of dementia, VaD, mixed dementia (VaD plus AD), LBD, frontotemporal dementia/degeneration (FTD), and dementia as a result of diseases such as type 2 diabetes, stroke, AIDS or multiple sclerosis. AD type of dementia is the most common cause of dementia and is rare before 60 years of age. VaD is known to involve the hardening of cerebral arteries leading to diffuse ischemia and neuronal loss. Molecular mechanisms of dementia are based on the accumulation of abnormal proteins (within or outside cells), mitochondrial dysfunction, increase in oxidative stress, dysregulation in calcium homeostasis, activation of microglia and astrocytes, and induction of neuroinflammation, along with early synaptic disconnection and late apoptotic cell death. At the molecular level, these processes are accompanied by the accumulation of abnormal proteins (Aβ, tau protein, α-syn), which induce the activation of microglia and astrocytes. A healthy lifestyle in dementia has been reported to decrease the rate of cognitive decline and help delay the onset of cognitive symptoms in the setting of age-associated diseases.

References

Aarsland, D., Ballard, C.G., Halliday, G., 2004. Are Parkinson's disease with dementia and dementia with Lewy bodies the same entity? J. Geriatr. Psychiatry Neurol. 17, 137–145.

Alzheimer's Association. What is dementia? Available online: http://www.alz.org/what-is-dementia.asp (accessed 15.12.16.).

Alzheimer's Australia tests used in diagnosing dementia. Available online: https://www.fightdementia.org.au/files/helpsheets/Helpsheet-DementiaQandA10-TestsUsedInDiagnosingDementia_english.pdf (accessed 15.12.16.).

Alzheimer's Society Mediterranean Diet. Available online: https://www.alzheimers.org.uk/info/20010/risk_factors_and_prevention/149/mediterranean_diet (accessed 15.12.16.).

American Psychiatric Association, 2013. Diagnostic and Statistical Manual of Mental Disorders, fifth ed. American Psychiatric Publishing, Arlington.

Aridi, Y.S., Walker, J.L., Wright, O.R.L., 2017. The association between the Mediterranean dietary pattern and cognitive health: a systematic review. Nutrients 9, 674.

Bamford, C., Olsen, K., Davison, C., et al., 2016. Is there a preference for PET or SPECT brain imaging in diagnosing dementia? The views of people with dementia, carers, and healthy controls. Int. Psychogeriatr. 28, 123–131.

Barnes, D.E., Yaffe, K., 2011. The projected impact of risk factor reduction on Alzheimer's disease prevalence. Lancet Neurol. 10, 819–828.

Baumgart, M., Snyder, H.M., Carrillo, M.C., Fazio, S., Kim, H., Johns, H., 2015. Summary of the evidence on modifiable risk factors for cognitive decline and dementia: a population-based perspective. Alzheimers Dement. 11, 718–726.

Bell, R.D., Winkler, E.A., Sagare, A.P., et al., 2010. Pericytes control key neurovascular functions and neuronal phenotype in the adult brain and during brain aging. Neuron 68, 409–427.

Bendiske, J., Bahr, B.A., 2003. Lysosomal activation is a compensatory response against protein accumulation and associated synaptopathogenesis--an approach for slowing Alzheimer disease? J. Neuropathol. Exp. Neurol. 62, 451–463.

Betarbet, R., Sherer, T.B., Greenamyre, J.T., 2002. Animal models of Parkinson's disease. BioEssays 24, 308–318.

Biomarkers Definitions Working Group, 2001. Biomarkers and surrogate endpoints: preferred definitions and conceptual framework. Clin. Pharmacol. Ther. 69, 89–95.

Borges, S.M., Radanovic, M., Forlenza, O.V., 2018. Correlation between functional mobility and cognitive performance in older adults with cognitive impairment. Neuropsychol. Dev. Cogn. B Aging Neuropsychol. Cogn. 25, 23–32.

Boya, P., Reggiori, F., Codogno, P., 2013. Emerging regulation and functions of autophagy. Nat. Cell Biol. 15, 713–720.

Bronstein, J., Carvey, P., Chen, H., Cory-Slechta, D., DiMonte, D., et al., 2009. Meeting report: Consensus statement—Parkinson's disease and the environment: Collaborative on health and the environment and Parkinson's Action Network (CHE PAN) conference, 26–28 June 2007. Environ. Health Perspect. 117, 117–121.

Brown, B.M., Peiffer, J.J., Taddei, K., et al., 2013. Physical activity and amyloid-β plasma and brain levels: results from the Australian imaging, biomarkers and lifestyle study of ageing. Mol. Psychiatry 18, 875–881.

Brown, W.M., 2015. Exercise-associated DNA methylation change in skeletal muscle and the importance of imprinted genes: a bioinformatics meta-analysis. Br. J. Sports Med. 49, 1567–1578.

Buntinx, F., De Lepeleire, J., Paquay, L., Iliffe, S., Schoenmakers, B., 2011. Diagnosing dementia: no easy job. BMC Fam. Pract. 12, 60.

Cai, W., Zhang, K., Li, P., Zhu, L., Xu, J., et al., 2017. Dysfunction of the neurovascular unit in ischemic stroke and neurodegenerative diseases: an aging effect. Ageing Res. Rev. 34, 77–87.

Camargo, L.H., Alves, F.H., Biojones, C., Correa, F.M., Resstel, L.B., et al., 2013. Involvement of N-methyl-d-aspartate glutamate receptor and nitric oxide in cardiovascular responses to dynamic exercise in rats. Eur. J. Pharmacol. 713, 16–24.

Cerejeira, J., Lagarto, L., Mukaetova-Ladinska, E.B., 2012. Behavioral and psychological symptoms of dementia. Front. Neurol. 3, 73.

Chen, C.-W., Lin, C.-C., Chen, K.-B., Kuo, Y.-C., Li, C.-Y., Chung, C.-J., 2013. Increased risk of dementia in people with previous exposure to general anesthesia: a nationwide population-based case control study. Alzheimers and Dement. 10, 196–204.

Cherra 3rd, S.J., Chu, C.T., 2008. Autophagy in neuroprotection and neurodegeneration: a question of balance. Future Neurol. 3, 309–323.

Chin, J., Massaro, C.M., Palop, J.J., et al., 2007. Reelin depletion in the entorhinal cortex of human amyloid precursor protein transgenic mice and humans with Alzheimer's disease. J Neurosci. 27, 2727–2733.

Chou, C.-H., Hwang, C.-L., Wu, Y.-T., 2012. Effect of exercise on physical function, daily living activities, and quality of life in the frail older adults: a meta-analysis. Arch. Phys. Med. Rehabil. 93, 237–244.

Craft, S., Foster, T.C., Landfield, P.W., Maier, S.F., Resnick, S.M., Yaffe, K., 2012. Session III: Mechanisms of age-related cognitive change and targets for intervention: inflammatory, oxidative, and metabolic processes. J. Gerontol. A Biol. Sci. Med. Sci. 67, 754–759.

Cuadrado-Tejedor, M., García-Osta, A., 2014. Current animal models of Alzheimer's Disease: challenges in translational research. Front. Neurol. 5, 182.

Cuervo, A.M., Stefanis, L., Fredenburg, R., Lansbury, P.T., Sulzer, D., 2004. Impaired degradation of mutant alpha-synuclein by chaperone-mediated autophagy. Science 305, 1292–1295.

Daviglus, M.L., Bell, C.C., Berrettini, W., Bowen, P.E., Connolly, E.S., et al., 2010. NIH State-of-the-Science Conference Statement: Preventing Alzheimer's Disease and Cognitive Decline. NIH Consens. State Sci. Statements 27, 1–30.

Daviglus, M.L., Plassman, B.L., Pirzada, A., et al., 2011. Risk factors and preventive interventions for Alzheimer disease: state of the science. Arch Neurol. 68, 1185–1190.

DeCarli, C., Kawas, C., Morrison, J.H., Reuter-Lorenz, P.A., Sperling, R.A., Wright, C.B., 2012. Session II: mechanisms of age-related cognitive change and targets for intervention: neural circuits, networks, and plasticity. J. Gerontol. A Biol. Sci. Med. Sci. 67, 747–753.

Dong, M.J., Peng, B., Lin, X.T., Zhao, J., Zhou, Y.R., Wang, R.H., 2007. The prevalence of dementia in the People's Republic of China: a systematic analysis of 1980-2004 studies. Age Ageing 36, 619–624.

Drapeau, E., Abrous, N.D., 2008. Stem cell review series: role of neurogenesis in age-related memory disorders. Aging Cell 7, 569–589.

Du, S.Q., Wang, X.R., Xiao, L.Y., Tu, J.F., Zhu, W., et al., 2017. Molecular mechanisms of vascular dementia: what can be learned from animal models of chronic cerebral hypoperfusion? Mol. Neurobiol. 54, 3670–3682.

Duncombe, J., Kitamura, A., Hase, Y., Ihara, M., Kalaria, R.N., Horsburgh, K., 2017. Chronic cerebral hypoperfusion: a key mechanism leading to vascular cognitive impairment and dementia. Closing the translational gap between rodent models and human vascular cognitive impairment and dementia. Clin. Sci. (Lond.) 131, 2451–2468.

Elder, G.A., Gama Sosa, M.A., De Gasperi, R., 2010. Transgenic mouse models of Alzheimer's disease. Mt. Sinai J. Med. 77, 69–81.

Eshkoor, S.A., Hamid, T.A., Mun, C.Y., Ng, C.K., 2015. Mild cognitive impairment and its management in older people. Clin. Interv. Aging 10, 687–693.

Farkas, E., De Jong, G.I., de Vos, R.A., Jansen Steur, E.N., Luiten, P.G., 2000. Pathological features of cerebral cortical capillaries are doubled in Alzheimer's disease and Parkinson's disease. Acta Neuropathol. 100, 395–402.

Farkas, E., Luiten, P.G., Bari, F., 2007. Permanent, bilateral common carotid artery occlusion in the rat: a model for chronic cerebral hypoperfusion-related neurodegenerative diseases. Brain Res. Rev. 54, 162–180.

Farooqui, A.A., 2010. Neurochemical Aspects of Neurotraumatic and Neurodegenerative Diseases. Springer, New York.

Farooqui, A.A., 2013. Metabolic Syndrome: An Important Risk Factor for Stroke, Alzheimer Disease, and Depression. Springer International Publishing, Switzerland.

Farooqui, A.A., 2014. Inflammation and Oxidative Stress in Neurological Disorders. Springer International Publishing, Switzerland.

Farooqui, A.A., 2015. High Calorie Diet and the HumanBrain. Springer International Publishing, Switzerland.

Farooqui, A.A., 2017. Neurochemical Aspects of Alzheimer's Disease. Academic Press, an imprint of Elsevier, San Diego, CA.

Finger, E.C., 2016. Frontotemporal dementias. Continuum (Minneap Minn) 22, 464–489.

Gill, R., Tsung, A., Billiar, T., 2010. Linking oxidative stress to inflammation: toll-like receptors. Free Radic. Biol. Med. 48, 1121–1132.

Girouard, H., Park, L., Anrather, J., Zhou, P., Iadecola, C., 2007. Cerebrovascular nitrosative stress mediates neurovascular and endothelial dysfunction induced by angiotensin ii. Arterioscler. Thromb. Vasc. Biol. 27, 303–309.

Goldman, J.G., Williams-Gray, C., Barker, R.A., Duda, J.E., Galvin, J.E., 2014. The spectrum of cognitive impairment in Lewy body diseases. Mov. Disord. 29, 608–621.

Gomez-Pinilla, F., Vaynman, S., Ying, Z., 2008. Brain-derived neurotrophic factor functions as a metabotrophin to mediate the effects of exercise on cognition. Eur. J. Neurosci. 28, 2278–2287.

Gordon-Salant, S., 2005. Hearing loss and aging: new research findings and clinical implications. J. Rehabil. Res. Dev 42 (4 Suppl. 2), 9–24.

Government of Canada, Statistics Canada, 2016. Census in brief: recent trends for the population aged 15 to 64 in Canada, census year 2016. Available online at: http://www12.statcan.gc.ca/census-recensement/2016/as-sa/98-200-x/2016003/98-200-x2016003-eng.cfm (accessed 22.06.17).

Graham, G., 2014. Population-based approaches to understanding disparities in cardiovascular disease risk in the United States. Int. J. Gen. Med. 7, 393–400.

Gurgel, R.K., Ward, P.D., Schwartz, S., Norton, M.C., Foster, N.L., et al., 2014. Relationship of hearing loss and dementia: a prospective, population-based study. Otol. Neurotol. 35, 775–781.

Gustafson, D., 2006. Adiposity indices and dementia. Lancet Neurol. 5, 713–720.

Gustafson, D.R., Backman, K., Waern, M., Ostling, S., Guo, X., et al., 2009. Adiposity indicators and dementia over 32 years in Sweden. Neurology 73, 1559–1566.

Hachinski, V.C., Lassen, N.A., Marshall, J., 1974. Multi-infarct dementia. A cause of mental deterioration in the elderly. Lancet 2, 207–210.

Hamer, M., Lavoie, K.L., Bacon, S.L., 2013. Taking up physical activity in later life and healthy ageing: the English longitudinal study of ageing. Br. J. Sports Med. 48, 239–243.

Hashimoto, M., Masliah, E., 1999. Alpha-synuclein in Lewy body disease and Alzheimer's disease. Brain Pathol. 9, 707–720.

Head, E., Corrada, M.M., Kahle-Wrobleski, K., et al., 2009. Synaptic proteins, neuropathology and cognitive status in the oldest-old. Neurobiol. Aging 30, 1125–1134.

Henriksen, K., O'Bryant, S.E., Hampel, H., et al., 2014. The future of blood-based biomarkers for Alzheimer's disease. Alzheimers Dement. 10, 115–131.

Herz, J., Chen, Y., 2006. Reelin, lipoprotein receptors and synaptic plasticity. Nat. Rev. Neurosci. 7, 850–859.

Hiesberger, T., Trommsdorff, M., Howell, B.W., et al., 1999. Direct binding of Reelin to VLDL receptor and ApoE receptor 2 induces tyrosine phosphorylation of disabled-1 and modulates tau phosphorylation. Neuron 24, 481–489.

Hillman, C.H., Erickson, K.I., Kramer, A.F., 2008. Be smart, exercise your heart: exercise effects on brain and cognition. Nat. Rev. Neurosci. 9, 58–65.

Hoe, H.S., Lee, K.J., Carney, R.S., et al., 2009. Interaction of reelin with amyloid precursor protein promotes neurite outgrowth. J. Neurosci. 29, 7459–7473.

Hugo, J., Ganguli, M., 2014. Dementia and cognitive impairment: epidemiology, diagnosis, and treatment. Clin. Geriatr. Med. 30, 421–442.

Hurd, M.D., Martorell, P., Delavande, A., Mullen, K.J., Langa, K.M., 2013. Monetary costs of dementia in the United States. N. Engl. J. Med. 368, 1326–1334.

Hurtig, H., Trojanowski, J., Galvin, J., Ewbank, D., Schmidt, M., et al., 2000. Alpha-synuclein cortical Lewy bodies correlate with dementia in Parkinson's disease. Neurology 54, 1916–1921.

Iadecola, C., 2013. The pathobiology of vascular dementia. Neuron 80, 844–866.

Iadecola, C., 2014. Hypertension and dementia. Hypertension 64, 3–5.

Iadecola, C., Park, L., Capone, C., 2009. Threats to the mind: aging, amyloid, and hypertension. Stroke 40, S40–S44.

Ironside, J.W., Bell, J.E., 2007. Infective dementias. Panminerva Med. 49, 209–225.

Jack, C.R., 2012. Alzheimer disease: new concepts on its neurobiology and the clinical role imaging will play. Radiology 263, 344–361.

Jack Jr., C.R., Knopman, D.S., Jagust, W.J., Shaw, L.M., Aisen, P.S., et al., 2010. Hypothetical model of dynamic biomarkers of the Alzheimer's pathological cascade. Lancet Neurol. 9, 119–128.

Jagust, W., 2006. Positron emission tomography and magnetic resonance imaging in the diagnosis and prediction of dementia. Alzheimers Dement. 2, 36–42.

Jankovic, J., 2008. Parkinson's disease: clinical features and diagnosis. J. Neurol. Neurosurg. Psychiatry 79, 368–376.

Jellinger, K.A., 2008. Morphologic diagnosis of "vascular dementia"—a critical update. J. Neurol. Sci. 270, 1–12.

Jiang, J., Dong, Y., Huang, W., Bao, M., 2017. General anesthesia exposure and risk of dementia: a meta-analysis of epidemiological studies. Oncotarget. 8, 59628–59637.

Jiwa, N.S., Garrard, P., Hainsworth, A.H., 2010. Experimental models of vascular dementia and vascular cognitive impairment: a systematic review. J. Neurochem. 115, 814–828.

Kabasakalian, A., Finney, G.R., 2009. Reversible dementias. Int. Rev. Neurobiol. 84, 283–302.

Kalaria, R.N., 2010. Vascular basis for brain degeneration: faltering controls and risk factors for dementia. Nutr. Rev. 68 (Suppl. 2), S74–S87.

Kales, H.C., Gitlin, L.N., Lyketsos, C.G., 2015. Assessment and management of behavioral and psychological symptoms of dementia. BMJ 350, h369.

Katusic, Z.S., Austin, S.A., 2014. Endothelial nitric oxide: protector of a healthy mind. Eur. Heart J. 35, 888–894.

Kazama, K., Anrather, J., Zhou, P., et al., 2004. Angiotensin ii impairs neurovascular coupling in neocortex through nadph oxidase-derived radicals. Circ. Res. 95, 1019–1026.

Kim, B.-K., Shin, M.-S., Kim, C.-J., Baek, S.-B., Ko, Y.-C., Kim, Y.-P., 2014. Treadmill exercise improves short-term memory by enhancing neurogenesis in amyloid beta-induced Alzheimer disease rats. J. Exerc. Rehabil. 10, 2–8.

Kim, S.K., Cho, K.O., Kim, S.Y., 2008. White matter damage and hippocampal neurodegeneration induced by permanent bilateral occlusion of common carotid artery in the rat: comparison between wistar and sprague-dawley strain. Korean J. Physiol. Pharmacol. 12, 89–94.

Kim, S.K., Cho, K.O., Kim, S.Y., 2009. The plasticity of posterior communicating artery influences on the outcome of white matter injury induced by chronic cerebral hypoperfusion in rats. Neurol. Res. 31, 245–250.

Knight, J., Nigam, Y., 2008. Exploring the anatomy and physiology of ageing. Part 1–The cardiovascular system. Nurs. Times 104, 26–27.

Koikkalainen, J., Rhodius-Meester, H., Tolonen, A., Barkhof, F., Tijms, B., et al., 2016. Differential diagnosis of neurodegenerative diseases using structural MRI data. Neuroimage Clin. 11, 435–449.

Kosik, K.S., Rapp, P.R., Raz, N., Small, S.A., Sweatt, J.D., Tsai, L.H., 2012. Mechanisms of age-related cognitive change and targets for intervention: epigenetics. J. Gerontol. A Biol. Sci. Med. Sci. 67, 741–746.

Kragh, C.L., Ubhi, K., Wyss-Coray, T., Masliah, E., 2012. Autophagy in dementias. Brain Pathol. 22, 99–109.

Kremen, W.S., Lachman, M.E., Pruessner, J.C., Sliwinski, M., Wilson, R.S., 2012. Mechanisms of age-related cognitive change and targets for intervention: social interactions and stress. J. Gerontol. A Biol. Sci. Med. Sci. 67, 760–765.

Kuhn, H.G., Dickinson-Anson, H., Gage, F.H., 1996. Neurogenesis in the dentate gyrus of the adult rat: age-related decrease of neuronal progenitor proliferation. J. Neurosci. 16, 2027–2033.

Lara, E., Koyanagi, A., Olaya, B., et al., 2016. Mild cognitive impairment in a Spanish representative sample: prevalence and associated factors. Int. J. Geriatr. Psychiatry 31, 858–867.

Lee, J.A., Gao, F.B., 2009. Inhibition of autophagy induction delays neuronal cell loss caused by dysfunctional ESCRT-III in frontotemporal dementia. J. Neurosci. 29, 8506–8511.

Levy, J.A., Chelune, G.J., 2007. Cognitive-behavioral profiles of neurodegenerative dementias: beyond Alzheimer's disease. J. Geriatr. Psychiatry Neurol. 20, 227–238.

Libon, D.J., Price, C.C., Heilman, K.M., Grossman, M., 2006. Alzheimer's "other dementia". Cogn. Behav. Neurol. 19, 112–116.

Lin, F.R., Yaffe, K., Xia, J., Xue, Q.L., Harris, T.B., et al., 2013. Hearing loss and cognitive decline in older adults. JAMA Intern. Med. 173, 293–299.

Llibre Rodriguez, J., Ferri, C., Acosta, D., et al., 2008. Prevalence of dementia in Latin America, India, and China: a population-based cross-sectional survey. Lancet 372, 464–474.

Luchsinger, J.A., Gustafson, D.R., 2009. Adiposity, type 2 diabetes, and Alzheimer's disease. J. Alzheimers Dis. 16, 693–704.

Lyketsos, C.G., Lopez, O., Jones, B., Fitzpatrick, A.L., Breitner, J., DeKosky, S., 2002. Prevalence of neuropsychiatric symptoms in dementia and mild cognitive impairment: results from the cardiovascular health study. JAMA 288, 1475–1483.

Marchesi, C., Paradis, P., Schiffrin, E.L., 2008. Role of the renin-angiotensin system in vascular inflammation. Trends Pharmacol. Sci. 29, 367–374.

Marosi, K., Bori, Z., Hart, N., et al., 2012. Long-term exercise treatment reduces oxidative stress in the hippocampus of aging rats. Neuroscience 226, 21–28.

Marstrand, J.R., Garde, E., Rostrup, E., Ring, P., Rosenbaum, S., Mortensen, E.L., 2002. Cerebral perfusion and cerebrovascular reactivity are reduced in white matter hyperintensities. Stroke 33, 972–976.

Marui, W., Iseki, E., Nakai, T., et al., 2002. Progression and staging of Lewy pathology in brains from patients with dementia with Lewy bodies. J. Neurol. Sci. 195, 153–159.

Masliah, E., Sisk, A., Mallory, M., Games, D., 2001. Neurofibrillary pathology in transgenic mice overexpressing V717F beta-amyloid precursor protein. J. Neuropathol. Exp. Neurol. 60, 357–368.

Matthews, F.E., Arthur, A., Barnes, L.E., Bond, J., Jagger, C., on behalf of the Medical Research Council Cognitive Function and Ageing Collaboration, et al., 2013. A two-decade comparison of prevalence of dementia in individuals aged 65 years and older from three geographical areas of England: results of the Cognitive Function and Ageing Study I and II. Lancet 382, 1405–1412.

Mattson, M.P., Magnus, T., 2006. Ageing and neuronal vulnerability. Nat. Rev. Neurosci. 2006 (7), 278–294.

Mattsson, N., Zetterberg, H., Hansson, O., et al., 2009. CSF biomarkers and incipient Alzheimer disease in patients with mild cognitive impairment. JAMA 302, 385–393.

Mayo, M.C., Bordelon, Y., 2014. Dementia with Lewy bodies. Semin. Neurol. 34, 182–188.

McKeith, I., 2009. Commentary: DLB and PDD: the same or different? Is there a debate? Int. Psychogeriatr. 21, 220–224.

McKhann, G.M., Knopman, D.S., Chertkow, H., et al., 2011. The diagnosis of dementia due to Alzheimer's disease: recommendations from the National Institute on Aging-Alzheimer's Association workgroups on diagnostic guidelines for Alzheimer's disease. Alzheimers Dement. 7, 263–269.

Mizushima, N., Yoshimori, T., Ohsumi, Y., 2011. The role of Atg proteins in autophagosome formation. Annu. Rev. Cell Dev. Biol. 27, 107–132.

Morrison, J.H., Baxter, M.G., 2012. The aging cortical synapse: hallmarks and implications for cognitive decline. Nat. Rev. Neurosci. 13, 240–250.

Mosconi, L., 2005. Brain glucose metabolism in the early and specific diagnosis of Alzheimer's disease FDG-PET studies in MCI and AD. Eur. J. Nucl. Med. Mol. Imaging 32, 486–516.

Mosconi, L., Murray, J., Tsui, W.H., Li, Y., Davies, M., Williams, S., et al., 2014. Mediterranean diet and magnetic resonance imaging-assessed brain atrophy in cognitively normal individuals at risk for alzheimer's disease. J. Prev. Alzheimers Dis. 1, 23–32.

Nagata, K., Saito, H., Ueno, T., et al., 2007. Clinical diagnosis of vascular dementia. J. Neurol. Sci. 257, 44–48.

Nakatogawa, H., Suzuki, K., Kamada, Y., Ohsumi, Y., 2009. Dynamics and diversity in autophagy mechanisms: lessons from yeast. Nat. Rev. Mol. Cell Biol. 10, 458–467.

Neeper, S.A., Gomez-Pinilla, F., Choi, J., Cotman, C.W., 1996. Physical activity increases mRNA for brain-derived neurotrophic factor and nerve growth factor in rat brain. Brain Res. 726, 49–56.

Nelson, A.R., Sweeney, M.D., Sagare, A.P., Zlokovic, B.V., 2016. Neurovascular dysfunction and neurodegeneration in dementia and Alzheimer's disease. Biochim. Biophys. Acta 1862, 887–900.

Niemantsverdriet, E., Feyen, B.F., Le Bastard, N., Martin, J.J., Goeman, J., et al., 2015. Overdiagnosing vascular dementia using structural brain imaging for dementia work-up. J. Alzheimer's Dis. 45, 1039–1043.

Niures, M., Matioli, P.S., Nitrini, R., 2015. Mechanisms linking brain insulin resistance to Alzheimer disease. Dement. Neuropsychol. 9, 96–102.

Nixon, R.A., 2007. Autophagy, amyloidogenesis and Alzheimer disease. J. Cell Sci. 120 (Pt 23), 4081–4091.

Okonkwo, O.C., Schultz, S.A., Oh, J.M., et al., 2014. Physical activity attenuates age-related biomarker alterations in preclinical AD. Neurology 83, 1753–1760.

Olney, N.T., Spina, S., Miller, B.L., 2017. Frontotemporal dementia. Neurol. Clin. 35, 339–374.

Ortman, J.M., Velkoff, V.A., Hogan, H., 2014. An Aging Nation: The Older Population in the United States. US Census Bureau, Washington, DC, pp. 25–1140.

Ostling, S., Gustafson, D., Waern, M., 2009. Psychotic and behavioural symptoms in a population-based sample of the very elderly subjects. Acta Psychiatr. Scand. 120, 147–152.

Park, L., Anrather, J., Girouard, H., Zhou, P., Iadecola, C., 2007. Nox2-derived reactive oxygen species mediate neurovascular dysregulation in the aging mouse brain. J. Cereb. Blood Flow Metab. 27, 1908–1918.

Park, L., Zhou, P., Pitstick, R., et al., 2008. Nox2-derived radicals contribute to neurovascular and behavioral dysfunction in mice overexpressing the amyloid precursor protein. Proc. Natl. Acad. Sci. U.S.A. 105, 1347–1352.

Perez, M., Ribe, E., Rubio, A., et al., 2005. Characterization of a double (amyloid precursor protein-tau) transgenic: Tau phosphorylation and aggregation. Neuroscience 130, 339–347.

Petrovitch, H., Ross, G.W., Steinhorn, S.C., et al., 2005. AD lesions and infarcts in demented and non-demented Japanese-American men. Ann. Neurol. 57, 98–103.

Planel, E., Richter, K.E.G., Nolan, C.E., et al., 2007. Anesthesia leads to tau hyperphosphorylation through inhibition of phosphatase activity by hypothermia. J. Neurosci. 27, 3090–3097.

Plassman, B.L., Langa, K.M., Fisher, G.G., et al., 2007. Prevalence of dementia in the United States: the aging, demographics, and memory study. Neuroepidemiology 29, 125–132.

Plassman, B.L., Williams Jr., J.W., Burke, J.R., Holsinger, T., Benjamin, S., 2010. Systematic review: factors associated with risk for and possible prevention of cognitive decline in later life. Ann. Intern. Med. 153, 182–193.

Prince, M., Prina, M., Guerchet, M., 2013a. World Alzheimer Report 2013. Journey of Caring. An Analysis of Long-Term Care for Dementia. Alzheimer's Disease International, London.

Prince, M., Bryce, R., Albanese, E., Wimo, A., Ribeiro, W., Ferri, C.P., 2013b. The global prevalence of dementia: a systematic review and meta-analysis. Alzheimer's Dement. 9, 63–75.

Ramos-Moreno, T., Galazo, M.J., Porrero, C., Martínez-Cerdeño, V., Clascá, F., 2006. Extracellular matrix molecules and synaptic plasticity: immunomapping of intracellular and secreted Reelin in the adult rat brain. Eur. J. Neurosci. 23, 401–422.

Rao, M.S., Hattiangady, B., Abdel-Rahman, A., Stanley, D.P., Shetty, A.K., 2005. Newly born cells in the ageing dentate gyrus display normal migration, survival and neuronal fate choice but endure retarded early maturation. Eur. J. Neurosci. 21, 464–476.

Rao, M.S., Hattiangady, B., Shetty, A.K., 2006. The window and mechanisms of major age-related decline in the production of new neurons within the dentate gyrus of the hippocampus. Aging Cell 5, 545–558.

Rasgon, N.L., Kenna, H.A., Wroolie, T.E., et al., 2010. Insulin resistance and hippocampal volume in women at risk for Alzheimer's disease. Neurobiol. Aging 32, 1942–1948.

Reagan, L.P., 2007. Insulin signaling effects on memory and mood. Curr. Opin. Pharmacol. 7, 633–637.
Ribe, E.M., Perez, M., Puig, B., Gich, I., Lim, F., Cuadrado, M., 2005. Accelerated amyloid deposition, neurofibrillary degeneration and neuronal loss in double mutant APP/tau transgenic mice. Neurobiol. Dis. 20, 814–822.
Ribeiro, F.M., Camargos, E.R., de Souza, L.C., Teixeira, A.L., 2013. Animal models of neurodegenerative diseases. Rev. Bras. Psiquiatr. 35 (Suppl. 2), S82–S91.
Risacher, S.L., Saykin, A.J., 2013. Neuroimaging biomarkers of neurodegenerative diseases and dementia. Semin. Neurol. 33, 386–416.
Rizzi, L., Rosset, I., Roriz-Cruz, M., 2014. Global epidemiology of dementia: Alzheimer's and vascular types. Biomed. Res. Int. 2014, 908915.
Román, G.C., Erkinjuntti, T., Wallin, A., Pantoni, L., Chui, H.C., 2002. Subcortical ischaemic vascular dementia. Lancet Neurol. 1, 426–436.
Rosano, C., Brach, J., Longstreth Jr, W.T., Newman, A.B., 2006. Quantitative measures of gait characteristics indicate prevalence of underlying subclinical structural brain abnormalities in high-functioning older adults. Neuroepidemiology 26, 52–60.
Sagare, A.P., Bell, R.D., Zhao, Z., et al., 2013. Pericyte loss influences Alzheimer-like neurodegeneration in mice. Nat. Commun. 4, 2932.
Savva, G.M., Arthur, A., 2015. Who has undiagnosed dementia? A cross-sectional analysis of participants of the Aging, Demographics and Memory Study. Age Ageing 44, 642–647.
Savva, G.M., Wharton, S.B., Ince, P.G., Forster, G., Matthews, F.E., Brayne, C., 2009. Age, neuropathology, and dementia. N. Engl. J. Med. 360, 2302–2309.
Scarmeas, N., Stern, Y., Mayeux, R., Luchsinger, J.A., 2006a. Mediterranean diet, Alzheimer disease, and vascular mediation. Arch. Neurol. 63, 1709–1717.
Scarmeas, N., Stern, Y., Mayeux, R., Luchsinger, J.A., 2006b. Mediterranean diet, Alzheimer disease, and vascular mediation. Arch. Neurol. 63, 1709–1717.
Schnaider Beeri, M., Haroutunian, V., Schmeidler, J., et al., 2012. Synaptic protein deficits are associated with dementia irrespective of extreme old age. Neurobiol. Aging 33, 1125.e1–e8.
Schreiber, S., Bueche, C.Z., Garz, C., Braun, H., 2013. Blood brain barrier breakdown as the starting point of cerebral small vessel disease? New insights from a rat model. Exp. Transl. Stroke Med. 5, 4.
Schuh, A.F., Rieder, C.M., Rizzi, L., Chaves, M., Roriz-Cruz, M., 2011. Mechanisms of brain aging regulation by insulin: implications for neurodegeneration in late-onset Alzheimer's Disease. ISRN Neurol. 2011, 306905.
Seppala, T.T., Nerg, O., Koivisto, A.M., et al., 2012. CSF biomarkers for Alzheimer disease correlate with cortical brain biopsy findings. Neurology 78, 1568–1575.
Seshadri, S., Beiser, A., Selhub, J., et al., 2002. Plasma homocysteine as a risk factor for dementia and Alzheimer's disease. N. Engl. J. Med. 346, 476–483.
Shi, Y., Wardlaw, J.M., 2016. Update on cerebral small vessel disease: a dynamic whole-brain disease. Stroke Vasc. Neurol. 1, 83–92.
Shintani, T., Klionsky, D.J., 2004. Autophagy in health and disease: a double-edged sword. Science 306, 990–995.
Simpson, J.E., Ince, P.G., Haynes, L.J., Theaker, R., Gelsthorpe, C., et al., 2010. Population variation in oxidative stress and astrocyte DNA damage in relation to Alzheimer-type pathology in the ageing brain. Neuropathol. Appl. Neurobiol. 36, 25–40.
Singh, B., Parsaik, A.K., Mielke, M.M., Erwin, P.J., Knopman, D.S., et al., 2014. Association of mediterranean diet with mild cognitive impairment and Alzheimer's disease: a systematic review and meta-analysis. J. Alzheimers Dis. 39, 271–282.
Smith, P.J., Blumenthal, J.A., 2016. Dietary factors and cognitive decline. J. Prev. Alzheimers Dis. 3, 53–64.
Sosa-Ortiz, A.L., Acosta-Castillo, I., Prince, M.J., 2012. Epidemiology of dementias and Alzheimer's disease. Arch. Med. Res. 43, 600–608.

Stomrud, E., Hansson, O., Blennow, K., et al., 2007. Cerebrospinal fluid biomarkers predict decline in subjective cognitive function over 3 years in healthy elderly. Dement. Geriatr. Cogn. Disord. 24, 118–124.

Stroth, S., Reinhardt, R.K., Thöne, J., Hille, K., Schneider, M., et al., 2010. Impact of aerobic exercise training on cognitive functions and affect associated to the COMT polymorphism in young adults. Neurobiol. Learn. Memory 94, 364–372.

Stukas, S., Robert, J., Wellington, C.L., 2014. High-density lipoproteins and cerebrovascular integrity in Alzheimer's disease. Cell Metab. 19, 574–591.

Takemori, K., Murakami, T., Kometani, T., Ito, H., 2013. Possible involvement of oxidative stress as a causative factor in blood-brain barrier dysfunction in stroke-prone spontaneously hypertensive rats. Microvasc. Res. 90, 169–172.

Tarantini, S., Tran, C.H.T., Gordon, G.R., Ungvari, Z., Csiszar, A., 2017. Impaired neurovascular coupling in aging and Alzheimer's disease: contribution of astrocyte dysfunction and endothelial impairment to cognitive decline. Exp. Gerontol. 94, 52–58.

The 2015 Ageing Report, 2015. The 2015 Ageing Report: Economic and Budgetary Projections for the 28 EU Member States (2013–2060). Publications Office, Luxembourg. Available online at: http://bookshop.europa.eu/uri?target = EUB:NOTICE:KCAR15003:EN:HTML

Toth, P., Tucsek, Z., Sosnowska, D., Gautam, T., Mitschelen, M., et al., 2013a. Age-related autoregulatory dysfunction and cerebromicrovascular injury in mice with angiotensin II-induced hypertension. J. Cereb. Blood Flow Metab. 33, 1732–1742.

Toth, P., Csiszar, A., Tucsek, Z., Sosnowska, D., Gautam, T., et al., 2013b. Role of 20-HETE, TRPC channels, and BKCa in dysregulation of pressure-induced Ca^{2+} signaling and myogenic constriction of cerebral arteries in aged hypertensive mice. Am. J. Physiol. Heart Circ. Physiol. 305, H1698–H1708.

Toth, P., Tarantini, S., Ashpole, N.M., et al., 2015. Igf-1 deficiency impairs neurovascular coupling in mice: implications for cerebromicrovascular aging. Aging Cell 14, 1034–1044.

Toth, P., Tarantini, S., Csiszar, A., Ungvari, Z., 2017. Functional vascular contributions to cognitive impairment and dementia: mechanisms and consequences of cerebral autoregulatory dysfunction, endothelial impairment, and neurovascular uncoupling in aging. Am. J. Physiol. Heart Circ. Physiol. 312, H1–H20.

United Nations, 2015. Department of Economic and Social Affairs, Population Division. World Population Ageing 2015 (ST/ESA/SER.A/390). United Nations, New York, NY.

Valkanova, V., Ebmeier, K.P., 2014. Neuroimaging in dementia. Maturitas 79, 202–208.

Van Den Wijngaart, M.A., Vernooij-Dassen, M.J., Felling, A.J., 2007. The influence of stressors, appraisal and personal conditions on the burden of spousal caregivers of persons with dementia. Aging Ment. Health 11, 626–636.

van Praag, H., Kempermann, G., Gage, F.H., 1999. Running increases cell proliferation and neurogenesis in the adult mouse dentate gyrus. Nat. Neurosci. 2, 266–270.

van Praag, H., Shubert, T., Zhao, C., Gage, F.H., 2005. Exercise enhances learning and hippocampal neurogenesis in aged mice. J. Neurosci. 25, 8680–8685.

Vas, C.J., Pinto, C., Panikkar, D., Noronha, S., 2001. Prevalence of dementia in an urban Indian population. Int. Psychogeriatr. 13, 389–393.

Venkat, P., Chopp, M., Chen, J., 2015. Models and mechanisms of vascular dementia. Exp. Neurol. 272, 97–108.

Verghese, J., LeValley, A., Hall, C.B., Katz, M.J., Ambrose, A.F., Lipton, R.B., 2006. Epidemiology of gait disorders in community-residing older adults. J. Am. Geriatr. Soc. 54, 255–261.

Verghese, J., Wang, C., Lipton, R.B., Holtzer, R., Xue, X., 2007. Quantitative gait dysfunction and risk of cognitive decline and dementia. J. Neurol. Neurosurg. Psychiatr. 78, 929–935.

Verghese, J., Wang, C., Lipton, R.B., Holtzer, R., 2013. Motoric cognitive risk syndrome and the risk of dementia. J. Gerontol. A Biol. Sci. Med. Sci. 68, 412–418.

Villemagne, V.L., Burnham, S., Bourgeat, P., et al., 2013. Amyloid β deposition, neurodegeneration, and cognitive decline in sporadic Alzheimer's disease: a prospective cohort study. Lancet Neurol. 12, 357–367.

WHO, 2016. Dementia. Available online at: http://www.who.int/mediacentre/factsheets/fs362/en/.

Wardlaw, J.M., Doubal, F., Armitage, P., et al., 2009. Lacunar stroke is associated with diffuse blood-brain barrier dysfunction. Ann. Neurol. 65, 194–202.

Wills, A.K., Hardy, R.J., Black, S., Kuh, D.J., 2010. Trajectories of overweight and body mass index in adulthood and blood pressure at age 53: the 1946 British birth cohort study. J. Hypertens. 28, 679–686.

Wimo, A., Guerchet, M., Ali, G.C., et al., 2017. The worldwide costs of dementia 2015 and comparisons with 2010. Alzheimers Dement. 13, 1.

Wongrakpanich, S., Petchlorlian, A., Rosenzweig, A., 2016. Sensorineural organs dysfunction and cognitive decline: a review article. Aging Dis. 7, 763–769.

World Health Organization, 2012. Dementia A Public Health Priority. World Health Organization.

Xie, Z., Culley, D.J., Dong, Y., et al., 2008. The common inhalation anesthetic isoflurane induces caspase activation and increases amyloid beta-protein level in vivo. Ann. Neurol. 64, 618–627.

Yaffe, K., Blackwell, T., Kanaya, A.M., Davidowitz, N., Barrett-Connor, E., Krueger, K., 2004. Diabetes, impaired fasting glucose, and development of cognitive impairment in older women. Neurology 63, 658–663.

Yang, Y., Rosenberg, G.A., 2011. Blood-brain barrier breakdown in acute and chronic cerebrovascular disease. Stroke 42, 3323–3328.

Yates, L.B., Djousse, L., Kurth, T., Buring, J.E., Gaziano, M., 2008. Exceptional longevity in men: modificable factors associated with survival and function to age 90 years. Arch. Intern. Med. 168, 284–290.

Zlokovic, B.V., 2008. The blood-brain barrier in health and chronic neurodegenerative disorders. Neuron 57, 178–201.

Further Reading

Ahmed, R.M., Paterson, R.W., Warren, J.D., Zetterberg, H., O'Brien, J.T., et al., 2017. Biomarkers in dementia: clinical utility and new directions. J. Neurol. Neurosurg. Psychiatry 85, 1426–1434.

Bowler, J.V., 2007. Modern concept of vascular cognitive impairment. Br. Med. Bull. 83, 291–305.

Marquis, S., Moore, M.M., Howieson, D.B., et al., 2002. Independent predictors of cognitive decline in healthy elderly persons. Arch. Neurol. 59, 601–606.

McAfoose, J., Baune, B.T., 2009. Evidence for a cytokine model of cognitive function. Neurosci. Biobehav. Rev. 33, 355–366.

Prince, M.J., Acosta, D., Castro-Costa, E., Jackson, J., Shaji, K.S., 2009. Packages of care for dementia in low- and middle-income countries. PLoS Med. 6, e1000176.

Waite, L.M., Broe, G.A., Grayson, D.A., Creasey, H., 2001. Preclinical syndromes predict dementia: the Sydney older persons study. J. Neurol. Neurosurg. Psychiatry 71, 296–302.

Wimo, A., Jönsson, L., Bond, J., Prince, M., Winblad, B., Alzheimer Disease International, 2013. The worldwide economic impact of dementia 2010. Alzheimers Dement. 9, 1–11.e3.

CHAPTER

2

Neurochemical Aspects of Poststroke Dementia

INTRODUCTION

Stroke is caused by the reduction or blockade of blood flow to the brain due to formation of a clot. In stroke, a decrease in blood flow leads not only to the deficiency of oxygen and reduction in glucose metabolism, but also a decrease in ATP production and breakdown of the blood–brain barrier (BBB) along with accumulation of toxic products (Strong et al., 2007; Farooqui, 2010, 2018). Stroke is the leading cause of physical and intellectual disability in adults and remains the major cause of mortality in the developed countries. Data from the World Health Organization (WHO) suggest that around 15 million people suffer stroke each year globally. Of these, 5 million die and another 5 million remain disabled permanently, putting a tremendous burden on the family and society. Stroke is an important risk factor for dementia and dementia predisposes to stroke. Dementia prevalence in subjects with stroke is comparable to that seen in stroke-free subjects who are 10 years older. Very little is known about the prevalence, time course, and risk factors for poststroke dementia (Pendlebury, 2009). However, recent studies on meta-analysis of pre- and poststroke dementia have indicated that there is considerable heterogeneity among individual studies and information from pooled dementia estimates have indicated that 1-in-10 patients become demented prior to first stroke, 1-in-10 develop new dementia soon after the first stroke, and over 1-in-3 develop dementia after a recurrent stroke. After the first year, the cumulative incidence of dementia is little greater than expected on the basis of recurrent stroke alone (Pendlebury, 2009). Significant information is available on the pathogenesis of stroke (Farooqui, 2018). Thus, at the molecular level, stroke is accompanied by the overstimulation of the NMDA-type of glutamate receptors, rapid influx of Ca^{2+} ions, and stimulation of

Molecular Mechanisms of Dementia
DOI: https://doi.org/10.1016/B978-0-12-816347-4.00002-7

Ca^{2+}-dependent enzymes, such as phospholipases A_2, C, and D (PLA_2, PLC, and PLD), calcium/calmodulin-dependent kinases (CaMKs), nitric oxide synthases (NOS), calpains, calcineurin, and endonucleases, extracellular signal-regulated kinase, p38, and c-Jun N-terminal kinase, and a disturbed docking of glutamate-containing vesicles (Farooqui, 2010, 2018). However, little is known about mechanisms of onset of poststroke dementia. It is suggested that vascular lesions of the brain contribute to the pathogenesis of poststroke dementia. Furthermore, poststroke dementia may be caused by asymptomatic Alzheimer pathology. In addition, white matter changes (WMCs) may also be associated with the pathogenesis of poststroke dementia because of the involvement of small-vessel disease and a higher risk of onset of stroke (Pasquier and Leys, 1997; Sun et al., 2014). Besides the onset of symptoms of stroke, there are also signs of memory disturbance and dementia. The common symptoms of poststroke dementia are slow thinking, forgetfulness, deficiencies in language, mood, and behavioral changes. The patients have reduced ability to perform their daily life activities (Xu and Shang, 2016; Lin et al., 2003). Risk factors for stroke are classified into two groups: (1) the nonmodifiable risk factors; and (2) modifiable risk factors. Nonmodifiable risk factors include age, sex, ethnicity, family history, and genetic predisposition. Modifiable risk factors include hypertension, diabetes, dyslipidemia, atrial fibrillation, obesity, smoking, and physical inactivity (Fig. 2.1). Risk factors of dementia after stroking are aging, low education level, diabetes mellitus, atrial fibrillation, myocardial infarction, hypertension, medial temporal lope atrophy, and WMCs (Sibolt et al., 2013; Pendlebury, 2009; Leys et al., 2005). Thus, poststroke dementia involves the activation of microglial cells and astrocytes, which secrete inflammatory cytokines and chemokines (TNF-α, interleukin (IL)-1β, IL-6, monocyte chemotactic protein (MCP)-1) contributing to neuroinflammation. Oxidative stress and neuroinflammation are closely interlinked processes and it is difficult to establish the temporal sequence of their relationship. For example, proinflammatory transcription factor, nuclear factor-κB (NF-κB), which modulates the expression of proinflammatory cytokines and chemokines, is redox sensitive. Thus, reactive oxygen species (ROS) modulate the expression and release of proinflammatory cytokines, which in turn enhance ROS production (Fig. 2.2) through the positive feedback stimulation of PLA_2 and the release of nitric oxide through the increased expression of nitric oxide synthase (Bryan et al., 2013), thus establishing a vicious circle. Onset of chronic neuroinflammation and oxidative stress not only leads to stroke, but also promotes the pathogenesis of chronic age-related neurodegenerative disorders, such as Alzheimer's disease (AD), Parkinson's disease (PD), and dementia (Farooqui et al., 2012; Farooqui, 2013, 2017, 2018). In addition to oxidative stress and neuroinflammation, stroke-mediated

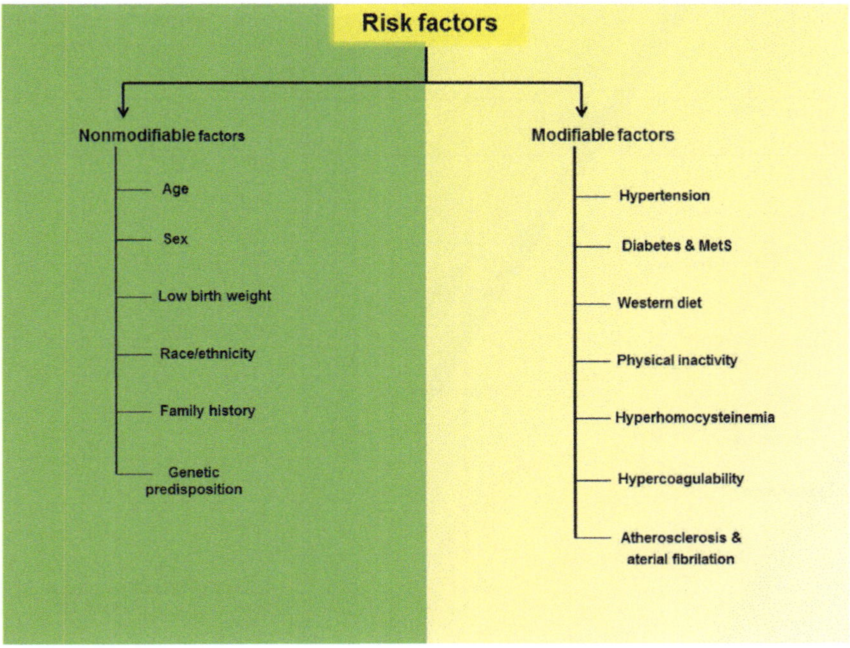

FIGURE 2.1 Risk factors for poststroke dementia.

brain damage also involves immunological alterations. Stroke-mediated immune system damage produces a powerful immunosuppressive effect that facilitates fatal intercurrent infections and threatens the survival of stroke patients. Ischemic stroke not only produces gray matter injury, but also elicits profound white matter injury. These changes are a risk factor for increased stroke incidence and poor neurological outcomes among stroke patients (Wang et al., 2016). The majority of damage caused by stroke and poststroke dementia is located in subcortical regions and, remarkably, white matter occupies nearly half of the average infarct volume. Indeed, white matter is more vulnerable to severe stroke than gray matter (Wang et al., 2016). Subacute phase complications of ischemic stroke include the management of neurological complications, which include brain edema, especially in large infarct volume, hemorrhagic transformation of ischemic lesions, and treatment of seizures (Jauch et al., 2013). Dysphagia is a common problem after both ischemic and hemorrhagic strokes and is a risk factor for pneumonia and urinary tract infections (UTIs). Management of stroke also involves the control of swallowing dysfunction, fever, and hyperglycemia (Middleton et al., 2011; Pollock et al., 2014). Finally, stroke patients should also undergo physical rehabilitation as soon as possible, since physical rehabilitation not only results in better outcome, but also

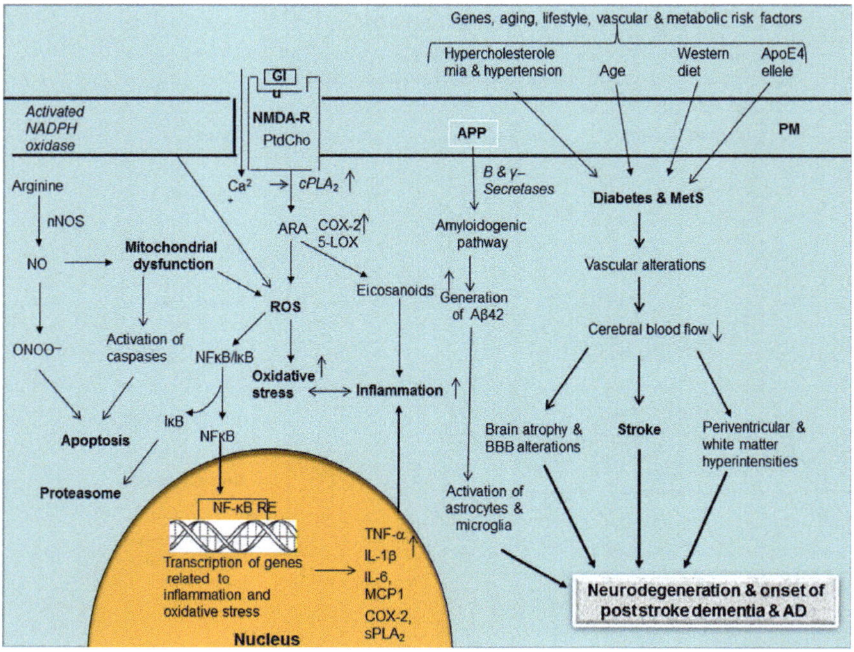

FIGURE 2.2 Mechanisms contributing to oxidative stress and neuroinflammation in stroke and poststroke dementia.
Aβ, β-amyloid; AD, Alzheimer disease; *APP*, amyloid precursor protein; *ARA*, arachidonic acid; *COX-2*, cyclooxygenase-2; *cPLA*$_2$, cytosolic phospholipase A_2; *Glu*, glutamate; *I-κB*, inhibitory subunit of NF-κB; *IL-1β*, interleukin-1β; *IL-6*, interleukin-6; *5-LOX*, 5-lipoxygenase; *MCP-1*, monocyte chemoattractant protein-1; *NF-κB*, nuclear factor-κB; *NF-κB-RE*, nuclear factor-κB-response element; *NMDA-R*, NMDA receptor; *NO*, nitric oxide; *ONOO −*, peroxynitrite; *PM*, plasma membrane; *PtdCho*, phosphatidylcholine; *ROS*, reactive oxygen species; *TNF-α*, tumor necrosis factor-α.

promotes a reduction in long-term disability (Pollock et al., 2014). Collectively, these studies suggest that oxidative stress, neuroinflammation, and immunological alterations contribute to the ischemic cascade from the early damaging events triggered by arterial occlusion, to the late regenerative processes underlying postischemic tissue repair (Iadecola and Anrather, 2011). Converging evidence suggests that multiple mechanisms contribute to neurodegeneration following stroke-mediated brain injury (Farooqui, 2018). Clinical aspects of poststroke vascular alterations are frequently dependent upon vascular risk factors or systemic vascular diseases. This condition leads to the development of large or small artery remodeling, which ultimately result in vascular brain lesions. The scenario is more complex when considering the potential direct impact of vascular or metabolic risk factors on cognition and the interaction between vascular load and neurodegenerative lesions such as AD-related pathology (Korczyn, 2002; Akinyemi et al., 2013). Cognitive impairment and dementia frequently occur following an acute stroke, and they are an important cause of stroke-related morbidity. Dementia may be related to a pure VaD or to a mixed form, which occurs after a stroke, or can represent the progression of prestroke vascular or degenerative-related cognitive impairment (Leys et al., 2005).

RISK FACTORS FOR POSTSTROKE DEMENTIA

Stroke is known to significantly increase the risk of dementia in subjects aged 55 years or more. One-third of stroke survivors develop dementia 5 years after a stroke (Mijajlović et al., 2017). This may be because of vascular risk factors such as hypertension, diabetes mellitus, hyperlipidemia, smoking, atrial fibrillation, myocardial infarctions, and smoking, which increase the risk of poststroke dementia (Mijajlović et al., 2017). Medial temporal lobe atrophy, female sex, and family history are more strongly associated with prestroke dementia, whereas the characteristics and complications of the stroke and the presence of multiple lesions in time and place are more strongly associated with poststroke dementia, indicating the likely impact of optimal acute stroke care and secondary prevention in reducing the burden of dementia (Pendlebury, 2009; Pendlebury and Rothwell, 2009; Surawan et al., 2017).

As stated above, the prevalence of the new-onset dementia in first stroke is about 10%, whereas in the recurrent stroke it is 30%. Stroke survivors have more than twice the risk of developing poststroke dementia compared with people who have never had a stroke (Patel et al., 2002). Stroke-mediated injury in the left hemisphere leads to problems in language and comprehension. This injury reduces the ability of patients to communicate (Pirmoradi et al., 2016). In contrast, stroke-induced damage in the right hemisphere produces alterations in intuitive thinking, reasoning, solving problems, and also perception, judgment and the visual–spatial functions can be impaired (Patel et al., 2002; Cumming et al., 2012; Sun et al., 2014; Harris et al., 2015; Save-Pédebos et al., 2016). The above changes in stroke patients lead to difficulties in locating objects, walking up or down stairs, or getting dressed. Consequently, cognitive disorders are one of the strongest predictors of the inability to return to work, thus contributing to the socioeconomic burden of stroke (Kauranen et al., 2013). However, stroke-mediated cognitive dysfunctions are often underestimated relative to motor impairments because they are confused with preexisting symptoms of age-related mild cognitive impairments (MCI) or AD (Sun et al., 2014; Corriveau et al., 2016). Furthermore, cognitive impairments are frequently associated with poor motor recovery (Patel et al., 2002; Leśniak et al., 2008; Rand et al., 2010), suggesting that stroke-mediated cognitive dysfunctions and decreases in neuroplasticity may produce changes not only in the stability and flexibility, but also in learning and memorizing of complex movements, which involve cognitive resources in older adults (Temprado et al., 2013; Cohen et al., 2016).

Neuroimaging studies have also indicated that patients with stroke and poststroke dementia show white matter hyperintensities, cerebrovascular lesions, small cerebral vessel disease, and cerebral amyloid angiopathy (Petrovitch et al., 2005). The progressive nature of white matter lesions often results in severe physical and mental disability. White matter lesions are characterized by myelin sheath loss and deformation, BBB disruption, and glial activation (Farkas et al., 2007). Traditional studies of white matter lesions have been focused on the oligodendrocytic death and axonal damage. However, multiple cell types and intercellular signaling cascades contribute to the maintenance of white matter integrity and connectivity (Hayakawa and Lo, 2016). Other ultrastructural abnormalities include changes in microvasculature, capillary wall deterioration, basement membrane thickening, and pericyte degeneration (Farkas et al., 2000) resulting in BBB permeability (Yang and Rosenberg, 2011), vascular cognitive impairment, and the genesis of cerebral microhemorrhages in the microvasculature (Toth et al., 2017). These processes not only impair the delivery of oxygen and glucose to the activated brain regions, but also decrease synaptic plasticity and long-term potentiation (LTP) producing MCI and ultimately dementia (Fig. 2.3) (Iadecola, 2014).

Molecular mechanisms associated with the above changes are not fully understood. However, it is suggested that the abovementioned changes in the aging brain are mediated and modulated by several conserved mechanisms, which not only control the aging process, but also are closely associated with the modulation of life span, and the onset of age-related diseases, through alterations in signal transduction processes involving insulin/insulin-like growth factor signaling, target of rapamycin signaling, sirtuins signaling, mitochondrial function, and caloric restriction. The cross-talk among the signaling pathways may modulate cognitive functions and neural cell longevity in the aging brain (Bishop et al., 2010). In addition to the above mechanisms, telomere shortening, mitochondrial oxidative damage, p53 activation, and reduced peroxisome proliferator-activated receptor gamma, coactivator 1 alpha and beta (PGC-1α and PGC-1β) (Sahin et al., 2011; Finck and Kelly, 2006) also modulate the integrity of genome and stability. These factors are major guarantors of viability and longevity. Stroke-mediated

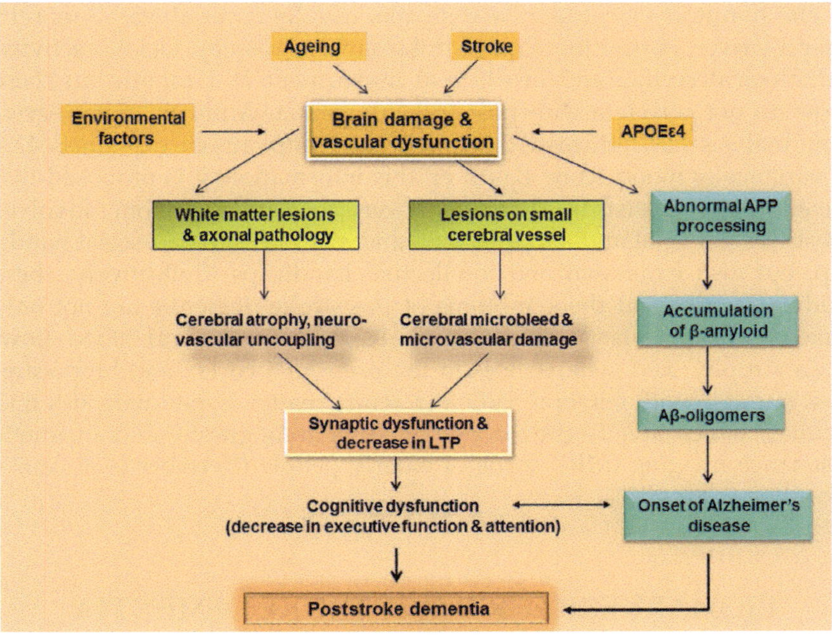

FIGURE 2.3 Risk factors and processes associated with the pathogenesis of poststroke dementia.
APOEε4, APOE epsilon 4 genotype; *LTP*, long-term potentiation.

damage in older adults with type 2 diabetes, atrial fibrillation, and small vessel disease are more severe than normal older subjects. These factors have also been found to be predictors of dementia (Andersen et al., 1995; de Leeuw et al., 2000, 2002; Farooqui, 2013; Pendlebury, 2009; Sibolt et al., 2013; Brainin et al., 2015). Poststroke dementia may also be associated with the AD pathology (Fig. 2.3). The reasons for such an association include: (1) some cases of dementia occurring after a stroke are progressive and AD is the most frequent cause of progressive dementia; (2) age and APOE epsilon 4 genotype, myocardial infarction, hypertension, and smoking are also risk factors for both AD and ischemic stroke; and (3) a vasculopathy is often associated with poststroke AD onset. Lastly, WMCs may also contribute to dementia because they not only indicate the presence of small-vessel disease, but also a higher risk of stroke recurrence, which may lead to further cognitive impairment. Finally, the summation of vascular lesions of the brain, WMCs, and AD pathology may lead to dementia, even when each type of lesion, on its own, is not severe enough to induce dementia (Surawan et al., 2017; Mijajlović et al., 2017).

The symptoms of poststroke dementia usually occur at least 3 months after a stroke. Furthermore, poststroke subjects display social inactivity, pathological crying, and intellectual impairment at 1 month but these signs do not correlate with poststroke dementia. A multivariate regression analysis has indicated that intellectual impairment explained 42% of variance of mood score. Based on this information, it is proposed that the etiology of poststroke dementia is very complex. It not only involves prestroke personal and social factors, and a stroke-induced social handicap, but also emotional and intellectual handicaps. Collectively, these studies suggest that the symptoms of poststroke dementia are not only linked with vascular dysfunction, a decrease in cerebral blood flow, brain atrophy, and synapse loss in the prefrontal cortex and hippocampus, but also with cerebral SVD and white matter hyperintensities and lacunes, which are diagnosed by computer tomography and magnetic resonance imaging (MRI) scans of elderly people (Vermeer et al., 2003; Wardlaw et al., 2013).

BIOMARKERS FOR POSTSTROKE DEMENTIA

Biomarkers are metabolites whose level, presence, and activity are closely associated with the pathogenesis of the disease processes. Clinically, biomarkers are used not only for early detection of the disease process (preclinical stage) and monitoring the disease progression, but also for following the treatment response more sensitively and objectively. The discovery of an ideal and specific biomarker will not only improve the differential diagnosis of poststroke dementia, but could also track the progression of poststroke dementia and neurodegenerative disease (AD) whose onset occurs after poststroke dementia, and be used to measure the efficacy of treatment. This means that there is an urgent need to develop biomarkers that are sensitive and specific to poststroke dementia pathology with positive and negative predictive value for the disorder (Grimes and Schulz, 2002). At present, the information on an ideal and specific biomarker for poststroke dementia is lacking. However, changes in levels of BACE1, soluble form of RAGE (sRAGE), Neutral endopeptidase (NEP), tau protein, and isoprostanes in CSF and serum are correlated with clinical manifestation of cognitive impairment after stroke. Furthermore, stroke severity and lesion volume did not significantly modify this relationship, suggesting that serum BACE1, sRAGE, NEP, tau protein, and isoprostanes may be suitable early biomarkers of cognitive impairment in poststroke dementia (Qian et al., 2012; Brainin et al., 2015; Tang et al., 2017). Other possible biomarkers of poststroke dementia are tumor necrosis factor-a, interleukin-1, interleukin-10, sE Selectin, vascular endothelial cell adhesion-1, and nerve microfilament proteins. It must be emphasized that the levels of these biomarkers are significantly increased in many neurotraumatic and neurodegenerative diseases (Farooqui, 2018). In the CSF, the quantification of these biomarkers by quantitative proteomics will advance this field.

Currently, the diagnosis of poststroke dementia is usually made on clinical manifestations, neuroimaging, and a battery of neuropsychological tests. The major imaging biomarkers for the diagnosis and prognosis of poststroke dementia include positron emission tomography (PET) neuroimaging of β-amyloid (Aβ) protein deposition, MRI of volume hippocampus and other brain structures, and in vivo imaging of insoluble Aβ species by fluorescent and near-infrared fluorescence imaging. Neurofunctional imaging modalities, such as FDG-PET, and regional cerebral blood flow imaging with single-photon emission computed tomography have been used to provide information about regional glucose metabolism and brain perfusion. Converging evidence suggests that neuroimaging methods are useful in the early diagnosis of poststroke dementia as well as AD (Westman et al., 2011). Neuroimaging techniques can also be used for the prediction of the conversion of MCI to dementia.

CELLULAR AND NEUROCHEMICAL CHANGES IN POSTSTROKE DEMENTIA

Neurochemical changes in poststroke dementia not only involve vascular factors (hypertension, coronary artery disease, insulin resistance, diabetes, and hyperlipidemia), but also WMCs, often after the age of 65 years, as well as pathogenesis of leukoaraiosis. The main vascular risk factor for small vessel disease is a decrease in cerebral blood flow due to the narrowing of cerebral blood vessels, leading to chronic hypertension. This decrease in cerebral blood flow leads to the activation and degeneration of astrocytes with the resulting fibrosis of the extracellular matrix (ECM). The fibrosis of cerebral vessels decreases elasticity leading to stiffening of blood vessels at the time of increased metabolic need (Rosenberg, 2017). Intermittent hypoxic/ischemic changes in poststroke dementia activate a molecular injury cascade, producing an incomplete infarction, which damages the deep layers of white matter due to the lack of cerebral blood flow. Neuroinflammation induced by ischemic injury activates microglia/macrophages to release proteases, ROS, and reactive nitrogen species (RNS) that perpetuate the damage over time to molecules in the ECM and the neurovascular unit (NVU).

Accumulations of ROS and RNS not only increase the susceptibility of brain tissue to ischemic damage but also trigger numerous molecular cascades, leading to increased BBB permeability, brain edema, hemorrhage and inflammation, and brain death (Pun et al., 2009). Activation of matrix metalloproteinases (MMPs) is a key step in the disruption of BBB. MMPs are proteolytic zinc-containing enzymes responsible for the degradation of the ECM around cerebral blood vessels and neurons. ROS are known to activate MMPs and subsequently induce the degradations of tight junctions (TJs), leading to BBB breakdown following ischemia–reperfusion injury and poststroke dementia. Breakdown of BBB promotes the entry of peripheral proinflammatory molecules into the brain and activates stress-activated pathways, thereby promoting the key pathological features of dementia/AD, such as mitochondrial dysfunction, and accumulation of neurotoxic beta-amyloid (Aβ) oligomers. These processes lead to synaptic loss, neuronal dysfunction, and cell death. Ceramides, an important molecule which forms the backbone of complex sphingolipids, can also pass the BBB, inducing proinflammatory reactions and oxidative stress. In a vicious circle, oxidative stress and the proinflammatory environment intensify, leading to further cognitive decline (Farooqui, 2014). Recent studies have revealed that caveolin-1, a membrane integral protein located at caveolae, can prevent the degradation of TJ proteins and protect the BBB integrity by inhibiting RNS production and MMPs activity. The interaction of caveolin-1 and RNS forms a positive feedback loop which provides amplified impacts on BBB dysfunction during cerebral ischemia–reperfusion injury (Gu et al., 2011). In addition, MMPs also contribute to vasogenic edema in white matter and vascular demyelination, which are the hallmarks of the subcortical ischemic vascular disease—the small vessel disease form of vascular cognitive impairment and dementia also called Binswanger's disease (Rosenberg, 2017). Poststroke dementia is also accompanied by lacunar infarct caused by occlusion of the penetrating small arteries that supply blood into the brain's deep structures. Cerebral microbleeds, which result from the impaired integrity of small vessels, contribute to either hypertensive vasculopathy or cerebral amyloid angiopathy in poststroke dementia. Microbleeds are commonly features of poststroke dementia and AD type of dementia after stroke (Yates et al., 2014). It is also reported that poststroke dementia can even be promoted by the onset of transient ischemic attack (TIA) (Bos et al., 2007). In addition, the strategic location of infarction in the brain may play an important role in the risk of developing poststroke dementia. These lesions include infarctions at the dominant thalamus, angular gyrus, deep areas of frontal lobe, medial temporal lobe, hippocampus, and left hemisphere, and can lead to multiple infarctions in both brain hemispheres (Grysiewicz and Gorelick, 2012).

In poststroke dementia, cyclin-dependent kinase 5 contributes to the cognitive impairment (Posada-Duque et al., 2015; Gutiérrez-Vargas et al., 2017). Administration of cyclin-dependent kinase 5 RNA interference results in the prevention of the impairment of reversal learning 4 months after ischemia and the decrease in neuronal loss, tauopathy, and microglial hyperreactivity (Posada-Duque et al., 2015; Gutiérrez-Vargas et al., 2017), supporting the view that inhibition of cyclin-dependent kinase 5 may not only prevent long-term postischemic brain damage and cognitive impairment, but also promote the maintenance of normal synaptic plasticity by stimulating the expression of brain-derived neurotrophic factor (BDNF) in the hippocampus (Posada-Duque et al., 2015; Gutiérrez-Vargas et al., 2017). Vascular factors also play a pathogenic role in the poststroke dementia. The contribution of vascular risk factors in poststroke dementia and AD is supported by recent findings on the involvement of the NVU, an entity which is composed of astrocytes, mural vascular smooth muscle cells, and pericytes, and endothelia. It regulates blood flow, controls the exchange across the BBB, contributes to immune surveillance in the brain, and provides trophic support to brain cells in the pathogenesis of poststroke dementia (Nelson et al., 2016). Aging is an important factor that influences the integrity of the NVU. The age-related physiological or pathological changes in the cellular components of the NVU increase the vulnerability of the NVU to ischemia/reperfusion injury resulting in brain damage (Cai et al., 2017). It is well known that the energy and O_2 demands of the brain tissue vary both spatially and temporally with changes in neuronal activity. This requires prompt adjustments of blood flow by regulating arteriolar resistance in a highly controlled fashion in order to maintain cellular homeostasis and function (Enager et al., 2009). There is compelling evidence that poststroke dementia patients exhibit significant impairment of neurovascular coupling responses (Rombouts et al., 2000). Hypertension-mediated alterations in NVU uncoupling is superimposed by beta-amyloid pathologies that not only exacerbates the dysregulation of cerebral blood, but also promotes cognitive decline. Preclinical and clinical studies have demonstrated that aging per se impairs neurovascular coupling responses (Toth et al., 2014; Balbi et al., 2015), suggesting that the combination of old age, amyloid pathologies, and hypertension likely results in a critical mismatch between supply and demand of oxygen and metabolic substrates in functioning cerebral tissue (Iadecola, 2009). Converging evidence suggests that stroke-mediated injury to the NVU alters cerebral blood flow regulation, depletes vascular reserves, disrupts the BBB, and reduces the brain's repair potential, inducing neurodegeneration and brain dysfunction (Iadecola, 2010). These processes accelerate the tempo of the dementia (Helzner et al., 2009). Poststroke dementia and AD patients with a reduced cerebrovascular reactivity to hypercapnia, an index of cerebrovascular function, have a more rapid cognitive decline (Silvestrini et al., 2006), linking disease progression with cerebrovascular dysfunction. Therefore, coexisting cerebrovascular disease or incident ischemic lesions may shorten the preclinical stage of poststroke dementia and AD, accelerating the disease progression.

OXIDATIVE STRESS-MEDIATED INJURY IN POSTSTROKE DEMENTIA

Oxidative stress is a threshold process that involves overwhelming the antioxidant defenses of the cells through the generation of ROS. This may be either due to an overproduction of ROS or to a failure of cell buffering mechanisms (Farooqui, 2010). ROS include superoxide anions ($O_2^{\bullet-}$), hydroxyl ($^{\bullet}OH$), alkoxyl, peroxyl radicals ($ROO^{\bullet}$), and hydrogen peroxide (H_2O_2). ROS are generated in mitochondria as a byproduct of oxidative phosphorylation. RNS such as nitric oxide ($^{\bullet}NO$) are also produced in mitochondria (Zorov et al., 2007). Among ROS, the $^{\bullet}OH$ radicals are very reactive. They have a very short in vivo half-life (approximately 10^{-9} seconds) (Sies, 1993). In contrast, $O_2^{\bullet-}$ radicals have low reactivity and can be detoxified by superoxide dismutase (SOD). The $O_2^{\bullet-}$ radical is frequently converted into H_2O_2, the most bioactive and stable form of ROS. It can diffuse from mitochondria into the cytosol and nucleus. H_2O_2 is detoxified by glutathione peroxidase in mitochondria and the cytosol and by catalase in peroxisomes (Zorov et al., 2007). $O_2^{\bullet-}$ reacts with the diffusible gas nitric oxide ($^{\bullet}NO$) to form the potent nucleophile oxidant and nitrating agent peroxynitrite ($ONOO^-$), which damages proteins by nitration (Mungrue et al., 2003; Beckman et al., 1993). $ONOO^-$ is genotoxic directly to neurons. It is not only capable of producing single- and double-strand breaks in DNA (Martin and Liu, 2002), but can produce apoptotic cell death in neural cells. Cu/ZnSOD can use $ONOO^-$ to catalyze the nitration (NO_2-Tyr) of mitochondrial protein tyrosine residues such as cyclophilin D (CyPD) and the adenine nucleotide translocator, which are core components of the mitochondrial permeability transition pore.

Low levels of ROS are essential for neuronal development and function, whereas excessive levels are hazardous. Under normal conditions, the deleterious effects of ROS production during aerobic metabolism are neutralized by the antioxidant system and in this manner the brain effectively regulates its oxygen consumption and redox generation capacity (Fig. 2.4). When ROS production exceeds the scavenging capacity of antioxidant response systems (superoxide dismutase, catalase, vitamin C, vitamin E, and reduced glutathione) then not only does

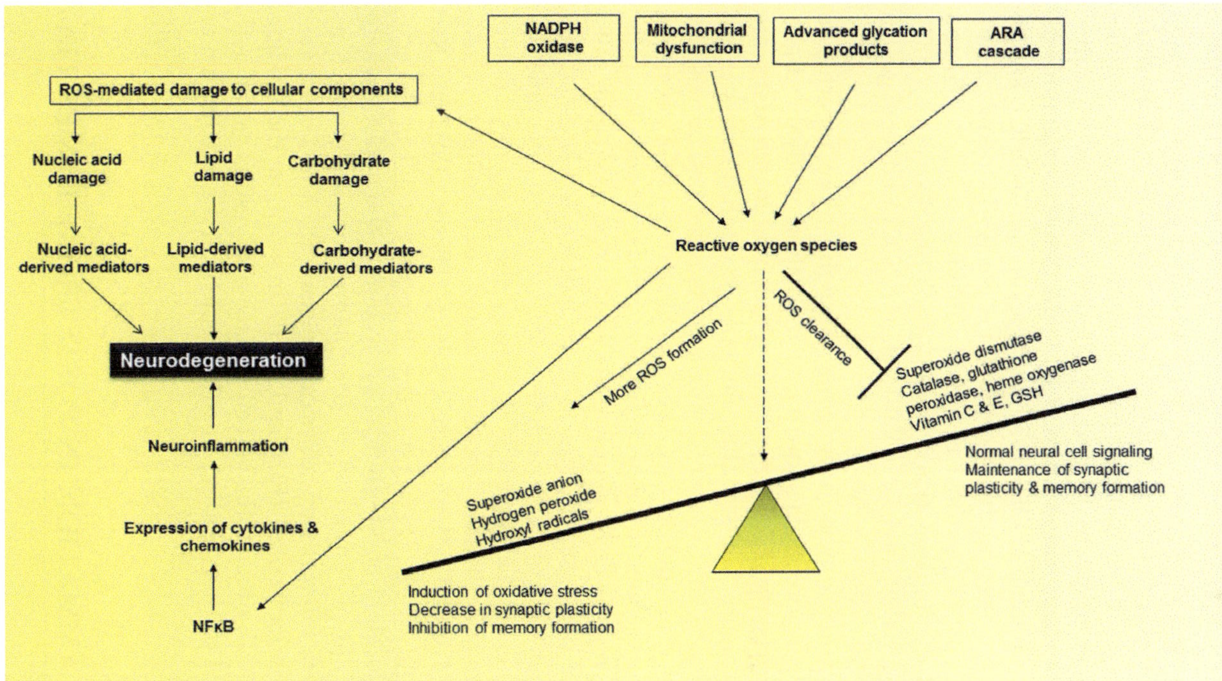

FIGURE 2.4 Mechanisms contributing to the induction of oxidative stress and their effects on neural cell components in the brain.

oxidative damage to cellular components and cellular degeneration occur, but also neural cell functional decline. The major sources of ROS in the brain include the mitochondrial respiratory chain, uncontrolled arachidonic acid (ARA) cascade, and activation of NADPH oxidase. ROS damage proteins, lipids, and nucleic acids. In poststroke dementia ROS-mediated damage to biomolecules not only produces high levels of oxidized proteins, lipid peroxidation end products (4-hydroxy-2,3-nonenal, acrolein, malondialdehyde and F_2-isoprostanes), and oxidative modifications in nuclear and mitochondrial DNA (8-hydroxyguanine (8-OHG), 8-hydroxyadenine (8-OHA), 5-hydroxycytosine (5-OHC), and 5-hydroxyuracil), but also produce membrane defects (Farooqui, 2010). These processes disrupt neuronal networks (Shankar et al., 2007), and induce neuronal dysfunction (Lacor et al., 2007; Shankar et al., 2007) resulting in impairment in LTP (Walsh et al., 2002) and changes in behavior (Ford et al., 2015).

In addition, patients who have gone through stroke/reperfusion injury and poststroke dementia show the accumulation of advanced glycation end products (AGEs) in the brain (Farooqui, 2010). AGEs are known to impair neural cell signaling not only through direct covalent cross-linking of AGEs with various domains of its receptors, but also by interfering signal transduction processes modulated by AGE receptors (RAGEs), which are found on macrophages, vascular endothelial cells, vascular smooth muscle cells, neurons, astrocytes, and microglial cells. RAGEs modulate many signal transductions pathways associated not only with the generation of more oxidative stress, but also inflammatory events. Interactions of AGE with RAGE increase the phosphorylation of p21ras, the mitogen-activated protein kinases, extracellular signal-regulated kinase 1/2 and p38, and also activate GTPases Cdc42 and Rac (Fig. 2.5) (Farooqui, 2010). These processes ultimately result in the activation and translocation of NF-κB from cytoplasm to the nucleus where it starts transcribing its target set of proinflammatory genes, such as TNF-α, IL-1β, IL-6, intercellular adhesion molecule (ICAM)-1, and vascular cell adhesion molecule-1 (Farooqui, 2010). In addition, the binding of AGEs with RAGE on endothelial cell surface also results in the activation of NADPH oxidase leading to enhancement in the production of ROS (Fig. 2.5) (Sun et al., 2007).

NEUROINFLAMMATION IN POSTSTROKE DEMENTIA

Neuroinflammation is a localized response to brain injury or infection which aids in the repair of damaged brain and/or destruction of the harmful agent. Classically, neuroinflammation is characterized by

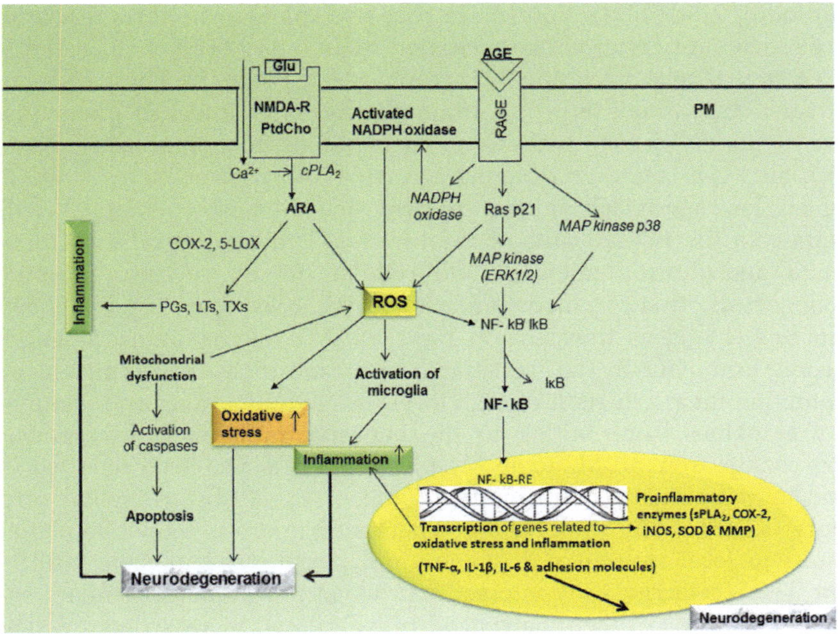

FIGURE 2.5 Interactions between AGE and AGE receptors (RAGE) along with the induction of ROS formation.

AGEs, advanced glycation products; *ARA*, arachidonic acid; *COX-2*, cyclooxygenase; *cPLA*$_2$, cytosolic phospholipase A_2; *IκB*, inhibitory subunit of NFκB; *IL-1β*, interleukin-1 beta; *IL-6*, interleukin-6; *iNOS*, inducible nitric oxide synthase; *5-LOX*, 5-lipoxygenase; *LTs*, leukotrienes; *MAPK*, mitogen-activated protein kinase; *MMPs*, matrix metalloproteinases; *NADP oxidase*, nicotinamide adenine dinucleotide phosphate oxidase; *NF-κB*, nuclear factor kappa-B; *NF-κB-RE*, nuclear factor kappa-B response element; *PGs*, prostaglandins; *PM*, plasma membrane; *PtdCho*, phosphatidylcholine; *PtdIns 3K*, phosphatidylinositol 3 kinase; *RAGE*, receptor for advanced glycation end products; *ROS*, reactive oxygen species; *SOD*, superoxide dismutase; *sPLA*$_2$, secretory phospholipase A_2; *TNF-α*, tumor necrosis factor-α; *TXs*, thromboxanes.

pain, heat, redness, swelling, and loss of function. Neuroinflammation protects and isolates the damaged brain tissue from the uninjured area. Neuroinflammation not only destroys injured cells, but also promotes neurorepair processes in the injured brain area (Minghetti et al., 2005). Neuroinflammation is orchestrated by activated microglia and astrocytes to reestablish homeostasis in the brain after injury. There are two types of neuroinflammation: (1) acute inflammation; and (2) chronic inflammation. Poststroke injury involves acute neuroinflammation and oxidative stress, whereas poststroke dementia and AD are accompanied with chronic neuroinflammation and oxidative stress

(Farooqui, 2010, 2018). Both these processes play an important role in secondary injury after stroke-mediated brain injury (Tao et al., 2017).

Microglial cells are resident macrophages of the brain. These cells are activated by various types of brain damage and undergo phenotype and functional transformations to maintain tissue homeostasis. Thus, microglial cells undergo different forms of polarized activation following stroke/reperfusion injury. Initially, activated M2 microglial cells migrate to the injured area of the brain. M2 microglial cells not only engulf and destroy microbes and cellular debris, but also promote angiogenesis, tissue remodeling, and repair (Gehrmann et al., 1995). But, this change is transient. In time, the M1 microglia/macrophages become dominant in the injury area and exacerbate brain damage by promoting the production of high levels of NO, ROS, and proinflammatory cytokines, contributing to the intensive inflammatory response. This change from an M2 to an M1 phenotype has also been reported in models of traumatic brain injury (Wang et al., 2013) and spinal cord injury (Kigerl et al., 2009). Stroke/reperfusion injury to the white matter results in local microglial activation and peripheral leukocyte infiltration. These cells mutually interact to propagate and intensify inflammatory injury through the involvement of NF-κB and increased expression of proinflammatory cytokine and chemokines (Wagner et al., 2006; Zhou et al., 2014). In addition, stroke/reperfusion injury also alters BBB permeability leading to the infiltration of more monocytes and the induction of mitochondrial dysfunction. Converging evidence suggests that the function of microglial cells is crucial for the homeostasis of the brain in health and disease, as they represent the first line of defense against pathogens and injuries, contributing to immune responses, but are also involved in tissue repair and remodeling (Fig. 2.6) (Lindsey et al., 1979; Correale and Villa, 2004). As stated above, the stimulation of microglial cells results in increased expression and the release of inflammation-mediating enzymes like MMPs, proinflammatory cytokines, and chemokines (IL-1β, IL-6, TNF-α, and MCP-1). A component of inflammasomes, nucleotide-binding domain, and leucine-rich repeat family, pyrin domain containing 3 (NLRP3), is also expressed by activated microglia. In the neural cells, the expression of these cytokines is regulated by NF-κB, a transcription factor, which is activated by ROS (Fig. 2.2). The generation of proinflammatory eicosanoids (prostaglandins, leukotrienes, and thromboxanes) via $cPLA_2$, COX-2, and LOX pathways is another mechanism, providing proinflammatory mediators (Phillis et al., 2006). In a brain that is damaged by stroke, the induction of sustained chronic neuroinflammation and the persistence of abnormalities in neuron–glia cross-talk may result in the loss of cellular homeostasis leading to compromised BBB and neurodegeneration (Raj et al., 2014).

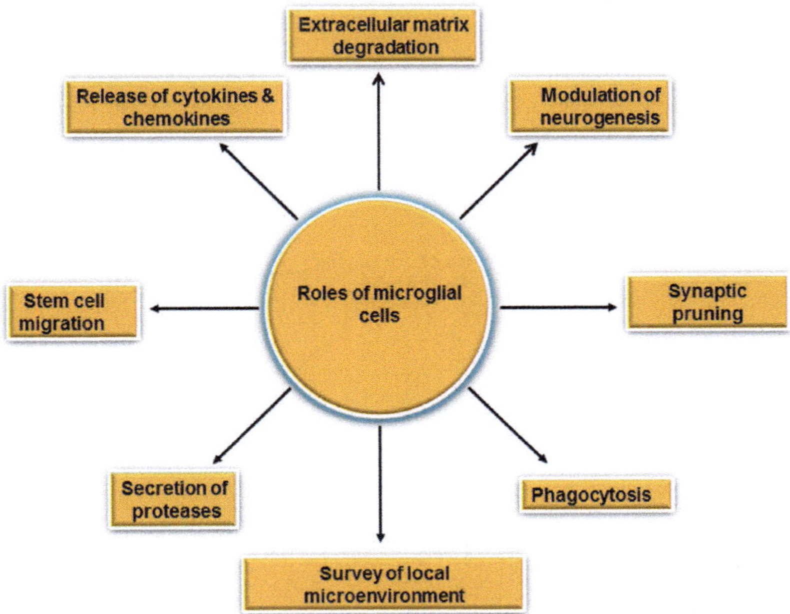

FIGURE 2.6 Roles of microglial cells in the brain.

Astrocytes are complex, highly differentiated, and the most abundant cells of the brain. They outnumber microglia within the central nervous system (CNS) parenchyma, and are the major components of the CNS innate immune system. Astrocytes perform multiple functions in the brain (Fig. 2.7). Thus, during neurogenesis and early development, astrocytes provide a scaffold for the correct migration of neurons and growth cones. Astrocytes also provide guidance cues and may also be associated with neuronal proliferation. In the adult brain, astrocytes maintain neuronal homeostasis and synaptic plasticity. In addition, astrocytes also secrete important neurotrophic factors, such as TGF-β, BDNF, nerve growth factor, and glial-derived neurotrophic factor (GDNF). Astrocytes not only modulate levels of extracellular glutamate, but also convert glucose into lactic acid, which is taken up by neurons and metabolized into pyruvate for energy metabolism (Allen and Barres, 2009; Maragakis and Rothstein, 2001). Astrocytes also play an important role in LTP, induction, and synaptic plasticity (Pita-Almenar et al., 2012). Astrocytes are electrically nonexcitable cells in which the onset of exocytosis depends on the release of Ca^{2+} from the internal stores. This suggests that there is a close relationship between the sites of Ca^{2+} release and the fusion process that occurs during exocytosis (Calì and Bezzi, 2010). Two distinct mechanisms

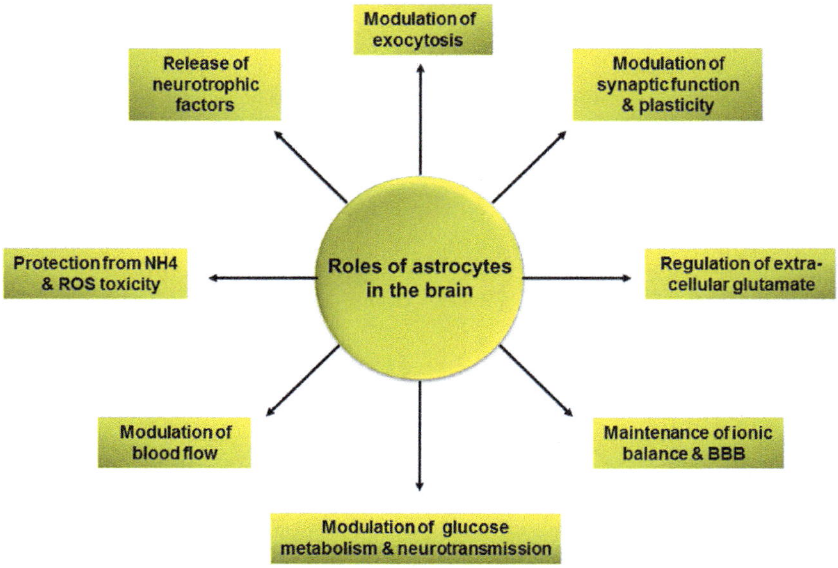

FIGURE 2.7 Roles of astrocytes in the brain.

contribute to the activation of astrocytes. The first mechanism involves the downregulation of gap junction proteins restricting the overall syncytia of astrocytes leading to alterations in the morphology and number of astrocytes–neuronal connections and the BBB (Brand-Schieber et al., 2005). The second mechanism is associated with changes in astrocyte morphology due to immune regulation and inflammation under pathological conditions (Leonard, 2010). Astrocytes react to the neuronal damage by not only over-expressing the GFAP, vimentin, nestin, and extracellular matrix (ECM) molecules, growth factors, and cytokines (IL-6, LIF, CNTF, TNFα, INFγ, Il1, Il10, TGFβ, FGF2, etc.), but also by releasing neurotransmitters and metabolites such as glutamate, noradrenaline, ATP, ROS, and NO. These metabolites are associated with the production of NH_4^+, which may contribute to systemic metabolic toxicity (Allaman et al., 2010; Sofroniew and Vinters, 2010). Astrocytes modulate astrogliosis and glial scar formation (Hwang et al., 2016). This process is supported by the release of NO, ROS, proinflammatory cytokines (such as TNFα, IL-1β, and IL-6), and eicosanoids, all of which at high concentrations can produce deleterious effects on neuronal function. Based on the above information, it is proposed that there may be a causal link between inflammation and dementia (Enciu and Popescu, 2013). Converging evidence thus suggests that activated microglia, astrocytes, leukocytes, neutrophils, macrophages, dendritic cells, and T lymphocytes interact with each other via intricate signaling pathways supporting and intensifying neuroinflammation. Although earlier studies indicated that neurons play a passive role in neuroinflammation, more recent studies have indicated that neurons contribute to neuroinflammation by providing many of their products (i.e., neuropeptides and transmitters), as well as the neuronal membrane proteins CD22, CD47, CD200, CX3CL1 (fractalkine), ICAM-5, neural cell adhesion molecule, semaphorins, and C-type lectins. All these neuronal factors regulate neuroinflammation (Tian et al., 2009). In addition, neurons express low levels of major histocompatibility complex (MHC) molecules and actively promote T-cell apoptosis via the Fas–Fas ligand pathway (CD95–CD95L). Interactions between oxidative stress, and inflammatory processes contribute to progressive neurodegeneration, which lead to the loss of synaptic connections in several interconnected brain regions, such as the prefrontal cortex and hippocampus (Kim et al., 2016). These regions are involved in learning and memory processes Grimm et al., 2016).

IMMUNE RESPONSES IN POSTSTROKE DEMENTIA

The BBB is one of the most essential protection mechanisms in the CNS. It selectively allows individual molecules, such as small lipid-soluble molecules, to pass through the capillary endothelial membrane while limiting the passage of pathogens or high molecular weight substances. The BBB is formed by specialized capillary endothelial cells, together with pericytes and perivascular glial cells. TJs between endothelial cells form a physical link and prevent the passage of molecules from the blood directly into the brain. It has been shown that both cell-to-cell interactions and diffusible cues from the CNS, primarily originating from pericytes and astrocytes, are necessary for cell polarity and the proper formation of TJs (Al Ahmad et al., 2011). Under normal conditions in a healthy human brain, endogenous macrophages and microglia present as immune cells. Occasionally, a T cell may enter the brain but due to the decreased expression of MHC molecules in the brain, the T cell leaves the brain within 24–48 hours (Miller, 1999). This makes the brain an immune-privileged organ, which is beneficial in protecting the brain from pathogens and high molecular weight substances from peripheral blood. Stroke is not only known to produce alterations in the immune responses, but also mediates a robust inflammatory response in the brain (Chapman et al., 2009). This response includes the infiltration of leukocytes and lymphocytes from the peripheral circulation into the ischemic brain as well as the activation of resident inflammatory cells (Fig. 2.8) (Iadecola and Anrather, 2011). Both the innate and the adaptive immune systems contribute to immune and inflammatory

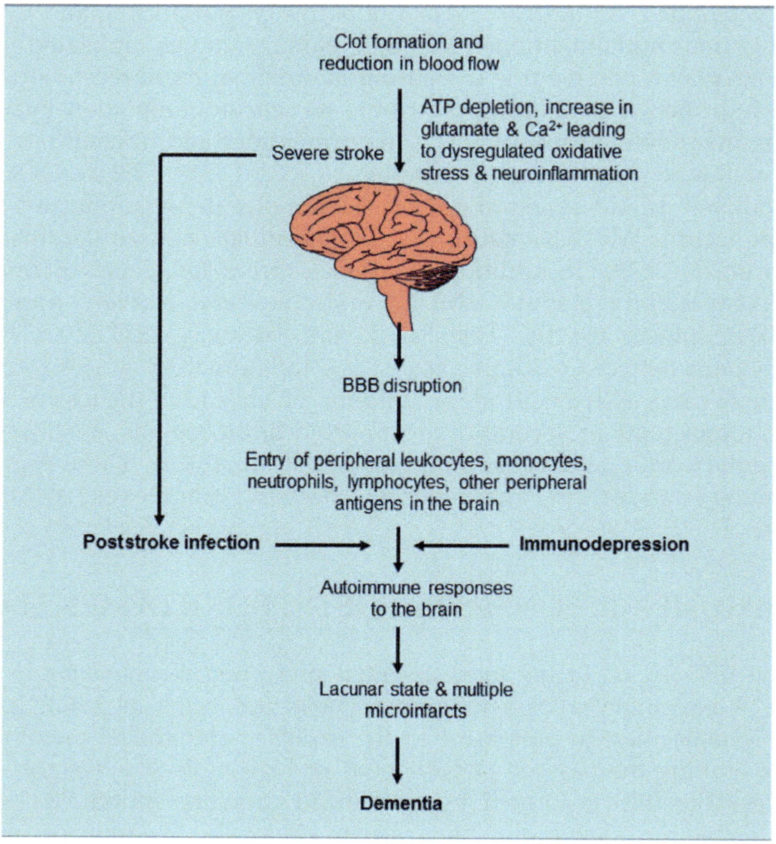

FIGURE 2.8 Effect of stroke on blood–brain barrier and autoimmune responses in the brain.

response following stroke-mediated brain injury. The innate system is germline-encoded, rapidly activated, and relies on low affinity receptors to gain wide-ranging target recognition. The adaptive system is based on high-affinity receptors, that is, T-cell receptors and immunoglobulins, which are randomly generated by somatic mutations. In contrast to the innate system, adaptive immunity needs antigen-driven clonal cell expansion, a process that requires several days, and retains a memory of this antigen exposure (Abbas, 2010). Although specific cell types are predominantly associated with one of the two types of immunities, there is considerable overlap between the role of these cells in innate and adaptive immunity. Inflammatory processes unresolved by innate immune mechanisms lead to recruitment of increased numbers of leukocytes, which modulate the adaptive arm of the immune system. Dendritic cells act as a communicative bridge, traveling from local inflammatory sites to lymph nodes, where they engage B and T cells. In contrast to the initial innate response, adaptive immunity recognizes specific molecular structures requiring clonal expansion of antigen-specific T and B cells—a process which takes several days and, therefore, requires longer durations of inflammation to initiate. In its most simplistic definition, chronic inflammation constitutes a persistence of inflammation beyond 6 weeks, but this is an arbitrary demarcation (Collins, 1999). Chronic inflammation not only damages neurons, but also activates glial cells, which express chemotaxic molecules. This process sends signals to the peripheral immune system that there has been an injury to the brain. The induction of cytokines and chemokines promote the upregulation of vascular adhesion molecules in endothelial cells and on immune cells. By disrupting the BBB, ischemic injury promotes the entry of peripheral immune cells into the brain (de Vries et al., 2012). In addition, stroke-mediated injury also influences immune cells in the peripheral circulation, possibly through increased activation of the sympathetic nervous system and the hypothalamic–pituitary–adrenal (HPA) axis (Prass et al., 2003; Haddad et al., 2002). This may not only result in a reduction in circulating immune cell counts, but also increases the risk of infectious complications (Prass et al., 2003). Furthermore, stroke also produces damage to the NVU through the activation of the innate and adaptive arms of the immune response system (Famakin, 2014). In stroke-mediated injury, self-epitopes, which are protected by the systemic immune system through different mechanisms may become open to adaptive immunity. This process may in turn modulate the immune system to respond to self-antigens in the brain thus leading to autoimmunity. Therefore, stroke-induced immune-suppression may help in preventing postinjury autoimmunity against CNS antigens (Kamel and Iadecola, 2012). It is also reported that the inflammatory mediators produced during the innate immune response in turn lead to recruitment of inflammatory cells and the production of more inflammatory mediators that result in activation of the adaptive immune response. Under normal conditions in the brain, neuroinflammation is known to promote neuroprotection by repairing damaged brain tissue. However, severe stroke not only prevents the development of autoimmune responses to brain antigens, but also predisposes to infections. The inflammatory response associated with infection overrides the systemic immunodepression and creates an environment that can support the successful activation of the immune response to self-antigens. It is proposed that infections occurring in the setting of immunodepression lead to an inflammatory response that is sufficient to allow for "bystander activation" of lymphocytes to brain antigens. In addition, the death of astrocytes, oligodendrocytes, and neurons exposes new and cryptic (intracellular) antigens to the immune system (Becker, 2012a,b). With ongoing tissue injury related to these immune responses, new antigens are constantly exposed, leading to the possibility of "epitope spreading." In a systematic review published on poststroke infection, it is estimated that approximately 30% of patients develop infection (Westendorp et al., 2011). The most common infections are pneumonias and UTIs, each occurring in about 10% of patients with stroke (Westendorp et al., 2011).

POSTSTROKE DEMENTIA AND COGNITIVE DYSFUNCTION

As stated above, stroke is accompanied by a decrease in cerebral blood flow. This reduction in cerebral blood low decreases oxygen and glucose delivery to the affected parts of the brain causing the activation of cellular anaerobic metabolism leading to the depletion of glucose, which is the only source of energy in the brain. Thus, the ischemic cascade produces neuronal damage and ionic pump failure in the brain due to energy depletion, and ultimately leads to necrosis and apoptosis of neurons and glial cells resulting in irreversible injury to core regions with partially reversible damage in the surrounding penumbra zone (Farooqui, 2018). Several studies have shown that there is increased neurogenic activity in the ischemic penumbra distant from the subventricular zone (SVZ) and in the neurogenic region of the lateral ventricular wall in the human brain after stroke (Jin et al., 2006; Nakayama et al., 2010). However, the limited number and capacity of neural stem cells due to stroke attacks and normal aging may lead to a decrease in the number and maturation of newly generated neurons in the ischemic penumbra of the cerebral cortex. In a rat model of neonatal ischemic injury, it is reported that the infusion of GDNF promoted endogenous self-repair by stimulating proliferation of glial progenitor cells derived from both the SVZ and white matter, activating their differentiation into more mature oligodendrocytes and raising the survival rate of these newly generated glial cells (Li et al., 2015). The molecular mechanisms contributing to endogenous self-repair process are not fully understood. However, it is well known that phosphatidylinositide 3-kinase (PtdIns 3K) is one of the well-established pathways affecting cell proliferation, growth, differentiation, motility, survival, and intracellular trafficking (Koh and Lo, 2015). It is likely that PtdIns 3K pathway may play an important role in the survival of neurons in penumbra.

In humans, cognitive dysfunction involves the loss of intellectual functions such as thinking, remembering, and reasoning that interfere with daily activities. Cognitive dysfunction not only depends on volume

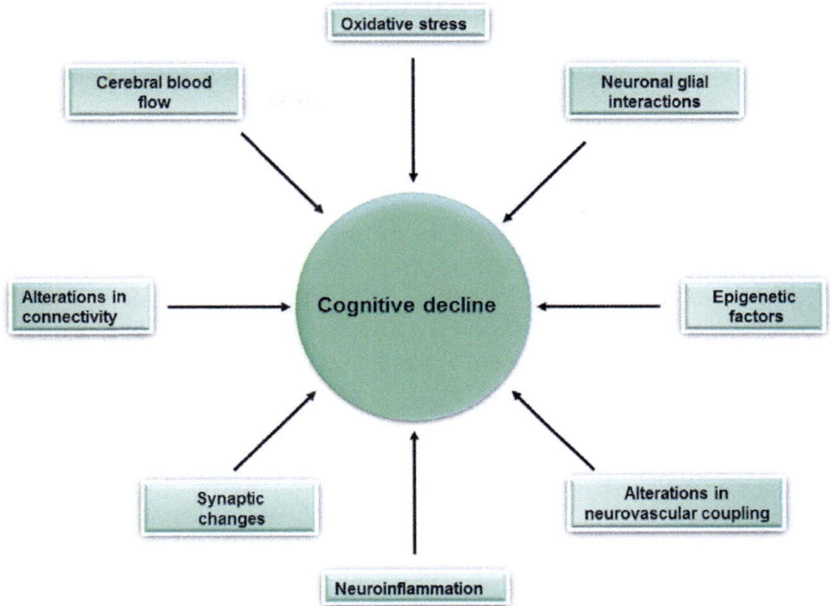

FIGURE 2.9 Factors modulating cognitive decline.

and strategic location of brain infarction, site and range of cerebral white matter injury bilaterality, but also on number of stroke lesions, and other coexistent pathologies (Grysiewicz and Gorelick, 2012). Cognitive function is regulated not only by the intensity of oxidative stress and neuroinflammation, neurochemical and intricate synaptic changes, and alterations in connectivity, but also by neuronal and glial interactions and epigenetic factors (Fig. 2.9) (Morrison and Baxter, 2012). Cognitive dysfunction is one of the primary disabilities of the aging process. It predisposes individuals for neurological and psychiatric disorders eventually affecting the quality of life. Cognitive decline during aging is a multifactorial process, which is controlled by several factors, such as genes for oxidative stress, neuroinflammation, immune response, mitochondrial functions, growth factors, neuronal survival, and calcium homeostasis (Lu et al., 2004; Loerch et al., 2008). The intensity of cognitive decline is markedly increased not only in patients with diabetes, metabolic syndrome, atrial fibrillation, stroke, and neurodegenerative diseases, but also in patients of neuropsychiatric diseases (Schuh et al., 2011; Farooqui, 2018). It is reported that 58% of stroke survivors have cognitive impairment with a quarter of them diagnosed with dementia (Sachdev et al., 2006). This observation supports the view that the onset of stroke doubles the risk of dementia (Leys et al., 2005). In a Latin American study, it was reported that 66% and 61% of stroke survivors develop cognitive impairment at 3 and 12 months of the study, respectively (39% cognitive impairment with no dementia; 22% dementia) (Delgado et al., 2010). The prevalence of cognitive impairment remains 21% at 3 months after stroke and after 14 years of follow-up period (Douiri et al., 2013). Prevalence of cognitive impairment varies due to differences in the study population with nonlacunar and lacunar stroke being common in hospital and community-based studies, respectively. There is a higher cognitive decline after lacunar stroke due to pathological causes where SVD affects a wide region of the brain compared to nonlacunar stroke that involves the extracranial region (Makin et al., 2013).

In poststroke subjects, stress response and genes related to inflammation and DNA repair are upregulated, while genes associated with neuronal growth and survival and mitochondrial functions are downregulated with advancing age (Yankner et al., 2008). Although some progress has been made on the molecular aspects of cognitive dysfunction in aging, significant work is still needed to understand the molecular mechanism(s) of cognitive dysfunction. Recent studies have indicated that peripheral proinflammatory mediators (e.g., IL-1β and IL-6) can cross the BBB to modulate central inflammatory processes that result in neurodegeneration and impairment of cognitive function (Poluektova et al., 2005; Richwine et al., 2008; Trapero and Cauli, 2014). In the brain, IL-1β and IL-6 enhance T and B lymphocyte proliferation and stimulate cytocidal activity to eliminate the injured cells or invading pathogen. IL-1β and IL-6 also induce the production of other cytokines, such as TNFα, which in turn has secondary effects on other cells. In rodents, chronic increases in peripheral inflammation mediators "primed" microglia to switch from a resting state to an activated state and to release a variety of inflammatory mediators including proinflammatory cytokines that organize host defense and restore homeostasis. These inflammatory mediators play a pathogenic role in age-related neurocognitive decline. Converging evidence suggests that in the aged and damaged brains, the neuroinflammatory response to a peripheral challenge is dysregulated, resulting in a potentiated proinflammatory cytokine response, whose source appears to be primed microglia. This exaggerated response is most prominent in the hippocampal formation, the critical brain region modulating contextual and spatial memory consolidation and may be associated with hippocampal memory impairments in aged individuals and victims of stroke. Proinflammatory cytokines such as IL-1β may affect cognitive processes by impairing synaptic plasticity through the activation of MAP kinases JNK and p38, and/or by inhibiting downstream mediators essential to hippocampal-dependent memory processes, such as BDNF and cytoskeletal-associated protein (Arc). In addition to inflammatory response, brain cytokines and other inflammatory molecules also modulate many processes such as fever, decrease in food/water intake, decrease in motor activity and social interaction, increase in slow-wave sleep, hyperalgesia, and HPA axis activation (Dantzer et al., 1998).

Blocking this exaggerated brain cytokine response pharmacologically, or through diet and exercise modifications, may effectively block the deleterious behavioral effects, not only suggesting that these may be useful therapeutic interventions, but also supporting the view that proinflammatory cytokines have a causal, rather than merely correlational relationship with impaired long-term memory in older individuals. In addition, during ischemic injury, microglial cells interact with neurons, possibly via P2x7 and NMDA receptors (Denes et al., 2007), to induce a neuroinflammatory response, which is characterized by an upregulation of cytokines. This process may alter the normal balance and physiological function of cytokines in synaptic plasticity and learning and memory. Alternatively, reduction in cerebral blood flow and tract-specific damage of white matter may decrease the expression of BDNF leading to network disruption due to stroke-mediated neuronal injury. These processes may eventually lead to cognitive decline in stroke and poststroke dementia patients (Fig. 2.10).

It should be noted that cognitive dysfunction exists in about 64% of patients with stroke, of which, up to one-third develop dementia (Nichols and Holmes, 2002). Poststroke dementia patients showed decreased cerebral blood flow and changes in white matter integrity caused either locally or remotely by ischemic injury. These contribute to cognitive deficits in stroke (Molko et al., 2002), resulting in the decreased ability to learn, recall, concentrate, and problem solve. Some studies have shown that cognitive dysfunction is correlated with white matter injury in the frontal lobes, basal ganglia, and thalamus of stroke patients (Cumming et al., 2013). The most common cognitive dysfunctions after stroke are aphasia (language impairment) and hemispatial neglect (failure to attend or respond to stimuli on the side contralateral to the stroke). Stroke may also cause hypoperfusion resulting in impairments in working memory, attention, learning, calculation, visual perception, or executive function (i.e., decision-making, organization, and problem solving). Stroke and poststroke dementia patients may also have ideomotor apraxia, an impairment in skilled movements in the absence of motor weakness or incoordination. Aphasia occurs in anywhere from 15% to one-third of patients with stroke (Inatomi et al., 2008; Engelter et al., 2006), depending on the population studied, the way language is tested, and when it is tested, and also typically occurs after left hemisphere stroke. Similar frequencies of occurrence have

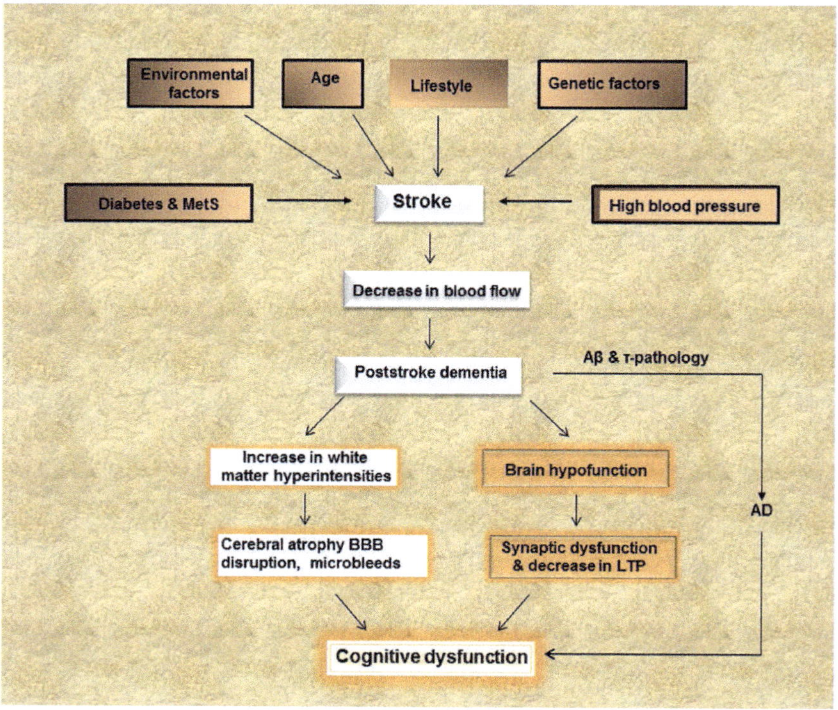

FIGURE 2.10 Changes associated with cognitive dysfunction in stroke and poststroke dementia.

been reported for hemispatial neglect, with rates above 40% among patients with right hemisphere stroke (Ringman et al., 2004). Collective evidence suggests that stroke is a risk factor for post-stroke dementia and vascular dementia is a risk factor for stroke. These pathological conditions predispose the human brain to injury, which is a major cause of disability and mortality throughout the world.

CONCLUSION

Poststroke cognitive impairment and dementia are major causes of long-term neurological disability. The prevalence of poststroke dementia and cognitive deficits varies between 20% and 80% depending on brain region, country, and diagnostic criteria. Unlike stroke resulting in physical disability, poststroke dementia is accompanied by the worsening of cognitive function over time and leads to detrimental impacts on the quality of life of survivors. The risk factors for poststroke dementia are multifactorial. They include genetic predisposition, older age, vascular risk factors, lower education status, prestroke cognitive and functional status, and prior TIA or stroke. Neuroimaging determinants not only include global cerebral atrophy, white matter lesions, and silent infarcts, but also lacunar infarcts and microbleeds. Common symptoms of poststroke dementia include slow thinking, forgetfulness, deficiencies in language, mood, and behavioral changes. The patients show reduced ability in their daily life until they no longer have any daily activities. The risk factors are diabetes mellitus, atrial fibrillation, myocardial infarction, hypertension, medial temporal lope atrophy, and WMCs. Molecular mechanisms of poststroke dementia involve oxidative stress, neuroinflammation, and immunological alterations. These mechanisms are frequently dependent upon vascular risk factors or systemic vascular diseases. This condition leads to the development of large or small artery remodeling, which ultimately result in vascular brain lesions. The scenario is more complex when considering the potential direct impact of vascular or metabolic risk factors on cognition and the interaction between vascular load and neurodegenerative lesions such as AD.

References

Abbas, A.K., 2010. Basic Immunology Updated Edition: Functions and Disorders of the Immune System. Saunders.

Akinyemi, R.O., Mukaetova-Ladinska, E.B., Attems, J., Ihara, M., Kalaria, R.N., 2013. Vascular risk factors and neurodegeneration in ageing related dementias: Alzheimer's disease and vascular dementia. Curr. Alzheimer. Res. 10, 642–653.

Al Ahmad, A., Taboada, C.B., Gassmann, M., Ogunshola, O.O., 2011. Astrocytes and pericytes differentially modulate blood–brain barrier characteristics during development and hypoxic insult. J. Cereb. Blood Flow Metab. 31, 693–705.

Allaman, I., Gavillet, M., Bélanger, M., Laroche, T., Viertl, D., Lashuel, H.A., et al., 2010. Amyloid-beta aggregates cause alterations of astrocytic metabolic phenotype: impact on neuronal viability. J. Neurosci. 30, 3326–3338.

Allen, N.J., Barres, B.A., 2009. Neuroscience: glia – more than just brain glue. Nature 457, 675–677.

Andersen, G., Vestergaard, K., Ingemann-Nielsen, M., Lauritzen, L., 1995. Risk factors for post-stroke depression and dementia. Acta Psychiatr. Scand. 92, 193–198.

Balbi, M., Ghosh, M., Longden, T.A., Jativa Vega, M., Gesierich, B., et al., 2015. Dysfunction of mouse cerebral arteries during early aging. J. Cereb. Blood Flow Metab. 35, 1445–1453.

Becker, K., 2012a. Autoimmune responses to brain following stroke. Transl. Stroke Res. 3, 310–317.

Becker, K.J., 2012b. Activation of immune responses to brain antigens after stroke. J. Neurochem. 123 (Suppl. 2), 148–155.

Beckman, J.S., Carson, M., Smith, C.D., Koppenol, W.H., 1993. ALS, SOD and peroxynitrite. Nature 364, 548.

Bishop, N.A., Lu, T., Yankner, B.A., 2010. Neural mechanisms of ageing and cognitive decline. Nature 8, 529–535.

Bos, M.J., van Rijn, M.J., Witteman, J.C., et al., 2007. Incidence and prognosis of transient neurological attacks. JAMA 298, 2877e2885.

Brainin, M., Tuomilehto, J., Heiss, W.D., Bornstein, N.M., Bath, P.M., Teuschl, Y., et al., 2015. Post-stroke cognitive decline: an update and perspectives for clinical research. Eur. J. Neurol. 22 (229–238), e13–e16.

Brand-Schieber, E., Werner, P., Iacobas, D.A., Iacobas, S., Beelitz, M., Lowery, S.L., et al., 2005. Connexin43, the major gap junction protein of astrocytes, is down-regulated in inflamed white matter in an animal model of multiple sclerosis. J. Neurosci. Res. 80, 798–808.

Bryan, S., Baregzay, B., Spicer, D., Singal, P.K., Khaper, N., 2013. Redox-inflammatory synergy in the metabolic syndrome. Can. J. Physiol. Pharmacol. 91, 22–30.

Cai, W., Zhang, K., Li, P., Zhu, L., Xu, J., et al., 2017. Dysfunction of the neurovascular unit in ischemic stroke and neurodegenerative diseases: an aging effect. Ageing Res. Rev. 34, 77–87.

Calì, C., Bezzi, P., 2010. CXCR4-mediated glutamate exocytosis from astrocytes. J. Neuroimmunol. 224, 13–21.

Chapman, K.Z., Dale, V.Q., Denes, A., Bennett, G., Rothwell, N.J., et al., 2009. A rapid and transient peripheral inflammatory response precedes brain inflammation after experimental stroke. J. Cereb. Blood Flow Metab. 29, 1764–1768.

Cohen, J.A., Verghese, J., Zwerling, J.L., 2016. Cognition and gait in older people. Maturitas 93, 73–77.

Collins, T., 1999. Acute and chronic inflammation. In: Cotran, R.S., KaTC, V. (Eds.), Pathologic Basis of Disease. WB Saunders, Philadelphia, pp. 50–88.

Correale, J., Villa, A., 2004. The neuroprotective role of inflammation in nervous system injuries. J. Neurol. 251, 1304–1316.

Corriveau, R.A., Bosetti, F., Emr, M., Gladman, J.T., Koenig, J.I., Moy, C.S., et al., 2016. The science of vascular contributions to cognitive impairment and dementia (VCID): a framework for advancing research priorities in the cerebrovascular biology of cognitive decline. Cell. Mol. Neurobiol. 36, 281–288.

Cumming, T.B., Tyedin, K., Churilov, L., Morris, M.E., Bernhardt, J., 2012. The effect of physical activity on cognitive function after stroke: a systematic review. Int. Psychogeriatr. 24, 557–567.

Cumming, T.B., Marshall, R.S., Lazar, R.M., 2013. Stroke, cognitive deficits, and rehabilitation: still an incomplete picture. Int. J. Stroke 8, 38–45.

Dantzer, R., Bluthé, R.M., Gheusi, G., Cremona, S., Layé, S., et al., 1998. Molecular basis of sickness behavior. Ann. N.Y. Acad. Sci. 856, 132–138.

de Leeuw, F.E., de Groot, J.C., Oudkerk, M., Kors, J.A., Hofman, A., van Gijn, J., et al., 2000. Atrial fibrillation and the risk of cerebral white matter lesions. Neurology 54, 1795–1801.

de Leeuw, F.E., de Groot, J.C., Oudkerk, M., Witteman, J.C., Hofman, A., van Gijn, J., et al., 2002. Hypertension and cerebral white matter lesions in a prospective cohort study. Brain 125 (Pt 4), 765–772.

de Vries, H.E., Kooij, G., Frenkel, D., Georgopoulos, S., Monsonego, A., et al., 2012. Inflammatory events at blood–brain barrier in neuroinflammatory and neurodegenerative disorders: implications for clinical disease. Epilepsia 53 (Suppl. 6), 45–52.

Delgado, C., Donoso, A., Orellana, P., Vásquez, C., Díaz, V., et al., 2010. Frequency and determinants of poststroke cognitive impairment at three and twelve months in Chile. Dement. Geriatr. Cogn. Disord. 29, 397–405.

Denes, A., Vidyasagar, R., Feng, J., et al., 2007. Proliferating resident microglia after focal cerebral ischaemia in mice. J. Cerebral Blood Flow Metab. 27, 1941–1953.

Douiri, A., Rudd, A.G., Wolfe, C.D.A., 2013. Prevalence of poststroke cognitive impairment: South London stroke register 1995–2010. Stroke 44, 138–145.

Enager, P., Piilgaard, H., Offenhauser, N., Kocharyan, A., Fernandes, P., et al., 2009. Pathway-specific variations in neurovascular and neurometabolic coupling in rat primary somatosensory cortex. J. Cereb. Blood Flow Metab. 29, 976–986.

Enciu, A.M., Popescu, B.O., 2013. Is there a causal link between inflammation and dementia? Biomed. Res. Int. 2013, 316495.
Engelter, S.T., Gostynski, M., Papa, S., et al., 2006. Epidemiology of aphasia attributable to first ischemic stroke: incidence, severity, fluency, etiology, and thrombolysis. Stroke 37, 1379–1384.
Famakin, B.M., 2014. The immune response to acute focal cerebral ischemia and associated post-stroke immunodepression: a focused review. Aging Dis. 5, 307–326.
Farkas, E., De Jong, G.I., de Vos, R.A., Jansen Steur, E.N., Luiten, P.G., 2000. Pathological features of cerebral cortical capillaries are doubled in Alzheimer's disease and Parkinson's disease. Acta Neuropathol. 100, 395–402.
Farkas, E., Luiten, P.G., Bari, F., 2007. Permanent, bilateral common carotid artery occlusion in the rat: a model for chronic cerebral hypoperfusion-related neurodegenerative diseases. Brain Res. Rev. 54, 162–180.
Farooqui, A.A., 2010. Neurochemical Aspects of Neurotraumatic and Neurodegenerative Diseases. Springer, New York.
Farooqui, A.A., 2013. Metabolic Syndrome an Important Risk Factor for Stroke, Alzheimer Disease and Depression. Springer, New York.
Farooqui, A.A., 2014. Inflammation and Oxidative Stress in Neurological Disorders. Springer, New York.
Farooqui, A.A., 2017. Neurochemical Aspects of Alzheimer's Disease. Academic Press, San Diego, CA.
Farooqui, A.A., 2018. Ischemic, Traumatic Brain, and Spinal Cord Injuries: Mechanisms and Potential Therapies. Academic Press.
Farooqui, A.A., Farooqui, T., Panza, F., Frisardi, V., 2012. Metabolic syndrome as a risk factor for neurological disorders. Cell. Mol. Life Sci. 69, 741–762.
Finck, B.N., Kelly, D.P., 2006. PGC-1 coactivators: inducible regulators of energy metabolism in health and disease. J. Clin. Invest. 116, 615–622.
Ford, L., Crossley, M., Williams, T., Thorpe, J.R., Serpell, L.C., et al., 2015. Effects of Abeta exposure on long-term associative memory and its neuronal mechanisms in a defined neuronal network. Sci. Rep. 5, 10614.
Gehrmann, J., Matsumoto, Y., Kreutzberg, G.W., 1995. Microglia: intrinsic immuneffector cell of the brain. Brain Res. Brain Res. Rev. 20, 269–287.
Grimes, D.A., Schulz, K.F., 2002. An overview of clinical research: the lay of the land. Lancet. 359, 57–61.
Grimm, A., Mensah-Nyagan, A.G., Eckert, A., 2016. Alzheimer, mitochondria and gender. Neurosci. Biobehav. Rev. 67, 89–101.
Grysiewicz, R., Gorelick, P.B., 2012. Key neuroanatomical structures for post-stroke cognitive impairment. Curr. Neurol. Neurosci. Rep. 12, 703–708.
Gu, Y., Dee, C.M., Shen, J., 2011. Interaction of free radicals, matrix metalloproteinases and caveolin-1 impacts blood-brain barrier permeability. Front Biosci. (Schol Ed) 3, 1216–1231.
Gutiérrez-Vargas, J.A., Moreno, H., Cardona-Gómez, G.P., 2017. Targeting CDK5 post-stroke provides long-term neuroprotection and rescues synaptic plasticity. J. Cereb. Blood Flow Metab. 37, 2208–2223.
Haddad, J.J., Saade, N.E., Safieh-Garabedian, B., 2002. Cytokines and neuro-immune-endocrine interactions: a role for the hypothalamic-pituitary-adrenal revolving axis. J. Neuroimmunol. 133, 1–19.
Harris, I.M., Wong, C., Andrews, S., 2015. Visual field asymmetries in object individuation. Conscious. Cogn. 37, 194–206.
Hayakawa, K., Lo, E.H., 2016. Brain-peripheral cell crosstalk in white matter damage and repair. Biochim. Biophys. Acta 1862, 901–908.

Helzner, E.P., Luchsinger, J.A., Scarmeas, N., Cosentino, S., Brickman, A.M., et al., 2009. Contribution of vascular risk factors to the progression in Alzheimer disease. Arch. Neurol. 66, 343–348.

Hwang, C.J., Park, M.H., Hwang, J.Y., Kim, J.H., Yun, N.Y., et al., 2016. CCR5 deficiency accelerates lipopolysaccharide-induced astrogliosis, amyloid-beta deposit and impaired memory function. Oncotarget 7, 11984–11999.

Iadecola, C., 2009. Bleeding in the brain: killer waves of depolarization in subarachnoid bleed. Nat. Med. 15, 1131–1132.

Iadecola, C., 2010. The overlap between neurodegenerative and vascular factors in the pathogenesis of dementia. Acta Neuropathol. 120, 287–296.

Iadecola, C., 2014. Hypertension and dementia. Hypertension 64, 3–5.

Iadecola, C., Anrather, J., 2011. The immunology of stroke: from mechanisms to translation. Nat. Med. 17, 796–808.

Inatomi, Y., Yonehara, T., Omiya, S., Hashimoto, Y., Hirano, T., Uchino, M., 2008. Aphasia during the acute phase in ischemic stroke. Cerebrovasc. Dis. 25, 316–323.

Jauch, E.C., Saver, J.L., Adams Jr, H.P., Bruno, A., Connors, J.J., et al., 2013. Guidelines for the early management of patients with acute ischemic stroke: a guideline for healthcare professionals from the American Heart Association/American Stroke Association. Stroke. 44, 870–947.

Jin, K., Wang, X., Xie, L., Mao, X.O., Zhu, W., Wang, Y., et al., 2006. Evidence for stroke-induced neurogenesis in the human brain. Proc. Natl. Acad. Sci. U.S.A. 103, 13198–13202.

Kamel, H., Iadecola, C., 2012. Brain-immune interactions and ischemic stroke: clinical implications. Arch. Neurol. 69, 576–581.

Kauranen, T., Turunen, K., Laari, S., Mustanoja, S., Baumann, P., Poutiainen, E., 2013. The severity of cognitive deficits predicts return to work after a first-ever ischaemic stroke. J. Neurol. Neurosurg. Psychiatry 84, 316–321.

Kigerl, K.A., Gensel, J.C., Ankeny, D.P., Alexander, J.K., Donnelly, D.J., et al., 2009. Identification of two distinct macrophage subsets with divergent effects causing either neurotoxicity or regeneration in the injured mouse spinal cord. J. Neurosci. 29, 13435–13444.

Kim, H.K., Nunes, P.V., Oliveira, K.C., Young, L.T., Lafer, B., 2016. Neuropathological relationship between major depression and dementia: a hypothetical model and review. Prog. Neuropsychopharmacol. Biol. Psychiatry 67, 51–57.

Koh, S.H., Lo, E.H., 2015. The role of the PI3K pathway in the regeneration of the damaged brain by neural stem cells after cerebral infarction. J. Clin. Neurol. 11, 297–304.

Korczyn, A.D., 2002. The complex nosological concept of vascular dementia. J. Neurol. Sci. 203–204, 3–6.

Lacor, P.N., Buniel, M.C., Furlow, P.W., Clemente, A.S., Velasco, P.T., et al., 2007. Abeta oligomer-induced aberrations in synapse composition, shape, and density provide a molecular basis for loss of connectivity in Alzheimer's disease. J. Neurosci. 27, 796–807.

Leonard, B.E., 2010. The concept of depression as a dysfunction of the immune system. Curr. Immunol. Rev. 6, 205–212.

Leśniak, M., Bak, T., Czepiel, W., Seniów, J., Członkowska, A., 2008. Frequency and prognostic value of cognitive disorders in stroke patients. Dement. Geriatr. Cogn. Disord. 2008 (26), 356–363.

Leys, D., Henon, H., Mackowiak-Cordoliani, M.A., Pasquier, F., 2005. Poststroke dementia. Lancet. Neurol. 4, 752–759.

Li, W.J., Mao, F.X., Chen, H.J., Qian, L.H., Buzby, J.S., 2015. Treatment with UDP-glucose, GDNF, and memantine promotes SVZ and white matter self-repair by endogenous glial progenitor cells in neonatal rats with ischemic PVL. Neuroscience 284, 444–458.

Lin, J.-H., Lin, R.-T., Tai, C.-T., et al., 2003. Prediction of poststroke dementia. Neurology 61, 343–348.

Lindsey, J.D., Landfield, P.W., Lynch, G., 1979. Early onset and topographical distribution of hypertrophied astrocytes in hippocampus of aging rats: a quantitative study. J. Gerontol. 34, 661–671.

Loerch, P.M., Lu, T., Dakin, K.A., Vann, J.M., Isaacs, A., et al., 2008. Evolution of the aging brain transcriptome and synaptic regulation. PLoS One 3, e3329.

Lu, T., Pan, Y., Kao, S.Y., Li, C., Kohane, I., Chan, J., et al., 2004. Gene regulation and DNA damage in the ageing human brain. Nature 429, 883–891.

Makin, S.D.J., Turpin, S., Dennis, M.S., Wardlaw, J.M., 2013. Cognitive impairment after lacunar stroke: systematic review and meta-analysis of incidence, prevalence and comparison with other stroke subtypes. J. Neurol. Neurosurg. Psychiatry 84, 893–900.

Maragakis, N.J., Rothstein, J.D., 2001. Glutamate transporters in neurologic disease. Arch. Neurol. 58, 365–370.

Martin, L.J., Liu, Z., 2002. DNA damage profiling in motor neurons: a single-cell analysis by comet assay. Neurochem. Res. 27, 1089–1100.

Middleton, S., McElduff, P., Ward, J., Grimshaw, J.M., Dale, S., et al., 2011. Implementation of evidence-based treatment protocols to manage fever, hyperglycaemia, and swallowing dysfunction in acute stroke (QASC): a cluster randomised controlled trial. Lancet 378, 1699–1706.

Mijajlović, M.D., Pavlović, A., Brainin, M., Heiss, W.D., Quinn, T.J., et al., 2017. Post-stroke dementia – a comprehensive review. BMC Med. 15, 11.

Miller, D.W., 1999. Immunobiology of the blood-brain barrier. J. Neurovirol. 5, 570–578.

Minghetti, L., Ajmone-Cat, M.A., De Berardinis, M.A., De Simone, R., 2005. Microglial activation in chronic neurodegenerative diseases: roles of apoptotic neurons and chronic stimulation. Brain Res. Rev. 48, 251–256.

Molko, N., Pappata, S., Mangin, J.F., Poupon, F., LeBihan, D., Bousser, M.G., et al., 2002. Monitoring disease progression in CADASIL with diffusion magnetic resonance imaging: a study with whole brain histogram analysis. Stroke 33, 2902–2908.

Morrison, J.H., Baxter, M.G., 2012. The aging cortical synapse: hallmarks and implications for cognitive decline. Nat. Rev. Neurosci. 13, 240–250.

Mungrue, I.N., Bredt, D.S., Stewart, D.J., Husain, M., 2003. From molecules to mammals: what's NOS got to do with it? Acta Physiol. Scand. 179, 123–135.

Nakayama, D., Matsuyama, T., Ishibashi-Ueda, H., Nakagomi, T., Kasahara, Y., Hirose, H., et al., 2010. Injury-induced neural stem/progenitor cells in post-stroke human cerebral cortex. Eur. J. Neurosci. 31, 90–98.

Nelson, A.R., Sweeney, M.D., Sagare, A.P., Zlokovic, B.V., 2016. Neurovascular dysfunction and neurodegeneration in dementia and Alzheimer's disease. Biochim. Biophys. Acta 1862, 887–900.

Nichols, T.E., Holmes, A.P., 2002. Nonparametric permutation tests for functional neuroimaging: a primer with examples. Hum. Brain. Mapp. 15, 1–25.

Pasquier, F., Leys, D., 1997. Why are stroke patients prone to develop dementia? J. Neurol. 244, 135–142.

Patel, M.D., Coshall, C., Rudd, A.G., Wolfe, C.D.A., 2002. Cognitive impairment after stroke: clinical determinants and its associations with long-term stroke outcomes. J. Am. Geriatr. Soc. 50, 700–706.

Pendlebury, S.T., 2009. Stroke-related dementia: rates, risk factors and implications for future research. Maturitas 64, 165–171.

Pendlebury, S.T., Rothwell, P.M., 2009. Prevalence, incidence, and factors associated with pre-stroke and post-stroke dementia: a systematic review and meta-analysis. Lancet Neurol. 8, 1006–1018.

Petrovitch, H., Ross, G.W., Steinhorn, S.C., Abbott, R.D., Markesbery, W., Davis, D., et al., 2005. AD lesions and infarcts in demented and non-demented Japanese-American men. Ann. Neurol. 57, 98–103.

Phillis, J.W., Horrocks, L.A., Farooqui, A.A., 2006. Cyclooxygenases, lipoxygenases, and epoxygenases in CNS: their role and involvement in neurological disorders. Brain Res. Rev. 52, 201–243.

Pirmoradi, M., Jemel, B., Gallagher, A., Tremblay, J., D'Hondt, F., et al., 2016. Verbal memory and verbal fluency tasks used for language localization and lateralization during magnetoencephalography. Epilepsy Res. 119, 1–9.

Pita-Almenar, J.D., Ranganathan, G.N., Koester, H.J., 2012. Impact of cortical plasticity on patterns of suprathreshold activity in the cerebral cortex. J. Neurophysiol. 107, 850–858.

Pollock, A., Baer, G., Campbell, P., Choo, P.L., Forster, A., et al., 2014. Physical rehabilitation approaches for the recovery of function and mobility following stroke. Cochrane Database Syst. Rev. 4, CD001920.

Poluektova, L., Meyer, V., Walters, L., Paez, X., Gendelman, H.E., 2005. Macrophage-induced inflammation affects hippocampal plasticity and neuronal development in a murine model of HIV-1 encephalitis. Glia 52, 344–353.

Posada-Duque, R.A., López-Tobón, A., Piedrahita, D., González-Billault, C., Cardona-Gomez, G.P., 2015. p35 and Rac1 underlie the neuroprotection and cognitive improvement induced by CDK5 silencing. J. Neurochem. 134, 354–370.

Prass, K., Meisel, C., Hoflich, C., Braun, J., Halle, E., et al., 2003. Stroke-induced immunodeficiency promotes spontaneous bacterial infections and is mediated by sympathetic activation reversal by poststroke T helper cell type 1-like immunostimulation. J. Exp. Med. 198, 725–736.

Pun, P.B., Lu, J., Moochhala, S., 2009. Involvement of ROS in BBB dysfunction. Free Radic. Res. 43, 348–364.

Qian, L., Ding, L., Cheng, L., Zhu, X., Zhao, H., et al., 2012. Early biomarkers for post-stroke cognitive impairment. J. Neurol. 259, 2111–2118.

Raj, D.D., Jaarsma, D., Holtman, I.R., Olah, M., Ferreira, F.M., Schaafsma, W., 2014. Priming of microglia in a DNA-repair deficient model of accelerated aging. Neurobiol. Aging 35, 2147–2160.

Rand, D., Eng, J.J., Liu-Ambrose, T., Tawashy, A.E., 2010. Feasibility of a 6-month exercise and recreation program to improve executive functioning and memory in individuals with chronic stroke. Neurorehabil. Neural. Repair 24, 722–729.

Richwine, A.F., Parkin, A.O., Buchanan, J.B., Chen, J., Markham, J.A., et al., 2008. Architectural changes to CA1 pyramidal neurons in adult and aged mice after peripheral immune stimulation. Psychoneuroendocrinology 33, 1369–1377.

Ringman, J.M., Saver, J.L., Woolson, R.F., Clarke, W.R., Adams, H.P., 2004. Frequency, risk factors, anatomy, and course of unilateral neglect in an acute stroke cohort. Neurology 63, 468–474.

Rombouts, S.A., Barkhof, F., Veltman, D.J., Machielsen, W.C., Witter, M.P., et al., 2000. Functional MR imaging in Alzheimer's disease during memory encoding. AJNR Am. J. Neuroradiol. 21, 1869–1875.

Rosenberg, G.A., 2017. Extracellular matrix inflammation in vascular cognitive impairment and dementia. Clin. Sci. (Lond.) 131, 425–437.

Sachdev, P.S., Brodaty, H., Valenzuela, M.J., et al., 2006. Clinical determinants of dementia and mild cognitive impairment following ischaemic stroke: the Sydney Stroke Study. Dement. Geriatr. Cogn. Disord. 21, 275–283.

Sahin, E., Colla, S., Liesa, M., Moslehi, J., Muller, F.L., Guo, M., 2011. Telomere dysfunction induces metabolic and mitochondrial compromise. Nature 470, 359–365.

Save-Pédebos, J., Pinabiaux, C., Dorfmuller, G., Sorbets, S.F., Delalande, O., et al., 2016. The development of pragmatic skills in children after hemispherotomy: contribution from left and right hemispheres. Epilepsy Behav. 55, 139–145.

Schuh, A.F., Rieder, C.M., Rizzi, L., Chaves, M., Roriz-Cruz, M., 2011. Mechanisms of brain aging regulation by insulin: implications for neurodegeneration in late-onset Alzheimer's Disease. ISRN Neurol. 2011, 306905.

Shankar, G.M., Bloodgood, B.L., Townsend, M., Walsh, D.M., Selkoe, D.J., et al., 2007. Natural oligomers of the Alzheimer amyloid-beta protein induce reversible synapse loss by modulating an NMDA-type glutamate receptor-dependent signaling pathway. J. Neurosci. 27, 2866–2875.

Sibolt, G., Curtze, S., Melkas, S., et al., 2013. Post-stroke dementia is associated with recurrent ischaemic stroke. J. Neurol. Neurosurg. Psych. 84, 722–726.

Sies, H., 1993. Strategies of antioxidant defense. Eur. J. Biochem. 215, 213–219.

Silvestrini, M., Pasqualetti, P., Baruffaldi, R., Bartolini, M., Handouk, Y., et al., 2006. Cerebrovascular reactivity and cognitive decline in patients with Alzheimer disease. Stroke 37, 1010–1015.

Sofroniew, M.V., Vinters, H.V., 2010. Astrocytes: biology and pathology. Acta Neuropathol. 119, 7–35.

Strong, A.J., Anderson, P.J., Watts, H.R., Virley, D.J., Lloyd, A., Irving, E.A., et al., 2007. Peri-infarct depolarizations lead to loss of perfusion in ischaemic gyrencephalic cerebral cortex. Brain. 130, 995–1008.

Sun, G.Y., Horrocks, L.A., Farooqui, A.A., 2007. The role of NADPH oxidase and phospholipases A_2 in mediating oxidative and inflammatory responses in neurodegenerative diseases. J. Neurochem. 103, 1–16.

Sun, J.-H., Tan, L., Yu, J.-T., 2014. Post-stroke cognitive impairment: epidemiology, mechanisms and management. Ann. Transl. Med. 2, 80.

Surawan, J., Areemit, S., Tiamkao, S., Sirithanawuthichai, T., Saensak, S., 2017. Risk factors associated with post-stroke dementia: a systematic review and meta-analysis. Neurol. Int. 9, 7261.

Tang, S.C., Yang, K.C., Hu, C.J., Chiou, H.Y., Wu, C.C., et al., 2017. Elevated plasma level of soluble form of RAGE in ischemic stroke patients with dementia. Neuromol. Med. 19, 579–583.

Tao, C., Hu, X., Li, H., You, C., 2017. White matter injury after intracerebral hemorrhage: pathophysiology and therapeutic strategies. Front. Hum. Neurosci. 11, 422.

Temprado, J.-J., Sleimen-Malkoun, R., Lemaire, P., Rey-Robert, B., Retornaz, F., et al., 2013. Aging of sensorimotor processes: a systematic study in Fitts' task. Exp. Brain Res. 228, 105–116.

Tian, L., Rauvala, H., Gahmberg, C.G., 2009. Neuronal regulation of immune responses in the central nervous system. Trends Immunol. 30, 91–99.

Toth, P., Tarantini, S., Tucsek, Z., Ashpole, N.M., Sosnowska, D., et al., 2014. Resveratrol treatment rescues neurovascular coupling in aged mice: role of improved cerebromicrovascular endothelial function and down-regulation of NADPH oxidase. Am. J. Physiol. Heart Circ. Physiol. 306, H299–H308.

Toth, P., Tarantini, S., Csiszar, A., Ungvari, Z., 2017. Functional vascular contributions to cognitive impairment and dementia: mechanisms and consequences of cerebral autoregulatory dysfunction, endothelial impairment, and neurovascular uncoupling in aging. Am. J. Physiol. Heart Circ. Physiol. 312, H1–H20.

Trapero, I., Cauli, O., 2014. Interleukin 6 and cognitive dysfunction. Metab. Brain. Dis. 29, 593–608.

Vermeer, S.E., Prins, N.D., den Heijer, T., Hofman, A., Koudstaal, P.J., et al., 2003. Silent brain infarcts and the risk of dementia and cognitive decline. N. Engl. J. Med. 348, 1215–1222.

Wagner, K.R., Beiler, S., Beiler, C., Kirkman, J., Casey, K., et al., 2006. Delayed profound local brain hypothermia markedly reduces interleukin-1β gene expression and vasogenic edema development in a porcine model of intracerebral hemorrhage. Acta Neurochir. Suppl. 96, 177–182.

Walsh, D.M., Klyubin, I., Fadeeva, J.V., Cullen, W.K., Anwyl, R., et al., 2002. Naturally secreted oligomers of amyloid beta protein potently inhibit hippocampal long-term potentiation in vivo. Nature 416, 535–539.

Wang, G., Zhang, J., Hu, X., Zhang, L., Mao, L., et al., 2013. Microglia/macrophage polarization dynamics in white matter after traumatic brain injury. J. Cereb. Blood Flow Metab. 33, 1864–1874.

Wang, Y., Liu, G., Hong, D., Chen, F., Ji, X., Cao, G., 2016. White matter injury in ischemic stroke. Prog. Neurobiol. 141, 45–60.

Wardlaw, J.M., Smith, E.E., Biessels, G.J., Cordonnier, C., Fazekas, F., et al., 2013. Neuroimaging standards for research into small vessel disease and its contribution to ageing and neurodegeneration. Lancet Neurol. 12, 822–838.

Westendorp, W.F., Nederkoorn, P.J., Vermeij, J.D., Dijkgraaf, M.G., van de Beek, D., 2011. Post-stroke infection: a systematic review and meta-analysis. BMC Neurol. 11, 110.

Westman, E., Wahlund, L.O., Foy, C., Poppe, M., Cooper, A., et al., 2011. Magnetic resonance imaging and magnetic resonance spectroscopy for detection of early Alzheimer's disease. J. Alzheimers Dis. 26 (Suppl. 3), 307–319.

Xu, Y.Y., Shang, J.H., 2016. Meta-analysis of risk factors for Parkinson's disease dementia. Transl. Neurodegener. 5, 1.

Yang, Y., Rosenberg, G.A., 2011. Blood-brain barrier breakdown in acute and chronic cerebrovascular disease. Stroke 42, 3323–3328.

Yankner, B.A., Lu, T., Loerch, P., 2008. The aging brain. Annu. Rev. Pathol. 3, 41–66.

Yates, P.A., Villemagne, V.L., Ellis, K.A., Desmond, P.M., Masters, C.L., Rowe, C.C., 2014. Cerebral microbleeds: a review of clinical, genetic, and neuroimaging associations. Front. Neurol. 4, 205.

Zhou, X., Xie, Q., Xi, G., Keep, R.F., Hua, Y., 2014. Brain CD47 expression in a swine model of intracerebral hemorrhage. Brain Res. 1574, 70–76.

Zorov, D.B., Isave, N.K., Plotnikov, E.Y., Zorova, L.D., Stelmashook, E.V., et al., 2007. The mitochondrion as Janus Bifrons. Biochemistry (Moscow) 72, 1115–1126.

Further Reading

Amor, S., Puentes, F., Baker, D., van der Valk, P., 2010. Inflammation in neurodegenerative diseases. Immunology 129, 154–169.

Dennis, N.A., Cabeza, R., 2008. Neuroimaging of healthy cognitive aging. In: Craik, F.I.M., Salthouse, T.A. (Eds.), The Handbook of Aging and Cognition, third ed Psychology Press, New York, pp. 1–54.

Dong, Y., Benveniste, E.N., 2001. Immune function of astrocytes. Glia 36, 180–190.

Pendlebury, S.T., 2012. Dementia in patients hospitalized with stroke: rates, time course, and clinico-pathologic factors. Int. J. Stroke 7, 570–581.

Risau, W., 1997. Mechanisms of angiogenesis. Nature 386, 671–674.

Soos, J.M., Ashley, T.A., Morrow, J., Patarroyo, J.C., Szente, B.E., Zamvil, S.S., 1999. Differential expression of B7 co-stimulatory molecules by astrocytes correlates with T cell activation and cytokine production. Int. Immunol. 11, 1169–1179.

CHAPTER

3

Neurochemical Aspects of Alzheimer's Type of Dementia

INTRODUCTION

Alzheimer's disease (AD) is the most common form of dementia. AD type of dementia is characterized by the accumulation of extracellular β-amyloid (Aβ) plaques (senile plaques) and intracellular neurofibrillary tangles (NFTs) composed of tau amyloid fibrils along with extensive neuronal death in selected brain regions (Hardy, 2009) (Fig. 3.1). As AD progresses, the NFTs affect more neocortical areas, resulting in deficits in other cognitive domains (Braak and Braak, 1991; Braak and Del Tredici, 2011). In contrast, amyloid plaques tend to accumulate more in the association cortices first and affect hippocampal structures only as the disease progresses (Braak and Braak, 1991; Thal et al., 2002). The main symptom of AD type of dementia is memory loss, which correlates with a decline of neuron population not only in the hippocampus, but also in the entorhinal cortex—the area of interface between the hippocampus and the neocortex (Jahn, 2013; Querfurth and Laferla, 2010). The decrease in cerebral blood flow (CBF) (Bangen et al., 2014) and accumulation of Aβ are independently linked with increased risk of developing memory loss and dementia (Rodrigue et al., 2012). Furthermore, AD is also accompanied by early cerebral vascular dysfunction (Zlokovic, 2011) and alterations in neurovascular function. These processes are closely linked with regulation of CBF by arterioles and the capillary neurovascular unit (Girouard and Iadecola, 2006). The neuropathology of AD develops decades prior to the initial cognitive symptoms in a preclinical or presymptomatic stage, in which accumulations of Aβ and tau start to occur in brain with the formation of amyloid plaques and NFTs (Sperling et al., 2011). As stated above, cerebrovascular function is also impaired in patients with early AD or at risk for AD, leading to a mismatch between the delivery of oxygen

Molecular Mechanisms of Dementia
DOI: https://doi.org/10.1016/B978-0-12-816347-4.00003-9

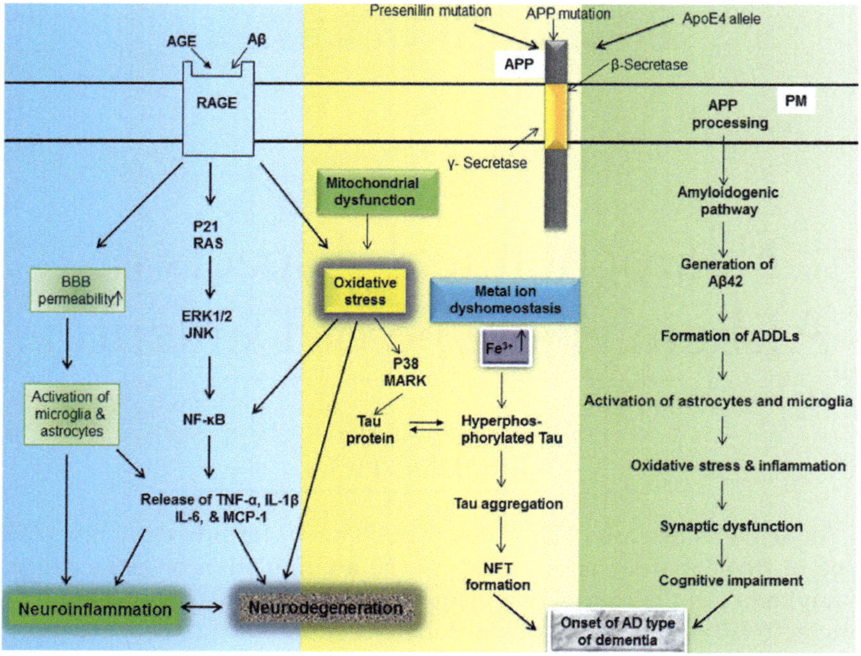

FIGURE 3.1 Processes contributing to the pathogenesis in the brain in AD patients. *Aβ*, beta-amyloid; *AD*, Alzheimer's disease; *ADDLs*, Aβ-derived diffusible ligands; *AGE*, advanced glycation end products; *APP*, amyloid precursor protein; *BBB*, blood–brain barrier; *ERK*, extracellular-signal-regulated kinase; *IL-1β*, interleukin-1β; *JNK*, jun amino-terminal kinases; *NFTs*, neurofibrillary tangles; *RAGE*, receptor for advanced glycation end products; *TNF-α*, tumor necrosis factor-α.

and glucose through blood flow and the energy demands of the active brain (Iadecola, 2013).

In AD type of dementia, the accumulation of Aβ plaques not only contributes to changes in the curvature of neurites and spine density (Meyer-Luehmann et al., 2008; Spires et al., 2005), but also to the inhibition of mitochondrial function and calcium homeostasis leading to rapid cell death in the vicinity of the plaques because of induction of oxidative stress (Xie et al., 2013). The growth of plaques occurs gradually over months, with a slower rate in older AD mice, and the degree of neuritic dystrophy correlates with the speed and extent of plaque enlargement (Condello et al., 2011). Collectively, these processes contribute to neurodegeneration and loss of synapses. Synapses are essential for transmitting, processing, and storing information from one cell to another in the brain. Synapses are composed of three main constituents: a presynaptic component (presynaptic ending, axon terminal); a synaptic cleft; and a postsynaptic component (dendritic spine). Two types of synapses (electrical and chemical) have been reported to occur in the brain tissue (Zoidl and Dermietzel, 2002). The loss of synapses in cortical and subcortical areas and the hippocampus leads to cognitive decline, along with progressive impairment of the activities of daily living and also often behavioral and physiological changes like apathy and depression. How these factors ultimately contribute to memory impairments and cognitive deficits is not fully understood. The molecular mechanisms of the pathogenesis of AD type of dementia still remain elusive. As AD type of dementia progresses, other debilitating noncognitive symptoms arise, including impaired sleep and appetite, and neuropsychiatric alterations (e.g., depression and apathy) (Ishii and Iadecola, 2015; Lanctôt et al., 2017).

The majority of AD cases (>90%–95%) are of sporadic (late-onset form). These patients are older than 65 years. After age 65, the risk of AD doubles every 5 years and after age 85, the risk reaches nearly 50%. The causes of the sporadic form of AD type of dementia are quite complex and may not only involve age-related alterations in metabolism, repair mechanisms, immune response, and the vascular system, but also may include exogenous factors such as brain trauma, obesity, insulin resistance, and diabetes (i.e., overall lifestyle) suggesting AD is a multifactorial disease that likely results from the complex interplay of multiple pathological processes, under the influence of internal and external determinants (Cohen and Dillin, 2008; Bishop et al., 2010; Farooqui, 2018). It is well known that abnormalities in energy metabolism are very frequently observed in AD patients. Thus, about 50%–60% of AD patients show abnormal eating behaviors (Ikeda et al., 2002) while 14%–80% of AD cases are of poor nutritional status (Droogsma et al., 2015). Furthermore, weight loss is an important clinical feature of AD in about 20%–45% of cases (Aziz et al., 2008). In AD type of dementia patients, cortical structures (e.g., parietal, posterior temporal, posterior cingulate–precuneal) show prominent hypometabolism. Whether and how this is related to weight loss remains unknown. However, the presence of senile plaques and NFTs has been reported in the hypothalamus at stages III and IV corresponding to early-moderate AD, and weight loss often occurs prior to cognitive impairments; factors other than tau and Aβ accumulation in the hypothalamus can contribute to such metabolic dysregulation. Furthermore, body mass index decline in older age is associated with increased risk of developing AD as well as with a faster rate of disease progression (Aziz et al., 2008).

Only 5%–7% cases are primarily genetic (early-onset familial form) involving apolipoprotein E (APOE), amyloid precursor protein (*APP*), presenilin 1 (*PS 1*), and presenilin 2 (*PS 2*) genes (van der Flier and Scheltens, 2005; Duthey, 2013). All the abovementioned AD related genes have been reported to upregulate the cerebral Aβ levels, with the majority of early-onset familial AD mutations increasing the ratio of Aβ42 to Aβ40, which enhances the oligomerization of Aβ into neurotoxic assemblies (Fig. 3.1) (Hardy and Selkoe, 2002).

Many hypotheses have been proposed to explain the pathogenesis of AD type of dementia including: (1) selective vulnerability of cholinergic neurons in the basal forebrain; (2) aluminum deposit hypothesis; (3) Aβ-cascade hypothesis; (4) reduction in neurotrophic factors; (5) protein misfolding and aggregation hypothesis; (6) amyloid cascade–inflammatory hypothesis; (7) neurovascular hypothesis; (8) insulin insensitivity hypothesis; and (9) dendritic hypothesis (Katzman and Saitoh, 1991; Castellani et al., 2009; Karran et al., 2011; McGeer and McGeer, 2013; Farooqui, 2013; Zlokovic, 2011; de la Monte and Tong, 2014; Cochran et al., 2014). Among the abovementioned hypotheses, the involvement of Aβ-cascade hypothesis in the pathogenesis of AD is the most popular. Aβ-cascade hypothesis is based on the generation and deposition of insoluble Aβ peptides and accumulation of Tau protein in the form of NFTs (Alzheimer's Association, 2012). Although, Aβ-cascade hypothesis does not explain all features of the AD type of dementia, it has dominated the research studies on AD type of dementia for the past 30 years since the hypothesis was proposed in the late 1980s (Allsop et al., 1988; Selkoe, 1989; Hardy and Higgins, 1992). Aβ is constantly synthesized from APP by the sequential action of two proteases, beta and gamma secretases (Querfurth and Laferla, 2010). The amount of Aβ in the cerebral tissue depends upon clearance mechanisms (Wang et al., 2017). Failure of any of these clearance mechanisms in the brain, at least partly due to the disruption of the phagocytic properties of glial cells and parenchymal neuroinflammation, leads to an Aβ overload, toxic oligomers accumulation, and formation of plaques (Heneka et al., 2015; Selkoe and Hardy, 2016). Tau protein contains more than 85 phosphorylated or phosphorylatable sites. These sites are phosphorylated by more than 30 kinases including glycogen synthase kinase 3β (GSK3β), cyclin dependent kinase 5 (cdk5), c-Jun N-terminal Kinase, microtubule affinity regulating kinase, extracellular-signal-regulated kinase 2, and Ca^{2+}/calmodulin-dependent protein kinase II, and 5′ adenosine monophosphate-activated protein kinase (Sergeant et al., 2008). Hyperphosphorylation of tau leads to conformational changes that notably impair its ability to bind to microtubules. Free monomers of misfolded tau then start to accumulate, oligomerize, and aggregate. Tau aggregates can deposit in NFTs that are observed early in life and increase during aging (Braak et al., 2011). According to Aβ-cascade hypothesis an imbalance between production and clearance of Aβ and its accumulation and aggregation in the brain is linked to the development of AD (Hardy, 2009). It should be noted that recent postmortem investigations in human AD have largely failed to provide unequivocal evidence in support of the original amyloid-cascade hypothesis (Koss et al., 2016). Conflicting evidence suggests, however, that Aβ plaques and NFTs, albeit to a lesser extent, are present in a substantial subset of nondemented individuals (Koss et al., 2016). Hence, a range of soluble tau and Aβ species have more recently been implicated as the disease-relevant toxic entities. Despite the incorporation of soluble proteins into a revised amyloid-cascade hypothesis, a detailed characterization of these species in the context of human AD onset, progression, and cognitive decline has been lacking. These observations along with the failure of some anti-Aβ therapies to preserve or rescue cognitive function suggests that Aβ may not be universally neurotoxic (Extance, 2010; Robakis, 2011; Tayeb et al., 2013), but other mechanisms directly or indirectly related to the oligomeric form of Aβ may contribute to the pathogenesis of AD type of dementia. Finally, neither Aβ plaques nor phospho-tau containing NFTs are specific for AD type of dementia. About 30% of normal aged people have as many Aβ plaques in their brains as in typical cases of AD (Dickson et al., 1992; van Duinen et al., 1987). Furthermore, other neuropathological conditions such as Parkinson's disease (Petrou et al., 2015), which is characterized by monoaminergic dysfunction, Lewy body pathology, and cerebrovascular disease along with cognitive impairment, and hereditary cerebral hemorrhage with amyloidosis of Dutch origin and sporadic cerebral amyloid angiopathy, which show amyloid pathology similar to AD without any dementia, suggesting that amyloid alone is insufficient to cause neuronal loss and cognitive symptoms observed in AD (Coria et al., 1987). Furthermore, another group of neurological disorders (stroke, head injury, and chronic traumatic encephalopathy) is also accompanied by the accumulation of Aβ along with other neurochemical changes in the brain suggesting that the accumulation of Aβ alone is not sufficient to explain the induction of neurodegeneration (Farooqui, 2017). Formation of NFTs with hyperphosphorylated tau is a hallmark of several neurodegenerative diseases called tauopathies which include frontotemporal dementia with parkinsonism linked to chromosome-17 tau, Pick disease, corticobasal degeneration, postsupranuclear palsy, dementia pugilistica/traumatic brain injury/chronic traumatic encephalopathy, and Guam parkinsonism–dementia complex. In addition, there is an important association between early-onset AD and Down's syndrome. From a neurobiological perspective, recognition of the role of APP overproduction due to increased gene dosage with trisomy 21 is an important clue for the involvement of the *APP* gene and the amyloid hypothesis of AD pathogenesis. Clinically, patients with Down's syndrome have increased risk of developing younger-onset dementia after the age of 35 years. AD-mediated changes at postmortem are essentially universal, while the prevalence of clinical dementia in individuals with Down's syndrome has been estimated as 15%–25% overall and increases steeply with increasing age (Nieuwenhuis-Mark, 2009; Castro et al., 2017). Clinical assessment of patients with these conditions is particularly challenging in this population, particularly as indices of executive and social and emotional functioning may be more important than tests of memory in indicating the onset of clinical dementia.

RISK FACTORS FOR ALZHEIMER TYPE OF DEMENTIA

AD is the most frequent cause of dementia, increasing in prevalence from less than 1% below the age of 60 years to more than 50% above 85 years of age. Aging is a major risk factor for AD. In addition, the gut microbiota, which comprises a complex community of microorganism species that reside in our gastrointestinal ecosystem and whose alterations influence not just various gut disorders, also play an important role in the pathogenesis of AD type of dementia (Jiang et al., 2017; Hoffman et al., 2017). The microbiota–gut–brain axis is a bidirectional communication system, which is associated with the functioning of neural, immune, endocrine, and metabolic pathways (Jiang et al., 2017; Hoffman et al., 2017). The increased permeability of the gut and blood–brain barrier (BBB) modulated by microbiota dysbiosis may promote the pathogenesis of AD type of dementia and other neurodegenerative disorders, especially those associated with aging (Jiang et al., 2017; Hoffman et al., 2017). In addition, bacteria populating the gut microbiota can secrete large amounts of amyloids and lipopolysaccharides, which may modulate signaling pathways and the production of proinflammatory cytokines associated with the pathogenesis of AD (Jiang et al., 2017; Hoffman et al., 2017). Moreover, imbalances in the gut microbiota can induce inflammation that is associated with the pathogenesis of obesity, type 2 diabetes mellitus, and AD (Jiang et al., 2017; Hoffman et al., 2017). Several conserved signaling pathways (insulin/IGF signaling and mitochondrial dysfunction) control the aging process. These pathways also control and modulate cognitive decline. However, the role of these conserved pathways in the onset and progression of AD type of dementia and other neurodegenerative disorders in humans is still unclear (Bishop et al., 2010). It is proposed that microbiota-derived metabolites such as the short-chain fatty acid (SCFA) butyrate are primary signals, which are linked to human health. This SCFA improves insulin resistance (Velasquez-Manoff, 2015). Orally consumed butyrate induces GLP-1 secretion, a hormone which supports the improvement of glucose tolerance and appetite control (Yadav et al.,

2013). Butyrate stimulates neurogenesis in the ischemic brain via brain-derived neurotrophic factor (BDNF) upregulation (Kim et al., 2009). It not only produces antidepressant-like effects (Yamawaki et al., 2012), but also inhibits NF-κB and increase I-κB levels as a countermeasure for improved long-term inflammatory control (Segain et al., 2000).

It is estimated that, in 2050 approximately 80 million people will suffer from AD worldwide. The initial phase of AD type of dementia is accompanied by a progressive deterioration of episodic memory. Other impairments in executive function may be entirely absent in the beginning of AD type of dementia. As dementia advances, impairment spreads to other aspects of memory and other domains of cognition (McKhann et al., 2011). Studies on patients suffering from the autosomal dominant version of AD and population studies that have followed patients or analyzed their performance retrospectively, suggest that the pathogenesis of AD probably starts 10–15 years earlier than when patients typically receive their diagnosis (Amieva et al., 2008; Bateman et al., 2012). This late diagnosis has created problems with the determination of the early mechanisms that contribute to the pathogenesis of AD and its progression.

Risk factors for AD type of dementia are classified into two groups: the nonmodifiable risk factors and modifiable risk factors. Nonmodifiable risk factors include age and genetic predisposition (ApoEε4). Modifiable risk factors include long-term consumption of the Western diet, physical inactivity, type 2 diabetes, traumatic brain injury, periodontitis, environmental factors, and deficiency of magnesium ions (Mg^{2+}) (Fig. 3.2). Among these factors, brain aging is a complex, inevitable, and undeniable factor, which is accompanied by a time-dependent

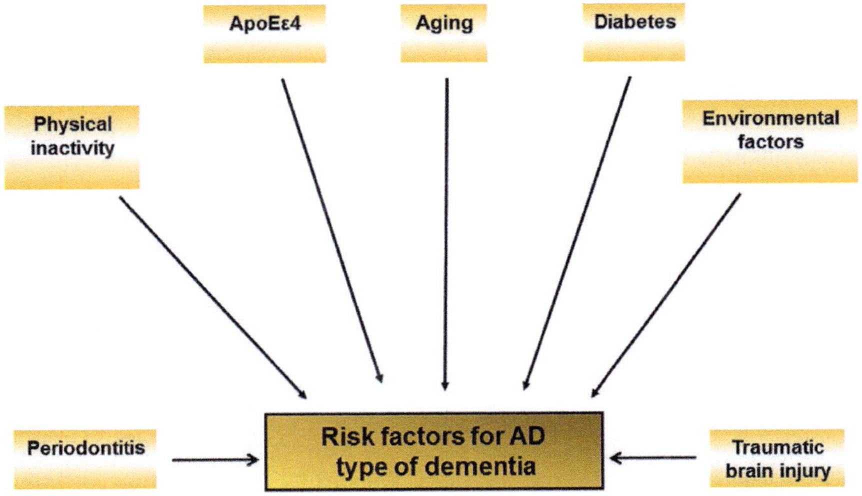

FIGURE 3.2 Risk factors for Alzheimer's disease (AD) type of dementia.

progressive decline of physiological function and tissue homeostasis leading to increased vulnerability to degeneration and death. Aging is the main risk factor for the advance of neurodegenerative diseases. At the molecular level, brain aging is not only accompanied by deregulated nutrient sensing, genomic instability, defects in nuclear architecture, telomere attrition, epigenetic alterations, and chromatin remodeling, but also by the induction of mitochondrial dysfunction, stem cell exhaustion, and altered intercellular communication (Lopez-Otin et al., 2013). In addition, brain aging also reduces brain weight and significantly decreases cognitive abilities (e.g., learning, memory formation, or executive control) producing gradual constraints in daily activities (Hasher and Zacks, 1988) with slower processing speed (Healey et al., 2008). Very little is known about the mechanisms, which contribute to cognitive decline in humans. Cognitive decline in aged individuals is associated not only with Mg^{2+} dis-homeostasis, but also with oxidative stress, mitochondrial dysfunction, and consequently energetic failure. Maintenance of cognitive function is crucial for conducting daily living activities such as attention, short-term and long-term memory, reasoning, coordination of movement, and planning of tasks (de Champlain et al., 2004). Interestingly, preclinical studies have shown that MMFS-01, a derivative compound of magnesium-L-threonate is effective in alleviating cognitive decline in aging rodents (Liu et al., 2015). As a synapse density enhancer (Zhou and Liu, 2015), the elevation of brain Mg^{2+} prevents synaptic loss and reverses cognitive deficits in AD mouse model (Li et al., 2014). Indeed, Mg^{2+} restoration attenuates memory impairment by activating protein kinase C in experimental animals (Libien et al., 2005). To this end earlier studies have revealed that Mg^{2+} treatment enhances clearance of Aβ in an APH-1α/1β-dependent manner in APP/PS1 transgenic (Tg) mice (Yu et al., 2010, 2015). It should be noted that cognitive changes in the aging brain are mediated and modulated by signal transduction processes, which not only control the aging process, but also modulate longevity. These signal transduction processes include insulin/insulin-like growth factor signaling, target of rapamycin signaling, sirtuins signaling, mitochondrial function, and caloric restriction (Bishop et al., 2010) (Fig. 3.3). In addition to the above mechanisms, telomere shortening, mitochondrial oxidative damage, p53 activation, and reduced peroxisome proliferator-activated receptor gamma, coactivator 1 α and β (PGC-1α and PGC-1β) (Sahin et al., 2011; Finck and Kelly, 2006) also modulate the integrity of the genome, its stability, and cellular longevity. Furthermore, the brain function is also modulated by CBF. Older age is accompanied by lower blood flow to the brain, probably due to the onset of atherosclerosis (Dennis and Cabeza, 2008). Prolonged reduction in blood flow due to aging and atherosclerosis not only results in hypertension-mediated damage in the

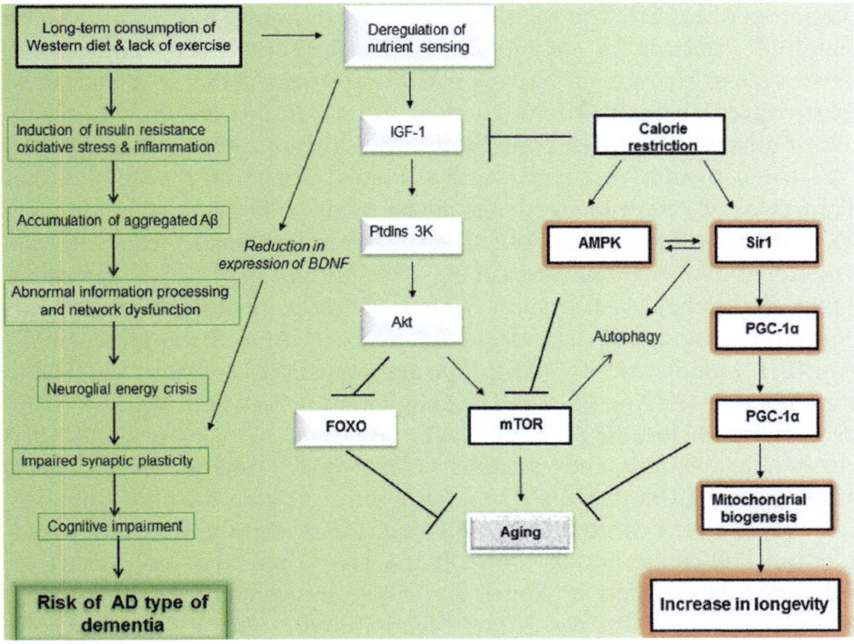

FIGURE 3.3 Signal transduction processes associated with aging process.
IGF-1, insulin/insulin-like growth factor; *PGC-1α*, peroxisome proliferator-activated receptor-γ coactivator-1α; *PtdIns 3K*, phosphatidylinositol 3-kinase; signaling, *mTOR*, target of rapamycin; *Sir*-1, sirtuins1.

occipitotemporal, prefrontal, and medial temporal lobe regions, but also impacts blood vessel function (Beason-Held et al., 2007). In fact, a mild chronic cerebrovascular hypoperfusion and hypometabolism caused by a decrease in CBF may lower metabolic rates of glucose, and oxygen consumption. It is proposed that this may be one of the very early events in the pathogenesis of AD (Iadecola, 2004). The reason for brain hypometabolism may include defects in glucose transport at the BBB, glycolysis, and/or mitochondrial dysfunction. The decrease in CBF due to atherosclerosis may negatively affect the synthesis of proteins required for learning and memory and eventually lead to neuritic injury and neuronal death. Thus, prolonged hypertension in old age contributes to cognitive dysfunction (Giordano et al., 2012) and dementia (Goldstein et al., 2013). This makes hypertension a major risk factor for vascular cognitive impairment in neurological disorders. Hypertension contributes to both the development and progression of cerebrovascular disease (MacMahon et al., 1990). There is growing evidence that hypertension is the most powerful modifiable risk factor for cerebral vessel dysfunction and it is closely associated with cognitive decline

(Nelson et al., 2011). The relationship between aging, hypertension, and cognitive function is complex and not completely understood. Nevertheless, changes in blood pressure are considered as a marker of cerebrovascular health (Barodka et al., 2011). In seniors, the consumption of an unhealthy diet containing high salt not only increases oxidative stress, insulin resistance, and elevates circulating catecholamines (Tran et al., 2009), and decreases nitric oxide (NO) bioavailability, but also enhances RAAS, and Ang II, levels (a potent vasoconstrictor) in a dose-dependent manner (Tran et al., 2009; Wright et al., 2013). As stated above, the brain RAAS system contains several functional components to produce the active ligands angiotensin II (Ang II), angiotensin III (Ang III), angiotensin IV (Ang IV), angiotensin 1–7 (Ang (1–7)), and angiotensin 3–7 (Ang (3–7)). These ligands interact with several receptor proteins including AT_1, AT_2, AT_4, and Mas, which are distributed within the central and peripheral nervous systems (Wright and Harding, 2010) and modulate blood pressure. Among the above ligands, Ang II is the best candidate for the mechanistic link between hypertension and AD (Kehoe et al., 2009). It is well known that vasoconstrictor and prooxidant effects of Ang II contribute to pathogenesis of essential hypertension (Coffman, 2011; Reudelhuber, 2013). Furthermore, increased circulating levels of Ang II accelerate development of AD pathology by promoting β-secretase activity (Cifuentes et al., 2015; Faraco et al., 2016). Existing literature supports the concept that Ang II increases superoxide production by activation of angiotensin II type 1 receptor (AT_1R) in the cerebral microvascular endothelium, thereby causing endothelial dysfunction (Xiao et al., 2015).

BIOMARKERS FOR ALZHEIMER'S TYPE OF DEMENTIA

As stated in Chapter 2, Neurochemical Aspects of Poststroke Dementia, biomarkers are metabolites or products whose level, presence, and activity are closely associated with the pathogenesis of the disease processes. Clinically, biomarkers are used not only for early detection of the disease process (preclinical stage) and monitoring the disease progression, but also for following pharmacological responses and therapeutic intervention (Biomarkers Definitions Working Group, 2001). An ideal biomarker is not only reproducible, stable over time, widely available, but also reflects directly the association with the pathogenesis of the disease (Biomarkers Definitions Working Group, 2001). For the AD type of dementia, biomarkers may be used to distinguish different aspects of the underlying pathology; detect presymptomatic pathological changes; predict decline or conversion between clinical disease states; and/or monitor disease progression and response to treatment.

Many investigators believe that AD type of dementia can be effectively monitored and treated in preclinical stages, before cognitive functions become impaired and neurons along with their synapses become damaged irreversibly (Golde et al., 2011; Kim and Hwang, 2016). Currently one major issue in identifying AD type of dementia patients is the difficulty in early and definitive diagnosis of this disease. However, a probable diagnosis of AD can be made with a confidence of >90%, based on clinical criteria, including medical history, physical examination, laboratory tests, neuroimaging, and neuropsychological evaluation. The main problem in discovering an ideal biomarker for AD type of dementia has not only been the slow understanding of pathogenesis of AD, unavailability of histopathological and biochemical diagnosis during patient lifetime, but also the occurrence of the large overlap with other types of dementia (dementia with Lewy bodies and vascular dementia) along with a lack of information on the treatment of the disease (Fjell and Walhovd, 2011).

Tentative biomarkers for AD type of dementia can be broadly classified into four types: neurochemical, neuroanatomical, genetic, and neuropsychological biomarkers (Fig. 3.4). Levels of t-tau, p-tau, or p-tau/Aβ have been determined by ELISA in in the cerebrospinal fluid but the data have led to a state of uncertainty not only because of the heterogeneity of the disease, but also because of different values in many studies. The heterogeneity observed in these studies awaits more

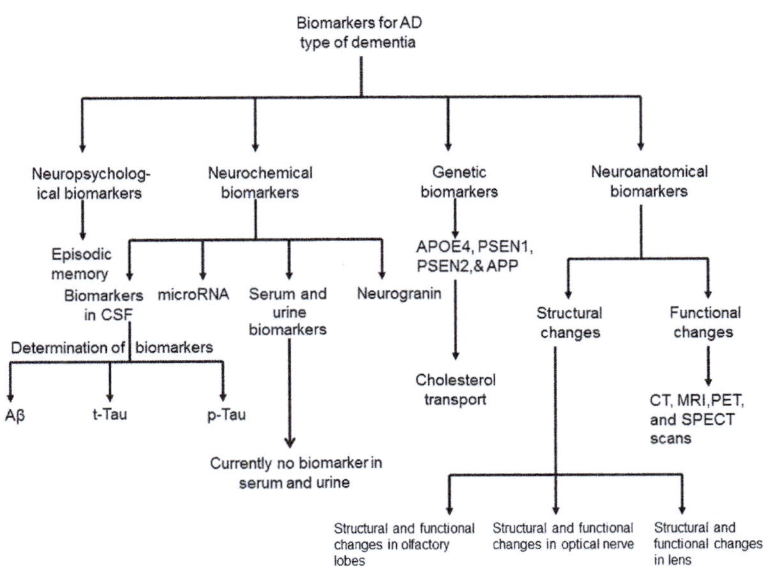

FIGURE 3.4 Biomarkers for Alzheimer's disease (AD) type of dementia.

information for clinical use. Functional imaging by Fluorine-18-Fluorodeoxyglucose Positron-Emission Tomography (FDG-PET) evaluates the CBF and glucose metabolism and is useful in the differentiation between vascular dementia and AD type of dementia. A large cohort study using PET shows that the increase in Pittsburg compound B retention and medial temporal atrophy is associated with AD type of dementia after stroke/TIA (Yang et al., 2015). Liu et al. used Pittsburgh Compound B positron-emission tomography (PiB-PET) to investigate patients with stroke or TIA who had a more progressive cognitive decline. They found that concurrent amyloid pathology was found in about one-fifth of the patients (Mok et al., 2016). It is also reported that cerebral PiB-PET examination is an important step in judging the abnormalities in amyloid metabolism in brains of AD patients (Gjedde et al., 2013). This can be used as a biomarker to detect AD type of dementia (McKhann et al., 2011; Rosenmann, 2012). Attempts have been made to detect Aβ in serum. However, these attempts have failed and data have been ambiguous. So currently, there are no blood-based or urine-based biomarkers available for routine clinical use. Magnetic resonance imaging (MRI) studies have been used to monitor structural abnormalities in early to late AD stages of AD type of dementia. Anatomical or volumetric MRI is the most widely used technique, which gives information on local and global measures of atrophy. According to the National Institute on Aging–Alzheimer's Association (NIA-AA) diagnostic guidelines (structural MRI), AD is diagnosed by structural MRI, which shows atrophy in the medial, basal, and lateral temporal lobes, as well as the medial parietal cortex. This can be used as a biomarker of neuronal degeneration or injury (Jack et al., 2011; McKhann, et al., 2011). MRI has also been used to assess vascular damage and white matter signal changes. Furthermore, one can also identify other neurodegenerative conditions such as spongiform and gliotic changes in prion disease on the basis of MRI (Ahmed et al., 2014). Furthermore, the use of advanced MRI analysis can provide information not only on the hippocampus, but also on alterations in fiber tract and neural network disintegration that may substantially contribute to early detection and the mapping of AD progression. The use of molecular in vivo amyloid imaging agents, such as the PIB and fluoro-2-deoxy-D-glucose (FDG), a marker for neurodegeneration, can provide information about the detection and neurodegeneration in earlier stages of AD. This information on AD biomarkers can be combined with other tests to increase accuracy or risk.

AD is accompanied by the chronic activation of microglia in the brain. The trigger for microglia activation is unclear, but the invasion of plaques by active microglia has been reported in AD transgenic mice models, when Aβ is injected into the brain or in in vitro experiments (Reed-Geaghan et al., 2009; Njie et al., 2012; Thanopoulou et al., 2010).

Activated microglia have the ability to sense neuronal activity. This allows them to regulate synaptic plasticity, learning and memory mechanisms, and hence determine cognitive abilities (Morris et al., 2013; Sipe et al., 2016). For instance, BDNF produced by microglia has been shown to modulate motor learning-dependent synapse formation in mice (Parkhurst et al., 2013). The activation of microglia by Aβ (Hashioka et al., 2008; Koenigsknecht-Talboo et al., 2008) results in cell transformation (Husemann et al., 2001). Activation of microglia in AD type of dementia can be determined by monitoring increased expression of a mitochondrial protein called 18-kDa translocator protein (TSPO). This protein can be imaged using (R)-[11C]PK11195 (Yokokura et al., 2011; Schuitemaker et al., 2013). More recently, several new TSPO ligands have been discovered (Venneti et al., 2013), and now TSPO has also been identified as a potential drug target (Chua et al., 2014). In particular, studies using [11C]PBR28 have shown a signal, which correlates with cognitive performance (Kreisl et al., 2013), providing a means for detecting changes early in the pathogenesis of AD type of dementia. However, a major disadvantage of the new TSPO ligands is genetic polymorphism (Owen et al., 2012), a subpopulation of subjects does not show binding. So, there is a need for novel TSPO ligands that provide a high signal, but are insensitive to this polymorphism.

NEUROCHEMICAL CHANGES IN ALZHEIMER'S TYPE OF DEMENTIA

Early events in the pathogenesis of AD include induction of oxidative stress, alterations in insulin and IGF signaling in the brain, accumulation of Aβ peptide, mitochondrial dysfunction, and tau hyperphosphorylation. Among these processes, the induction of mitochondrial dysfunction is a major feature of AD, which may be one of the early events that trigger downstream principal events including the loss of synapses in brains of AD patients. This suggestion is supported by a substantial decrease in the number of dendritic spines in brains of patients with AD (Penzes et al., 2011). The reduction in dendritic spines correlates with loss of learning and memory (Alvarez and Sabatini, 2007). Accumulation of Aβ leads to its oligomerization and formation of Aβ plaques, deposition of NFTs, and impaired glucose and insulin tolerance (Farooqui, 2010; Pedrós et al., 2014; Farooqui, 2016). These processes may also contribute to cognitive impairment and amyloid plaque load in AD patients (Nelson et al., 2012). Collectively, these studies suggest that the pathogenesis of AD is not only multifactorial and progressive, but also complex, and irreversible. It is driven by metabolic

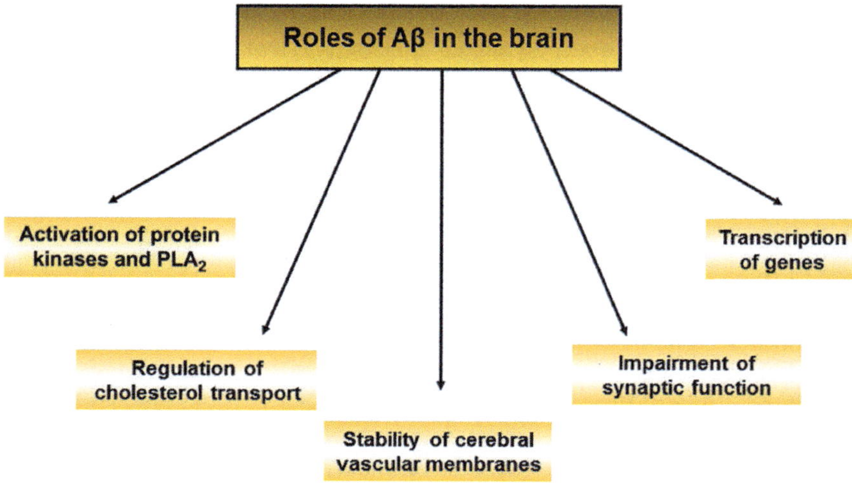

FIGURE 3.5 Roles of beta-amyloid (Aβ) in the brain.

(type 2 diabetes, hypertension, and neuroinflammation), environmental (exposure to neurotoxins), and genetic factors (ApoE, PS1, and PS2) (Shinohara et al., 2014).

As stated above, the cleavage of APP by β-secretase (BACE1) and γ-secretase results in formation of Aβ peptide (Hardy and Selkoe, 2002; De Strooper, 2010). Under physiological conditions, Aβ contributes to a variety of important functions in healthy subjects (Lahiri and Maloney, 2010). These functions include activation of kinases, regulation of cholesterol transport, modulation of synaptic plasticity, and proinflammatory activity (Fig. 3.5) (Tabaton et al., 2010; Igbavboa et al., 2009; Soscia et al., 2010). In addition, Aβ inhibits the activity of ubiquitin C-terminal hydrolase L1 (Uch-L1), a neuronal enzyme, which plays an important role in the elimination of misfolded proteins (Guglielmotto et al., 2017). The inhibition of Uch-L1 by BACE1results not only in its upregulation, but also in the induction of neuronal apoptosis in the control as well as in the transgenic AD mouse model. This process is supported by the activation of the NF-κB pathway as well as impairment of its lysosomal degradation (Guglielmotto et al., 2012).

In AD type of dementia, Aβ undergoes oligomerization and forms Aβ-derived diffusible ligands (ADDLs), which are considered to be an initiator of AD type of dementia not only due to inducing synaptic loss and progressive cognitive decline, but also by mediating the development of tau pathology and synaptic dysfunction (Fig. 3.6) (Tu et al., 2014; Viola and Klein, 2015; Selkoe, 2008; Bloom, 2014). In axons, ADDLs impair the transport of cargoes such as mitochondria and vesicles containing BDNF (Decker et al., 2010). Increasing evidence suggests that ADDLs may be the

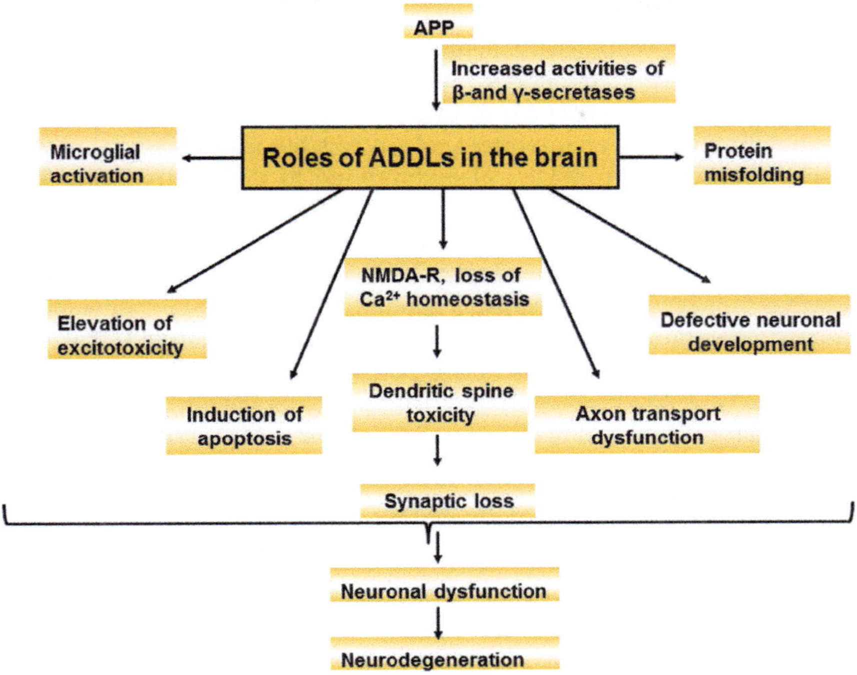

FIGURE 3.6 Role of ADDLs in the brain.
APP, amyloid precursor protein; *ADDLs*, Aβ-derived diffusible ligands; *PLA_2*, phospholipase A_2.

primary cause of AD because ADDLs have a greater correlation with dementia than insoluble Aβ (Viola and Klein, 2015). These ADDLs act by binding to a putative receptor and activating the receptor tyrosine kinase EphA4 and Fyn. ADDLs' binding triggers aberrant activation of NMDARs and abnormal increase in postsynaptic Ca^{2+} (Luine and Frankfurt, 2012). Based on these studies it is proposed that impairment in transport of mitochondria and BDNF may contribute to synaptic dysfunction in AD type of dementia (Scharfman and Chao, 2013). ADDLs may also disrupt mitochondrial membrane function by inserting into the membrane and creating calcium-permeable channels (Reddy, 2009; Kawahara, 2010; Farooqui, 2010). ADDL has the ability to incorporate into neuronal membranes, leading to the dysregulation of calcium homeostasis in the neuron (Kawahara, 2010). Other investigators have proposed that Aβ forms fibrils. The Aβ fibrils form pores in neurons producing a calcium influx and the neuronal death associated with AD type of dementia (Demuro et al., 2011). Collective evidence suggests that reactive oxygen species (ROS)-mediated increases in ADDL not only induce oxidative stress, but also accelerate the progression of AD type of dementia. The build-up of high amounts of ADDLs induces excitotoxicity, neuroinflammation, causes oxidative damage, and negatively influences multiple signal transduction pathways including the activation of GSK-3β (Muyllaert et al., 2008), a pivotal kinase, which not only plays an important role in AD, but is also associated with memory consolidation. ADDs also contribute to the disruption of BBB (Zlokovic, 2011). Aβ peptide also interacts with for advanced glycation end products receptor (RAGE) and promote Aβ transport across the BBB. This may contribute to the deposition of Aβ in the brain (Fig. 3.1). However, molecular mechanisms underlying Aβ−RAGE interaction-induced alterations in the BBB have not been fully identified. It is proposed that Aβ not only enhances permeability and disrupts zonula occludin-1 (ZO-1) expression in the plasma membrane, but also increases intracellular calcium and matrix metalloproteinase (MMP) secretion in cultured endothelial cells. Neutralizing antibodies against RAGE and inhibitors of calcineurin and MMPs prevents Aβ-mediated changes in ZO-1, suggesting that Aβ−RAGE interactions alter TJ proteins through the Ca^{2+}-calcineurin pathway. Consistent with these in vitro findings, it is suggested that disruption of microvessels near Aβ plaque-deposited areas, elevates RAGE expression, and enhances MMP secretion in microvessels of the brains of 5XFAD mice, an animal model for AD. Kook et al. (2012) supported the view that the accumulation of Aβ, disruption of BBB along with hyperphosphorylation, and accumulation of NFTs may be closely associated with the pathogenesis of AD (Fig. 3.1). Finally, ADDL can also induce neuronal apoptosis through the activation of neutral sphingomyelinase (N-SMase) (Jana and Pahan, 2004). Treatment of human primary neurons with ADDLs promote the formation of ceramide and further activation of N-SMase. Treatment of neuronal cultures with antisense of N-SMase protects neurons from ADDL-induced apoptosis and cell death. These studies support the view that ADDL may cause neuronal damage through the stimulation of N-SMase (Jana and Pahan, 2004) and the sphingomyelin cycle may play an important role in neurodegeneration in the AD brain.

OXIDATIVE STRESS IN ALZHEIMER'S TYPE OF DEMENTIA

The major sources of ROS in brain are mitochondrial respiratory chain, uncontrolled arachidonic acid (ARA) cascade, and activation of NADPH oxidase. ARA is a constituent of neural membrane glycerophospholipids. It is released by the action of cytosolic phospholipase A_2 ($cPLA_2$) and oxidized by cyclooxygenase (COX), lipoxygenase, and epoxygenase. Translocation and activation of NADPH oxidase in plasma membranes generates superoxide radical by the one-electron reduction of oxygen, using NADPH as the electron donor. The mitochondrial electron transport chain consists of several complexes containing multiple redox centers that normally facilitate transfer of electrons to their final acceptor, molecular oxygen, which is reduced by four electrons to water at complex IV. Thus, over 90% of ROS production occurs in mitochondria during metabolism of oxygen when some of the electrons passing "down" the electron transport chain leak away from the main path and go directly to reduce oxygen molecules to the superoxide anion (Pieczenik and Neustadt, 2007). In the presence of high levels of metal ions, such as Fe^{2+} and Cu^{2+}, H_2O_2 is converted into $^{\bullet}OH$ through the Fenton reaction. Hydroxyl radicals can attack polyunsaturated fatty acids in neural membrane phospholipids forming $ROO^{\bullet}$ and then can propagate the chain reaction of lipid peroxidation. Under physiological conditions, low levels of ROS are associated with growth and adaptation responses. At high levels, ROS contribute to neural membrane damage. Thus, at high levels, ROS promote the translocation of NF-κB from cytoplasm to the nucleus, where it interacts with NF-κB response element to facilitate the expression of proinflammatory enzymes ($sPLA_2$, COX-2, iNOS), cytokines (TNF-α, IL-1β, IL-6, IL-12), chemokines (MIP-1α, MCPP1), growth factors, cell cycle regulatory molecules, adhesion molecule leading to inflammation (ICAM, VCAM, and E-selectin), and antiinflammatory molecules and adhesion molecules (Farooqui, 2014). At high levels, ROS also attack cellular components (nucleic acids, lipids, and proteins) and produce changes in metabolism. This process impairs membrane integrity and produces changes in neural membrane functions causing neuronal cell death (Farooqui, 2010). In AD, the increase in ROS does not only produce impairment in mitochondrial function, but also dysregulates levels of important biological metals, such as iron and copper (Kawahara, 2010) supporting the view that the above mechanisms are major contributors of high levels of ROS in AD.

Tau is an essential protein, which is predominantly expressed in neurons (Avila et al., 2004). Physiologically, it promotes the assembly and stabilization of microtubules, and participates in neuronal development, axonal transport, and neuronal polarity. However, in AD, tau undergoes pathological modifications in which soluble tau assembles into insoluble filaments, leading not only to synaptic failure and neurodegeneration (Kolarova et al., 2012), but also inhibits apoptotic death (Li et al., 2007), particularly in axons (Rodríguez-Martín et al., 2013). Hyperphosphorylation of tau destabilizes microtubules by decreasing the binding affinity of tau for filament proteins. This process modulates axonal transport of tau and results in its aggregation in NFTs

(Rodríguez-Martín et al., 2013), which are composed of paired helical filaments (PHFs). PHFs are enriched in hyperphosphorylated tau. The role of tau in the pathogenesis of AD is unclear. However, it is proposed that hyperphosphorylation, oligomerization, fibrillization, and propagation of tau pathology are likely to be the pathological processes that induce the loss of function or gain of tau toxicity leading to neurodegeneration in AD (Yoshiyama et al., 2013). Overexpression of human wild-type full-length tau (termed hTau) produces memory deficits with decrease in synaptic plasticity. Both in vivo and in vitro data show that hTau accumulation produces remarkable dephosphorylation of cAMP response element binding protein (CREB) in the nuclear fraction (Yin et al., 2016). Simultaneously, the Ca^{2+}-dependent protein phosphatase calcineurin (CaN) is upregulated, and the calcium/calmodulin-dependent protein kinase IV (CaMKIV) is suppressed. Furthermore, the activation of CaN activation results in dephosphorylation of CREB and CaMKIV, whereas the effect of CaN on CREB dephosphorylation is independent of CaMKIV inhibition (Yin et al., 2016). Finally, inhibition of CaN attenuates the hTau-mediated CREB dephosphorylation with improvement in synapse and memory functions. Collective evidence suggests that the hTau accumulation impairs synapse and memory by CaN-mediated suppression of nuclear CaMKIV/CREB signaling (Yin et al., 2016).

Tau can be located both in pre- and postsynaptic compartments, and the number of synaptosomes containing tau did not differ between control and AD human brains; however, a particular form of phosphorylated tau (pS396/pS404) and tau oligomers are specifically found in AD synaptosomes (Tai et al., 2012). Very little is known on the link between tau and synaptic activity. Synaptic activation has been reported to enhance the secretion of tau in vitro and in vivo (Pooler et al., 2013; Yamada et al., 2014). Synaptic activity is also shown to induce tau translocation to excitatory synapses—to be precise in dendritic spines and postsynaptic compartments—in wild-type neurons (Frandemiche et al., 2014). In the same study, investigators have indicated that Aβ oligomers induce tau localization to synapses; intriguingly, such translocation requires the residue S404 of tau to be phosphorylated, the same observed specifically in AD synaptosomes (Tai et al., 2012). Synaptic activation induces tau phosphorylation on residue T205; however, this phosphorylation is not mandatory for tau translocation to synapses (Frandemiche et al., 2014).

The relationship and underlying mechanisms between oxidative stress and tau hyperphosphorylation remain elusive. Fatty acid oxidative products provide a direct link between the mechanisms of how oxidative stress induces the formation of NFTs in AD (Patil and Chan, 2005). Chronic oxidative stress is known to increase the levels of tau phosphorylation at PHF-1 epitope (serine 396/404). This phosphorylation is inhibited by glutathione synthesis with buthionine sulfoximine. Similarly, the treatment of primary rat cortical neuron cultures with cuprizone, a copper chelator, in combination with oxidative stress (Fe^{2+}/H_2O_2), significantly increases the aberrant tau phosphorylation due to elevation in GSK-3 activity (Lovell et al., 2004). In addition, treatment of rat hippocampal cells and SHSY5Y human neuroblastoma cells with H_2O_2 also results in dephosphorylation of tau at the tau1 epitope by CDK5 via PP1 activation suggesting that phosphorylation of tau protein may contribute to neurodegeneration (Zambrano et al., 2004). Furthermore, it is also reported that oxidative stress is a causal factor in tau-induced neurodegeneration in *Drosophila* (Dias-Santagata et al., 2007; Frost et al., 2014). Several studies on various cellular or animal models of tauopathies have indicated that the overexpression of mutant forms of human tau increases both the expression of oxidative stress markers and the sensitivity of neurons to H_2O_2 or paraquat (Alavi Naini and Soussi-Yanicostas, 2015), supporting the view that oxidative stress modulates the phosphorylation of tau protein.

Advanced glycation end products (AGEs) are also formed in the brain of AD patients. They may cause the accumulation of oxidized glycated proteins in the senile plaques (Durany et al., 1999). The microtubuli-associated protein tau is also subject to intracellular AGE formation. AGEs participate in neuronal death causing direct (chemical) radical production: glycated proteins produce nearly 50-fold more radicals than nonglycated proteins, and indirect (cellular) radical production: interaction of AGEs with cells increases oxidative stress (Durany et al., 1999). During aging, cellular defense mechanisms weaken and the damages to cell constituents accumulate leading to loss of function and finally cell death (Durany et al., 1999).

NEUROINFLAMMATION IN ALZHEIMER'S TYPE OF DEMENTIA

As stated in Chapter 2, Neurochemical Aspects of Poststroke Dementia, activation of microglia and astrocytes contribute to neuroinflammation, which is mediated by elevation in levels of proinflammatory lipid mediators (prostaglandins, leukotrienes, and thromboxanes) (Fig. 3.7) and increased expression of proinflammatory cytokines (TNF-α, IL-1β, and IL-6) and chemokines (MCP-1). It is modulated by interactions between neurons and microglia by several molecular and cellular pathways (Farooqui, 2014). The dysregulation of these pathways often produces neurobiological consequences, including aberrant neuronal

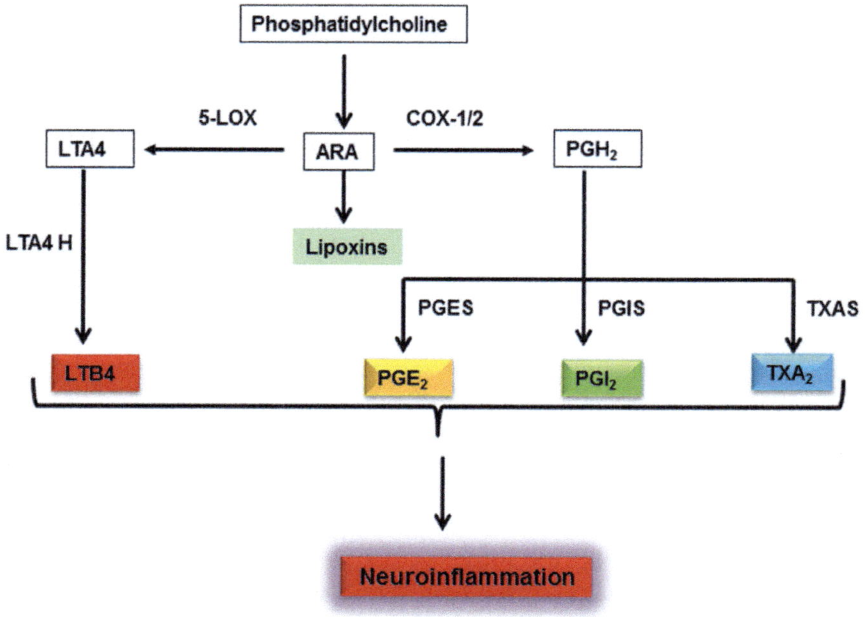

FIGURE 3.7 Generation of inflammatory mediators from arachidonic acid.
ARA, arachidonic acid; *COX-1*, cyclooxygenase-1; *COX-2*, cyclooxygenase-2; *5-LOX*, 5-lipoxygenase; *LTA_4*, leukotriene A_4; *LTA_4H*, leukotriene A_4 hydrolase; *LTB_4*, leukotriene B_4; *PGE_2*, prostaglandin E_2; *PGES*, prostaglandin E_2 synthase; *PGI_2*, prostaglandin I_2; *PGIS*, prostaglandin I_2 synthase; *PLA_2*, phospholipase A_2; *TXA_2*, thromboxane A_2; *TXAS*, thromboxane A_2 synthase.

responses and microglia activation (Farooqui, 2014). Functional changes in microglia are indicative of an immune state termed parainflammation in which tissue-resident macrophages (i.e., microglia) respond to malfunctioning cells by initiating modest inflammation in an attempt to restore homeostasis (Farooqui, 2014). Aβ oligomers have been implicated in initiating the inflammatory processes (Minter et al., 2016). As stated above, Aβ oligomers not only activate microglia and astrocytes, but also promote changes in astrocytes and microglial cell metabolism (Walker et al., 2015) by increasing the Ca^{2+} concentration in the postsynapse leading to neuroinflammation and cell death through the activation of NMDA and RAGE receptors (Dinamarca et al., 2012; Walker et al., 2015). In addition, there is mounting evidence to indicate that the disruption of BBB also potentiates the neuroinflammatory cycle. In mouse models of AD, microglia are clustered around Aβ plaques, leaving the surrounding tissue covered by fewer processes than usually occurs in age-matched wild-type mice (Baron et al., 2014). In addition, the number of fine processes on microglial cells is significantly decreased in older mice compared to adult mice, and the overexpression of the human mutated APP and the deposition of plaques in the brain significantly accelerates this reduction (Baron et al., 2014).

Furthermore, in mouse model of AD, a significant reduction has been observed in the number of microglial processes surrounding Aβ plaques (Baron et al., 2014). Microglial accumulation near Aβ plaques may thus not only shift the molecular and cellular milieu to one that can enhance neurotoxicity (Varvel et al., 2012; Heneka and Kummer et al., 2013), but it also causes a progressive decrease in microglial process complexity, which may not only impair the clearance of Aβ oligomers, but also contribute to alterations in the synaptic network or neuronal repair processes.

Cerebral endothelial cells and astrocytes are among the key players in the human brain inflammatory responses, initiated by inflammatory events in the brain's environment. It is reported that cells of the BBB are highly responsive to the neuroinflammatory processes and can be modulated by neuroinflammatory mediators (proinflammatory eicosanoids, cytokines, chemokines, and high levels of ROS) of both the systemic and central nervous systems. A major consequence of chronic neuroinflammation is the loss of barrier integrity. In AD, many proinflammatory mediators such as TNFα or IL-1β induce loss of "tightness" that increases BBB permeability (Abbott, 2000; Farooqui, 2014). This increase in BBB permeability allows immune cells to enter the brain parenchyma and worsen pathology. Collectively, these studies indicate that there is a link among Aβ, cytokine release, the BBB, and AD progression.

Astrocytes are the most abundant cells of the brain. They play critical roles in neuronal homeostasis through their physical properties and neuron–glia signaling pathways. Astrocytes become reactive in response to neuronal injury and this process is called reactive astrogliosis. Reactive astrogliosis represents a continuum of pathobiological processes and is associated with morphological, functional, and gene expression changes of varying degrees. Changes in astrocyte function have been observed in brains from individuals with AD, as well as in AD in vitro and in vivo animal models. The presence of Aβ has been shown to disrupt neurotransmission, neurotransmitter uptake, and alter calcium signaling in astrocytes. Furthermore, astrocytes express APOE and are involved in the production, degradation, and removal of Aβ. In addition to microglia, astrocytes also contribute to the pathological characteristics of AD (González-Reyes et al., 2017). Astrocytes participate in the inflammatory/immune responses of the central nervous system. The presence of Aβ activates different cell receptors and intracellular signaling pathways, mainly the advanced glycation end products receptor/nuclear factor kappa-light-chain-enhancer of activated B cells (NF-κB) pathway. As stated above this pathway is responsible for the transcription of proinflammatory cytokines and chemokines in astrocytes. The release of these proinflammatory agents may not only induce cellular damage, but also may stimulate the production of Aβ in astrocytes. Additionally, Aβ-mediated production of ROS and RNS in astrocytes is aided not only by an increase in intracellular calcium and NADPH oxidase, but NF-κB signaling, and the onset of excitotoxicity and mitochondrial function. The Aβ/NF-κB interaction in astrocytes may play a central role in these inflammatory and oxidative stress-mediated changes in AD (González-Reyes et al., 2017).

Converging evidence suggests that chronic oxidative stress and neuroinflammation are interrelated processes. In AD, neuroinflammatory changes are caused by the activation of microglia, astrocytes, and macrophages, particularly in the area where amyloid deposition occurs (Heneka and O'Banion, 2007). In AD, neurodegenerative process also results in the release of large amounts of proinflammatory mediators, including cytokines, chemokines, eicosanoids, and nitric oxide (NO), all of which increase the generation of insoluble ADDLs (Velez-Pardo et al., 2002). The Aβ-mediated respiratory burst in microglia produces ROS and tumor necrosis factor alpha (TNF-α), which aggravates Aβ deposition and further neuronal dysfunction and eventual death (Liu et al., 2002). The potentially significant contribution of inflammatory mechanisms in AD has prompted consideration of antiinflammatory treatment strategies (Kim et al., 2010). Collective evidence suggests that neurodegeneration in vulnerable regions of the brain in AD may contribute to the release of the abovementioned inflammatory mediators and activated complement components (Agostinho et al., 2010; Farooqui, 2010). Induction of neuroinflammation and oxidative stress in AD brain is also supported by excitotoxicity, a process which is closely associated with the pathogenesis of AD. Induction of excitotoxicity is not only accompanied by the production of ROS, but also by the stimulation of NF-κB and the increased expression of proinflammatory cytokines in brains of AD patients (Farooqui et al., 2008).

IMMUNE RESPONSES IN ALZHEIMER'S TYPE OF DEMENTIA

It is well known that immune responses commonly occur in the brain, despite the perception that it is an immune-privileged site. Brain mast cells are the first responders before microglia in brain injuries since mast cells can release prestored mediators (Hendriksen et al., 2017). Immune responses in the brain are mediated and modulated by interplay among resident microglia, astrocytes, and mast cells resulting in the release of mediators such as cytokines, proteases, proinflammatory eicosanoids, ROS, and nitric oxide, along with neurotrophic factors and complement factors, which may mediate neuroprotective and neurotoxic effects (Sofroniew and Vinters, 2010; Farina et al., 2007). Several molecules have been reported to contribute to immune function in the brain. They include MHC class I (Huh et al., 2000), neuronal pentraxins (Bjartmar et al., 2006), and complement (Stevens et al., 2007), and these mediate synaptic remodeling in the developing mouse brain, yet very little is known about the signals regulating the expression and function at the developing synapses. Classical complement cascade proteins are components of the innate immune system that control and modulate synaptic pruning, a process which is critical for the establishment of precise synaptic circuits. Complement proteins, C1q and C3, are expressed in the adult brain (Stevens et al., 2007). C1q is the recognition domain of the initiating protein called C1, in the classical complement cascade. It is a large secreted protein, which is composed of C1qA, C1qB, and C1qC peptide chains. In the immune system, interactions of C1q to apoptotic cell membranes or pathogens triggers a proteolytic cascade of downstream complement proteins, resulting in C3 opsonization and phagocytosis by macrophages that express complement receptors. The function of complement proteins in the brain appears analogous to their immune system function: clearance of cellular material that has been "tagged" for elimination (Blalas and Stevens, 2013). Consistent with the well-ascribed role of complement proteins as opsonins or "eat me" signals, C1q and C3 localize to retinogeniculate synapses, and presynaptic terminals of retinal ganglion cells are similarly eliminated by phagocytic microglia expressing complement receptors. Genetic deletion of C1q, C3, or the microglia-specific complement receptor, CR3 (CD11b) results in sustained defects in eye-specific segregation, suggesting that these molecules function in a common pathway to refine synaptic circuits (Stevens et al., 2007; Schafer et al., 2012). Importantly, microglial engulfment of retinogeniculate inputs occurs during a narrow window of postnatal development (P5–P8) coincident with retinal C1q expression (Stevens et al., 2007), suggesting that complement-dependent synaptic pruning is initiated by C1q. Collective evidence suggests that in the brain immune reactions often take place in virtual isolation from the innate/adaptive immune interplay that characterizes peripheral immunity.

Mast and microglial cells can detect amyloid plaque formation during pathogenesis of AD. During immune responses, Aβ acts as an upstream activator of astroglial NF-κB. This results in the release of complement protein C3, which acts on the neuronal C3a receptor (C3aR) to disrupt dendritic morphology and network function influencing cognitive function (Fig. 3.8) (Lian et al., 2016). The activation of

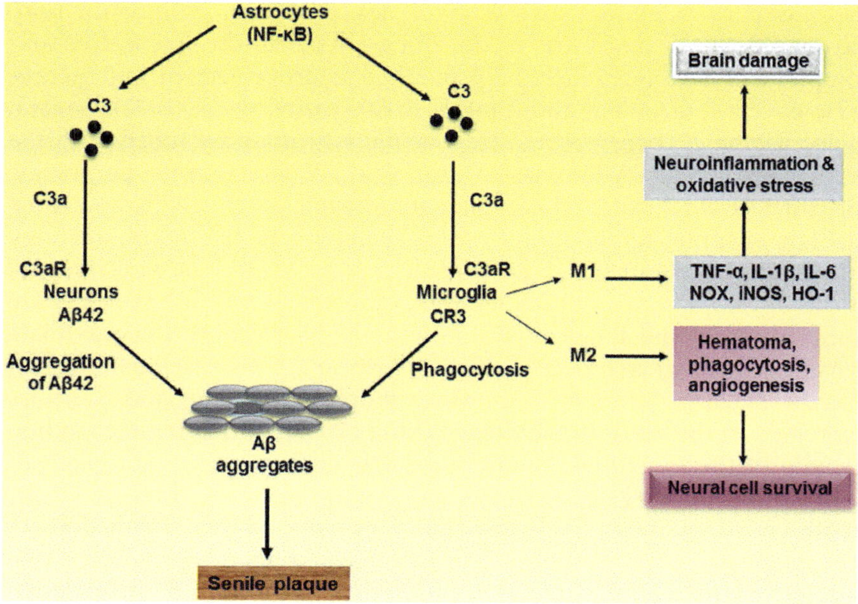

FIGURE 3.8 NF-κB- and C3-mediated astrocyte–neuron and astrocyte–microglia signaling network that induces neuronal homeostasis in Alzheimer's disease type of dementia.

astrocytic complement regulates Aβ dynamics not only in in vitro experiments, but also affects amyloid pathology in AD mouse models through microglial C3aR involvement. In primary microglial cultures an acute C3 or C3a activation promotes microglial phagocytosis, whereas chronic C3/C3a activation attenuates microglial phagocytosis. The chronic C3 activation can be blocked by cotreatment with a C3aR antagonist and by genetic deletion of C3aR (Lian et al., 2016). It is interesting to note that Aβ pathology and neuroinflammation in APP transgenic mice are worsened by astroglial NF-κB hyperactivation and resulting C3 elevation. The treatment of APP transgenic mice with the C3aR antagonist (C3aRA) ameliorates plaque load and microgliosis. Based on these studies it is proposed that there is a complement-dependent intercellular cross-talk in which neuronal overproduction of Aβ activates astroglial NF-κB to elicit extracellular release of C3. This promotes a pathogenic cycle by which C3 in turn interacts with neuronal and microglial C3aR to alter cognitive function and impair Aβ phagocytosis. This feedforward loop can be effectively blocked by C3aR inhibition, supporting the therapeutic potential of C3aR antagonists under chronic neuroinflammation conditions (Lian et al., 2016). Furthermore, TGF-β, an important transcription factor, which promotes the formation of Aβ plaques in AD. It is localized in senile plaques (Wyss-Coray et al., 1997). The blocking of TGF-β and Smad2/3 signaling mitigates plaque formation in mouse models of AD (Town et al., 2008). C1q associates with plaques in AD as well (Afagh et al., 1996), and in mouse models of Alzheimer's, C1q-deficiency has been shown to be neuroprotective (Fonseca et al., 2004). Synapse loss and/or dysfunction have emerged as early hallmarks of AD, supporting the view that aberrant complement upregulation may reactivate the developmental synapse elimination pathway in AD to promote synapse loss. It is proposed that a new link between TGF-β signaling, complement, and synapse elimination may open up new avenues of research into the role of regulatory mechanism for C1q in these disorders and in other regions of the healthy CNS. Collective evidence suggests that both C1q and C3 complement proteins perform the role of critical mediators of synaptic refinement and plasticity via C3-dependent microglial phagocytosis of synapses.

The adaptive immune system also modulates neuroinflammation. Brain immune cells (microglia, astrocytes, and mast cells) also engage in significant cross-talk with brain-infiltrating T cells and other components of the innate immune system (Ransohoff and brown, 2012). T cells also interact with dendritic cells. and initiate an immune response, as is seen in adaptive immune responses elsewhere in the body (Matyszak, 1998; Charo and Ransohoff, 2006). This fundamental difference represents the cellular basis of immune privilege of the CNS. Thus, T cells respond in the periphery and traffic to the CNS to respond to the disease process (Charo and Ransohoff, 2006). This "efferent" system by which immune cells respond to an antigen depot in the brain is efficient and implies immunosurveillance. Collectively, these studies suggest that the impact of the immune system responses on the brain is profound (Ransohoff et al., 2015) and infiltrating peripheral immune cells (macrophages) can access the brain under certain circumstances. Thus, in APP/PS1 mice, alterations in BBB are accompanied by increased infiltration of macrophages (Minogue et al., 2014), which phagocytize Aβ. It is suggested that infiltrating macrophages from aged animals and APP/PS1 mice are more responsive to inflammatory stimuli (Barrett et al., 2015a,b) than microglial cells. Therefore, when infiltrating macrophages encounter the inflammatory environment that exists in these mice, they have the potential to exacerbate the already-existing neuroinflammation. Infiltrating macrophages may also be the source of the increase in IFNγ observed with age and in APP/PS1 mice (Minogue et al., 2014). It is interesting to note that IFNγ inhibits long-term potentiation (LTP) (Kelly et al., 2013) and synergizes with Aβ to increase microglial activation (Jones et al., 2015). Infiltration of T cells also occurs not only in aged mice, but also in APP/PS1 mice, and it is proposed that Th1 and also Th17 cells activate microglia in vitro and in vivo while their presence in the brain of APP/PS1 mice negatively impacts hippocampal-dependent cognitive function (Browne et al., 2013; McManus et al., 2014, 2015) suggesting that in the brain immune responses contribute to modulation of neuroinflammation.

COGNITIVE DYSFUNCTION IN ALZHEIMER'S TYPE OF DEMENTIA

Cognitive impairment is an essential part of the diagnostic criteria for dementia, and it may indicate the initiation of AD or other types of dementia (Cui et al., 2011; Cheng et al., 2014).

Cognitive dysfunction in AD type dementia is associated with the loss of intellectual functions such as thinking, remembering, and reasoning that interfere with daily functioning. In older adults, obesity is associated with increased risk for cognitive and functional decline (Kharabian et al., 2016). Persistent cognitive dysfunction not only depends on volume and strategic location of brain infarction, site and range of cerebral white matter injuries, but also on number of stroke lesions, and other coexistent pathologies, which promote behavioral disturbances that interfere with independence and daily functioning (Grysiewicz and Gorelick, 2012). Cognitive dysfunction is regulated not only by neurochemical and intricate synaptic changes, but also by neuronal and glial interactions (Morrison and Baxter, 2012). It predisposes individuals for neurological and psychiatric changes, which compromise neuronal and glial function, with a reduction in neurotransmitter homeostasis and induction of neuroinflammation and oxidative stress. These neurochemical alterations promote the accumulation of Aβ oligomer in the form of plaques that are neurotoxic. Additionally, there is generation and accumulation of hyperphosphorylated insoluble fibrillar tau which can exacerbate cytoskeletal collapse and synaptic disconnection resulting in the onset of AD type of dementia (Schuh et al., 2011; Farooqui, 2018). Synaptic dysfunction/degeneration is considered one of the most reliable markers of cognitive impairment in AD type of dementia, which can be detected very early on in the progression of AD, as early as on the onset of mild cognitive impairment (MCI) (Arendt, 2009). It has been demonstrated that there is up to an 18% loss of synapses in the CA1 hippocampal region of MCI patients, which progressed to a 55% loss in mild AD cases. In AD type of dementia, the loss of synapses is not always confined to degenerating neurons but can also occur in surviving neurons (Coleman and Yao, 2003). Analysis of synaptic components from brains of AD type of dementia patients shows significant reduction in levels of synaptophysin as well as synaptopodin, and PSD-95 proteins indicating that presynaptic and postsynaptic proteins are critically involved in AD progression. Importantly, the loss of synapses is confined to brain regions affected by AD type of dementia and closely correlates with NFT counts indicating a link between synaptic dysfunction and tangle formation in AD patients (Reddy and Beal, 2008).

The molecular mechanism associated with loss of synapses in AD is not fully understood. However, it has been proposed that the accumulation of ADDL induces the early synaptic disruptions by stimulating $cPLA_2$ isoforms. This enzyme releases ARA and other fatty acids from sn-2 position of neural membrane phospholipids (Farooqui, 2011, 2014). ARA is as associated with the regulation of LTP (Volterra et al., 1992; Williams et al., 1989). ARA is converted into cannabinoids and eicosanoids. These metabolites are involved in synaptic signaling and neuroinflammation (Sheinin et al., 2008; Farooqui, 2011). ARA also inhibits presynaptic and postsynaptic channels as well as the formation and recycling of synaptic vesicles (Marza et al., 2008). Upregulation of ARA metabolism has been reported to occur in hAPP-J20 AD mice (Sanchez-Mejia et al., 2008) as well as in the postmortem brain from AD patients, particularly in regions reported to have high densities of senile (neuritic) plaques with activated microglia (Esposito et al., 2008). Enhancement of the ARA cascade by interleukin-1β is closely associated with the working memory impairment (Matsumoto et al., 2004). Furthermore, ARA has been shown to inhibit ligand binding to several types of G protein-coupled receptors, such as muscarinic acetylcholine receptor subtypes (Bordayo et al., 2005). It is well known that cognitive dysfunction is related to diminished cholinergic function, which can be treated with the stimulation of central cholinergic activity which results in the improvement of cognitive performances (Bhattacharya et al., 1993). The loss of neuronal cholinergic observed in the hippocampal area is responsible for the major characteristic of AD. In treating AD type senile dementia, it is suggested to improve the central cholinergic system. Administration of nootropic agent increases the level of ACh and promotes the upregulation of receptor binding for cholinergic in the frontal cortex and hippocampus (Bhattacharya et al., 2000). Downregulation of noradrenergic function has been shown to diminish the behavioral impairment due to degeneration of the cholinergic system (Sara, 1989) and subsequently the reduction in cholinergic function may lead to upregulation of ACh expression in the brain. Thus, good agents of nootropics are able to decrease norepinephrine (NE) and elevate the 5-hydroxytryptamine (5-HT) expression observed in the central cortex, hippocampus, and hypothalamus (Singh and Dhawan, 1997).

Patients with AD type of dementia also show a decrease in CBF, along with changes in white matter integrity, caused due to either local or remote ischemic injury (Cumming et al., 2012). White matter changes involve axonal loss, through Wallerian-like degeneration, cortical phosphorylated tau burden, small vessel disease, hypoperfusion, and demyelination in patients with AD, along with vasculopathy and ischemia (McAleese et al., 2017). These processes may contribute to cognitive deficits, which may be associated with a decrease in ability to learn, recall, concentrate, and problem solve. A decrease in CBF in AD type of dementia may also cause hypoperfusion resulting in impairments in working memory, attention, learning, calculation, visual perception, or executive function (i.e., decision-making, organization, and problem solving). Another underexplored brain structures in aging and dementia is the BBB, a complex cellular entity, which tightly regulates the transport of molecules into and out of the brain. Disruption of BBB is now increasingly documented not only in brain vascular diseases but also in aging and neurodegenerative disorders. It has been proposed that there is a possible causal link among the disruption of BBB, CBF, and cognitive decline. These processes increase the oxidative stress and neuroinflammation predisposing human subjects to loss of memory and cognitive dysfunction.

CONCLUSION

AD is by far the most common cause of dementia in the elderly. The characteristic clinical phenotype of AD is a gradual and progressive loss of memory and cognition. Aβ is formed in a two-step cleavage of the transmembrane protein APP by proteases called secretases. The APP is first cleaved by either secretase α or β, and then by γ. Thus, the accumulation of abnormally folded Aβ peptide in extracellular plaques and hyperphosphorylated tau proteins in intracellular tangles are two major pathological hallmarks of AD. The accumulation of ADDLs reflect the imbalance between their production and their elimination from the brain. ADDLs have been found in mitochondrial membranes, where they interact with mitochondrial proteins, induce free radical production, alter mitochondrial enzymes, disrupt the electron transport chain, inhibit adenosine triphosphate production, and damage mitochondria. A prominent criticism of the Aβ hypothesis has been the lack of association of total plaque load with cognitive status, which is in contrast to the more robust and graded correlation of tau pathology to neuronal loss and symptomatic presentation. A growing body of electrophysiological, biochemical, and behavioral evidence suggests that mitochondrial dysfunction, synaptic damage, and neuronal network disorganization underlie the progressive cognitive manifestations of the clinical AD occurring before the onset of symptoms.

References

Abbott, N.J., 2000. Inflammatory mediators and modulation of blood-brain barrier permeability. Cell Mol. Neurobiol. 20, 131–147.

Afagh, A., Cummings, B.J., Cribbs, D.H., Cotman, C.W., Tenner, A.J., 1996. Localization and cell association of C1q in Alzheimer's disease brain. Exp. Neurol. 138, 22–32.

Agostinho, P., Cunha, R.A., Oliveira, C., 2010. Neuroinflammation, oxidative stress and the pathogenesis of Alzheimer's disease. Curr. Pharm. Des. 16, 2766–2778.

Ahmed, R.M., Murphy, E., Davagnanam, I., et al., 2014. A practical approach to diagnosing adult onset leukodystrophies. J. Neurol. Neurosurg. Psychiatry 85, 770–781.

Alavi Naini, S.M., Soussi-Yanicostas, N., 2015. Tau hyperphosphorylation and oxidative stress, a critical vicious circle in neurodegenerative tauopathies? Oxid. Med. Cell Longev. 2015, 151979.

Allsop, D., Wong, C.W., Ikeda, S., Landon, M., Kidd, M., Glenner, G.G., 1988. Immunohistochemical evidence for the derivation of a peptide ligand from the amyloid beta-protein precursor of Alzheimer disease. Proc. Natl. Acad. Sci. U.S.A. 85, 2790–2794.

Alvarez, V.A., Sabatini, B.L., 2007. Anatomical and physiological plasticity of dendritic spines. Annu. Rev. Neurosci. 30, 79–97.

Alzheimer's Association, 2012. Alzheimer's disease facts and figures. Alzheimers Dement. 8, 131–168.

Amieva, H., Le Goff, M., Millet, X., Orgogozo, J.M., Peres, K., et al., 2008. Prodromal Alzheimer's disease: successive emergence of the clinical symptoms. Ann. Neurol. 2008 (64), 492–498.

Arendt, T., 2009. Synaptic degeneration in Alzheimer's disease. Acta Neuropathol. 118, 167–179.

Avila, J., Lucas, J.J., Pérez, M., Hernández, F., 2004. Role of Tau protein in both physiological and pathological conditions. Physiol. Rev. 84, 361–384.

Aziz, N.A., van der Marck, M.A., Pijl, H., Olde Rikkert, M.G.M., Bloem, B.R., Roos, R.A.C., 2008. Weight loss in neurodegenerative disorders. J. Neurol. 255, 1872–1880.

Bangen, K.J., Nation, D.A., Clark, L.R., Harmell, A.L., Wierenga, C.E., et al., 2014. Interactive effects of vascular risk burden and advanced age on cerebral blood flow. Front. Aging Neurosci. 6, 159.

Barodka, V.M., Joshi, B.L., Berkowitz, D.E., Hogue, C.W., Nyhan, D., 2011. Review article: implications of vascular aging. Anesth. Analg. 112, 1048–1060.

Baron, R., Babcock, A.A., et al., 2014. Accelerated microglial pathology is associated with Abeta plaques in mouse models of Alzheimer's disease. Aging Cell 13, 584–595.

Barrett, J.P., Costello, D.A., O'Sullivan, J., Cowley, T.R., Lynch, M.A., et al., 2015a. Bone marrow-derived macrophages from aged rats are more responsive to inflammatory stimuli. J. Neuroinflamm. 12, 67.

Barrett, J.P., Minogue, A.M., Ones, R.S., Ribeiro, C., Kelly, R.J., Lynch, M.A., et al., 2015b. Bone marrow-derived macrophages from AbetaPP/PS1 mice are sensitized to the effects of inflammatory stimuli. J. Alzheimers Dis. 44, 949–962.

Bateman, R.J., Xiong, C., Benzinger, T.L., Fagan, A.M., Goate, A., et al., 2012. Clinical and biomarker changes in dominantly inherited Alzheimer's disease. N. Engl. J. Med. 367, 795–804.

Beason-Held, L.L., Moghekar, A., Zonderman, A.B., Kraut, M.A., Resnick, S.M., 2007. Longitudinal changes in cerebral blood flow in the older hypertensive brain. Stroke 38, 1766–1773.

Bhattacharya, S.K., Upadhyay, S.N., Jaiswal, A.K., 1993. Effect of piracetam on electroshock induced amnesia and decrease in brain acetylcholine in rats. Indian J. Exp. Biol. 31, 822–824.

Bhattacharya, S.K., Bhattacharya, A., Kumar, A., Ghosal, S., 2000. Antioxidant activity of *Bacopa monniera* in rat frontal cortex, striatum and hippocampus. Phytother. Res. 14, 174–179.

Biomarkers Definitions Working Group, 2001. Biomarkers and surrogate endpoints: preferred definitions and conceptual framework. Clin. Pharmacol. Ther. 69, 89–95.

Bishop, N.A., Lu, T., Yankner, B.A., 2010. Neural mechanisms of ageing and cognitive decline. Nature 8, 529–535.

Bjartmar, L., Huberman, A.D., Ullian, E.M., Rentería, R.C., Liu, X., et al., 2006. Neuronal pentraxins mediate synaptic refinement in the developing visual system. J. Neurosci. 26, 6269–6281.

Blalas, A.R., Stevens, B., 2013. TGF-β signaling regulates neuronal C1q expression and developmental synaptic refinement. Nat. Neurosci. 16, 1773–1782.

Bloom, G.S., 2014. Amyloid-β and Tau: the trigger and Bullet in Alzheimer Disease Pathogenesis. JAMA Neurol. 71, 505–508.

Bordayo, E.Z., Fawcett, J.R., Lagalwar, S., Svitak, A.L., Frey Jr., W.H., 2005. Inhibition of ligand binding to G protein-coupled receptors by arachidonic acid. J. Mol. Neurosci. 27, 185–194.

Braak, H., Braak, E., 1991. Neuropathological stageing of Alzheimer-related changes. Acta Neuropathol. (Berl.) 82, 239–259.

Braak, H., Del Tredici, K., 2011. Alzheimer's pathogenesis: is there neuron-to-neuron propagation? Acta Neuropathol. (Berl.) 121, 589–595.

Braak, H., Thal, D.R., Ghebremedhin, E., Del Tredici, K., 2011. Stages of the pathologic process in Alzheimer disease: age categories from 1 to 100 years. J. Neuropathol. Exp. Neurol. 70, 960–969.

Browne, T.C., McQuillan, K., McManus, R.M., O'Reilly, J.A., Mills, K.H., et al., 2013. IFN-gamma Production by amyloid beta-specific Th1 cells promotes microglial activation and increases plaque burden in a mouse model of Alzheimer's disease. J. Immunol. 190, 2241–2251.

Castellani, R.J., Zhu, X., Lee, H.-G., Smith, M.A., Perry, G., 2009. Molecular pathogenesis of Alzheimer's disease: reductionist versus expansionist approaches. Int. J. Mol. Sci. 10, 1386–1406.

Castro, P., Zaman, S., Holland, A., 2017. Alzheimer's disease in people with Down's syndrome: the prospects for and the challenges of developing preventative treatments. J. Neurol. 264, 804–813.

Charo, I.F., Ransohoff, R.M., 2006. The many roles of chemokines and chemokine receptors in inflammation. N. Engl. J. Med. 354, 610–621.

Cheng, Q., Sun, H.X., Ye, F.L., Wang, G., Ling, H.W., et al., 2014. Dementia among elderly in Shanghai suburb: a rural community survey. J. Alzheimer's Dis. 39, 883–889.

Chua, S.W., Kassiou, M., Ittner, L.M., 2014. The translocator protein as a drug target in Alzheimer disease. Expert Rev. Neurother. 14, 439–448.

Cifuentes, D., Poittevin, M., Dere, E., Broquères-You, D., Bonnin, P., et al., 2015. Hypertension accelerates the progression of Alzheimer-like pathology in a mouse model of the disease. Hypertension 65, 218–224.

Cochran, J.N., Hall, A.M., Roberson, E.D., 2014. The dendritic hypothesis for Alzheimer's Disease pathophysiology. Brain Res. Bull. 103, 18–28.

Coffman, T.M., 2011. Under pressure: the search for the essential mechanisms of hypertension. Nat. Med. 17, 1402–1409.

Cohen, E., Dillin, A., 2008. The insulin paradox: aging, proteotoxicity and neurodegeneration. Nat. Rev. Neurosci. 8, 759–767.

Coleman, P.D., Yao, P.J., 2003. Synaptic slaughter in Alzheimer's disease. Neurobiol. Aging 24, 1023–1027.

Condello, C., Schain, A., Grutzendler, J., 2011. Multicolor time-stamp reveals the dynamics and toxicity of amyloid deposition. Sci. Rep. 1, 19.

Coria, F., Castaño, E.M., Frangione, B., 1987. Brain amyloid in normal aging and cerebral amyloid angiopathy is antigenically related to Alzheimer's disease beta-protein. Am. J. Pathol. 129, 422–428.

Cui, G.H., Yao, Y.H., Xu, R.F., Tang, H.D., Jiang, G.X., et al., 2011. Cognitive impairment using education-based cutoff points for CMMSE scores in elderly Chinese people of agricultural and rural Shanghai China. Acta Neurol. Scand 124, 361–367.

Cumming, T.B., Marshall, R.S., Lazar, R.M., 2012. Stroke, cognitive deficits and rehabilitation: still an incomplete picture. Int. J. Stroke 8, 38–45.

de Champlain, J., Wu, R., Girouard, H., Karas, M., Elm, A., Laplante, M.A., et al., 2004. Oxidative stress in hypertension. Clin. Exp. Hypertens. 26, 593–601.

de la Monte, S.M., Tong, M., 2014. Brain metabolic dysfunction at the core of Alzheimer's disease. Biochem Pharmacol. 88, 548–559.

De Strooper, B., 2010. Proteases and proteolysis in Alzheimer disease: a multifactorial view on the disease process. Physiol. Rev. 90, 465–494.

Decker, H., Lo, K.Y., Unger, S.M., Ferreira, S.T., Silverman, M.A., 2010. Amyloid-beta peptide oligomers disrupt axonal transport through an NMDA receptor-dependent mechanism that is mediated by glycogen synthase kinase 3beta in primary cultured hippocampal neurons. J. Neurosci. 30, 9166–9171.

Demuro, A., Smith, M., Parker, I., 2011. Single-channel Ca^{2+} imaging implicates Aβ1-42 amyloid pores in Alzheimer's disease pathology. J. Cell Biol. 195, 515–524.

Dennis, N.A., Cabeza, R., 2008. Neuroimaging of healthy cognitive aging. In: Craik, F.I.M., Salthouse, T.A. (Eds.), The Handbook of Aging and Cognition, third ed. Psychology Press, New York, pp. 1–54.

Dias-Santagata, D., Fulga, T.A., Duttaroy, A., Feany, M.B., 2007. Oxidative stress mediates tau-induced neurodegeneration in Drosophila. J. Clin. Invest. 117, 236–245.

Dickson, D.W., Crystal, H.A., Mattiace, L.A., Masur, D.M., Blau, A.D., Davies, P., et al., 1992. Identification of normal and pathological aging in prospectively studied nondemented elderly humans. Neurobiol. Aging 13, 179–189.

Dinamarca, M.C., Rios, J.A., Inestrosa, N.C., 2012. Postsynaptic receptors for amyloid-beta oligomers as mediators of neuronal damage in Alzheimer's disease. Front. Physiol. 3, 464.

Droogsma, E., van Asselt, D., De Deyn, P.P., 2015. Weight loss and undernutrition in community-dwelling patients with Alzheimer's dementia. Z. Gerontol. Geriatr. 48, 318–324.

Durany, N., Münch, G., Michel, T., Riederer, P., 1999. Investigations on oxidative stress and therapeutical implications in dementia. Eur Arch Psychiatry Clin Neurosci. 249 (Suppl 3), 68–73.

Duthey, B., 2013. Background Paper 6. 11. Alzheimer Disease and Other Dementias 5, 4–74.

Esposito, G., Giovacchini, G., Liow, J.S., Bhattacharjee, A.K., Greenstein, D., et al., 2008. Imaging neuroinflammation in Alzheimer's disease with radiolabeled arachidonic acid and PET. J. Nucl. Med. 49, 1414–1421.

Extance, A., 2010. Alzheimer's failure raises questions about disease-modifying strategies. Nat. Rev. Drug Discov. 9, 749–751.

Faraco, G., Park, L., Zhou, P., Luo, W., Paul, S.M., et al., 2016. Hypertension enhances Aβ-induced neurovascular dysfunction, promotes β-secretase activity, and leads to amyloidogenic processing of APP. J. Cereb. Blood Flow Metab. 36, 241–252.

Farina, C., Aloisi, F., Meinl, E., 2007. Astrocytes are active players in cerebral innate immunity. Trends Immunol. 28, 138–145.

Farooqui, A.A., 2010. Neurochemical Aspects of Neurotraumatic and Neurodegenerative Diseases. Springer, New York.

Farooqui, A.A., 2011. Lipid Mediators and Their Metabolism in the Brain. Springer, New York.

Farooqui, A.A., 2013. Metabolic Syndrome: An Important Risk Factor for Stroke, Alzheimer, and Depression. Spinger Science-Business, New York.

Farooqui, A.A., 2014. Inflammation and Oxidative Stress in Neurological Disorders. Springer International Publishing, Switzerland.
Farooqui, A.A., 2016. Therapeutic Potentials of Curcumin for Alzheimer Disease. Springer International Publishing, Switzerland.
Farooqui, A.A., 2017. Neurochemical Aspects of Alzheimer's Disease. Academic Press, an imprint of Elsevier, San Diego, CA.
Farooqui, A.A., 2018. Ischemic, Traumatic Brain, and Spinal Cord Injuries: Mechanisms and Potential Therapies. Academic Press an Imprint of Elsevier, San Diego, CA.
Farooqui, A.A., Ong, W.Y., Horrocks, L.A., 2008. Neurochemical Aspects of Excitotoxicity. Springer, New York.
Finck, B.N., Kelly, D.P., 2006. PGC-1 coactivators: inducible regulators of energy metabolism in health and disease. J. Clin. Invest. 116, 615–622.
Fjell, A.M., Walhovd, K.B., 2011. New tools for the study of Alzheimer's disease: what are biomarkers and morphometric markers teaching us? Neuroscientist 17, 592–605.
Fonseca, M.I., Zhou, J., Botto, M., Tenner, A.J., 2004. Absence of C1q leads to less neuropathology in transgenic mouse models of Alzheimer's disease. J. Neurosci. 24, 6457–6465.
Frandemiche, M.L., De Seranno, S., Rush, T., Borel, E., et al., 2014. Activity-dependent tau protein translocation to excitatory synapse is disrupted by exposure to amyloid-beta oligomers. J. Neurosci. 34, 6084–6097.
Frost, B., Hemberg, M., Lewis, J., Feany, M.B., 2014. Tau promotes neurodegeneration through global chromatin relaxation. Nat. Neurosci. 17, 357–366.
Giordano, N., Tikhonoff, V., Palatini, P., Bascelli, A., Boschetti, G., et al., 2012. Cognitive functions and cognitive reserve in relation to blood pressure components in a population-based cohort aged 53 to 94 years. Int. J. Hypertens. 2012, 274851.
Girouard, H., Iadecola, C., 2006. Neurovascular coupling in the normal brain and in hypertension, stroke, and Alzheimer disease. J. Appl. Physiol. 100, 328–335.
Gjedde, A., Aanerud, J., Braendgaard, H., Rodell, A.B., 2013. Blood-brain transfer of Pittsburgh compound B in humans. Front. Aging Neurosci. 5, 70.
Golde, T.E., Schneider, L.S., Koo, E.H., 2011. Anti-aβ therapeutics in Alzheimer's disease: the need for a paradigm shift. Neuron 69, 203–213.
Goldstein, F.C., Levey, A.I., Steenland, N.K., 2013. High blood pressure and cognitive decline in mild cognitive impairment. J. Am. Geriatr. Soc. 61, 67–73.
González-Reyes, R.E., Nava-Mesa, M.O., Vargas-Sánchez, K., Ariza-Salamanca, D., Mora-Muñoz, L., 2017. Involvement of Astrocytes in Alzheimer's Disease from a neuroinflammatory and oxidative stress perspective. Front. Mol. Neurosci. 10, 427.
Grysiewicz, R., Gorelick, P.B., 2012. Key neuroanatomical structures for post-stroke cognitive impairment. Curr. Neurol. Neurosci. Rep. 12, 703–708.
Guglielmotto, M., Monteleone, D., Boido, M., Piras, A., Giliberto, L., Borghi, R., et al., 2012. Aβ1-42-mediated down-regulation of Uch-L1 is dependent on NFκB activation and impaired BACE1 lysosomal degradation. Aging Cell 11, 834–844.
Guglielmotto, M., Monteleone, D., Vasciaveo, V., Repetto, I.E., Manassero, G., Tabaton, M., et al., 2017. The decrease of Uch-L1 activity is a common mechanism responsible for Aβ 42 accumulation in Alzheimer's and vascular disease. Front. Aging Neurosci. 9, 320.
Hardy, J., 2009. The amyloid hypothesis for Alzheimer's disease: a critical reappraisal. J. Neurochem. 110, 1129–1134.
Hardy, J., Selkoe, D.J., 2002. The amyloid hypothesis of Alzheimer's disease: progress and problems on the road to therapeutics. Science 297, 353–356.
Hardy, J.A., Higgins, G.A., 1992. Alzheimer's disease: the amyloid cascade hypothesis. Science 256, 184–185.
Hasher, L., Zacks, R.T., 1988. In: Bower, G.H. (Ed.), The Psychology of Learning and Motivation, vol. 22. Academic Press, pp. 193–225.

Hashioka, S., Miklossy, J., Schwab, C., Klegeris, A., Mcgeer, P.L., 2008. Adhesion of exogenous human microglia and THP-1 cells to amyloid plaques of postmortem Alzheimer's disease brain. J. Alzheimers Dis. 14, 345–352.

Healey, M.K., Campbell, K.L., Hasher, L., 2008. Progress in brain research. In: Sossin, W., Lacaille, J.C., Castellucci, V.F., Belleville, S. (Eds.), The Essence of Memory, vol. 169. Elsevier.

Hendriksen, E., van Bergeijk, D., Oosting, R.S., Redegeld, F.A., 2017. Mast cells in neuroinflammation and brain disorders. Neurosci. Biobehav. Rev. 79, 119–133.

Heneka, M.T., O'Banion, M.K., 2007. Inflammatory processes in Alzheimer's disease. J. Neuroimmunol. 184, 69–91.

Heneka, M.T., Kummer, M.P., et al., 2013. NLRP3 is activated in Alzheimer's disease and contributes to pathology in APP/PS1 mice. Nature 493, 674–678.

Heneka, M.T., Carson, M.J., El Khoury, J., Landreth, G.E., Brosseron, F., Feinstein, D.L., et al., 2015. Neuroinflammation in Alzheimer's disease. Lancet Neurol. 14, 388–405.

Hoffman, J.D., Parikh, I., Green, S.J., Chlipala, G., Mohney, R.P., et al., 2017. Age drives distortion of brain metabolic, vascular and cognitive functions, and the gut microbiome. Front. Aging Neurosci. 9, 298.

Huh, G.S., Boulanger, L.M., Du, H., Riquelme, P.A., Brotz, T.M., et al., 2000. Functional requirement for class I MHC in CNS development and plasticity. Science 290, 2155–2159.

Husemann, J., Loike, J.D., Kodama, T., Silverstein, S.C., 2001. Scavenger receptor class B type I (SR-BI) mediates adhesion of neonatal murine microglia to fibrillar beta-amyloid. J. Neuroimmunol. 114, 142–150.

Iadecola, C., 2004. Neurovascular regulation in the normal brain and in Alzheimer's disease. Nat. Rev. Neurosci. 5, 347–360.

Iadecola, C., 2013. The pathobiology of vascular dementia. Neuron 80, 844–866.

Igbavboa, U., Sun, G.Y., Weisman, G.A., He, Y., Wood, W.G., 2009. Amyloid beta-protein stimulates trafficking of cholesterol and caveolin-1 from the plasma membrane to the Golgi complex in mouse primary astrocytes. Neuroscience 162, 328–338.

Ikeda, M., Brown, J., Holland, A.J., Fukuhara, R., Hodges, J.R., 2002. Changes in appetite, food preference and eating habits in frontotemporal dementia and Alzheimer's disease. J. Neurol. Neurosurg. Psychiatry 73, 371–376.

Ishii, M., Iadecola, C., 2015. Metabolic and non-cognitive manifestations of Alzheimers disease: the hypothalamus as both culprit and target of pathology. Cell Metab. 22, 761–776.

Jack, C.R., Albert, M.S., Knopman, D.S., McKhann, G.M., Sperling, R.A., et al., 2011. Introduction to the recommendations from the National Institute on Aging-Alzheimer's Association workgroups on diagnostic guidelines for Alzheimer's disease. Alzheimers Dement. 7, 257–262.

Jahn, H., 2013. Memory loss in Alzheimer's disease. Dialogues Clin. Neurosci. 15, 445–454.

Jana, A., Pahan, K., 2004. Fibrillar amyloid-beta peptides kill human primary neurons via NADPH oxidase-mediated activation of neutral sphingomyelinase. Implications for Alzheimer's disease. J. Biol. Chem. 279, 51451–51459.

Jiang, C., Li, G., Huang, P., Liu, Z., Zhao, B., 2017. The gut microbiota and Alzheimer's Disease. J. Alzheimers Dis. 58, 1–15.

Jones, R.S., Minogue, A.M., Fitzpatrick, O., Lynch, M.A., et al., 2015. Inhibition of JAK2 attenuates the increase in inflammatory markers in microglia from APP/PS1 mice. Neurobiol. Aging 36, 2716–2724.

Karran, E., Mercken, M., Strooper, B.D., 2011. The amyloid cascade hypothesis for Alzheimer's disease: an appraisal for the development of therapeutics. Nat. Rev. Drug Discov. 10, 698–712.

Katzman, R., Saitoh, T., 1991. Advances in Alzheimer's disease. FASEB J. 5, 278–286.

Kawahara, M., 2010. Neurotoxicity of beta-amyloid protein: oligomerization, channel formation and calcium dyshomeostasis. Curr. Pharm. Des. 16, 2779–2789.

Kehoe, P.G., Miners, S., Love, S., 2009. Angiotensins in Alzheimer's disease – friend or foe? Trends Neurosci. 32, 619–628.

Kelly, R.J., Minogue, A.M., Lyons, A., Jones, R.S., Browne, T.C., et al., 2013. Glial activation in a beta PP/PS1 mice is associated with infiltration of IFN gamma-producing cells. J. Alzheimers Dis. 37, 63–75.

Kharabian, M.S., Arelin, K., Horstmann, A., Lampe, L., Kipping, J.A., et al., 2016. Higher body mass index in older adults is associated with lower gray matter volume: implications for memory performance. Neurobiol. Aging 40, 1–10.

Kim, H.J., Leeds, P., Chuang, D.M., 2009. The HDAC inhibitor, sodium butyrate, stimulates neurogenesis in the ischemic brain. J. Neurochem. 110, 1226–1240.

Kim, J., Lee, H.J., Lee, K.W., 2010. Naturally occurring phytochemicals for the prevention of Alzheimer's disease. J. Neurochem. 112, 1415–1430.

Kim, M., Hwang, D., 2016. Network-based protein biomarker discovery platforms. Genomics Inform. 14, 2–11.

Koenigsknecht-Talboo, J., Meyer-Luehmann, M., Parsadanian, M., Garcia-Alloza, M., Finn, M.B., et al., 2008. Rapid microglial response around amyloid pathology after systemic anti-Abeta antibody administration in PDAPP mice. J. Neurosci. 28, 14156–14164.

Kolarova, M., García-Sierra, F., Bartos, A., Ricny, J., Ripova, D., 2012. Structure and pathology of tau protein in Alzheimer disease. Int. J. Alzheimer's Dis 2012, 13.

Kook, S.Y., Hong, H.S., Moon, M., Ha, C.M., Chang, S., et al., 2012. $A\beta_{1-42}$-RAGE interaction disrupts tight junctions of the blood-brain barrier via Ca^{2+}-calcineurin signaling. J. Neurosci. 32, 8845–8854.

Koss, D.J., Jones, G., Cranston, A., Gardner, H., Kanaan, N.M., et al., 2016. Soluble prefibrillar tau and β-amyloid species emerge in early human Alzheimer's disease and track disease progression and cognitive decline. Acta Neuropathol. 132, 875–895.

Kreisl, W.C., Lyoo, C.H., McGwier, M., Snow, J., Jenko, K.J., Kimura, N., et al., 2013. In vivo radioligand binding to translocator protein correlates with severity of Alzheimer's disease. Brain 136, 2228–2238.

Lahiri, D.K., Maloney, B., 2010. Beyond the signaling effect role of amyloid-ß42 on the processing of APP, and its clinical implications. Exp. Neurol. 225, 51–54.

Lanctôt, K.L., Amatniek, J., Ancoli-Israel, S., Arnold, S.E., Ballard, C., Cohen-Mansfield, J., et al., 2017. Neuropsychiatric signs and symptoms of Alzheimer's disease: new treatment paradigms. Alzheimer's Dement. Transl. Res. Clin. Interv 3, 440–449.

Li, H.-L., Wang, H.-H., Liu, S.-J., et al., 2007. Phosphorylation of tau antagonizes apoptosis by stabilizing beta-catenin, a mechanism involved in Alzheimer's neurodegeneration. Proc. Natl. Acad. Sci. U.S.A. 104, 3591–3596.

Li, W., Yu, J., Liu, Y., Huang, X., Abumaria, N., Zhu, Y., et al., 2014. Elevation of brain magnesium prevents synaptic loss and reverses cognitive deficits in Alzheimer's disease mouse model. Mol. Brain 7, 65.

Lian, H., Litvinchuk, A., Chiang, A.C., Aithmitti, N., Jankowsky, J.L., et al., 2016. Astrocyte-microglia cross talk through complement activation modulates amyloid pathology in mouse models of Alzheimer's Disease. J. Neurosci. 36, 577–589.

Libien, J., Sacktor, T.C., Kass, I.S., 2005. Magnesium blocks the loss of protein kinase C, leads to a transient translocation of PKCα and PKCα and improves recovery after anoxia in rat hippocampal slices. Mol. Brain Res. 136, 104–111.

Liu, H., Deng, Y., Gao, J., Liu, Y., Li, W., Shi, J., et al., 2015. Sodium hydrosulfide attenuates β-amyloid-induced cognitive deficits and neuroinflammation via modulation of MAPK/NF-κB pathway in rats. Curr. Alzheimer Res. 12, 673–683.

Liu, Y., Qin, L., Wilson, B.C., An, L., Hong, J.-S., Liu, B., 2002. Inhibition by naloxone stereoisomers of β-amyloid peptide (1-42)-induced superoxide production in microglia

and degeneration of cortical and mesencephalic neurons. J. Pharmacol. Exp. Ther. 302, 1212–1219.
Lopez-Otin, C., Blasco, M.A., Partridge, L., Serrano, M., Kroemer, G., 2013. The hallmarks of aging. Cell 153, 1194–1217.
Lovell, M.A., Xiong, S., Xie, C., Davies, P., Markesbery, W.R., 2004. Induction of hyperphosphorylated tau in primary rat cortical neuron cultures mediated by oxidative stress and glycogen synthase kinase-3. J. Alzheimer's Dis. 6, 659–671.
Luine, V., Frankfurt, M., 2012. Interactions between estradiol, BDNF and dendritic spines in promoting memory. Neuroscience 239, 34–45.
MacMahon, S., Peto, R., Cutler, J., Collins, R., Sorlie, P., Neaton, J., et al., 1990. Blood pressure, stroke, and coronary heart disease. Part 1, prolonged differences in blood pressure: prospective observational studies corrected for the regression dilution bias. Lancet 335, 765–774.
Marza, E., Long, T., Saiardi, A., Sumakovic, M., Eimer, S., et al., 2008. Polyunsaturated fatty acids influence synaptojanin localization to regulate synaptic vesicle recycling. Mol. Biol. Cell 19, 833–842.
Matsumoto, D., LeRoux, J.A., Bernhard, R., Gray, H., 2004. Personality and behavioral correlates of intercultural adjustment potential. Int. J. Intercultural Relations 28, 281–309.
Matyszak, M.K., 1998. Inflammation in the CNS: balance between immunological privilege and immune responses. Prog. Neurobiol. 56, 19–35.
McAleese, K.E., Walker, L., Graham, S., Moya, E.L.J., Johnson, M., Erskine, D., et al., 2017. Parietal white matter lesions in Alzheimer's disease are associated with cortical neurodegenerative pathology, but not with small vessel disease. Acta Neuropathol. 134, 459–473.
McGeer, P.L., McGeer, E.G., 2013. The amyloid cascade-inflammatory hypothesis of Alzheimer disease: implications for therapy. Acta Neuropathol. 126, 479–497.
McKhann, G., Knopman, D., Chertkow, H., Hyman, B., Jack, C.J., et al., 2011. The diagnosis of dementia due to Alzheimer's disease: recommendations from the National Institute on Aging-Alzheimer's Association workgroups on diagnostic guidelines for Alzheimer's disease. Alzheimers Dement. 7, 263–269.
McManus, R.M., Higgins, S.C., Mills, K.H., Lynch, M.A., 2014. Respiratory infection promotes T cell infiltration and amyloid-beta deposition in APP/PS1 mice. Neurobiol. Aging 35, 109–121.
McManus, R.M., Mills, K.H., Lynch, M.A. T., 2015. Cells-protective or pathogenic in Alzheimer's Disease? J. Neuroimmune Pharmacol. 10, 547–560.
Meyer-Luehmann, M., Spires-Jones, T.L., Prada, C., Garcia-Alloza, M., de Calignon, A., et al., 2008. Rapid appearance and local toxicity of amyloid-beta plaques in a mouse model of Alzheimer's disease. Nature 451, 720–724.
Minogue, A.M., Jones, R.S., Kelly, R.J., McDonald, C.L., Connor, T.J., et al., 2014. Age-associated dysregulation of microglial activation is coupled with enhanced blood-brain barrier permeability and pathology in APP/PS1 mice. Neurobiol. Aging 35, 1442–1452.
Minter, M.R., Taylor, J.M., Crack, P.J., 2016. The contribution of neuro-inflammation to amyloid toxicity in Alzheimer's disease. J. Neurochem. 136, 457–474.
Mok, V.C., Liu, W.Y., Wong, A., 2016. Detection of amyloid plaques in patients with post-stroke dementia. Hong Kong Med. J. 22 (Suppl. 2), S40–S42.
Morris, G.P., Clark, I.A., Zinn, R., Vissel, B., 2013. Microglia: a new frontier for synaptic plasticity, learning and memory, and neurodegenerative disease research. Neurobiol. Learn. Mem. 105, 40–53.
Morrison, J.H., Baxter, M.G., 2012. The aging cortical synapse: hallmarks and implications for cognitive decline. Nat. Rev. Neurosci. 13, 240–250.
Muyllaert, D., Kremer, A., Jaworski, T., Borghgraef, P., Devijver, H., et al., 2008. Glycogen synthase kinase-3beta, or a link between amyloid and tau pathology? Genes Brain Behav. 7 (Suppl. 1), 57–66.

Nelson, P.T., Head, E., Schmitt, F.A., Davis, P.R., Neltner, J.H., Jicha, G.A., et al., 2011. Alzheimer's disease is not "brain aging": neuropathological, genetic, and epidemiological human studies. Acta Neuropathol. 121, 571–587.

Nelson, P.T., Alafuzoff, I., Bigio, E.H., Bouras, C., Braak, H., et al., 2012. Correlation of Alzheimer disease neuropathologic changes with cognitive status: a review of the literature. J. Neuropathol. Exp. Neurol. 71, 362–381.

Nieuwenhuis-Mark, R.E., 2009. Diagnosing Alzheimer's dementia in Down syndrome: problems and possible solutions. Res. Dev. Disabil. 30, 827–838.

Njie, E.G., Boelen, E., Stassen, F.R., Steinbusch, H.W., Borchelt, D.R., et al., 2012. Ex vivo cultures of microglia from young and aged rodent brain reveal age-related changes in microglial function. Neurobiol. Aging 195, e1–e12.

Owen, D.R., Yeo, A.J., Gunn, R.N., Song, K., Wadsworth, G., et al., 2012. An 18-kDa Translocator Protein (TSPO) polymorphism explains differences in binding affinity of the PET radioligand PBR28. J. Cereb. Blood Flow Metab. 32, 1–5.

Parkhurst, C.N., Yang, G., Ninan, I., Savas, J.N., Yates, J.R., Lafaille, J.J., et al., 2013. Microglia promote learning-dependent synapse formation through brain-derived neurotrophic factor. Cell 155, 1596–1609.

Patil, S., Chan, C., 2005. Palmitic and stearic fatty acids induce Alzheimer-like hyperphosphorylation of tau in primary rat cortical neurons. Neurosci. Lett. 384, 288–293.

Pedrós, I., Petrov, D., Allgaier, M., Sureda, F., Barroso, E., et al., 2014. Early alterations in energy metabolism in the hippocampus of APPswe/PS1dE9 mouse model of Alzheimer's disease. Biochim. Biophys. Acta 1842, 1556–1566.

Penzes, P., Cahill, M.E., Jones, K.A., VanLeeuwen, J.-E., Woolfrey, K.M., 2011. Dendritic spine pathology in neuropsychiatric disorders. Nat. Neurosci. 14, 285–293.

Petrou, M., Dwamena, B.A., Foerster, B.R., MacEachern, M.P., Bohnen, N.I., Müller, M.L., et al., 2015. Amyloid deposition in Parkinson's disease and cognitive impairment: a systematic review. Mov. Disord. 30, 928–935.

Pieczenik, S.R., Neustadt, J., 2007. Mitochondrial dysfunction and molecular pathways of disease. Exp. Mol. Pathol. 83, 84–92.

Pooler, A.M., Phillips, E.C., Lau, D.H., Noble, W., Hanger, D.P., 2013. Physiological release of endogenous tau is stimulated by neuronal activity. EMBO Rep. 14, 389–394.

Querfurth, H.W., Laferla, F.M., 2010. Alzheimer's disease. N. Engl. J. Med. 362, 329–344.

Ransohoff, R.M., Brown, M.A., 2012. Innate immunity in the central nervous system. J. Clin. Invest. 122, 1164–1171.

Ransohoff, R.M., Schafer, D., Vincent, A., Blachère, N.E., Bar-Or, A., 2015. Neuroinflammation: ways in which the immune system affects the brain. Neurotherapeutics 12, 896–909.

Reddy, P.H., 2009. Amyloid beta, mitochondrial structural and functional dynamics in Alzheimer's disease. Exp. Neurol. 218, 286–292.

Reddy, P.H., Beal, M.F., 2008. Amyloid beta, mitochondrial dysfunction and synaptic damage: implications for cognitive decline in aging and Alzheimer's disease. Trends Mol. Med. 14, 45–53.

Reed-Geaghan, E.G., Savage, J.C., Hise, A.G., Landreth, G.E., 2009. CD14 and toll-like receptors 2 and 4 are required for fibrillar Abeta-stimulated microglial activation. J. Neurosci. 29, 11982–11992.

Reudelhuber, T.L., 2013. Where hypertension happens. J. Clin. Invest. 123, 1934–1936.

Robakis, N.K., 2011. Mechanisms of AD neurodegeneration may be independent of Abeta and its derivatives. Neurobiol. Aging 32, 372–379.

Rodrigue, K.M., Kennedy, K.M., Devous Sr, M.D., Rieck, J.R., Hebrank, A.C., et al., 2012. beta-Amyloid burden in healthy aging: regional distribution and cognitive consequences. Neurology 78, 387–395.

Rodríguez-Martín, T., Cuchillo-Ibáñez, I., Noble, W., Nyenya, F., Anderton, B.H., Hanger, D.P., 2013. Tau phosphorylation affects its axonal transport and degradation. Neurobiol. Aging 34, 2146–2157.

Rosenmann, H., 2012. CSF biomarkers for amyloid and tau pathology in Alzheimer's disease. J. Mol. Neurosci. 47, 1–14.

Sahin, E., Colla, S., Liesa, M., Moslehi, J., Muller, F.L., Guo, M., 2011. Telomere dysfunction induces metabolic and mitochondrial compromise. Nature 470, 359–365.

Sanchez-Mejia, R.O., Newman, J.W., Toh, S., Yu, G.Q., Zhou, Y., et al., 2008. Phospholipase A_2 reduction ameliorates cognitive deficits in a mouse model of Alzheimer's disease. Nat. Neurosci. 11, 1311–1318.

Sara, S.J., 1989. Noradrenergic-cholinergic interaction: its possible role in memory dysfunction associated with senile dementia. Arch. Gerontol. Geriatr. 8 (1), 99–108.

Schafer, D.P., Lehrman, E.K., Kautzman, A.G., Koyama, R., Mardinly, A.R., et al., 2012. Microglia sculpt postnatal neural circuits in an activity and complement-dependent manner. Neuron 74, 691–705.

Scharfman, H.E., Chao, M.V., 2013. The entorhinal cortex and neurotrophin signaling in Alzheimer's disease and other disorders. Cogn. Neurosci 4, 123–135.

Schuh, A.F., Rieder, C.M., Rizzi, L., Chaves, M., Roriz-Cruz, M., 2011. Mechanisms of brain aging regulation by insulin: implications for neurodegeneration in late-onset Alzheimer's Disease. ISRN Neurol. 2011, 306905.

Schuitemaker, A., Kropholler, M.A., Boellaard, R., van der Flier, W.M., Kloet, R.W., et al., 2013. Microglial activation in Alzheimer's disease: an (R)-[(1)(1)C]PK11195 positron emission tomography study. Neurobiol. Aging 34, 128–136.

Segain, J., De La Blétiere, D.R., Bourreille, A., Leray, V., Gervois, N., Rosales, C., et al., 2000. Butyrate inhibits inflammatory responses through NFkappa B inhibition: implications for Crohn's disease. Gut 47, 397–403.

Selkoe, D.J., 1989. Amyloid beta protein precursor and the pathogenesis of Alzheimer's disease. Cell 58, 611–612.

Selkoe, D.J., 2008. Soluble oligomers of the amyloid β-protein impair synaptic plasticity and behavior. Behav. Brain Res. 192, 106–113.

Selkoe, D.J., Hardy, J., 2016. The amyloid hypothesis of Alzheimer's disease at 25 years. EMBO Mol. Med. 8, 595–608.

Sergeant, N., Bretteville, A., Hamdane, M., Caillet-Boudin, M.L., Grognet, P., Bombois, S., et al., 2008. Biochemistry of Tau in Alzheimer's disease and related neurological disorders. Expert Rev. Proteomics 5, 207–224.

Sheinin, A., Talani, G., Davis, M.I., Lovinger, D.M., 2008. Endocannabinoid- and mGluR5-dependent short-term synaptic depression in an isolated neuron/bouton preparation from the hippocampal CA1 region. J. Neurophysiol. 100, 1041–1052.

Shinohara, M., Sato, N., Shimamura, M., Kurinami, H., Hamasaki, T., et al., 2014. Possible modification of Alzheimer's disease by statins in midlife: interactions with genetic and non-genetic risk factors. Front. Aging Neurosci. 6, 71.

Singh, H.K., Dhawan, B.N., 1997. Neuropsychopharmacological effects of the Ayurvedic nootropic *Bacopa monniera* Linn. (Brahmi). Indian J. Pharmacol. 29, 359–365.

Sipe, G.O., Lowery, R.L., Tremblay, M.-È., Kelly, E.A., Lamantia, C.E., et al., 2016. Microglial P2Y12 is necessary for synaptic plasticity in mouse visual cortex. Nat. Commun. 7.

Sofroniew, M.V., Vinters, H.V., 2010. Astrocytes: biology and pathology. Acta Neuropathol. 119, 7–35.

Soscia, S.J., Kirby, J.E., Washicosky, K.J., Tucker, S.M., Ingelsson, M., et al., 2010. The Alzheimer's disease-associated amyloid beta-protein is an antimicrobial peptide. PLoS One 5, e9505.

Sperling, R.A., Aisen, P.S., Beckett, L.A., Bennett, D.A., Craft, S., Fagan, A.M., et al., 2011. Toward defining the preclinical stages of Alzheimer's disease: recommendations from the National Institute on Aging-Alzheimer's Association workgroups on diagnostic guidelines for Alzheimer's disease. Alzheimers Dement. 7, 280–292.

Spires, T.L., Meyer-Luehmann, M., Stern, E.A., McLean, P.J., Skoch, J., Nguyen, P.T., et al., 2005. Dendritic spine abnormalities in amyloid precursor protein transgenic mice demonstrated by gene transfer and intravital multiphoton microscopy. J. Neurosci. 25, 7278–7287.

Stevens, B., Allen, N.J., Vazquez, L.E., Howell, G.R., Christopherson, K.S., et al., 2007. The classical complement cascade mediates CNS synapse elimination. Cell 131, 1164–1178.

Tabaton, M., Zhu, X., Perry, G., Smith, M.A., Giliberto, L., 2010. Signaling effect of amyloid-beta(42) on the processing of AbetaPP. Exp. Neurol. 221, 18–25.

Tai, H.C., Serrano-Pozo, A., Hashimoto, T., Frosch, M.P., Spires-Jones, T.L., et al., 2012. The synaptic accumulation of hyperphosphorylated tau oligomers in Alzheimer disease is associated with dysfunction of the ubiquitin-proteasome system. Am. J. Pathol. 181, 1426–1435.

Tayeb, H.O., Murray, E.D., Price, B.H., Tarazi, F.I., 2013. Bapineuzumab and solanezumab for Alzheimer's disease: is the 'amyloid cascade hypothesis' still alive? Expert Opin. Biol. Ther. 13, 1075–1084.

Thal, D.R., Rüb, U., Orantes, M., Braak, H., 2002. Phases of Aβ-deposition in the human brain and its relevance for the development of AD. Neurology 58, 1791–1800.

Thanopoulou, K., Fragkouli, A., Stylianopoulou, F., Georgopoulos, S., 2010. Scavenger receptor class B type I (SR-BI) regulates perivascular macrophages and modifies amyloid pathology in an Alzheimer mouse model. Proc. Natl. Acad. Sci. U.S.A. 107, 20816–20821.

Town, T., Laouar, Y., Pittenger, C., Mori, T., Szekely, C.A., et al., 2008. Blocking TGF-β–Smad2/3 innate immune signaling mitigates Alzheimer-like pathology. Nat. Med. 14, 681–687.

Tran, L.T., MacLeod, K.M., McNeill, J.H., 2009. Chronic etanercept treatment prevents the development of hypertension in fructose-fed rats. Mol. Cell Biochem. 330, 219–228.

Tu, S., Okamoto, S., Lipton, S.A., Xu, H., 2014. Oligomeric Aβ-induced synaptic dysfunction in Alzheimer's disease. Mol. Neurodegener. 9, 48.

van der Flier, W.M., Scheltens, P., 2005. Epidemiology and risk factors of dementia. J. Neurol. Neurosurg. Psychiatry 76, 2–7.

van Duinen, S.G., Castano, E.M., Prelli, F., Bots, G.T., Luyendijk, W., Frangione, B., 1987. Hereditary cerebral hemorrhage with amyloidosis in patients of Dutch origin is related to Alzheimer disease. Proc. Natl. Acad. Sci. U.S.A. 84, 5991–5994.

Varvel, N.H., Grathwohl, S.A., Baumann, F., Liebig, C., Bosch, A., et al., 2012. Microglial repopulation model reveals a robust homeostatic process for replacing CNS myeloid cells. Proc. Natl. Acad. Sci. U.S.A. 109, 18150–18155.

Velasquez-Manoff, M., 2015. Gut microbiome: the peacekeepers. Nature 518, S3–S11.

Velez-Pardo, C., Ospina, G.G., Jimenez del Rio, M., 2002. Abeta[25-35] peptide and iron promote apoptosis in lymphocytes by an oxidative stress mechanism: involvement of H2O2, caspase-3, NF-kappaB, p53 and c-Jun. Neurotoxicology 23, 351–365.

Venneti, S., Lopresti, B.J., Wiley, C.A., 2013. Molecular imaging of microglia/macrophages in the brain. Glia. 61, 10–23.

Viola, K.L., Klein, W.L., 2015. Amyloid β oligomers in Alzheimer's disease pathogenesis, treatment, and diagnosis. Acta Neuropathol. 129, 183–206.

Volterra, A., Trotti, D., Cassutti, P., Tromba, C., Galimberti, R., et al., 1992. A role for the arachidonic acid cascade in fast synaptic modulation: ion channels and transmitter uptake systems as target proteins. Adv. Exp. Med. Biol. 318, 147–158.

Walker, D., Lue, L.F., Paul, G., Patel, A., Sabbagh, M.N., 2015. Receptor for advanced glycation endproduct modulators: a new therapeutic target in Alzheimer's disease. Expert Opin. Invest. Drugs 24, 393–399.

Wang, J., Gu, B.J., Masters, C.L., Wang, Y.J., 2017. A systemic view of Alzheimer disease—insights from amyloid-β metabolism beyond the brain. Nat. Rev. Neurol. 13, 612–623.

Williams, J.H., Errington, M.L., Lynch, M.A., Bliss, T.V., 1989. Arachidonic acid induces a long-term activity-dependent enhancement of synaptic transmission in the hippocampus. Nature 341, 739–742.

Wright, J.W., Harding, J.W., 2010. The brain RAS and Alzheimer's disease. Exp. Neurol. 223, 326–333.

Wright, J.W., Kawas, L.H., Harding, J.W., 2013. A role for the brain RAS in Alzheimer's and Parkinson's Diseases. Front. Endocrinol. (Lausanne). 4, 158.

Wyss-Coray, T., Masliah, E., Mallory, M., McConlogue, L., Johnson-Wood, K., et al., 1997. Amyloidogenic role of cytokine TGF-β1 in transgenic mice and in Alzheimer's disease. Nature 389, 603–606.

Xiao, X., Zhang, C., Ma, X., Miao, H., Wang, J., et al., 2015. Angiotensin-(1-7) counteracts angiotensin II-induced dysfunction in cerebral endothelial cells via modulating Nox2/ROS and PI3K/NO pathways. Exp. Cell Res. 336, 58–65.

Xie, H., Hou, S., Jiang, J., Sekutowicz, M., Kelly, J., Bacskai, B.J., 2013. Rapid cell death is preceded by amyloid plaque-mediated oxidative stress. Proc. Natl. Acad. Sci. U.S.A. 110, 7904–7909.

Yadav, H., Lee, J.H., Lloyd, J., Walter, P., Rane, S.G., 2013. Beneficial metabolic effects of a probiotic via butyrate-induced GLP-1 hormone secretion. J. Biol. Chem. 288, 25088–25097.

Yamada, K., Holth, J.K., Liao, F., Stewart, F.R., Mahan, T.E., et al., 2014. Neuronal activity regulates extracellular tau in vivo. J. Exp. Med. 211, 387–393.

Yamawaki, Y., Fuchikami, M., Morinobu, S., Segawa, M., Matsumoto, T., Yamawaki, S., 2012. Antidepressant-like effect of sodium butyrate (HDAC inhibitor) and its molecular mechanism of action in the rat hippocampus. World J. Biol. Psychiatry 13, 458–467.

Yang, J., Wong, A., Wang, Z., et al., 2015. Risk factors for incident dementia after stroke and transient ischemic attack. Alzheimers Dement. 11, 16–23.

Yin, Y., Gao, D., Wang, Y., Wang, Z.H., Wang, X., et al., 2016. Tau accumulation induces synaptic impairment and memory deficit by calcineurin-mediated inactivation of nuclear CaMKIV/CREB signaling. Proc. Natl. Acad. Sci. U.S.A. 113, E3773–E3781.

Yokokura, M., Mori, N., Yagi, S., Yoshikawa, E., Kikuchi, M., et al., 2011. In vivo changes in microglial activation and amyloid deposits in brain regions with hypometabolism in Alzheimer's disease. Eur. J. Nucl. Med. Mol. Imaging 38, 343–351.

Yoshiyama, Y., Lee, V.M.Y., Trojanowski, J.Q., 2013. Therapeutic strategies for tau mediated neurodegeneration. J. Neurol. Neurosurg. Psychiatry 84, 784–795.

Yu, J., Sun, M., Chen, Z., Lu, J., Liu, Y., Zhou, L., et al., 2010. Magnesium modulates amyloid-β protein precursor trafficking and processing. J. Alzheimers Dis. 20, 1091–1106.

Yu, X., Guan, P.P., Guo, J.W., Wang, Y., Cao, L.L., Xu, G.B., et al., 2015. By suppressing the expression of anterior pharynx-defective-1α and -1β and inhibiting the aggregation of β-amyloid protein, magnesium ions inhibit the cognitive decline of amyloid precursor protein/presenilin 1 transgenic mice. FASEB J. 29, 5044–5058.

Zambrano, C.A., Egaña, J.T., Núñez, M.T., Maccioni, R.B., González-Billault, C., 2004. Oxidative stress promotes τ dephosphorylation in neuronal cells: the roles of cdk5 and PP1. Free Radic. Biol. Med. 36, 1393–1402.

Zhou, H., Liu, G., 2015. Regulation of density of functional presynaptic terminals by local energy supply. Mol. Brain 8, 45.

Zlokovic, B.V., 2011. Neurovascular pathways to neurodegeneration in Alzheimer's disease and other disorders. Nat. Rev. Neurosci. 12, 723–738.
Zoidl, G., Dermietzel, R., 2002. On the search for the electrical synapse: a glimpse at the future. Cell Tissue Res. 310, 137–142.

Further Reading

Alzheimer's Disease International. World Alzheimer Report, 2009.
Holtzman, D.M., Morris, J.C., Goate, A.M., 2011. Alzheimer's disease the challenge of the second century. Sci. Transl. Med. 3, 77sr1.
Mosconi, L., Pupi, A., De Leon, M.J., 2008. Brain glucose hypometabolism and oxidative stress in preclinical Alzheimer's disease. Ann. N.Y. Acad. Sci. 1147, 180–195.
Prokop, S., Miller, K.R., Heppner, F.L., 2013. Microglia actions in Alzheimer's disease. Acta Neuropathol. 126, 461–477.

CHAPTER

4

Neurochemical Aspects of Lewy Body Dementia

INTRODUCTION

Parkinson's disease (PD) is a chronic, multifactorial and progressive neurological disorder characterized by the selective loss of dopaminergic neurons of the substantia nigra pars compacta as well as the formation of intracellular inclusion bodies, also known as Lewy bodies (LBs). PD is often associated with cognitive impairment that can progress to dementia. Its incidence in the human population ranges from approximately 1% in people over 65 years old to 4% in people over 86 years old. PD is the second most common neurodegenerative disorder after Alzheimer's disease (AD) (Stuendl et al., 2016). In the United States, more than 1 million people suffer from PD (Goldenberg, 2008). PD is more common in men (about 1.5 times) than in women (Davie, 2008), and higher incidences of PD have been reported in developed countries (Bove et al., 2005), due to an increase in the aged population (Cannon and Greenamyre, 2011). Eighty percent of PD patients develop dementia as PD progresses (Hely et al., 2008). Lewy body dementia (LBD), PD, and Parkinson's disease dementia (PDD) have been grouped under the umbrella term of LBD spectrum due to the overlap in symptom profile, similar treatment response, and common underlying neuropathology (Francis, 2009). Collective evidence suggests that LBD, PD, and PDD patients share the presence of α-synuclein aggregates in LBs and neuritis, and different timing in the onset of cognitive and motor manifestations may reflect the diverse regional burden and cerebral distribution of the pathology. Moreover, β-amyloid deposition is a frequent feature of LBD strongly affecting clinical manifestations (Merdes et al., 2003). In PDD, duration of parkinsonism before dementia is associated with different patterns of brain pathology and neurochemical abnormalities (Aarsland et al., 2006). Pathologically, LBD is characterized by LB and

Molecular Mechanisms of Dementia
DOI: https://doi.org/10.1016/B978-0-12-816347-4.00004-0

TABLE 4.1 Symptoms of Parkinson Disease

Primary motor symptoms	Primary nonmotor symptoms
Tremor	Depression
Rigidity	Dementia
Bradykinesia	Sleep disturbance
Postural instability	Fatigue and loss of energy
Secondary motor symptoms	Secondary nonmotor symptoms
Difficulty in swallowing and chewing	Sweating and urinary problems
Muscle cramps dystonia	Hypotension
Sexual dysfunction	Emotional changes

Lewy neurites in the brainstem, limbic system, and cortical areas (Fujishiro et al., 2008; Ince, 2011). In contrast, PD is accompanied by additional atypical features defining the Parkinson-plus syndromes, like multiple system atrophy (dysautonomia and/or cerebellar signs), progressive supranuclear palsy (impaired vertical eye movements and prominent postural instability), and corticobasal degeneration (apraxia).

The molecular mechanisms contributing to the pathogenesis of LBD, PD, and PDD, remains unknown. However, based on earlier investigations, it is suggested that the neurodegeneration of dopaminergic neurons result in the depletion of dopamine leading to abnormal dopaminergic neurotransmission in the basal ganglia motor circuit, not only causing resting tremor, muscular rigidity, akinesia, bradykinesia, posture, but also producing ambulation difficulty, sleep disorder, depression, dementia, and gastrointestinal dysfunction (Table 4.1) (Jankovic, 2008; Maiti et al., 2017). The symptoms of PD, LBD, and PDD include cognitive impairment, hallucinations, depression, intermittent confusion, and PD-like motor signs (bradykinesia, rigidity, and myoclonus). Among the above symptoms, akinesia and bradykinesia are assumed to be the result of a disruption of motor cortex activity (Jankovic, 2008), tremor, and rigidity. These processes are related to nigrostriatal dopaminergic deficits (Hornykiewicz, 2008). Patients with LBD not only show the accumulation of LBs, which are enriched in α-synuclein, but also signs of cerebral angiopathy, and deposition of β-amyloid and hyperphosphorylation of tau (Fig. 4.1). Postural instability is a major component of functional mobility. It is often overlooked by both clinicians and patients with LBD, PD, and parkinson disease dementia (PDD). Balance problems and resulting falls are major factors determining quality of life, morbidity, and mortality in individuals with LBD,

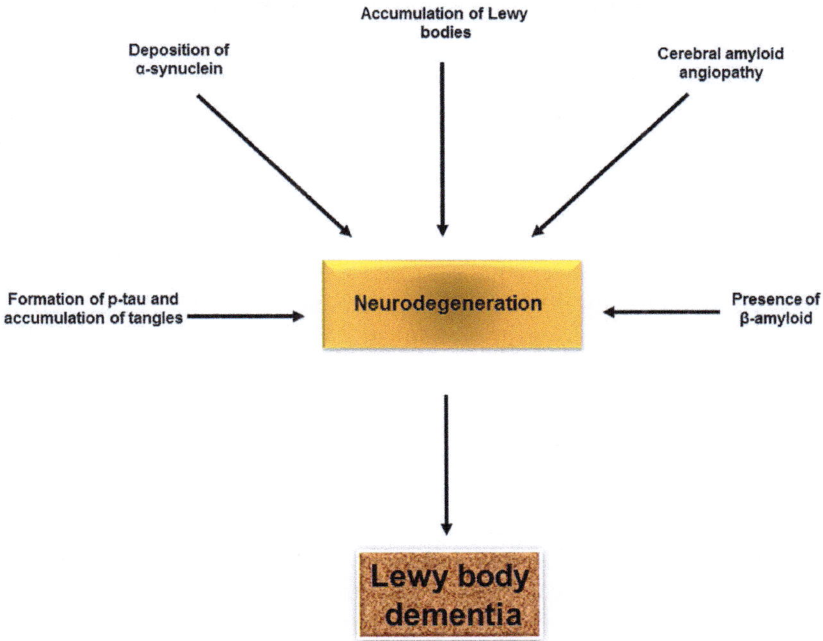

FIGURE 4.1 Neuropathological processes associated with the pathogenesis of Lewy body dementia.

PD, and PDD (Stolze et al., 2004). According to one study of 489 LBD, PD, and PDD patients who had been admitted to a hospital, approximately 60% of PD patients reported at least one fall in the previous 12 months (Stolze et al., 2004). In LBD, PD, and PDD patients falls contribute to the increased risk of bone fracture, impairment of mobility, low body mass index, and low bone mineral density (Sato et al., 2001). In LBD, PD, and PDD patients with disturbances of gait due to parkinsonism, fractures are more common than in those with gait disturbances due to other neurological conditions such as peripheral neuropathies (Syrjala et al., 2003). Most of above problems start with dopaminergic neuronal loss in substantia nigra pars compacta. The molecular mechanisms of dopaminergic neuronal loss are not fully understood. However, earlier investigations indicate that the degeneration of dopaminergic neurons in the substantia nigra pars compacta is due to monoamine oxidase (MAO)-mediated abnormal dopamine metabolism and hydrogen peroxide generation leading to oxidative stress. However, only 5%–10% of PD patients are known to have monogenic forms of PD. The majority of patients have sporadic PD, which may be induced by complex interactions among genetic factors, environmental exposures to toxins (paraquat, rotenone, herbicide, and insecticide), and aging of genetic variants with environmental risk factors (Lesage and Brice, 2009). There is a large variability in the onset and course of the disease.

It is interesting to note that almost half of PDD patients show significant Alzheimer's type pathologies, including the presence of amyloid-β (Aβ) plaques and Neurofibrillary tangles (NFTs). Recent studies have indicated that patients with PDD are prone to comorbid insulin resistance (Bosco et al., 2012; Ashraghi et al., 2016), even in the absence of type 2 diabetes. Furthermore, it is also demonstrated that insulin treatment not only normalizes the production and functionality of dopamine, but also ameliorates motor impairments in 6-OHDA-induced rat PD models. GSK3β, a downstream substrate of PtdIns 3K/Akt signaling following induction by insulin and IGF-1, exerts an influence on PD physiopathology. The genetic overexpression of GSK3β in cortex and hippocampus results in signs of neurodegeneration and spatial learning deficits in in vivo models of PD (Lucas et al., 2001). Accordingly, insulin- or IGF-1-activated PtdIns 3K/Akt/GSK3β signaling may contribute to the pathogenesis of PDD. In contrast, mutations in several genes have been implicated in familial and sporadic forms of PD, and their impact on DA neuronal cell death is slowly emerging. These genes include α-synuclein, Parkin, PTEN-induced putative kinase 1 (PINK1), protein DJ-1 (DJ1), leucine-rich repeat kinase 2 (LRRK2), and ubiquitin carboxyl-terminal hydrolase isozyme 1 (UCHL1) (Maiti et al., 2017). Little is known about the involvement of these genes and their proteins in the pathogenesis of LBD, PD, and PDD. However, PINK1 and Parkin have important roles in mitophagy, a cellular process associated with clearance of damaged mitochondria. PINK1 activates Parkin to ubiquitinate outer mitochondrial membrane proteins to induce a selective degradation of damaged mitochondria by autophagy (Hu and Wang, 2016). LRRK2 is a large multidomain protein bearing GTPase and kinase activity, and mutations in this gene represent one of the stronger risk factors for the development of PD (Cookson, 2010; Liu et al., 2012). Although the underlying pathogenesis of PD remains poorly understood, increased LRRK2 kinase activity, which is caused by the G2019S mutation, is thought to be associated with LRRK2-linked PD (Schwab and Ebert, 2015). Several studies have shown dopaminergic neurodegeneration from cultured dopaminergic neurons of pluripotent stem cells from PD patients harboring the LRRK2-G2019S mutation and human LRRK2-G2019S-expressing transgenic mice (Ramonet et al., 2011). The potential relationship between LRRK2, α-synuclein, and tau in inducing PD pathogenesis has been suggested (Cookson, 2010; Liu et al., 2012). LBs, eosinophilic proteinaceous round-shaped inclusions and Lewy neurites, enlarged aberrant thread-containing neuritic structures, are mainly composed of α-synuclein. LBs are key hallmarks of PD. Among PD associated proteins, α-synuclein and PTEN-induced putative kinase (PINK1) are two critical proteins associated with the pathogenesis of PD. α-synuclein induces mitochondrial deficits and apoptosis. PINK1 alleviates α-synuclein-mediated toxicity. However, the mechanistic details remain obscure. PINK1 interacts with α-synuclein mainly in the cytoplasm, where it initiates autophagy, a process which targets long-lived cytosolic proteins and damaged organelles. It involves a sequential set of events including double membrane formation, elongation, vesicle maturation, and finally delivery of the targeted materials to the lysosome. These interactions depend on the kinase activity of PINK1 and are abolished by deletion of the kinase domain or a G309D point mutation, an inactivating mutation in the kinase domain (Zaltieri et al., 2015; Liu et al., 2017). Interaction between PINK1 and α-synuclein stimulate the removal of excess α-synuclein, which prevents mitochondrial deficits and apoptosis (Liu et al., 2017).

Another protein whose mutations have been found to induce rare forms of autosomal recessive parkinsonism is DJ-1. This protein acts as a redox-sensitive molecular chaperone, whose loss of function may induce oxidative stress and consequently mitochondrial damage (Trancikova et al., 2012; Meulener et al., 2006). DJ-1 knockout mice show an enhanced sensitivity to the exposure of mitochondrial toxins; however, they do not develop PD-like pathological alterations per se. Instead, expression of mutant forms of LRRK2, that are the most common cause for the onset of familial PD, only produces subtle alterations in mitochondria morphology and integrity in vivo, although it has been hypothesized that the protein may also regulate mitochondrial dynamics (Trancikova et al., 2012). Interactions among these genes and their proteins are closely associated with neurodegeneration in PD. A hypothetical diagram showing interactions among these proteins is shown in Fig. 4.2.

Sporadic PD cases may be mediated by the environmental and genetic risk factors provoking oxidative stress, excitotoxicity, mitochondrial dysfunction, energy failure, neuroinflammation, misfolding and aggregation of α-synuclein, impairment of protein clearance pathways, cell-autonomous mechanisms, and deficits in proteasomal function or autophagy-lysosomal degradation of defective proteins (e.g., α-synuclein) (Fig. 4.3) (Alexander, 2004; Davie, 2008; Michel et al., 2016; Si et al., 2017; Maiti et al., 2017; Franco-Iborra et al., 2016; Truban et al., 2017; Moors et al., 2016). Among these processes, protein misfolding and subsequent accumulation of misfolded proteins in intracellular spaces has become a leading hypothesis for PD and LBD (Martin et al., 2011; Chauhan and Jeans, 2015). Misfolded α-synuclein not only undergoes phosphorylation, nitration, and truncatation, but also has abnormal solubility and has the ability to prompt the production of oligomeric species, aggregates into fibrils, and is ubiquitinated (Hashimoto et al.,

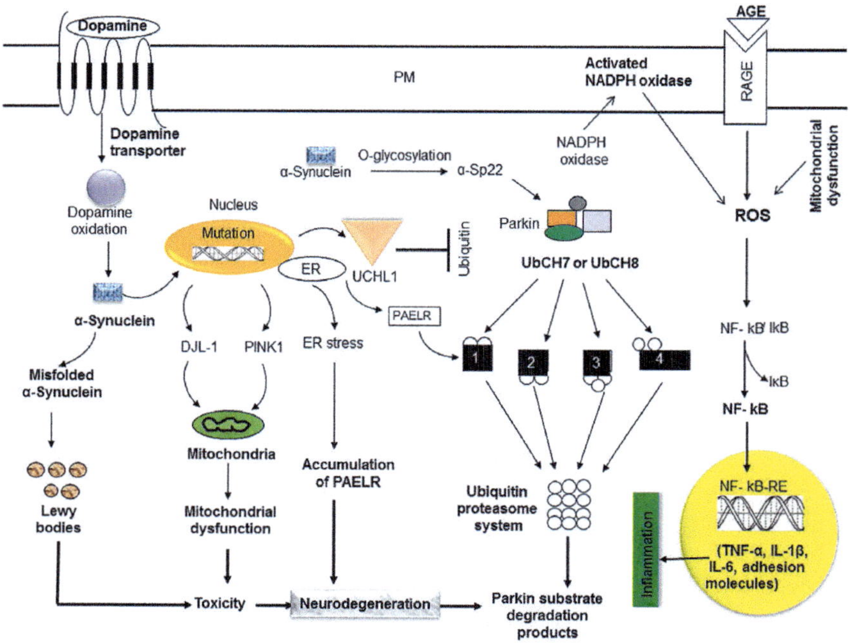

FIGURE 4.2 Hypothetical diagram showing interactions among genes and their proteins associated with neurodegeneration in familial form of PD.
DJ-1, protein DJ-1; *ER*, endoplasmic reticulum; *PAELR*, Pael receptor; *PINK1*, PTEN-induced putative kinase 1; *PM*, plasma membrane; *UCHL1*, ubiquitin carboxyl-terminal hydrolase isozyme 1; PAELR (1); cyclin (2); other Parkin substrate (3); synaptotagmin (4).

2004; Mukaetova-Ladinska and McKeith, 2006). Like misfolded Aβ protein inclusion in AD, the intracellular spaces of substantia nigra pars compacta neurons in LBD and PD contain aggregated α-synuclein (Alexander, 2004; Chauhan and Jeans, 2015; Berg, 2008; Maiti et al., 2017). In addition, substantia nigra pars compacta neurons also contain intracytoplasmic inclusions called LBs and Lewy neurites which contain several misfolded amyloid proteins, including aggregated and nitrated α-synuclein, phosphorylated tau (p-tau), and Aβ protein (Chauhan and Jeans, 2015; Kim et al., 2014).

α-SYNUCLEIN AND LBD SPECTRUM DISORDERS

α-Synuclein is a 140-amino acid soluble protein (mol mass 14 kDa) found predominantly within the brain. It is also enriched in the peripheral nervous system and circulating erythrocytes (Barbour et al., 2008; Maiti et al., 2017). α-Synuclein is encoded by a single gene consisting of

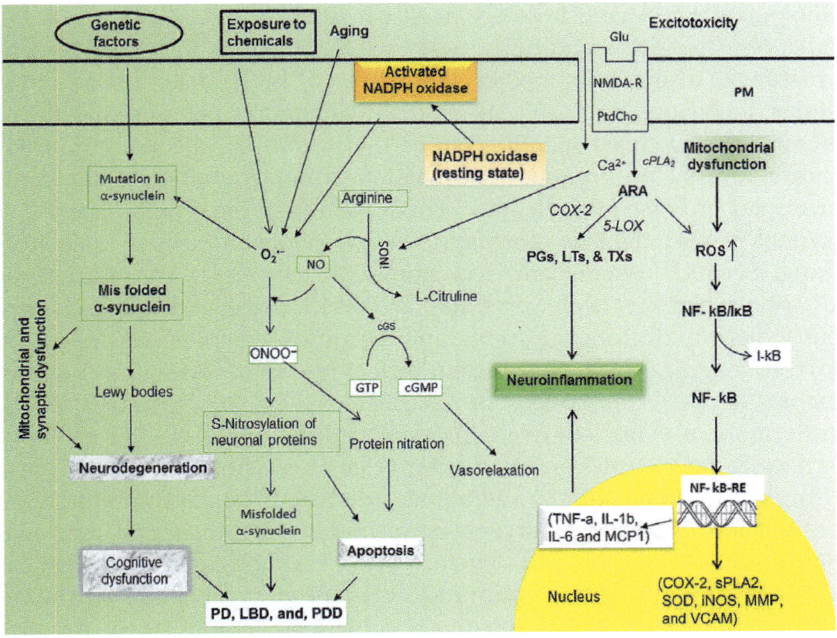

FIGURE 4.3 Hypothetical diagram showing molecular mechanisms contributing to the pathogenesis of Sporadic PD.
ARA, arachidonic acid; *COX-2*, cyclooxygenase-2; *cPLA$_2$*, cytosolic phospholipase A$_2$; *Glu*, glutamate; *IL-1β*, interleukin-1β; *IL-6*, interleukin-6; *iNOS*, inducible nitric oxide synthase; *LOX*, lipoxygenase; *LTs*, leukotriens; *NF-κB*, nuclear factor-κB; *NF-κB-RE*, nuclear factor-κB response element; *NMDA-R*, N-methyl-D-aspartate receptor; *NO$^•$*, nitric oxide; $O_2^{•-}$, superoxide; *ONOO$^-$*, peroxinitrite; *PGs*, prostaglandin; *PM*, plasma membrane; *PtdCho*, phosphatidylcholine; *ROS*, reactive oxygen species; *TNF-α*, tumor necrosis factor-α; *TXs*, thromboxanes.

seven exons located in chromosome 4 (Chen et al., 1995). α-Synuclein is primarily localized at the presynaptic terminals of neurons (Iwai et al., 1995). This protein lacks both cysteine and tryptophan residues. α-Synuclein is present in high concentration at presynaptic terminals and is found in both soluble and membrane-associated fractions of the brain (Lee et al., 2002). α-Synuclein is composed of three distinct regions: (1) an amino terminus (residues 1–60), containing apolipoprotein lipid-binding motifs, which contribute to the generation of amphiphilic helices involved in the formation of α-helical structures on membrane binding; (2) a central hydrophobic region (61–95) called NAC (non-Aβ component), which confers the ability to form β-sheets; and (3) a carboxyl terminus that is highly negatively charged, and is prone to be unstructured (Lee et al., 2011). There are at least two shorter alternatively spliced variants of the α-synuclein gene transcript, but their physiological and pathological roles have not been well characterized (Ueda et al., 1993). Under physiological conditions, α-synuclein functions in its native conformation as a soluble monomer. However, PD is characterized by intracellular inclusions of insoluble fibrils. Oligomers and protofibrils of α-synuclein have been identified to be the most toxic species, with their accumulation at presynaptic terminals affecting several steps of neurotransmitter release (Bridi and Hirth, 2018). First, high levels of α-synuclein alter the size of synaptic vesicle pools and impair their trafficking. Second, α-synuclein overexpression can either misregulate or redistribute proteins of the presynaptic SNARE complex leading to deficient tethering, docking, priming, and fusion of synaptic vesicles at the active zone (AZ). Third, α-synuclein inclusions are found within the presynaptic AZ, accompanied by a decrease in AZ protein levels. Furthermore, α-synuclein overexpression reduces the endocytic retrieval of synaptic vesicle membranes during vesicle recycling (Bridi and Hirth, 2018). These presynaptic alterations mediated by accumulation of α-synuclein, together impair neurotransmitter exocytosis and neuronal communication (Bridi and Hirth, 2018). Although α-synuclein is expressed throughout the brain and enriched at presynaptic terminals, dopaminergic neurons are the most vulnerable in PD, likely because α-synuclein directly regulates dopamine levels. Indeed, evidence suggests that α-synuclein is a negative modulator of dopamine by inhibiting enzymes responsible for its synthesis. In addition, α-synuclein is able to interact with and reduce the activity of VMAT2 and DAT (Bridi and Hirth, 2018). The resulting dysregulation of dopamine levels directly contributes to the formation of toxic α-synuclein oligomers resulting in a vicious cycle of α-synuclein accumulation and deregulation of dopamine resulting in synaptic dysfunction and impaired neuronal communication. Collective evidence suggests that native α-synuclein plays an important role in the regulation of synaptic vesicle release and trafficking, maintenance of synaptic vesicle pools, fatty acid binding, neurotransmitter release, synaptic plasticity, and neuronal survival (Bridi and Hirth, 2018) (Fig. 4.4). It should be noted that α-synuclein-induced neurodegeneration involves mitochondrial thiol oxidation and activation of caspases downstream of mitochondrial outer membrane permeabilization, leading to apoptosis-like cell death execution with some unusual aspects (Tolö et al., 2018). The overexpression of α-synuclein is not influenced by neurotrophic factors, calpain inhibition, and increased lysosomal protease capacity. In contrast, Bcl-Xl almost completely blocks neuron death. However, Bcl-Xl does not prevent mitochondrial thiol oxidation. Importantly, α-synuclein reduces excitability of neurons by external stimuli and robust impairments in endogenous neuronal network activity by decreasing the frequency of action potentials generated without external stimulation. This finding suggests that α-synuclein can

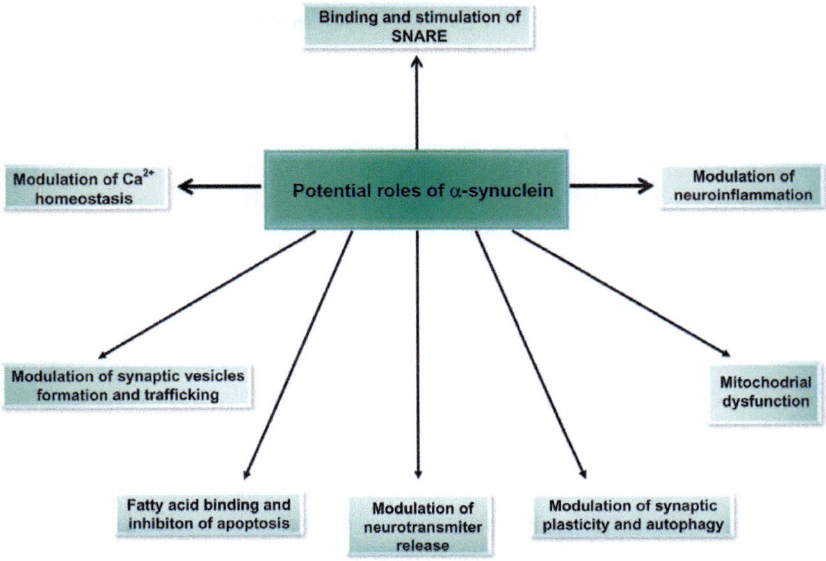

FIGURE 4.4 Potential roles of α-synuclein in the brain.

induce neuronal dysfunction independent from its induction of neurotoxicity and may be responsible for functional deficits that precede neurodegeneration in synucleopathies like PD or LBD (Tolö et al., 2018). Finally, by interacting with the SNARE-protein synaptobrevin-2/VAMP2, α-synuclein promotes the formation of SNARE-complex. This complex plays an important role at the presynaptic terminal during aging (Burré et al., 2010).

The self-aggregation of α-synuclein is accelerated not only in the presence of calcium, dopamine, proteins, and lipids, but also through posttranslational modifications, and oxidative stress. Other factors, which promote α-synuclein aggregation in vitro include subtle changes in the environment (i.e., increase in temperature, decrease in pH), addition of amphipathic molecules, such as herbicides, presence of external metal ions (industrial pollutants), and the interactions with membranes and other proteins (Bisaglia et al., 2009; Uversky et al., 2001a,b). Oxidative stress also upregulates the expression of α-synuclein, and promotes its fibrillization and aggregation (Vila et al., 2000). Conversely, a high degree of fibrillization and aggregation of α-synuclein results in an increase of reactive oxygen species (ROS) and neurotoxicity (Hsu et al., 2000). Thus, oxidative stress is one of the basic mechanisms that contributes to neurodegeneration in LBD, PD, and PDD (Clark et al., 2010). This vicious cycle between nitrated and aggregated α-synuclein and oxidative stress may not only contribute to the progression of loss of substantia nigra pars compacta dopaminergic neurons in PD, but may also result in the stimulation of microglial cells, activation of the innate immune system, production of inflammatory cytokines and chemokines, and subsequent neurodegeneration. It is also reported that as LBD, PD, and PDD progress, secretions from α-synuclein-activated microglia can engage neighboring glial cells in a cycle of autocrine and paracrine amplification of neurotoxic immune products. Such pathogenic processes affect the balance between a microglial neurotrophic and neurotoxic signature. Detailed investigations have demonstrated that interactions between nitrated and aggregated α-synuclein and microglia result in secretion of inflammatory, regulatory, redox-active, enzymatic, and cytoskeletal proteins (Reynolds et al., 2008; Reynolds et al., 2009). An increase in extracellular glutamate and cysteine and diminished intracellular glutathione and secreted exosomal proteins have also been demonstrated. An increase in redox-active proteins suggests regulation of microglial responses by misfolded, nitrated α-synuclein. These changes are linked with the discontinuous cystatin expression, cathepsin activity, and nuclear factor-kappa B (NF-κB) activation. Inhibition of cathepsin B attenuates, in part, the neurotoxicity of nitrated α-synuclein.on microglia (Reynolds et al., 2008, 2009). Extracellular α-synuclein is known to induce neuroinflammatory reactions in glial cells leading to neurodegeneration. Radiolabeled α-synuclein has been demonstrated to move across the blood–brain barrier (BBB) in both directions and this movement can have important therapeutic significance (Sui et al., 2014). High levels of α-synuclein in blood can contribute to CNS pathology, since the plasma levels of α-synuclein are significantly higher than in the colony-stimulating factor (CSF) levels (Shi et al., 2014). α-Synuclein from neurons enters into the glial cells and induces the expression and secretion of proinflammatory cytokines and chemokines; moreover, this release is directly proportional to the amount of α-synuclein in the glial cells (Lee et al., 2010). Recently, a neuron-to-neuron transfer of α-synuclein aggregates has also been reported in the cell culture system as well as in transgenic mice with neuronal progenitor cell grafts (Desplats et al., 2009). The transferred α-synuclein induces LBs-like inclusion and apoptotic changes in the recipient neurons.

RISK FACTORS FOR LEWY BODY DEMENTIA, PARKINSON'S DISEASE, AND PDD

Risk factors for PD and PDD include advanced age, older age of disease onset, limited cognitive reserve, hallucinations, and predominant gait dysfunction (Hely et al., 2008). With advancing age, a number of processes essential for the function of substantia nigra pars compacta neurons, including dopamine metabolism, wild-type mitochondrial DNA copy number, and protein degradation, decline (Reeve et al., 2014). A decline in wild-type mtDNA copy number may lead to a decrease in ATP production and a reduction in efficient protein degradation will affect the function of neurons (Subramaniam and Chesselet, 2013). Cognitive deficits in PD, LBD, and PDD typically affect executive functions, attention, visuospatial function, and processing speed (Williams-Gray et al., 2007). The pattern of cognitive impairment varies, however, in not only the extent to which different cognitive domains are affected but also which domains are affected first. Other risk factors for the LBD, PD, and PDD include exposure to herbicides and pesticides, high calorie intake, drug abuse by methamphetamine/amphetamine, traumatic brain injury (TBI), chronic traumatic encephalopathy (CTE), and drug-mediated mitochondrial dysfunction (Logroscino, 2005). Among above risk factors, TBI, CTE, and exposure to environmental toxins are major risk factors (Fig. 4.5). In addition, genetic mutations in DJ-1, PINK1, Parkin (PARK2), α-synuclein, and LRRK2 genes also contribute to LBD, PD, and PDD. These genes impact in complex ways on mitochondrial function leading to exacerbation of ROS

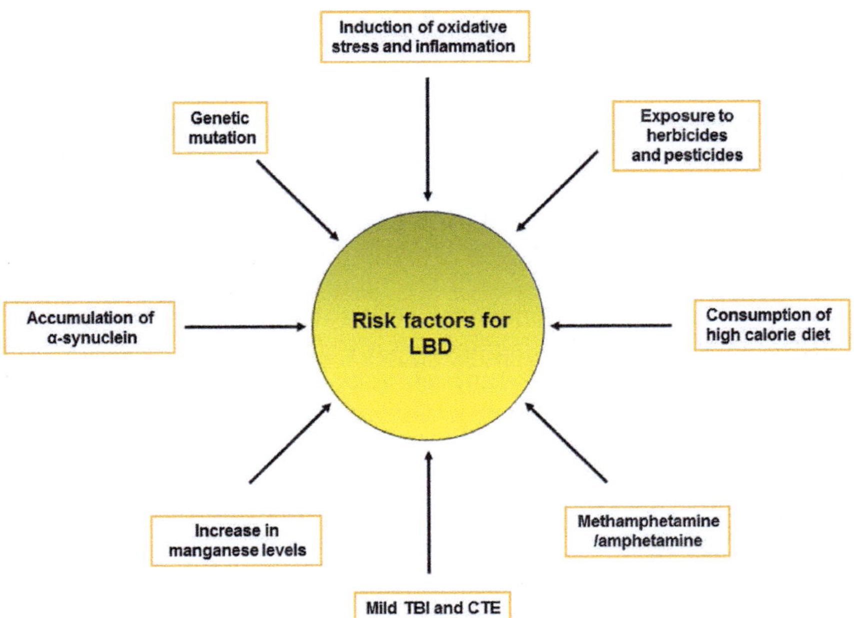

FIGURE 4.5 Risk factors for Lewy body dementia.

generation and susceptibility to oxidative stress. Additionally, cellular homeostatic processes including the ubiquitin–proteasome system and mitophagy are impacted by oxidative stress. It is apparent that the interplay between these various mechanisms contributes to neurodegeneration in LBD, PD, and PDD. Premature death of LBD, PD, and PDD patients often results due to complications such as movement impairment-related injuries and pneumonia (DeMaagd and Philip, 2015). LBD, PD, and PDD also cause cognitive, psychiatric, autonomic, and sensory disturbances. Cognitive impairments are common in a large fraction of patients with LBD, PD, and PDD at the initial diagnosis and afflict a majority of patients as the disease progresses. The secondary manifestation includes anxiety, insecurity, stress, confusion, memory loss, constipation, depression, difficulty in swallowing and excessive salivation, diminished sense of smell, increased sweating, erectile dysfunction, skin problems, and a monotone voice (Savitt et al., 2006).

DIAGNOSIS AND BIOMARKERS FOR LEWY BODY DEMENTIA, PARKINSON'S DISEASE, AND PDD

Differential diagnosis of LBD is quite difficult, especially in early stages of the disease, because specific biomarkers are not known and there are a lot of overlaps between the clinical and neuropathological characteristics among LBD, PD, PDD, and AD. The neuropathological diagnosis of LBD, PD, and PDD is based on the detection and quantification of LBs (Beach et al., 2009). As stated above, LBs are insoluble protein aggregates forming fibrils, which are composed of α-synuclein (Wakabayashi et al., 2007). In PD, LBs are mainly found at predilection sites of neuronal loss, that is, the substantia nigra pars compacta and locus coeruleus supporting the view that LBs somehow contribute to nerve cell loss in LBD, PD, and PDD. The number of LBs in patients with mild to moderate loss of neurons in the substantia nigra pars compacta is higher than in patients with severe neuronal depletion indicating that Lewy body-containing neurons are degenerating (Wakabayashi et al., 2007). Attempts have also been made to correlate the density of LBs in the cortex or brain stem with clinical symptoms of the disease (presence or absence of cognitive dysfunction, visual hallucinations, delusions, recurrent falls, severity of parkinsonism) in LBD, PD, and PDD, but these attempts have not been successful (Gómez-Tortosa et al., 2000). It is well known that nigrostriatal dopamine levels are depleted in LBD, PD, and PDD. Similarly, CSF levels of 3,4-dihydroxyphenylacetic acid (DOPAC), are also decreased in these patients. Whether low CSF DOPAC is associated specifically with parkinsonism has been unclear. In the neuronal cytoplasm, dopamine undergoes not only enzymatic oxidation to form DOPAC but also spontaneous oxidation to form 5-S-cysteinyl-dopamine (Cys-DA). Theoretically, oxidative stress or decreased activity of aldehyde dehydrogenase (ALDH) in the residual nigrostriatal dopaminergic neurons should increase CSF Cys-DA levels with respect to DOPAC levels. It is reported that in CSF Cys-DA/DOPAC ratio is substantially increased in LBD, PD, and PDD indicating that in LBD, PD, and PDD an elevated CSF Cys-DA/DOPAC ratio may provide a specific biomarker for these pathological conditions (Goldstein et al., 2016). Recent studies have indicated that there are specific microRNAs that correlate with LBD, PD, and PDD progression, and since microRNAs have been shown to be involved in the maintenance of neuronal development, mitochondrial dysfunction, and oxidative stress, there is a strong possibility that these microRNAs can be potentially used to differentiate among subsets of PD patients. PD is mainly diagnosed at the late stage, when almost the majority of the dopaminergic neurons are lost. Therefore, identification of molecular biomarkers for early detection of PD is important. Given that miRNAs are crucial in controlling the gene expression, these regulatory microRNAs and their target genes can be used as biomarkers for early diagnosis of PD (Arshad et al., 2017; Wang et al., 2017; Li and Le, 2017; Teixeira Santos et al., 2016).

In addition, α-synuclein can be detected in CSF and blood (Gao et al., 2015). However, at present, there are no reliable blood or CSF markers for Lewy Body Dementia (DLB), PD, and PDD that can be used for diagnosis, to follow disease progression, or as an outcome parameter for therapeutic interventions in DLB, PD, and PDD. The use of α-synuclein as a potential biomarker for DLB, PD, and PDD has been controversial (Mollenhauer et al., 2008; Ohrfelt et al., 2009). However, measurement of oligomeric α-synuclein in CSF and brain tissue by a specific enzyme-linked immunosorbent assay procedure has been used to distinguish between LBD and PD patients and age-matched control subjects (Tokuda et al., 2010; Paleologou et al., 2009). These results need replication and confirmation in a larger cohort. According to the latest diagnostic guidelines, DLB can be diagnosed by the absence or minimal atrophy of the medial temporal lobe on MRI (McKeith et al., 2017). At autopsy, the hippocampus shows the presence of tangles rather than plaques or Lewy body-associated pathology (Burton et al., 2009). Still, some investigators are using the presence of α-synuclein in CSF and blood along with neuroimaging data (PET, single-photon emission computerized tomography, magnetic resonance imaging (MRI)) to diagnose PD, PDD, and LBD, but these diagnostic tests are quite costly and are not always available in many hospitals (Walker et al., 2015). Furthermore, the frequent presence of concomitant AD pathology in LBD patients renders amyloid markers and MRI information less discriminative (Walker et al., 2015; Nedelska et al., 2015). In contrast, electroencephalography (EEG) has been proposed as a low-cost and readily available diagnostic tool to distinguish between LBD and AD (Lee et al., 2015; Roks et al., 2008). At present, in a clinical setting, data from patient history and the abovementioned diagnostic tests are weighted differently in each individual patient to make a diagnosis (Van Der Flier et al., 2014). The exact contribution of the (combinations of) EEG and other diagnostic tests to the differential diagnosis of DLB and AD remains unclear. Among LBD genetic markers, it is reported that there is a significant association between glucocerebrosidase (GBA1) mutation carrier status and LBD, and the GBA1 may be linked with PD (Sohma et al., 2013). Using 150 AD patients, 50 LBD patients, and 279 healthy elderly controls, it was reported that annexin A5 (a calcium and phospholipid binding protein) and ApoE ε4 are common plasma markers for AD and LBD (Sohma et al., 2013). In addition, familial LBD has been strongly associated to a region of chromosome 2, 2q35-q36 (Meeus et al., 2010).

NEUROCHEMICAL CHANGES IN LEWY BODY DEMENTIA, PARKINSON'S DISEASE, AND PDD

It has been hypothesized that the high concentration and aggregation of α-synuclein is not only linked with the dendritic spine degeneration and neurodegeneration in LBD, PD and PDD, but also with mitochondrial dysfunction. Dopaminergic neurons have a high energy demand that relies on the efficiency of the mitochondria respiratory chain (Bose and Beal, 2016). Mitochondrial dysfunction not only causes bioenergetic defects, mutations in mitochondrial DNA, and nuclear DNA gene mutations linked to mitochondria, but also changes mitochondrial dynamics such as fusion or fission, changes in size and morphology, alterations in trafficking or transport, altered movement of mitochondria, impairment of transcription, and the presence of mutated proteins associated with mitochondria in LBD, PD, and PDD (Bose and Beal, 2016). This suggestion is supported by the physiological action of α-synuclein relevant for mitochondrial homeostasis. The pathological aggregation of α-synuclein can negatively impinge on mitochondrial function. Thus, imbalances in the equilibrium between the reciprocal modulatory action of mitochondria and α-synuclein can contribute to PD onset by inducing neuronal impairment (Fujita et al., 2012; Faustini et al., 2017). Signal transduction pathways contributing to the above processes are not clearly understood. However, Bose and Beal (2016) have suggested that a new signaling pathway called the retromer-trafficking pathway may contribute to the pathogenesis of PD. This pathway includes bioenergetic defects, mutations in mitochondrial DNA, nuclear DNA gene mutations, alterations in mitochondrial dynamics, alterations in trafficking/transport and mitochondrial movement, impairment of transcription, and the presence of mutated proteins associated with mitochondria (Bose and Beal, 2016). It is also reported that the aggregated α-synuclein may produce toxicity through a gain-of-function mechanism (Benskey et al., 2016; Fujita et al., 2012). α-Synuclein contributes to a diverse range of essential cellular processes such as the regulation of neurotransmission and response to cellular stress. Another alternative hypothesis is that the aggregation of α-synuclein results in toxicity because of a toxic loss of necessary α-synuclein functions leading to degeneration. The possibility that presynaptic aggregated α-synuclein interferes with the release of neurotransmitter is supported by the observation that neurochemical changes in C57/Bl6 mice slices are closely associated not only with the depletion of dopamine, but also with progressive impairments in neuronal excitability and connectivity. These changes lead to profound loss of dendritic spines (Day et al., 2006). The imbalance of dendritic spine changes in relation to the relative preservation of presynaptic terminals may be explained by the finding that the bidirectional synaptic plasticity is based on the morphological plasticity of the dendritic spines (Nagerl et al., 2004). This link between α-synuclein aggregation, synaptic pathology, and mitochondrial dysfunction paves the way towards explaining the clinical symptoms of PD, PDD, and LBD. It also serves as the basis for understanding the effect of L-DOPA therapy at the beginning of symptoms and its failure later in the disease process. Moreover, the neurodegeneration produces loss of neuronal cell function. This process may be responsible for the clinical symptoms in DLB, PD, and PDD. This makes the treatment of these pathological conditions very difficult. Oxidative stress is not only known to produce nuclear membrane modifications, but also to promote the translocation of α-synuclein from cytoplasm to the nucleus, where it can form complexes with histones leading to its oligomerization into insoluble fibrils (Zhou et al., 2013). As stated above, aggregation and high levels of α-synuclein have been shown to induce oxidant production or increase the level of oxidative stress. Within cells, α-synuclein normally adopts an α-helical conformation. However, under high levels of oxidative stress α-synuclein undergoes a profound conformational transition to a β-sheet-rich structure that polymerizes to form toxic oligomers. Involvement of soluble oligomeric and protofibrillar forms of α-synuclein aggregates in the pathogenesis of PD is not only supported by the consistent detection of α-synuclein deposits in affected brain areas, but also by pathogenic mutations affecting the α-synuclein gene in familial PD and association of the α-synuclein locus with idiopathic

PD in genome-wide association studies. Furthermore, in vitro studies on cell culture systems and animal models also support association of α-synuclein with PD (Irvine et al., 2008; Simón-Sánchez et al., 2009). Recent studies on the neurodegenerative potency of α-synuclein fibrils have indicated that the toxicity of α-synuclein fibrils may be due to their ability to penetrate neural cell membranes (Volles et al., 2001; Pieri et al., 2012). Thus, compounds that inhibit α-synuclein aggregation and fibrilization and stabilize it in a nontoxic state can therefore serve as therapeutic molecules for both prevention of accumulation of aggregated α-synuclein and maintenance of normal physiological concentrations of α-synuclein (Li et al., 2004).

Studies on patients with Gaucher disease have indicated that deficiency of lysosomal hydrolase β-glucocerebrosidase also plays an important role in the development of synucleinopathies, such as PD and LBD (Blanz and Saftig, 2016; Stojkovska et al., 2018). The decrease in β-glucocerebrosidase activity leads to the accumulation of glucosylceramide and related lipid metabolites. Glucosylceramide is known to stabilize toxic oligomeric forms of α-synuclein, which not only effect β-glucocerebrosidase activity, but partially block the newly synthesized β-glucocerebrosidase from the endoplasmic reticulum amplifying the pathological effects of α-synuclein and ultimately resulting in neuronal cell death. This pathogenic molecular feedback loop and most likely other factors (such as impaired endoplasmic reticulum-associated degradation, activation of the unfolded protein response, and dysregulation of calcium homeostasis mediated by misfolded GC mutants) are involved in shifting the cellular homeostasis from monomeric α-synuclein towards oligomeric neurotoxic and aggregated forms, which contribute to the progression of PD. The molecular mechanism for the association between Gaucher disease and PD is not fully understood. However, it is proposed that GBA1-related neuronal death and α-synuclein accumulation, including disruptions in lipid metabolism, protein trafficking, and impaired protein quality control, may be an important link (Blanz and Saftig, 2016; Stojkovska et al., 2018). Based on several studies, it is proposed that lysosomal dysfunction may also contribute to the pathogenesis of PD. Mutations in the lysosomal hydrolase β-glucocerebrosidase (GBA1) may be a major risk factor for the development of PD and LBD (Blanz and Saftig, 2016; Stojkovska et al., 2018).

ANIMAL MODELS FOR PARKINSON'S DISEASE

Several environmental toxins are associated with sporadic PD, which can be partially mimicked in experimental animal models of PD, such

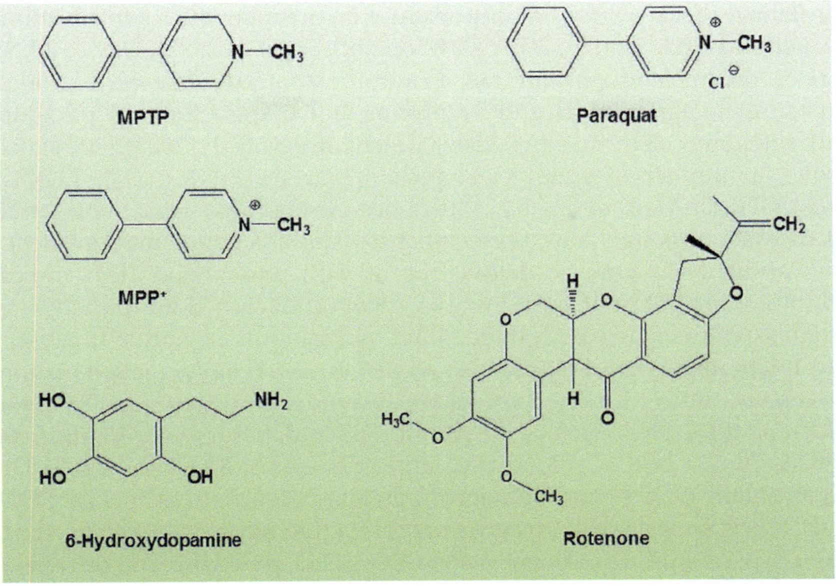

FIGURE 4.6 Chemical structures of compounds used for inducing PD in animal models.

as the use of 1-methyl-4-phenyl-1,2,3,6-tetrahydropyridine (MPTP), 6-hydroxydapoamine (6-OHDA), paraquat, rotenone, and other pesticides and herbicides (Fig. 4.6) (Bove et al., 2005; Bus and Gibson, 1984; Greenamyre et al., 2003; Srinivas et al., 2008). Unlike sporadic PD, familial cases of PD are rare, and do not follow the prescribed symptoms of PD, which makes it more difficult to understand the pathogenesis of PD (Martin et al., 2011; Klein and Westenberger, 2012). These neurotoxins provoke oxidative stress and impair mitochondrial respiration and energy metabolism, which in turn results in neurodegeneration. Postmortem tissues from PD patients have revealed a significant insight into the failure of complex I in the substantia nigra pars compacta. Complex I is a component of the mitochondrial electron-transport chain, and 30%–40% decrease in activity may be the central prognosis of sporadic PD (Dawson and Dawson, 2003). The decrease in the activity of complex I can result in self-inflected oxidative damage, underproduction of certain, complex I subunits, and may be due to complex I disassembly (Keeney et al., 2006). Immunocytochemical confirmation of protein glycation and nitration in substantia nigra pars compacta region of human PD brain revealed oxidative damage to DNA and protein resulting from persistent oxidative trauma (Floor and Wetzel, 1998).

PD is classified into two major subtypes: rare familial forms resulting from the inheritance of single gene mutations and the common sporadic disease with important environmental contributions (Dauer and Przedborski, 2003; Dawson and Dawson, 2003; Davie, 2008). Age is the most important risk factor for sporadic disease, although exposure to agricultural and environmental toxins, such as paraquat and rotenone, also increases risk (Greenamyre et al., 2003). As stated above, a number of genes (α-synuclein, Parkin, PINK1, DJ1, LRRK2, and UCHL1) (Maiti et al., 2017) have been found associated with familial disease, although how these inherited disease genes may influence development of sporadic disease is not well understood. The clinical symptoms in LBD not only involve synaptic dysfunction, induction of progressive dementia with deficits in attention and executive functions, and fluctuating cognition, but also recurrent visual hallucinations before or concurrently with the parkinsonian syndrome (McKeith, 2007). In DLB and PDD, extrastriatal dopaminergic and particularly cholinergic deficits play a central role in mediating dementia (McKeith, 2007). The involvement of presynaptic neurotransmitter deficiencies in PD, PDD, and DLB is supported by in vivo neuroimaging studies (Nikolaus et al., 2009). These studies indicate that in PD, PDD, and DLB the degenerative process is located at the presynapse (Linazasoro, 2007) and results in a neurotransmitter deficiency syndrome.

OXIDATIVE STRESS IN LEWY BODY DEMENTIA, PARKINSON'S DISEASE, AND PDD

As stated in chapters 2 and 3, oxidative stress occurs when the level of prooxidants exceeds the level of antioxidants in cells resulting in redox imbalance between prooxidants and antioxidants in favor of the former ones, leading to oxidation of cellular components and consequent loss of cellular function. ROS include superoxide anions ($O_2^{\bullet -}$), hydroxyl, alkoxyl, and peroxyl radicals ($^{\bullet}OH$ and $ROO^{\bullet}$), and nonradical hydrogen peroxide (H_2O_2). The initial product, $O_2^{\bullet -}$ is generated through mitochondrial dysfunction, uncontrolled ARA cascade, and activation of NADPH oxidases (Sun et al., 2007). $O_2^{\bullet -}$ are readily transformed by oxidoreduction reactions with transition metals or other redox cycling compounds into more aggressive radical species ($OH^{\bullet}$ and H_2O_2) (Fig. 4.7) (Hancock et al., 2001; Beal, 2005). A number of sources and mechanisms control the generation of ROS in the brain including the metabolism of dopamine itself, mitochondrial dysfunction, iron, neuroinflammatory cells, calcium, and aging. PD producing gene products, including DJ-1, PINK1, parkin, α-synuclein, and LRRK2, also impact in complex ways on mitochondrial function leading to exacerbation of ROS generation and susceptibility to oxidative stress

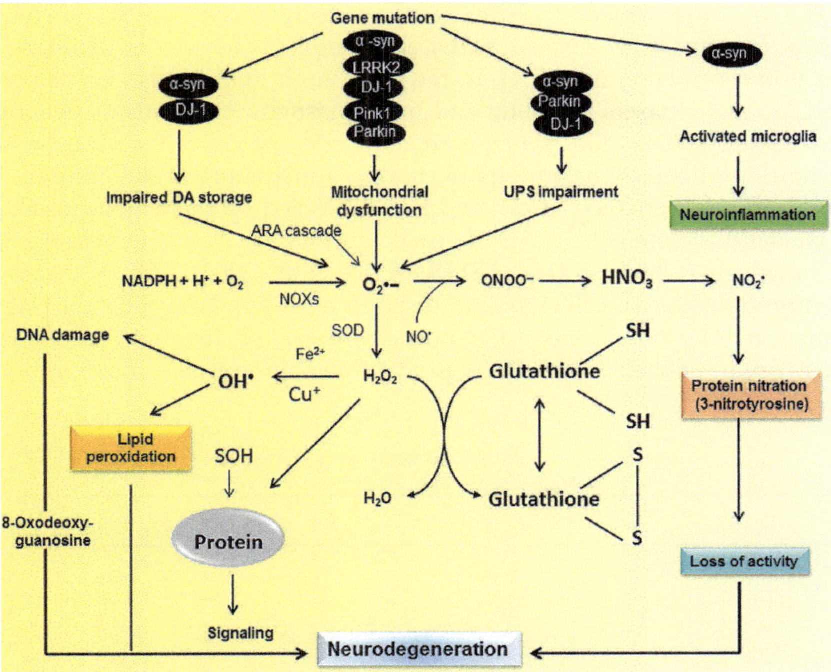

FIGURE 4.7 Generation of ROS/RNS and coupling and neutralization of oxidative stress by glutathione.
H_2O_2, hydrogen peroxide; *$NO^{\bullet}$*, nitric oxide; *$O_2^{\bullet-}$*, superoxide; *$OH^{\bullet}$*, hydroxyl radical; *$ONOO^-$*, peroxynitrite; *SOD*, superoxide dismutase; *α-Syn*, α-synuclein.

(Dias et al., 2013). In addition, cellular homeostatic processes including the ubiquitin–proteasome system and mitophagy are impacted by oxidative stress. It is apparent that the interplay among these various mechanisms contributes to neurodegeneration in PD as a feed-forward scenario where primary insults lead to oxidative stress, which damages key cellular pathogenetic proteins that in turn cause more ROS production (Dias et al., 2013). Two neuroprotective mechanisms operate in the brain to tackle the threat posed by ROS: (1) the antioxidant enzyme system; and (2) the low-molecular-weight antioxidants (Kohen et al., 1999). The antioxidant enzyme system includes superoxide dismutase (SOD), glutathione reductase, glutathione peroxidase, and catalase (CAT) (Griendling et al., 2000). The low-molecular-weight antioxidants include glutathione, uric acid, ascorbic acid, and melatonin, which offer neutralizing functions by causing chelation of transition metals (Halliwell, 2006). Low levels of intracellular ROS are needed to maintain normal cellular functions (proliferation, migration, and survival) and redox signaling (Forman et al., 2004). However, an excess of ROS results in oxidative stress, which involves damage to cellular components, such as lipids, proteins, and nucleic acids, and leads to the loss of biological function (Forman et al., 2004). In the brain high ROS levels also diminish LTP and synaptic signaling and brain plasticity mechanisms (Knapp and Klann, 2002). This is regarded as a state of oxidative stress and becomes particularly hazardous for normal functioning of the brain.

$NO^{\bullet}$ is another free radical, and is synthesized by the action of nitric oxide synthase on arginine. $NO^{\bullet}$ reacts with $O_2^{\bullet -}$ to form the neurotoxic peroxynitrite ($ONOO^-$) (Fig. 4.8) (Bal-Price et al., 2002). ROS/RNS play an important role in cell signaling through redox signaling. To maintain proper cellular homeostasis and normal neural cell function, a balance must occur between ROS/RNS production and oxygen consumption

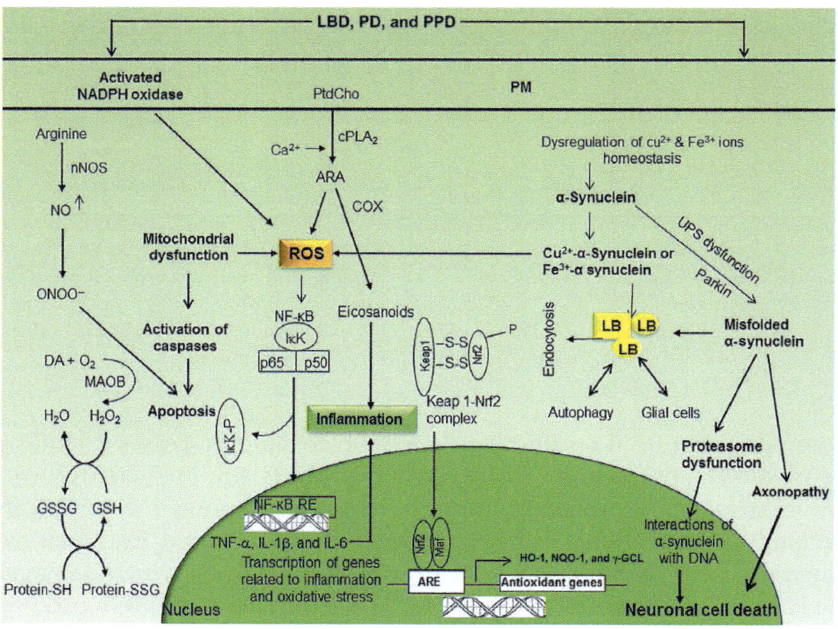

FIGURE 4.8 Neurochemical mechanisms contributing to the pathogenesis of LBD, PD, and PDD.

ARA, arachidonic acid; *cPLA₂*, cytosolic phospholipase A_2; *DA*, dopamine; *DJ-1*, neuroprotective protein DJ-1; *IL-1β*, interleukin-1beta; *IL-6*, interleukin-6; *LOX*, cyclooxygenase; *NF-κB*, nuclear factor-kappa B; *NF-κB-RE*, nuclear factor-kappa B response element; *PINK1*, PTEN-induced putative kinase 1; *PM*, plasma membrane; *PtdCho*, phosphatidylcholine; *ROS*, reactive oxygen species; *TNF-α*, tumor necrosis factor-alpha; *UPS*, ubiquitin-proteasome system; *MAOB*, monoaminoxidase B; *GSH*, reduced glutathione; *GSSG*, oxidized glutathione; *Nrf2*, nuclear factor E2-related factor 2; *Keap1*, kelch-like ECH-associated protein 1; *ARE*, antioxidant response element; *Maf*, small leucine zipper proteins; *HO-1*, heme oxygenase; *NQO-1*, NADPH quinine oxidoreductase; *γ-GCL*, γ-glutamate cystein ligase; *LB*, Lewy bodies.

and $NO^{\bullet}$. It has been hypothesized that $ONOO^{-}$ contributes to PINK1/Parkin-mediated mitophagy activation via the triggering of dynamin-related protein 1 (Drp1) recruitment leading to mitochondrial damage. Excessive ROS/RNS have to be either quenched by converting them into metabolically nondestructive molecules or be scavenged/neutralized right after their formation. This protective mechanism is called the antioxidant defence system preventing ROS/RNS-mediated damage of cells leading to various diseases and aging (Yu, 1994; Winterbourn and Hampton, 2008). Another mechanism of redox signaling is through the involvement of the glutathione thiol/disulfide redox couple (GSH/GSSG) system. This pathway is another predominant mechanism for maintaining the intracellular microenvironment in a highly reduced state that is essential for antioxidant/detoxification capacity, redox enzyme regulation, cell cycle progression, and transcription of antioxidant response elements (ARE) (Fig. 4.8) (Biswas et al., 2006; Fratelli et al., 2005).

$2GSH + O_2 \rightarrow GSSG + 2H_2O_2$
$2GSH + 2H_2O_2 \rightarrow GSSG + 2H_2O$ (GSH peroxidase)
$2GSSG + NADH \rightarrow 2GSH + NADP$ (GSSG reductase)

In response to oxidative and nitrosative stress, neural cells increase their antioxidant defenses through activation of nuclear factor erythroid 2-related factor (Nrf2), an important transcription factor (Maes et al., 2011). Nrf2 is a key component of this control system and recognizes the antioxidant response element (ARE) found in the promoter regions of many genes that encode antioxidants and detoxification enzymes, such as heme oxygenase 1 (HO-1), NAD(P)H dehydrogenase quinone 1, SOD1, glutathione peroxidase 1 (GPx1), and CAT (Itoh et al., 1997).

As stated above, deposition of misfolded α-synuclein, mitochondrial dysfunction, and induction of oxidative stress are closely associated with the pathogenesis of LBD, PD, and PDD (Blesa et al., 2015). Oxidative stress results in the generation of higher levels of cholesterol hydroperoxide, MDA, 4-HNE, and OH8dG. One of the suggested causes of induction of oxidative stress in the substantia nigra pars compacta is the production of ROS during normal DA metabolism. In human substantia nigra pars compacta, the oxidation products of DA (mainly 6-hydroxydopamine) may polymerize to form neuromelanin, which may also be toxic by inducing apoptosis (Berman and Hastings, 1999). Furthermore, postmortem studies have indicated a decrease in GSH levels and an increase in GSSG levels in the substantia nigra pars compacta. This can be a critical primary event that weakens or abrogates the natural antioxidant defence mechanisms of neural cells, thereby triggering degeneration of the nigral neurons and promoting the pathogenesis of LBD, PD, and PDD (Gu et al., 2015. Since dysregulation of metal ion homeostasis is a potential catalyst to further production of ROS, the highly oxidative environment for DA interaction with α-synuclein, and the resulting oxidant-mediated toxicity and protein aggregation, is one of the most likely underlying mechanisms for LBD, PD, and PDD. It is proposed that neurodegeneration in PD may occur as a result of self-propagating reactions that involve not only DA, α-synuclein, and mitochondrial dysfunction, but also involve the participation of redox-active metals (Carboni and Lingor, 2015). Furthermore, PD causing gene products, including DJ-1, PINK1, parkin, alpha-synuclein, and LRRK2, also impact on mitochondrial function in complex ways leading to the exacerbation of ROS generation and susceptibility to oxidative stress. Additionally, cellular homeostatic processes, including the ubiquitin–proteasome system and mitophagy, are impacted by oxidative stress. It is apparent that the interplay between these various mechanisms contributes to neurodegeneration in PD as a feed-forward scenario where primary insults lead to oxidative stress, which damages key cellular pathogenetic proteins that in turn cause more ROS production.

Advanced glycation end products (AGEs) are proteins or lipids that become glycated after exposure to sugars. The formation of AGEs promotes the deposition of proteins due to the protease-resistant cross-linking between the peptides and proteins. PD, PDD, and LBD are characterized by the abnormal accumulation or aggregation of proteins such as amyloid β, tau, and α-synuclein, which become glycated and the extent of glycation is correlated with the pathologies of PD, PDD, and LBD (Fig. 4.2) (Li et al., 2012). It is reported that AGE-mediated modification of glycated proteins triggers the sustained local oxidative stress and inflammatory response, eventually contributing to the pathological and clinical aspects of PD, PDD, and LBD.

NEUROINFLAMMATION IN LEWY BODY DEMENTIA, PARKINSON'S DISEASE, AND PDD

It is well known that neuroinflammation is closely associated with the pathogenesis of LBD, PD, and PDD (Farooqui, 2014). Neuroinflammation not only involves resident cells (microglia, astrocytes, neurons) of the brain, but also the cells and humoral factors of the peripheral immune system that penetrate into the brain (Phani et al., 2012; Farooqui, 2014). In LBD, PD, and PDD, the onset of neuroinflammation is initially associated with the cleaning up of dead neurons to control the severity and progression of the disease. However, neuroinflammation also acts as a double-edged sword (Kielian, 2016). On the one hand, neuroinflammation induces and/or aggravates neurodegeneration in the brain, while on the other hand, it promotes neural cell homeostasis by the induction of resolution and removal of the degenerating neurons (Lucas et al., 2006).

Furthermore, chronic neuroinflammation not only promotes regeneration to some extent (Bollaerts et al., 2017), but also induces cytotoxic effects, which increase the severity of neurodegenerative disease symptoms. Thus, in AD, LBD, PD, and PDD, the neurodegenerative process is mediated by inflammatory and neurotoxic mediators, such as TNF-α, interleukin-1beta (IL-1β), IL-6, IL-8, IL-33, chemokine (C-C motif) ligand 2 (CCL2), CCL5, matrix metalloproteinase (MMPs), granulocyte macrophage colony-stimulating factor (GM-CSF), glia maturation factor (GMF), substance P, ROS, RNS) mast cells-mediated histamine and proteases, protease activated receptor-2 (PAR-2), CD40, CD40L, CD88, intracellular Ca^+ elevation, and activation of mitogen-activated protein kinases (MAPKs) and NF-kB (Farooqui, 2014; Kempuraj et al., 2016, 2018). Under physiological conditions, the quiescent state of microglia is maintained by a variety of immunomodulators, such as CX3CL1, CD200, CD22, CD47, CD95, and neural cell adhesion molecule (NCAM), which are produced mainly by neuronal cells (Sheridan and Murphy, 2013). Interestingly, the receptors for these molecules are almost exclusively expressed by microglia in the CNS, indicating the critical role of neuron–microglia interactions in the regulation of neuroinflammation (Sheridan and Murphy, 2013). As stated above, activated microglia, astrocytes, neurons,T cells, and mast cells release the above inflammatory mediators and induce and support neuroinflammation by crossing the defective BBB (Table 4.2)

TABLE 4.2 Inflammatory Mediators, Which Are Release by Microglia, Astrocytes, and Mast Cells in LBD, PD, and PDD

Inflammatory mediators	Mast cells	Microglia	Astrocytes
CRH and CRH-R	Elevated	CRH-R elevated	Altered BBB permeability
Histamine and tryptase	Elevated	–	–
TNF-α, Il-1β, IL-6, IL-18, Il-33, Il-36, CCL2	Elevated	Elevated	Elevated
VEGF	Elevated	Elevated	–
ROS and NO	Elevated	Elevated	Elevated
MMPs	Elevated	Elevated	–
CD40L	Elevated	Elevated	–
PGD2 and LT4	Elevated	Elevated	–
PAF	Elevated	Elevated	–

CRH, Corticotropin-releasing hormone; *CRH-R*, corticotropin-releasing hormone receptor; *BBB*, blood–brain barrier; *TNF-α*, tumor necrosis factor-alpha; *IL-33*, interleukin-33; *VEGF*, vascular endothelial growth factor; *ROS*, reactive oxygen species; *CCL2*, chemokine (C-C motif) ligand 2; *MMPs*, matrix metalloproteinase; *PGD_2*, prostaglandin D_2; *LT4*, leukotriene 4; *PAF*, platelet-activating factor.

(Farooqui, 2014; Kempuraj et al., 2016, 2018). In LBD, PD, and PDD, the T cell and mast cell interaction with glial cells and neurons results in neuroinflammation. The onset of neuroinflammation in LBD, PD, and PDD has not only been confirmed by the elevated levels of cytokines, chemokines, and proinflammatory eicosanoids, but also by in vivo studies using PET imaging (Surendranathan et al., 2015). In LBD, PD, and PDD, the onset of the chronic inflammatory process is linked with the accumulation of misfolded α-synuclein, activation of microglia, cognitive dysfunction, neuronal loss (Surendranathan et al., 2015; Streit and Xue, 2016), and with a broad range of components of the innate and adaptive immune systems (Surendranathan et al., 2015). Evidence of neuroinflammation in LBD, PD, and PDD is further supported by pathological and biomarker studies. Furthermore, genetic and epidemiological studies also support a role for neuroinflammation in LBD, PD, and PDD (Wang et al., 2015).

IMMUNE RESPONSES IN LEWY BODY DEMENTIA, PARKINSON'S DISEASE, AND PDD

Microglia are the resident immune cells of the brain. They are the primary contributor to innate immunity in the brain (Katsumoto et al., 2014). They play important roles not only in exploring the cellular environment and phagocytosis, but also in antigen processing and presentation, and production of cytokines and chemokines (Katsumoto et al., 2014). T cell infiltration and glial cell activation are common features of both human PD patients and animal models of PD, playing vital roles in the degeneration of DA neurons (Hirsch et al., 2012). Detailed investigations on human PD patients have indicated that a sustained long-term increase in levels of proinflammatory cytokines and chemokines along with elevated levels of eicosanoids contribute to chronic inflammatory responses (Fig. 4.9) (Hirsch et al., 2012). In PD, under pathological conditions (aging, protein aggregation, gene mutations, environmental factors), M1 microglia become activated due to the infiltration of T cells. The proinflammatory mediators from M1 microglia activate astrocytes, leading to elevated production of proinflammatory factors, nitric oxide and superoxide radical, contributing to the degeneration of DA neurons. The molecules released from degenerative DA neurons can further cause the activation of glia and enhance the inflammatory response. At a certain stage of PD, a subpopulation of microglia may become the activated M2 phenotype releasing antiinflammatory factors, including TGF-β, and exert a neuroprotective effect in PD (Wang et al., 2015).

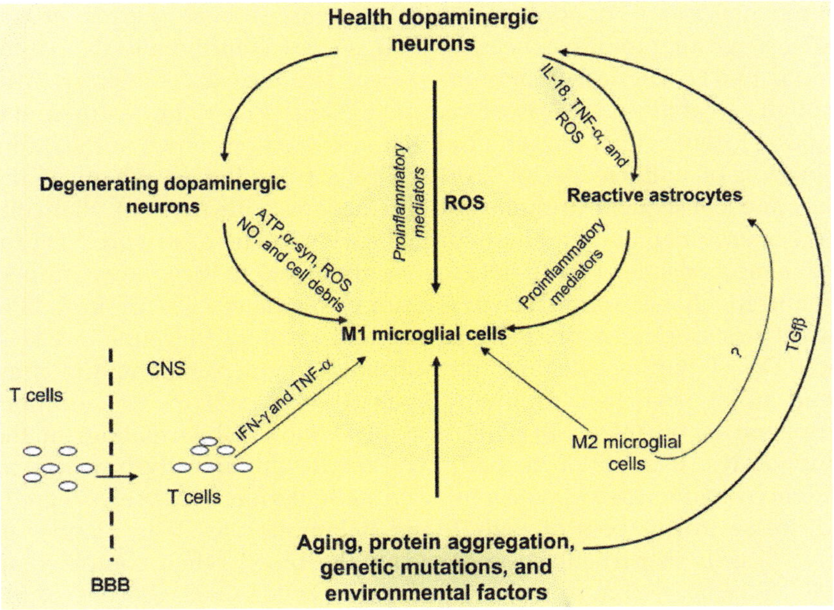

FIGURE 4.9 Hypothetical diagram showing inflammatory mechanisms contributing to the pathogenesis of LBD, PD, and PDD.
IL-1β, interleukin-1β; *IL-18*, interleukin-18; *IFN-γ*, interferon-gamma; *ROS*, reactive oxygen species; *TNF-α*, Tumor necrosis factor-α.

There is an intimate relation between α-synuclein and the immune system (Bandres-Ciga and Cookson, 2017). In a rat model of PD, overexpression of α-synuclein results in upregulation of TNF-α, IL-1β, and IFN-γ in the striatum (Chung et al., 2009). A recent study has also shown that T cells from PD patients recognize α-synuclein peptides as antigenic epitopes which may explain the association of PD with specific major histocompatibility complex alleles (Sulzer et al., 2017). Major histocompatibility complex I is expressed in human substantia nigra pars compacta neurons and can be induced in human stem cell-derived dopaminergic neurons. Dopaminergic neurons internalize foreign ovalbumin and display antigen derived from this protein by major histocompatibility complex I to activate T cells, resulting in autoimmune responses and the death of dopaminergic neurons (Cebrian et al., 2014). α-Synuclein promotes the entry of proinflammatory peripheral CCR^{2+} monocytes into substantia nigra pars compacta, inducing the expression of major histocompatibility complex II and the subsequent degeneration of dopaminergic neurons (Harms et al., 2018). Genetic allelic variants of Mhc2ta, the major regulator of major histocompatibility complex II expression, regulates α-synuclein-induced microglial activation and the neurodegeneration of dopaminergic neurons (Jimenez-Ferrer et al., 2017), suggesting immunochemical changes may contribute to the pathogenesis of PD. The innate immune system is linked to its adaptive arm through the abilities to provide required "signals" for antigen presentation and to act as final effectors byT cell-mediated responses in the brain. Migration of antigen-specific $CD4^+$ T cells from the periphery to the brain and consequent immune cell interactions with resident glial cells affect neuroinflammation and neuronal survival. The destructive or protective mechanisms of these interactions are linked to the relative numerical and functional dominance of effector or regulatory T cells. In the PD patient postmortem brain, there is a 10-fold greater infiltration of $CD4^+$ and $CD8^+$ T lymphocytes into the substantia nigra pars compacta compared to age-matched controls (Brochard et al., 2009). In addition, peripheral blood from LBD, PD, and PDD shows abnormalities in the lymphocytes. In addition to peripheral innate immune dysfunction, as evidenced by increased neutrophils and natural killer (NK) cells, there is a decrease in numbers of both T and B lymphocytes. The number of $CD4^+$ T cells is particularly reduced in the blood when compared to $CD8^+$ T cells, which are unchanged (Baba et al., 2005; Stevens et al., 2012). The reduction in $CD4^+$ T cells is correlated with UPDRS III performance in PD patients (Baba et al., 2005). Characterization of $CD4^+$ peripheral T cells from PD patients shows that they are likely to be Th1 cells, as the ratio of IFN-γ:IL-4-producing cells is increased (Baba et al., 2005). Furthermore, various surface markers are altered and correlate with disease state as assessed by UPDRS III (Saunders et al., 2012). These results suggest that peripheral T lymphocytes in PD are activated effector/memory cells with a Th1 phenotype. Furthermore, these T cells are likely undergoing activation-induced Fas-mediated apoptosis, leading to their decrease in number (Saunders et al., 2012). Decreases in $\alpha4\beta7$ integrin on $CD4^+$ T cells can signal a relative increase in brain-homing function or an active immune response in the gut sequestering $CD4^+$ $\alpha4\beta7^+$ T cells from peripheral blood (Saunders et al., 2012). Indeed, T cell protein expression may be used as a biomarker for PD in the future; a panel of 13 proteins expressed in T lymphocytes can be quantified by multiple reaction monitoring and be validated as PD-specific in a small blinded cohort (Alberio et al., 2014). Collective evidence implicates the activation of both the innate and adaptive immune systems in PD (Reish and Standaert, 2005). This inflammatory response plays an essential role in neurodegeneration. The evidence reviewed in this chapter implicates α-synuclein itself as the primary trigger of the immune response in PD. While modification of α-synuclein by nitration elicits a stronger immune response than aggregation alone, it is likely that aggregation is sufficient to induce the inflammation seen in PD. This concept has implications for both the prevention and treatment of PD.

COGNITIVE DYSFUNCTION IN LEWY BODY DEMENTIA, PARKINSON'S DISEASE, AND PDD

The prevalence of LBD, PD, and PDD increases with age, disease duration, motor severity, postural instability/gait disorder phenotype, baseline cognitive impairment, and presence of other nonmotor and neuropsychiatric issues (Litvan et al., 2011). In these pathological conditions, the onset of mild cognitive impairment may progress to dementia rapidly. In LBD, PD, and PDD, the rate of prevalence of cognitive dysfunction is increased four to six times compared to age-matched controls. Social isolation, depression, and medical illness may worsen cognition in general and in PD. Even after accounting for these factors, however, cognitive function varies among individuals. The mechanisms contributing to cognitive dysfunction in LBD, PD, and PDD are not fully understood. However, it is proposed that cognitive decline is not only accompanied by α-synuclein-mediated induction of oxidative stress and neuroinflammation, but also due to cortical thinning, hypometabolism, white matter changes, dopaminergic/cholinergic dysfunction, and increased α-synuclein burden (Jellinger, 2012; Hanganu et al., 2013). These alterations can not only cause destruction of essential neuronal networks, but also synaptic rarefaction, leading to cognitive dysfunction (Jellinger, 2012). These cognitive impairments are severe enough to impair their everyday functional abilities in LBD, PD, and PDD. Furthermore, there is emerging evidence that healthy lifestyles may decrease the rate of cognitive decline seen with aging and help delay the onset of cognitive symptoms in age-associated diseases (Farooqui, 2012, 2018). Some investigators have implicated LBs in the neocortex, other investigators have pointed out that α-synuclein pathology in the hippocampus contributes to cognitive impairment and depression in LBD, PD, and PDD (Yang and Yu, 2017). A major question of whether and how much AD pathology and α-synuclein pathology contribute to cognitive decline in PD remains disputable (Mollenhauer et al., 2011). A majority of LBD patients show an increase in cortical ^{11}C-PIB binding, similar to AD (Edison et al., 2008), suggesting that LBD is actually a dementia associated with both α-synuclein and Aβ pathology, thereby possibly explaining its aggressive nature. In contrast, brain tissue from PD and PDD patients shows a reduction in the prevalence of amyloid plaques and lower levels of cortical ^{11}C-PIB binding than LBD (Edison et al., 2008; Jokinen et al., 2010). This finding suggests that the LBD is more likely due to a specific α-synuclein pathology rather than only an overlap of other pathologies. This is in agreement with postmortem observations (Kramer and Schulz-Schaeffer, 2007). The cognitive decline in LBD, PD, and PDD can be probed clinically by comprehensive neuropsychological assessment (Pievani et al., 2011). Neuroimaging studies in these disorders have also revealed the presence of cortical atrophy, hypometabolism, white matter changes, and dopaminergic/cholinergic dysfunction. Combined analysis of neuroimaging and cerebrospinal fluid markers (tau, Aβ-42, and α-synuclein) is the most promising method for identifying cognitive dysfunction in LBD and PDD (Mak et al., 2015). A detailed investigation on systematic examination of hippocampal Lewy pathology and its distribution in hippocampal subfields in 95 clinically and neuropathologically characterized LBD patients has indicated that α-synuclein pathology is highest in two hippocampal-related subregions: the CA2 subfield and the entorhinal cortex (EC) (Adamowicz et al., 2017; Yang et al., 2018). While EC had numerous classic somatic LBs, CA2 contained mainly Lewy neurites in presumed axon terminals, suggesting the involvement of the EC → CA2 circuitry in the pathogenesis of LBD symptoms. There is a correlation between the measurements of verbal and visual memory with EC, but not the CA2 subfield. However, this subfield does not contribute to memory deficits. Lewy pathology in the CA1 subfield—the main output region for CA2—correlates best with results from memory testing despite a milder pathology, indicating that CA1 may be more functionally relevant than CA2 in the context of memory impairment in LBD. These correlations remain significant after controlling for several factors, including concurrent Alzheimer's pathology (neuritic plaques and NFTs) and the interval between time of testing and time of death (Adamowicz et al., 2017; Yang et al., 2018). It is also reported that α-synuclein-containing inclusions (small α-synuclein aggregates or oligomers) found in the hippocampus may be the real culprit responsible for causing deficits in neurotransmission and neurogenesis. This may constitute the major mechanism for the hippocampal dysfunctions and associated neuropsychiatric manifestations in various synucleinopathies (Adamowicz et al., 2017; Yang et al., 2018).

CONCLUSION

LBD, PD, and PDD are progressive neurodegenerative disorders that are characterized by progressive decline of motor and nonmotor functions such as bradykinesia, rigidity, tremor, and postural instability. LBD, PD, and PDD are multifactorial diseases, in which age, genetics, and environmental toxins are all considered significant risk factors. The overexpression or mutation of α-synuclein has been identified as a major genetic factor associated with PD. Under physiological conditions, α-synuclein functions in its native conformation as a soluble monomer. However, brains from LBD, PD, and PDD patients are characterized by intracellular inclusions of insoluble α-synuclein fibrils. Yet, oligomers and protofibrils of α-synuclein have been identified to be the most toxic species, with their accumulation at presynaptic terminals affecting several steps of neurotransmitter release. These presynaptic alterations induced by the accumulation of α-synuclein, together impair dopamine exocytosis and neuronal communication. Although α-synuclein is expressed throughout the brain and enriched at presynaptic terminals, dopamine neurons are the most vulnerable in LBD, PD, and PDD because α-synuclein directly regulates dopamine levels. Indeed, evidence suggests that α-synuclein is a negative modulator of dopamine by inhibiting enzymes responsible for its synthesis. There are a number of neurologic conditions that mimic these diseases, making it difficult to diagnose LBD, PD, and PDD in their early stages. Deposition of β-amyloid is a frequent feature of DLB strongly affecting clinical manifestations. In PDD, the duration of parkinsonism before dementia is associated with different patterns of brain pathology and neurochemical abnormalities. Furthermore, inflammation, which occurs in LBD, PD, and PDD is induced by elevated levels of proinflammatory cytokines and chemokines. At the molecular level these conditions (LBD, PD, and PDD) are not only accompanied by the accumulation of α-synuclein, but also induction of mitochondrial dysfunction-mediated oxidative stress, and neuroinflammation, which play causative roles in LBD, PD, and PDD. Collective evidence suggests that LBD, PD, and PDD are clinically similar neurological disorders, distinguished on the basis of the relative timing of dementia and parkinsonism (the 1-year rule). In view of the heterogeneity of the clinical course and symptomatology, these disorders share the same neurochemistry and pathophysiology. More studies are needed on biomarkers, new molecular imaging tracers, and multimodal imaging to understand their pathophysiology.

References

Aarsland, D., Perry, E., Perry, R., Larsen, J., Ballard, C., 2006. Duration of Parkinsonism prior to dementia is associated with a different pattern of neuropathological and neurochemical substrates in DLB and PDD (Poster Sessions: Clinical Science). Mov. Disord. 21, S96.

Adamowicz, D.H., Roy, S., Salmon, D.P., Galasko, D.R., Hansen, L.A., et al., 2017. Hippocampal α-synuclein in dementia with lewy bodies contributes to memory impairment and is consistent with spread of pathology. J. Neurosci. 37, 1675–1684.

Alberio, T., McMahon, K., Cuccurullo, M., Gethings, L.A., Lawless, C., et al., 2014. Verification of a Parkinson's disease protein signature in T-lymphocytes by multiple reaction monitoring. J. Proteome Res. 13, 3554–3561.

Alexander, G.E., 2004. Biology of Parkinson's disease: pathogenesis and pathophysiology of a multisystem neurodegenerative disorder. Dialogues Clin. Neurosci. 6, 259–280.

Arshad, A.R., Sulaiman, S.A., Saperi, A.A., Jamal, R., Mohamed Ibrahim, N., et al., 2017. MicroRNAs and target genes as biomarkers for the diagnosis of early onset of parkinson disease. Front. Mol. Neurosci. 10, 352.

Ashraghi, M.R., Pagano, G., Polychronis, S., Niccolini, F., Politis, M., 2016. Parkinson's disease, diabetes and cognitive impairment. Recent Pat. Endocr. Metab. Immune Drug Discov. 10, 11–21.

Baba, Y., Kuroiwa, A., Uitti, R.J., Wszolek, Z.K., Yamada, T., 2005. Alterations of T-lymphocyte populations in Parkinson disease. Parkinsonism Relat. Disord. 11, 493–498.

Bal-Price, A., Matthias, A., Brown, G.C., 2002. Stimulation of the NADPH oxidase in activated rat microglia removes nitric oxide but induces peroxynitrite production. J. Neurochem. 80, 73–80.

Bandres-Ciga, S., Cookson, M.R., 2017. Alpha-synuclein triggers T-cell response. Is Parkinson's disease an autoimmune disorder? Mov. Disord. 32, 1327.

Barbour, R., Kling, K., Anderson, J., Banducci, K., Cole, T., et al., 2008. Red blood cells are the major source of α-synuclein in blood. Neurodegener. Dis. 5, 55–59.

Beach, T.G., Adler, C.H., Lue, L., et al., 2009. Unified staging system for Lewy body disorders: correlation with nigrostriatal degeneration, cognitive impairment and motor dysfunction. Acta Neuropathol. 117, 613–634.

Beal, M.F., 2005. Mitochondria take center stage in aging and neurodegeneration. Ann. Neurol. 58, 495–505.

Benskey, M.J., Perez, R.G., Manfredsson, F.P., 2016. The contribution of alpha synuclein to neuronal survival and function - implications for Parkinson's disease. J. Neurochem. 137, 331–359.

Berg, D., 2008. Biomarkers for the early detection of Parkinson's and Alzheimer's disease. Neurodegener Dis. 5, 133–136.

Berman, S.B., Hastings, T.G., 1999. Dopamine oxidation alters mitochondrial respiration and induces permeability transition in brain mitochondria: implications for Parkinson's disease. J. Neurochem. 73, 1127–1137.

Bisaglia, M., Tessari, I., Mammi, S., Bubacco, L., 2009. Interaction between a-synuclein and metal ions still looking for a role in the pathogenesis of Parkinson's disease. Neuro Mol. Med. 11, 239–251.

Biswas, S., Chida, A.S., Rahman, I., 2006. Redox modifications of protein-thiols: emerging roles in cell signaling. Biochem. Pharmacol. 71, 551–564.

Blanz, J., Saftig, P., 2016. Parkinson's disease: acid-glucocerebrosidase activity and alpha-synuclein clearance. J. Neurochem. 139 (Suppl. 1), 198–215.

Blesa, J., Trigo-Damas, I., Quiroga-Varela, A., Jackson-Lewis, V.R., 2015. Oxidative stress and Parkinson's disease. Front. Neuroanatomy 9.

Bollaerts, I., Van Houcke, J., Andries, L., De Groef, L., Moons, L., 2017. Neuroinflammation as fuel for axonal regeneration in the injured vertebrate central nervous system. Med. Inflamm. 9478542.

Bosco, D., Plastino, M., Cristiano, D., Colica, C., Ermio, C., et al., 2012. Dementia is associated with insulin resistance in patients with Parkinson's disease. J. Neurol. Sci. 315, 39–43.

Bose, A., Beal, M.F., 2016. Mitochondrial dysfunction in Parkinson's disease. J. Neurochem. 139 (Suppl 1), 216–231.

Bove, J., Prou, D., Perier, C., Przedborski, S., 2005. Toxin-induced models of Parkinson's disease. NeuroRx. 2, 484–494.

Bridi, J.C., Hirth, F., 2018. Mechanisms of α-synuclein induced synaptopathy in Parkinson's disease. Front. Neurosci. 12, 80.

Brochard, V., Combadiere, B., Prigent, A., Laouar, Y., Perrin, A., et al., 2009. Infiltration of CD4 + lymphocytes into the brain contributes to neurodegeneration in a mouse model of Parkinson disease. J. Clin. Invest. 119, 182–192.

Burré, J., Sharma, M., Tsetsenis, T., Buchman, V., Etherton, M.R., et al., 2010. Alpha-synuclein promotes SNARE-complex assembly in vivo and in vitro. Science 329, 1663–1667.

Burton, E.J., Barber, R., Mukaetova-Ladinska, E.B., Robson, J., Perry, R.H., et al., 2009. Medial temporal lobe atrophy on MRI differentiates Alzheimer's disease from dementia with Lewy bodies and vascular cognitive impairment: a prospective study with pathological verification of diagnosis. Brain 132, 195–203.

Bus, J.S., Gibson, J.E., 1984. Paraquat: model for oxidant-initiated toxicity. Environ. Health Perspect. 55, 37–46.

Cannon, J.R., Greenamyre, J.T., 2011. The role of environmental exposures in neurodegeneration and neurodegenerative diseases. Toxicol. Sci. 124, 225–250.

Carboni, E., Lingor, P., 2015. Insights on the interaction of alpha-synuclein and metals in the pathophysiology of Parkinson's disease. Metallomics 7, 395–404.

Cebrian, C., Zucca, F.A., Mauri, P., Steinbeck, J.A., Studer, L., et al., 2014. MHC-I expression renders catecholaminergic neurons susceptible to T-cell-mediated degeneration. Nat. Commun. 5, 3633.

Chauhan, A., Jeans, A.F., 2015. Is Parkinson's disease truly a prion-like disorder? An appraisal of current evidence. Neurol. Res. Int. 2015, 345285.

Chen, X., de Silva, H.A., Pettenati, M.J., Rao St, P.N., George-Hyslop, P., et al., 1995. The human NACP/alpha-synuclein gene: chromosome assignment to 4q21.3–q22 and TaqI RFLP analysis. Genomics 26, 425–427.

Chung, C.Y., Koprich, J.B., Siddiqi, H., Isacson, O., 2009. Dynamic changes in presynaptic and axonal transport proteins combined with striatal neuroinflammation precede dopaminergic neuronal loss in a rat model of AAV alpha-synucleinopathy. J. Neurosci. 29, 3365–3373.

Clark, T.A., Lee, H.P., Rolston, R.K., et al., 2010. Oxidative stress and its implications for future treatments and management of Alzheimer disease. Int. J. Biomed. Sci. 6, 225–227.

Cookson, M.R., 2010. The role of leucine-rich repeat kinase 2 (LRRK2) in Parkinson's disease. Nat. Rev. Neurosci. 11, 791–797.

Dauer, W., Przedborski, S., 2003. Parkinson's disease: mechanisms and models. Neuron 39, 889–909.

Davie, C.A., 2008. A review of Parkinson's disease. Br. Med. Bull. 86, 109–127.

Dawson, T.M., Dawson, V.L., 2003. Molecular pathways of neurodegeneration in Parkinson's disease. Science 302, 819–822.

Day, M., Wang, Z., Ding, J., et al., 2006. Selective elimination of glutamatergic synapses on striatopallidal neurons in Parkinson disease models. Nat. Neurosci. 9, 251–259.

DeMaagd, G., Philip, A., 2015. Parkinson's disease and its management: part 1: disease entity, risk factors, pathophysiology, clinical presentation, and diagnosis. Pharm. Therap. 40, 504–532.

Desplats, P., Lee, H.-J., Bae, E.-J., Patrick, C., Rockenstein, E., et al., 2009. Inclusion formation and neuronal cell death through neuron-to-neuron transmission of α-synuclein. Proc. Natl. Acad. Sci. USA 106, 13010–13015.

Dias, V., Junn, E., Mouradian, M.M., 2013. The role of oxidative stress in Parkinson's disease. J. Parkinsons Dis. 3, 461–491.

Edison, P., Rowe, C.C., Rinne, J.O., Ng, S., Ahmed, I., et al., 2008. Amyloid load in Parkinson's disease dementia and Lewy body dementia measured with [11C]PIB positron emission tomography. J. Neurol. Neurosurg. Psych. 79, 1331–1338.

Farooqui, A.A., 2012. Phytochemicals, Signal Transduction, and Neurological Disorders. Springer, New York.

Farooqui, A.A., 2014. Inflammation and Oxidative Stress in Neurological Disorders. Springer International Publishing, Switzerland.

Farooqui, A.A., 2018. Neurochemical Aspects of Alzheimer's Disease. Academic Press, an imprint of Elsevier, San Diego, CA.

Faustini, G., Bono, F., Valerio, A., Pizzi, M., Spano, P., et al., 2017. Mitochondria and α-synuclein: friends or foes in the pathogenesis of Parkinson's disease? Genes (Basel) 8, E377.

Floor, E., Wetzel, M.G., 1998. Increased protein oxidation in human substantia nigra pars compacta in comparison with basal ganglia and prefrontal cortex measured with an improved dinitrophenylhydrazine assay. J. Neurochem. 70, 268–275.

Forman, H.J., Fukuto, J.M., Torres, M., 2004. Redox signaling: thiol chemistry defines which reactive oxygen and nitrogen species can act as second messengers. Am. J. Physiol. Cell Physiol. 287, C246–C256.

Francis, P.T., 2009. Biochemical and pathological correlates of cognitive and behavioural change in DLB/PDD. J. Neurol. 256, 280–285.

Franco-Iborra, S., Vila, M., Perier, C., 2016. The Parkinson disease mitochondrial hypothesis: where are we at? Neuroscientist. 22, 266–277.

Fratelli, M., Goodwin, L.O., Ørom, U.A., et al., 2005. Gene expression profiling reveals a signaling role of glutathione in redox regulation. Proc. Natl. Acad. Sci. USA. 102, 13998–14003.

Fujishiro, H., Ferman, T.J., Boeve, B.F., Smith, G.E., Graff-Radford, N.R., et al., 2008. Validation of the neuropathologic criteria of the third consortium for dementia with Lewy bodies for prospectively diagnosed cases. J. Neuropathol. Exp. Neurol. 67, 649–656.

Fujita, M., Sekigawa, A., Sekiyama, K., Takamatsu, Y., Hashimoto, M., 2012. Possible alterations in β-synuclein, the non-amyloidogenic homologue of α-synuclein, during progression of sporadic α-synucleinopathies. Int. J. Mol. Sci. 13, 11584–11592.

Gao, L., Tang, H., Nie, K., Wang, L., Zhao, J., et al., 2015. Cerebrospinal fluid alpha-synuclein as a biomarker for Parkinson's disease diagnosis: a systematic review and meta-analysis. Int. J. Neurosci. 125, 645–654.

Goldenberg, M.M., 2008. Medical management of Parkinson's disease. P & T: Peer-Rev. J. Formulary Manag. 33, 590–606.

Goldstein, D.S., Holmes, C., Sullivan, P., Jinsmaa, Y., Kopin, I.J., et al., 2016. Elevated cerebrospinal fluid ratios of cysteinyl-dopamine/3,4-dihydroxyphenylacetic acid in parkinsonian synucleinopathies. Parkinsonism Relat. Disord. 31, 79–86.

Gómez-Tortosa, E., Irizarry, M.C., Gómez-Isla, T., Hyman, B.T., 2000. Clinical and neuropathological correlates of dementia with Lewy bodies. Ann. N Y Acad. Sci 920, 9–15.

Greenamyre, J.T., Betarbet, R., Sherer, T.B., 2003. The rotenone model of Parkinson's disease: genes, environment and mitochondria. Parkinsonism Relat. Disord 9 (Suppl. 2), S59–S64.

Griendling, K.K., Sorescu, D., Lassègue, B., Ushio-Fukai, M., 2000. Modulation of protein kinase activity and gene expression by reactive oxygen species and their role in vascular physiology and pathophysiology. Arterioscler Thromb. Vasc. Biol. 20, 2175–2183.

Gu, F., Chauhan, V., Chauhan, A., 2015. Glutathione redox imbalance in brain disorders. Curr. Opin. Clin. Nutr. Metab. Care. 18, 89–95.

Halliwell, B., 2006. Reactive species and antioxidants. Redox biology is a fundamental theme of aerobic life. Plant Physiol. 141, 312–322.

Hancock, J.T., Desikan, R., Neill, S.J., 2001. Role of reactive oxygen species in cell signalling pathways. Biochem. Soc. Transac. 29, 345–350.

Hanganu, A., Bedetti, C., Jubault, T., Gagnon, J.F., Mejia-Constain, B., et al., 2013. Mild cognitive impairment is linked with faster rate of cortical thinning in patients with Parkinson's disease longitudinally. Mov. Disord. 28, 1360–1369.

Harms, A.S., Thome, A.D., Yan, Z., Schonhoff, A.M., Williams, G.P., et al., 2018. Peripheral monocyte entry is required for alpha-synuclein induced inflammation and neurodegeneration in a model of Parkinson disease. Exp. Neurol. 300, 179–187.

Hashimoto, M., Kawahara, K., Bar-On, P., Rockenstein, E., Crews, L., et al., 2004. The role of alpha-synuclein assembly and metabolism in the pathogenesis of Lewy body disease. J. Mol. Neurosci. 24, 343–352.

Hely, M.A., Reid, W.G.J., Adena, M.A., Halliday, G.M., Morris, J.G.L., 2008. The Sydney multicenter study of Parkinson's disease: the inevitability of dementia at 20 years. Mov. Disord. 23, 837–844.

Hirsch, E.C., Vyas, S., Hunot, S., 2012. Neuroinflammation in Parkinson's disease. Parkinsonism Relat. Disord. 18, S210–S212.

Hornykiewicz, O., 2008. Basic research on dopamine in Parkinson's disease and the discovery of the nigrostriatal dopamine pathway: the view of an eyewitness. Neurodegener Dis. 5, 114–117.

Hsu, L.J., Sagara, Y., Arroyo, A., Rockenstein, E., Sisk, A., et al., 2000. Alpha-synuclein promotes mitochondrial deficit and oxidative stress. Am. J. Pathol 157, 401–410.

Hu, Q., Wang, G., 2016. Mitochondrial dysfunction in Parkinson's disease. Transl. Neurodegener. 5, 14.

Ince, P.G., 2011. Dementia with Lewy bodies and Parkinson's disease with dementia. In: Dickson, D.W., Weller, R.O. (Eds.), Neurodegeneration, the Molecular Pathology of Dementia and Movement Disorders. Wiley-Blackwell, Oxford, pp. 224–237.

Irvine, G.B., El-Agnaf, O.M., Shankar, G.M., Walsh, D.M., 2008. Protein aggregation in the brain: the molecular basis for Alzheimer's and Parkinson's diseases. Mol. Med. 14, 451–464.

Itoh, K., Chiba, T., Takahashi, S., Ishii, T., Igarashi, K., et al., 1997. An Nrf2/small maf heterodimer mediates the induction of phase II detoxifying enzyme genes through antioxidant response elements. Biochem. Biophys. Res. Commun. 236, 313–322.

Iwai, A., Masliah, E., Yoshimoto, M., Ge, N., Flanagan, L., et al., 1995. The precursor protein of non-A beta component of Alzheimer's disease amyloid is a presynaptic protein of the central nervous system. Neuron 14, 467–475.

Jankovic, J., 2008. Parkinson's disease: clinical features and diagnosis. J. Neurol. Neurosurg. Psych. 79, 368–376.

Jellinger, K.A., 2012. Neurobiology of cognitive impairment in Parkinson's disease. Expert Rev. Neurother. 12, 1451–1466.

Jimenez-Ferrer, I., Jewett, M., Tontanahal, A., Romero-Ramos, M., Swanberg, M., 2017. Allelic difference in Mhc2ta confers altered microglial activation and susceptibility to alpha-synuclein-induced dopaminergic neurodegeneration. Neurobiol. Dis. 106, 279–290.

Jokinen, P., Scheinin, N., Aalto, S., Nagren, K., Savisto, N., et al., 2010. [(11)C]PIB-, [(18)F] FDG-PET and MRI imaging in patients with Parkinson's disease with and without dementia. Parkinsonism Relat. Disord. 16, 666–670.

Katsumoto, A., Lu, H., Miranda, A.S., Ransohoff, R.M., 2014. Ontogeny and functions of central nervous system macrophages. J. Immunol. 193, 2615–2621.

Keeney, P.M., Xie, J., Capaldi, R.A., Bennett Jr., J.P., 2006. Parkinson's disease brain mitochondrial complex I has oxidatively damaged subunits and is functionally impaired and misassembled. J. Neurosci. 26, 5256–5564.

Kempuraj, D., Thangavel, R., Natteru, P.A., Selvakumar, G.P., Saeed, D., et al., 2016. Neuroinflammation induces neurodegeneration. J. Neurol. Neurosurg. Spine. 1, 1003.

Kempuraj, D., Selvakumar, G.P., Zaheer, S., Thangavel, R., Ahmed, M.E., et al., 2018. Cross-talk between glia, neurons and mast cells in neuroinflammation associated with Parkinson's disease. J. Neuroimmune Pharmacol. 13, 100–112.

Kielian, T., 2016. Multifaceted roles of neuroinflammation: the need to consider both sides of the coin. J. Neurochem. 136, 5–9.

Kim, W.S., Kagedal, K., Halliday, G.M., 2014. Alpha-synuclein biology in Lewy body diseases. Alzheimers Res. Ther. 6, 73.

Klein, C., Westenberger, A., 2012. Genetics of Parkinson's disease. Cold Spring Harb. Perspect. Med. 2, a008888.

Knapp, L.T., Klann, E., 2002. Role of reactive oxygen species in hippocampal long-term potentiation: contributory or inhibitory? J. Neurosci. Res. 70, 1–7.

Kohen, R., Beit-Yannai, E., Berry, E.M., Tirosh, O., 1999. Overall low molecular weight antioxidant activity of biological fluids and tissues by cyclic voltammetry. Methods Enzymol. 300, 285–296.

Kramer, M.L., Schulz-Schaeffer, W.J., 2007. Presynaptic alpha-synuclein aggregates, not Lewy bodies, cause neurodegeneration in dementia with Lewy bodies. J. Neurosci. 27, 1405–1410.

Lee, H., Brekelmans, G.J., Roks, G., 2015. The EEG as a diagnostic tool in distinguishing between dementia with Lewy bodies and Alzheimer's disease. Clin. Neurophysiol. 126, 1735–1739.

Lee, H.J., Choi, C., Lee, S.J., 2002. Membrane-bound α-synuclein has a high aggregation propensity and the ability to seed the aggregation of the cytosolic form. J. Biol. Chem. 277, 671–678.

Lee, H.J., Suk, J.E., Patrick, C., Bae, E.J., Cho, J.H., et al., 2010. Direct transfer of α-synuclein from neuron to astroglia causes inflammatory responses in synucleinopathies. J. Biol. Chem. 285, 9262–9272.

Lee, H.J., Kang, S.J., Lee, K., Im, H., 2011. Human α-synuclein modulates vesicle trafficking through its interaction with prenylated rab acceptor protein 1. Biochem. Biophys. Res. Commun. 412, 526–531.

Lesage, S., Brice, A., 2009. Parkinson's disease: from monogenic forms to genetic susceptibility factors. Hum. Mol. Genet. 18, R48–R59.

Li, J., Zhu, M., Rajamani, S., Uversky, V.N., Fink, A.L., 2004. Rifampicin inhibits alpha-synuclein fibrillation and disaggregates fibrils. Chem. Biol. 11, 1513–1521.

Li, J., Liu, D., Sun, L., Lu, Y., Zhang, Z., 2012. Advanced glycation end products and neurodegenerative diseases: mechanisms and perspective. J. Neurol. Sci. 317, 1–5.

Li, S., Le, W., 2017. Biomarker discovery in Parkinson's disease: present challenges and future opportunities. Neurosci. Bull. 33, 481–482.

Linazasoro, G., 2007. Classical Parkinson disease versus Parkinson complex—reflections against staging and in favour of heterogeneity. Eur. J. Neurol. 14, 721–728.

Litvan, I., et al., 2011. MDS Task Force on mild cognitive impairment in Parkinson's disease: critical review of PD-MCI. Mov. Disord. 26, 1814–1824.

Liu, G.H., Qu, J., Suzuki, K., Nivet, E., Li, M., Montserrat, N., et al., 2012. Progressive degeneration of human neural stem cells caused by pathogenic LRRK2. Nature 491, 603–607.

Liu, J., Wang, X., Lu, Y., Duan, C., Gao, G., et al., 2017. Pink1 interacts with α-synuclein and abrogates α-synuclein-induced neurotoxicity by activating autophagy. Cell Death Dis. 8, e3056.

Logroscino, G., 2005. The role of early-life environmental risk factors in Parkinson disease: what is the evidence? Environ. Health Perspect 113, 1234–1238.

Lucas, J.J., Hernández, F., Gómez-Ramos, P., Morán, M.A., Hen, R., et al., 2001. Decreased nuclear beta-catenin, tau hyperphosphorylation and neurodegeneration in GSK-3beta conditional transgenic mice. EMBO J. 20, 27–39.

Lucas, S.M., Rothwell, N.J., Gibson, R.M., 2006. The role of inflammation in CNS injury and disease. Br. J. Pharmacol. 147, S232–S240.

Maes, M., Galecki, P., Chang, Y.S., Berk, M., 2011. A review on the oxidative and nitrosative stress (O&NS) pathways in major depression and their possible contribution to the (neuro)degenerative processes in that illness. Prog. Neuropsychopharmacol. Biol. Psychiatry 35, 676–692.

Maiti, P., Manna, J., Dunbar, G.L., 2017. Current understanding of the molecular mechanisms in Parkinson's disease: targets for potential treatments. Transl. Neurodegener. 6, 28.

Mak, E., Su, L., Williams, G.B., O'Brien, J.T., 2015. Neuroimaging correlates of cognitive impairment and dementia in Parkinson's disease. Parkinsonism Relat. Disord. 21, 862–870.

Martin, I., Dawson, V.L., Dawson, T.M., 2011. Recent advances in the genetics of Parkinson's disease. Ann. Rev. Genomics Hum. Genet. 12, 301–325.

McKeith, I., 2007. Dementia with Lewy bodies. In: Koller WC, Melamed E (eds.), Parkinson's Disease and Related Disorders, Part II. Handb. Clin. Neurol. 84:531–548.

McKeith, I.G., Boeve, B.F., Dickson, D.W., Halliday, G., Taylor, J.P., Weintraub, D., et al., 2017. Diagnosis and management of dementia with Lewy bodies: fourth consensus report of the DLB consortium. Neurology 89, 88–100.

Meeus, B., Nuytemans, K., Crosiers, D., Engelborghs, S., Peeters, K., et al., 2010. Comprehensive genetic and mutation analysis of familial dementia with Lewy bodies linked to 2q35-q36. J. Alzheimers Dis. 20, 197–205.

Merdes, A.R., Hansen, L.A., Jeste, D.V., Galasko, D., Hofstetter, C.R., et al., 2003. Influence of Alzheimer pathology on clinical diagnostic accuracy in dementia with Lewy bodies. Neurology 60, 1586–1590.

Meulener, M.C., Xu, K., Thompson, L., Ischiropoulos, H., Bonini, N.M., 2006. Mutational analysis of DJ-1 in Drosophila implicates functional inactivation by oxidative damage and aging. Proc. Natl. Acad. Sci. USA 103, 12517–12522.

Michel, P.P., Hirsch, E.C., Hunot, S., 2016. Understanding dopaminergic cell death pathways in Parkinson disease. Neuron 90, 675–691.

Mollenhauer, B., Cullen, V., Kahn, I., Krastins, B., Outeiro, T.F., Pepivani, I., et al., 2008. Direct quantification of CSF alpha-synuclein by ELISA and first cross-sectional study in patients with neurodegeneration. Exp. Neurol. 213, 315–325.

Mollenhauer, B., Schulz-Schaeffer, W., Schlossmacher, M., 2011. Synaptic alpha-synuclein pathology as the likely cause of Parkinson's disease dementia. Lancet Neurol. 10, 68–69.

Moors, T., Paciotti, S., Chiasserini, D., Calabresi, P., Parnetti, L., et al., 2016. Lysosomal dysfunction and alpha-synuclein aggregation in Parkinson's disease: diagnostic links. Mov. Disord. 31, 791–801.

Mukaetova-Ladinska, E.B., McKeith, I.G., 2006. Pathophysiology of synuclein aggregation in Lewy body disease. Mech. Ageing Dev. 127, 188–202.

Nagerl, U.V., Eberhorn, N., Cambridge, S.B., Bonhoeffer, T., 2004. Bidirectional activity-dependent morphological plasticity in hippocampal neurons. Neuron. 44, 759–767.

Nedelska, Z., Ferman, T.J., Boeve, B.F., Przybelski, S.A., Lesnick, T.G., et al., 2015. Pattern of brain atrophy rates in autopsy-confirmed dementia with Lewy bodies. Neurobiol. Aging. 36, 452–461.

Nikolaus, S., Antke, C., Muller, H.W., 2009. In vivo imaging of synaptic function in the central nervous system: I. Movement disorders and dementia. Behav. Brain Res. 204, 1–31.

Ohrfelt, A., Grognet, P., Andreasen, N., Wallin, A., Vanmechelen, E., Blennow, K., et al., 2009. Cerebrospinal fluid alpha-synuclein in neurodegenerative disorders-a marker of synapse loss? Neurosci. Lett. 450, 332–335.

Paleologou, K.E., Kragh, C.L., Mann, D.M., Salem, S.A., Al-Shami, R., et al., 2009. Detection of elevated levels of soluble α-synuclein oligomers in post-mortem brain extracts from patients with dementia with Lewy bodies. Brain 132, 1093–1101.

Phani, S., Loike, J.D., Przedborski, S., 2012. Neurodegeneration and inflammation in Parkinson's disease. Parkinsonism Relat. Disord. 18, 207–209.

Pieri, L., Madiona, K., Bousset, L., Melki, R., 2012. Fibrillar α-synuclein and huntingtin exon 1 assemblies are toxic to the cells. Biophys. J. 102, 2894–2905.

Pievani, M., de Haan, W., Wu, T., Seeley, W.W., Frisoni, G.B., 2011. Functional network disruption in the degenerative dementias. Lancet Neurol. 10, 829–843.

Ramonet, D., Daher, J.P., Lin, B.M., Stafa, K., Kim, J., Banerjee, R., et al., 2011. Dopaminergic neuronal loss, reduced neurite complexity and autophagic abnormalities in transgenic mice expressing G2019S mutant LRRK2. PLoS One 6, e18568.

Reeve, A., Simcox, E., Turnbull, D., 2014. Ageing and Parkinson's disease: why is advancing age the biggest risk factor? Ageing Res. Rev. 14, 19–30.

Reish, A., Standaert, D., 2005. Role of α-synuclein in inducing innate and adaptive immunity in Parkinson disease. Parkinson Dis. 5, 1–19.

Reynolds, A.D., Kadiu, I., Garg, S.K., Glanzer, J.G., Nordgren, T., et al., 2008. Nitrated alpha-synuclein and microglial neuroregulatory activities. J. Neuroimmune Pharmacol. 3, 59–74.

Reynolds, A.D., Stone, D.K., Mosley, R.L., Gendelman, H.E., 2009. Proteomic studies of nitrated alpha-synuclein microglia regulation by CD4 + CD25 + T cells. J. Proteome Res. 8, 3497–3511.

Roks, G., Korf, E.S., van der Flier, W.M., Scheltens, P., Stam, C.J., 2008. The use of EEG in the diagnosis of dementia with Lewy bodies. J. Neurol. Neurosurg. Psychiatry 79, 377–380.

Sato, Y., Kaji, M., Tsuru, T., et al., 2001. Risk factors for hip fracture among elderly patients with Parkinson's disease. J. Neurol. Sci. 182, 89–93.

Saunders, J.A., Estes, K.A., Kosloski, L.M., Allen, H.E., Dempsey, K.M., et al., 2012. CD4 + regulatory and effector/memory T cell subsets profile motor dysfunction in Parkinson's disease. J. Neuroimmune Pharmacol. 7, 927–938.

Savitt, J.M., Dawson, V.L., Dawson, T.M., 2006. Diagnosis and treatment of Parkinson disease: molecules to medicine. J. Clin. Invest. 116, 1744–1754.

Schwab, A.J., Ebert, A.D., 2015. Neurite aggregation and calcium dysfunction in iPSC-derived sensory neurons with Parkinson's disease-related LRRK2 G2019S mutation. Stem. Cell Reports 5, 1039–1052.

Sheridan, G.K., Murphy, K.J., 2013. Neuron-glia crosstalk in health and disease: fractalkine and CX3CR1 take centre stage. Open Biol. 3, 130181.

Shi, M., Liu, C., Cook, T.J., Bullock, K.M., Zhao, Y., et al., 2014. Plasma exosomal α-synuclein is likely CNS-derived and increased in Parkinson's disease. Acta Neuropathol. 128, 639–650.

Si, X., Pu, J., Zhang, B., 2017. Structure, distribution, and genetic profile of α-synuclein and their potential clinical application in Parkinson's disease. J. Mov. Disord. 10, 69–79.

Simón-Sánchez, J., Schulte, C., Bras, J.M., Sharma, M., Gibbs, J.R., et al., 2009. Genome-wide association study reveals genetic risk underlying Parkinson's disease. Nat. Genet. 41, 1308–1312.

Sohma, H., Imai, S.-I., Takei, N., Honda, H., Matsumoto, K., et al., 2013. Evaluation of annexin A5 as a biomarker for Alzheimer's disease and dementia with lewy bodies. Front. Aging Neurosci. 5, 15.

Srinivas, B.M.M., Mythri, R., Jagatha, B., Vali, S., 2008. Neuroprotective effect of curcumin against inhibition of mitochondrial complex I in vitro and in vivo. Implications for Parkinson's disease explained via in silico studies. J. Neurochem. 106, 9–28.

Stevens, C.H., Rowe, D., Morel-Kopp, M.C., Orr, C., Russell, T., et al., 2012. Reduced T helper and B lymphocytes in Parkinson's disease. J. Neuroimmunol. 252, 95–99.

Stojkovska, I., Krainc, D., Mazzulli, J.R., 2018. Molecular mechanisms of α-synuclein and GBA1 in Parkinson's disease. Cell Tissue Res. 373, 51–60.

Stolze, H., Klebe, S., Zechlin, C., Baecker, C., Friege, L., et al., 2004. Falls in frequent neurological diseases--prevalence, risk factors and aetiology. J. Neurol. 251, 79–84.

Streit, W.J., Xue, Q.S., 2016. Microglia in dementia with Lewy bodies. Brain Behav. Immun. 55, 191–201.

Stuendl, A., Kunadt, M., Kruse, N., Bartels, C., Moebius, W., Danzer, K.M., et al., 2016. Induction of α-synuclein aggregate formation by CSF exosomes from patients with Parkinson's disease and dementia with Lewy bodies. Brain 139 (Pt 2), 481–494.

Subramaniam, S.R., Chesselet, M.F., 2013. Mitochondrial dysfunction and oxidative stress in Parkinson's disease. Prog. Neurobiol. 106–107, 17–32.

Sui, Y.T., Bullock, K.M., Erickson, M.A., Zhang, J., Banks, W.A., 2014. α synuclein is transported into and out of the brain by the blood-brain barrier. Peptides 62, 197–202.

Sulzer, D., Alcalay, R.N., Garretti, F., Cote, L., Kanter, E., et al., 2017. T cells from patients with Parkinson's disease recognize alpha-synuclein peptides. Nature 546, 656–661.

Sun, G.Y., Horrocks, L.A., Farooqui, A.A., 2007. The roles of NADPH oxidase and phospholipases A_2 in oxidative and inflammatory responses in neurodegenerative diseases. J. Neurochem. 103, 1–16.

Surendranathan, A., Rowe, J.B., O'Brien, J.T., 2015. Neuroinflammation in Lewy body dementia. Parkinsonism Relat. Disord. 21, 1398–1406.

Syrjala, P., Luukinen, H., Pyhtinen, J., Tolonen, U., 2003. Neurological diseases and accidental falls of the aged. J. Neurol. 250, 1063–1069.

Teixeira Dos Santos, M.C., Bell, R., da Costa, A.N., 2016. Recent developments in circulating biomarkers in Parkinson's disease: the potential use of miRNAs in a clinical setting. Bioanalysis 8, 2497–2518.

Tokuda, T., Qureshi, M.M., Ardah, M.T., Varghese, S., Shehab, S.A., et al., 2010. Detection of elevated levels of α-synuclein oligomers in CSF from patients with Parkinson disease. Neurology 75, 1766–1772.

Tolö, J., Taschenberger, G., Leite, K., Stahlberg, M.A., Spehlbrink, G., et al., 2018. Pathophysiological consequences of neuronal α-synuclein overexpression: impacts on ion homeostasis, stress signaling, mitochondrial integrity, and electrical activity. Front. Mol. Neurosci. 11, 49.

Trancikova, A., Tsika, E., Moore, D.J., 2012. Mitochondrial dysfunction in genetic animal models of Parkinson's disease. Antioxidants Redox Signaling. 16, 896–919.

Truban, D., Hou, X., Caulfield, T.R., Fiesel, F.C., Springer, W., 2017. PINK1, parkin, and mitochondrial quality control: what can we learn about Parkinson's disease pathobiology? J. Parkinsons Dis. 7, 13–29.

Ueda, K., Fukushima, H., Masliah, E., Xia, Y., Iwai, A., et al., 1993. Molecular cloning of cDNA encoding an unrecognized component of amyloid in Alzheimer disease. Proc. Natl. Acad. Sci. 90, 11282–11286.

Uversky, V.N., Li, J., Fink, A.L., 2001a. Metal-triggered structural transformations, aggregation, and fibrillation of human α-synuclein. A possible molecular NK between Parkinson's disease and heavy metal exposure. J. Biol. Chem. 276, 44284–44296.

Uversky, V.N., Li, J., Fink, A.L., 2001b. Pesticides directly accelerate the rate of alpha-synuclein fibril formation a possible factor in Parkinson's disease. FEBS Lett. 500, 105–108.

Van Der Flier, W.M., Pijnenburg, Y.A., Prins, N., Lemstra, A.W., Bouwman, F.H., et al., 2014. Optimizing patient care and research: the Amsterdam dementia cohort. J. Alzheimers Dis. 41, 313–327.

Vila, M., Vukosavic, S., Jackson-Lewis, V., Neystat, M., Jakowec, M., Przedborski, S., 2000. α-Synuclein up-regulation in substantia nigra dopaminergic neurons following administration of the parkinsonian toxin MPTP. J. Neurochem. 74, 721–729.

Volles, M.J., Lee, S.J., Rochet, J.C., Shtilerman, M.D., Ding, T.T., et al., 2001. Vesicle permeabilization by protofibrillar alpha-synuclein: implications for the pathogenesis and treatment of Parkinson's disease. Biochemistry 40, 7812–7819.

Wakabayashi, K., Tanji, K., Mori, F., Takahashi, H., 2007. The Lewy body in Parkinson's disease: molecules implicated in the formation and degradation of alpha-synuclein aggregates. Neuropathology 27, 494–506.

Walker, Z., Possin, K.L., Boeve, B.F., Aarsland, D., 2015. Lewy body dementias. Lancet 386, 1683–1697.
Wang, Q., Liu, Y., Zhou, J., 2015. Neuroinflammation in Parkinson's disease and its potential as therapeutic target. Transl. Neurodegener. 4, 19.
Wang, Y., Yang, Z., Le, W., 2017. Tiny but mighty: promising roles of MicroRNAs in the diagnosis and treatment of Parkinson's disease. Neurosci. Bull. 33, 543–551.
Williams-Gray, C.H., Foltynie, T., Brayne, C.E., Robbins, T.W., Barker, R.A., 2007. Evolution of cognitive dysfunction in an incident Parkinson's disease cohort. Brain 130, 1787–1798.
Winterbourn, C.C., Hampton, M.B., 2008. Thiol chemistry and specificity in redox signaling. Free Radic. Biol. Med. 45, 549–561.
Yang, L., Wang, H., Liu, L., Xie, A., 2018. The role of insulin/IGF-1/PI3K/Akt/GSK3β signaling in Parkinson's disease dementia. Front. Neurosci. 12, 73.
Yang, W., Yu, S., 2017. Synucleinopathies: common features and hippocampal manifestations. Cell Mol. Life Sci. 74, 1485–1501.
Yu, B.P., 1994. Cellular defenses against damage from reactive oxygen species. Physiol. Rev. 74, 139–162.
Zaltieri, M., Longhena, F., Pizzi, M., Missale, C., Spano, P., et al., 2015. Mitochondrial dysfunction and α-synuclein synaptic pathology in Parkinson's disease: who's on first? Parkinsons Dis. 2015, 108029.
Zhou, M., Xu, S., Mi, J., Uéda, K., Chan, P., 2013. Nuclear translocation of alpha-synuclein increases susceptibility of MES23.5 cells to oxidative stress. Brain Res. 1500, 19–27.

Further Reading

Beach, T.G., White, C.L., Hamilton, R.L., et al., 2008. Evaluation of alpha-synuclein immunohistochemical methods used by invited experts. Acta Neuropathol. 116, 277–288.
Cromarty, R.A., Elder, G.J., Graziadio, S., Baker, M., Bonanni, L., et al., 2016. Neurophysiological biomarkers for Lewy body dementias. Clin. Neurophysiol. 127, 349–359.
Emre, M., Aarsland, D., Brown, R., Burn, D.J., Duyckaerts, C., Mizuno, Y., et al., 2007. Clinical diagnostic criteria for dementia associated with Parkinson's disease. Mov. Disord. 22, 1689–1707.
Gomperts, S.N., Rentz, D.M., Moran, E., Becker, J.A., Locascio, J.J., et al., 2008. Imaging amyloid deposition in Lewy body diseases. Neurology. 71, 903–910.
Krogsgaard, M., Davis, M.M., 2005. How T cells 'see' antigen. Nat. Immunol. 6, 239–245.
Mosley, R.L., Hutter-Saunders, J.A., Stone, D.K., Gendelman, H.E., 2012. Inflammation and adaptive immunity in Parkinson's disease. Cold Spring Harb. Perspect. Med. 2, 009381.
Nalls, M.A., Duran, R., Lopez, G., Kurzawa-Akanbi, M., McKeith, I.G., et al., 2013. A multicenter study of glucocerebrosidase mutations in dementia with Lewy bodies. JAMA Neurol. 70, 727–735.

CHAPTER

5

Neurochemical Aspects of Vascular Dementia

INTRODUCTION

Vascular dementia is a heterogeneous and progressive neurocognitive disorder caused by reduction in blood flow to the brain. It is caused by a reduction in cerebral blood flow due to hemorrhagic, ischemic, and hypoxic injuries. Subtypes of vascular dementia include multiinfarct dementia (characterized by multiple small strokes), single infarct dementia (caused by a single major stroke that damages the hippocampus), small vessel disease (SVD), and vasculitic dementia in which patients additionally suffer from migraine-like headaches caused by inflammation of blood vessels (Fig. 5.1) (Venkat et al., 2015). Vascular dementia is not only accompanied by behavioral symptoms and locomotor abnormalities, but also autonomic dysfunction. In contrast, vascular cognitive impairment (VCI) is a group of cognitive disorders with vascular causes. VCI is caused by irreversible structural damage to the vascular system in the brain (Marshall and Lazar, 2011). Recent studies indicate that cerebral hypoperfusion can hinder the function of the brain before structural damage occurs (Marshall and Lazar, 2011). The latter is supported by the finding that nondemented patients with cardiovascular disease show cognitive decline (Okonkwo et al., 2010) and that in patients with heart failure cognitive functioning can be enhanced by improving cardiac function (Zuccala et al., 2005). The term VCI has generally superseded the term vascular dementia. The combination of Alzheimer's disease (AD) and vascular dementia pathological changes in the brain of older people is extremely common, making mixed dementia probably the most common type of dementia (Langa et al., 2004; Moorhouse and Rockwood, 2008). Vascular dementia is the second most common type of dementia, accounting for approximately 15%–20% of all dementia patients (Venkat et al., 2015; O'Brien and Thomas, 2015).

Molecular Mechanisms of Dementia
DOI: https://doi.org/10.1016/B978-0-12-816347-4.00005-2

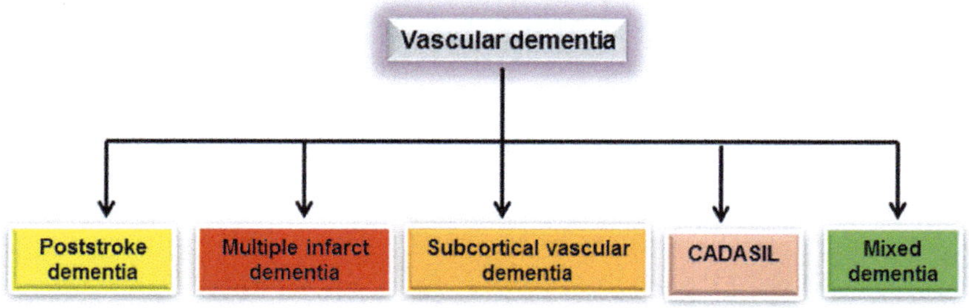

FIGURE 5.1 Classification of vascular dementia.

As stated above, vascular dementia results from inadequate blood supply to the brain caused by occlusion or rupture of cerebral arteries (Venkat et al., 2015), leading to impairment in cognitive abilities. Vascular dementia is accompanied by slow thinking, forgetfulness, depression and anxiety, disorientation, loss of executive functions, and induction of neurochemical changes that occur in AD and other neurodegenerative pathologies (Gorelick et al., 2011; Toledo et al., 2013). It is well known that there is a close link between cardiac and nervous system functions and under physiological conditions, cerebral blood flow is regulated by three major regulatory mechanisms: (1) cerebral autoregulation; (2) endothelium-dependent vasomotor function; and (3) neurovascular coupling response. Interactions among these mechanisms provide moment-to-moment adjustment of cerebral blood flow, preventing both cerebral hypo- and hyperperfusions in order to ensure adequate delivery of oxygen and nutrients to the brain. Given the high metabolic demand and insufficient energy reserve of neuron, it is not surprising that age-mediated cerebrovascular dysfunction or vascular injury can lead to significant consequences on brain functions and cognitive impairments (Tucsek et al., 2014). Thus, in cardiac surgery patients postoperative brain injury is a major concern. Postoperative brain injury may contribute to increased morbidity and mortality not only due to microemboli, increase in white matter lesions, SVD, microbleeds, cerebral infarcts, gray matter atrophy, and regional structural alterations, but also due to the induction in cerebral hypoperfusion, neuroinflammation, and increased amyloid disposition (Leritz et al., 2011; Richardson et al., 2012). Induction of the above changes in the brain contribute to the progression of cognitive decline (Snyder et al., 2015). In addition, vascular risk factors (hypertension, obesity, hyperlipidemia, diabetes, and the metabolic syndrome, MetS) also contribute to cognitive decline even in asymptomatic individuals (Friedman et al., 2014). Furthermore, genetic marker studies have indicated that ε4 allele of the apolipoprotein E (*APOE*) gene is a risk factor for both AD and

CVD (Adluru et al., 2014). These findings are supported by autopsy studies, which indicate that there is a synergistic link among heart disease, AD, vascular pathologies, and dementia (Kalaria, 2010). It is tempting to speculate that CVD leads to vascular dementia through an integrated, complex network of vascular, metabolic, and neural changes, which regulate not only cerebral blood flow, but may also contribute to the onset of dementia and cognitive decline.

New experimental findings have revealed the involvement of functional and pathogenic synergy between neurons, glia, and vascular cells, particularly endothelial cells (Iadecola, 2010; Quaegebeur et al., 2011; Zlokovic, 2011). Recently, the occurrence of the neurovascular unit (NVU) has been described in the literature. This unit comprises brain endothelial cells, pericytes or vascular smooth muscle cells, astrocytes, and neurons (Iadecola, 2010). It controls blood–brain barrier (BBB) permeability, cerebral blood flow, and maintains the chemical composition of the neuronal "milieu," which is required for proper functioning of neuronal circuits (Iadecola, 2010). The disrupted BBB may promote the leakage of plasma components and blood cells, eventually leading to perivascular inflammation, demyelination and gliosis. These findings provide the opportunity to reevaluate the possibility that alterations in cerebral blood vessels can contribute to new mechanisms of the neuronal dysfunction and cognitive impairment. Converging evidence suggests that vascular dementia not only leads to neuronal dysfunction, and neurodegeneration, but may also contribute to the development of SVDs and cerebrovascular storage disorders, such as cerebral β-amyloidosis and cerebral amyloid angiopathy (CAA), which are caused by accumulation of the peptide amyloid-β in the brain and the vessel wall, respectively, and are features of AD (Zlokovic, 2008). The presence of SVDs frequently occurs in the brains of elderly individuals and they become more prevalent and severe with advancing age (Jellinger and Attems, 2012). There are less common forms of cerebral SVDs, including various types of vasculitis and inherited diseases that affect vessel integrity, some of which are associated with the development of dementia in the absence of AD, for example, cerebral autosomal dominant arteriopathy with subcortical infarcts and leukoencephalopathy (CADASIL)—a monogenic form of cerebral SVD caused by mutations in the *Notch3* gene and associated with recurrent lacunar strokes and cognitive decline leading to dementia. (Chabriat et al., 2009; Jellinger, 2013; Kalaria, 2017). It should be noted that even in CADASIL patients cognitive deficits are detected even before the onset of stroke and dementia (Amberla et al., 2004), particularly in areas of attention, processing speed, and executive functions (Amberla et al., 2004; Buffon et al., 2006).

A reduction of greater than 80% in cerebral blood flow results in neuronal death due to ischemia (Moskowitz et al., 2010). Because of its diverse etiology, vascular dementia can be classified into various subtypes, including hypoperfusion dementia, subcortical vascular dementia, multiinfarct dementia, and strategic infarct dementia (O'Brien and Thomas, 2015; Iadecola, 2013; Jellinger, 2013). Another form of dementia called hereditary form of vascular dementia is caused by gene mutations. Factors defining subtypes of vascular dementias include the nature and extent of vascular pathologies, degree of involvement of extra and intracranial vessels and the anatomical location of tissue changes as well as the time after the initial vascular event (Lopez et al., 2005; Ferrari et al., 2017).

SMALL VESSEL DISEASE AND VASCULAR DEMENTIA

SVD is a group of intracranial disorders affecting the small arteries, arterioles, venules, and capillaries of the brain. SVD is not only a major contributor to stroke in humans but also an important cause of vascular and mixed dementia (Pantoni, 2010). The clinical manifestations of SVD include a wide range of symptoms including signs typical of stroke onset, neurological deficits ranging from mild to progressive cognitive decline, dementia, depression, and physical disabilities (Wardlaw et al., 2013a). Mechanisms underlying cognitive impairment in SVD remain largely unknown. However, it is proposed that SVD-related lesions such as small subcortical infarct, lacunes, white matter hyperintensities (WMHs), prominent perivascular spaces, cerebral microbleeds (CMBs), and atrophy affect structural brain connectivity and thereby modulate the efficiency of the brain network to process information (Wardlaw et al., 2013b). Several studies have indicated that the global network efficiency of the brain network to process information is related to the reduced processing speed and executive functioning in patients with SVD (Reijmer et al., 2015; Lawrence et al., 2014; Tuladhar et al., 2016). In these studies, associations between network efficiency and cognition are found to be stronger than between individual MRI markers of SVD and cognition (Patel and Markus, 2011; Sun et al., 2014). It should be noted that there is significant overlap in risk factors for AD and SVD. This makes their clinical differentiation often challenging (Rincon and Wright, 2014); thus, the estimated proportion of dementia caused by SVD ranges between 36% and 67% (Chui, 2001). Vascular risk factors contribute to the pathogenesis of SVD and hence, the development of cognitive impairment (Imamine et al., 2011; Iadecola, 2014). Among them, hypertension emerges as a major modifiable risk factor for cerebral complications. Pathological changes in blood pressure (BP) have been directly linked to cognitive decline. This finding initiates the controversial discussions about BP control as a potential therapeutic strategy to achieve optimal brain perfusion and thus, reduce the occurrence of a mild stage of cognitive impairment preceding both AD and vascular dementia (Chan et al., 2013; Lopez et al., 2003). Yet, the underlying mechanisms linking hypertension to cognitive decline and specifically to SVD have not yet been fully elucidated, which makes the search for effective therapies quite difficult.

In recent years, SVD has been recognized as an important substrate for cognitive impairment and vascular dementia. SVD is not only characterized by arteriolosclerosis, lacunar infarcts, and cortical and subcortical microinfarcts, but also by diffuse white matter changes, which are associated with the loss of myelin and axonal abnormalities (Kalaria, 2017). The presence of lacunar infarcts and leukoaraiosis is another feature of vascular dementia. It is associated with arterial stiffness. Patients with leukoaraiosis have a higher pulse wave velocity which transmits an increased pulse pressure into the brain through the middle cerebral artery (Webb et al., 2012). These changes not only result in a decrease in motor performance and early impairment of attention and executive function, but also the slowing of information processing due to the development of lacunar infarcts or multiple microinfarcts in the basal ganglia, thalamus, and brainstem. Similar to AD, vascular dementia is also accompanied by lobal brain atrophy and focal degeneration of the cerebrum including medial temporal lobe atrophy. Studies on hereditary arteriopathies have provided insights into the mechanisms of vascular dementia, particularly how arteriolosclerosis, a major contributor of SVD, promotes cognitive impairment. Recently described validated neuropathology guidelines indicate that the best predictors of VCI are small or lacunar infarcts, microinfarcts, perivascular space dilation, myelin loss, arteriolosclerosis, and leptomeningeal CAA. While these substrates do not suggest high specificity, vascular dementia is likely defined by key neuronal and dendro-synaptic changes, resulting in executive dysfunction and related cognitive deficits. Collective evidence suggests that SVD produces brain injury in both the cortical and subcortical gray and white matter. It often coexists with atherosclerosis involving large extracranial vessels and embolic disease (Li et al., 2015). The heterogeneity of cerebrovascular disease makes it challenging to elucidate the neuropathological substrates and mechanisms of vascular dementia as well as VCI. Subcortical small vessel disease (SSVD) refers to pathological processes affecting a spectrum of subcortical vascular changes visible on Computed Tomography/Magnetic Resonance Imaging (CT/MRI) as white matter lesions, lacunes, and CMBs. Underlying vascular pathologies associated with SSVD include arteriolosclerosis, lipohyalinosis, fibroid necrosis, edema, and damage to the blood–cerebrospinal fluid and BBB. These changes contribute to chronic leakage of fluid and macromolecules in the white matter leading to neuroinflammation (Kalaria, 2016). Based on this information, it is proposed that more studies should be performed on molecular mechanisms and neuropathological changes to clearly define molecular mechanisms of microvascular diseases and their vascular substrates of various subtypes of vascular dementia (Kalaria, 2017).

RISK FACTORS FOR VASCULAR DEMENTIA

Many risk factors have been described for the pathogenesis of vascular dementia. Nonmodifiable risk factors include age, family history, sex, and genetics (presence of Apolipoprotein E ε4 gene). Modifiable risk factors are smoking, long-term consumption of a Western diet, high body mass index, physical inactivity, high BP, low levels of uric acid, high cholesterol, hypertension, type 2 diabetes, and MetS (Fig. 5.2) (Saunders et al., 1993; Farrer et al., 1997; Xu et al., 2016). Heart failure and atrial fibrillation are other risk factors for vascular dementia. Cardiac disease can cause or worsen cerebral hypoperfusion, creating a cellular energy crisis setting off a cascade of events leading not only to hypertension, but also to the production of toxic proteins (de la Torre, 2012). These risk factors not only trigger peripheral and neuroinflammation and oxidative/nitrosative

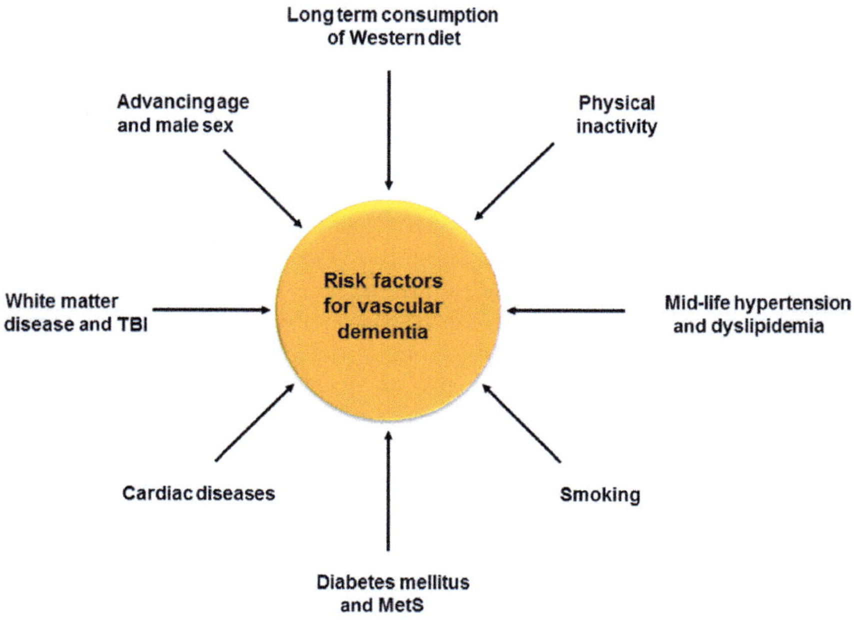

FIGURE 5.2 Risk factors contributing to vascular dementia and cognitive dysfunction in vascular dementia.

stress that in turn decrease nitric oxide and enhance endothelin, but also promote accumulation of Aβ, CAA, and BBB disruption (Farooqui, 2017). Induction of TNF-α, IL-1β, IL-6, endothelin-1, and induction of oxidative/nitrosative stress triggers several pathological feedforward and feedback loops. These upstream factors persist in the brain for decades, upregulating amyloid and tau, before the cognitive decline. These cascades not only promote neuronal Ca^{2+} increase, induction of neurodegeneration, but also a decline in cognitive function and memory, leading to dementia and AD (Farooqui, 2017). Collective evidence suggests that vascular dementia shares multiple risk factors with post-stroke dementia and AD type of dementia. Both these conditions frequently occur in seniors who have hypertension. This makes the differential diagnosis of these conditions difficult. The onset of vascular dementia is more sudden than AD and stroke-linked dementia. Among the above risk factors, chronic arterial hypertension is a major contributor to cognitive impairment (Gorelick et al., 2011). Hypertension affects an estimated 80 million people in the United States and 1 billion individuals worldwide (Mozaffarian et al., 2015). It is associated with insulin resistance, diabetes, and MetS, which are leading causes of global disease burden and overall health loss among seniors (Lim et al., 2012). The brain is one of the main target organs affected by hypertension. Thus, as stated above, hypertension is the most important risk factor for cerebrovascular pathology leading to stroke, vascular dementia, and AD. The harmful effects of hypertension on cognitive function were recognized in the early 1960s on the psychomotor speed of air traffic controllers and pilots, where individuals with hypertension show reduced performance (Elias et al., 2012). Hypertension is not only known to contribute to the executive dysfunction and slowing of mental processing speed, but also to memory deficits (Gąsecki et al., 2013). National Institutes of Health (NIH, 2010) organized a conference on risk factors for vascular dementia and AD and it was reported that there is sufficient evidence on a clinical level to support the association of any modifiable risk factors and vascular dementia (Farooqui, 2013). However, the evidence in many human studies (particularly with respect to dementia as opposed to cognitive decline) is inconclusive due in large part to the limited data collected to date and the limited number of clinical studies involving specific interventions (NIH, 2010). However, there is strong evidence from a population-based perspective, to conclude: (1) regular exercise and management of cardiovascular risk factors (diabetes, obesity, smoking, and hypertension) have been shown to reduce the risk of cognitive decline and may reduce the risk of vascular dementia; and (2) a healthy diet (original Mediterranean diet) and lifelong learning/cognitive training may also reduce the risk of vascular dementia and cognitive decline (Farooqui, 2013, 2015; Farooqui and Farooqui, 2018). The Institute of Medicine panel of distinguished researchers in the field has reached a virtually identical conclusion (Institute of Medicine, 2015; Baumgart et al., 2015). Collective evidence suggests that hypertension damages the cerebral tissues resulting in subcortical white matter lesions (leukoaraiosis), which not only contribute to the risk of vascular dementia, but also stroke. Increase in BP also contributes to more severe periventricular and subcortical white matter lesions (ischemic damage), and poorly controlled hypertension has an even higher risk of white matter lesions and thus cognitive impairment than those without hypertension, controlled hypertension, or untreated hypertension (van Dijk et al., 2004). In contrast, chronic hypotension (low BP) is accompanied by a variety of complaints including fatigue, reduced drive, faintness, dizziness, headaches, palpitations, and increased pain sensitivity (Duschek and Schandry, 2007). Physicians have generally ignored the effect of hypotension on cognition in clinical practice. One reason for this attitude is the current dogma that low systemic BP does not cause brain dysfunction because compensatory cerebral autoregulation prevents brain hypoperfusion from being activated (Duschek and Schandry, 2007). However, studies have confirmed, particularly in the elderly, that cerebral autoregulation does not necessarily protect the brain from chronic low BP and low cardiac output, an outcome that can result in cerebral blood flow insufficiency and its accompanying consequences (Kennelly and Collins, 2012).

The pathogenesis of vascular dementia remains unknown. However, it is becoming increasingly evident that multiple causes, including cerebrovascular disease and coexisting cardiovascular risk factors, such as aging of blood vessels, hypertension, atherosclerosis, insulin resistance, dyslipidemia, increased waist circumference, and stroke, play important roles in the etiopathogenesis of vascular dementia (Craft, 2009; Fillit et al., 2008; Abraham et al., 2016; Dearborn et al., 2015). There are few concerted studies on the protein and lipid chemistry of vascular dementia. It is suggested that alterations in vasculature, oxidative stress, neuroinflammation, apoptosis, and autophagy play a causative role in the pathogenesis of vascular dementia by virtue of their involvement in cerebral ischemia (Mulugeta et al., 2008; Li et al., 2014; Kalaria, 2016; Förstermann et al., 2017). It is also reported that the monocyte chemoattractant protein-1 and interleukin (IL)-6 concentrations are significantly decreased in the frontal lobe of vascular and mixed dementia subjects, suggesting that the induction of these changes have a vascular basis rather than being due to AD pathology.

Although many studies have been published on biomarkers of AD in CSF (Farooqui, 2017), such information is still lacking for vascular dementia. However, CSF has been used to determine elevated CSF/blood albumin ratio, alterations in CSF matrix metalloproteinase (MMP) activity, blood, and CSF inflammatory cytokines and adhesion molecules and it is reported that multimodal biomarkers are needed to identify vascular dementia (Rosenberg, 2016; Wallin et al., 2017). The multimodal approach involves the use of biochemical, neuroimaging, and clinical markers. Using this approach, it is reported that the CSF/blood albumin ratio is increased, the activity of MMP-2 is decreased, and levels of neurofilament (κL) are increased in the CSF of vascular dementia patients (Fig. 5.3) (Rosenberg, 2016; Wallin et al., 2017). However, CSF/blood albumin ratio may be nonspecific and may not distinguish vascular dementia from AD (Leblanc et al., 2006). An increase in MMP-2 activity in CSF indicates changes in the extracellular matrix (ECM) associated with vascular diseases with inflammation. Furthermore, the increase in neurofibrillary tangle levels indicate the axonal degeneration and the extent of white matter damage, which are characteristic of vascular dementia (Leblanc et al., 2006; Wallin and Sjögren, 2001). Sulfatide, a marker for myelin, has been used to identify the extent of demyelination in the white matter and it is found to be elevated in vascular dementia (Fredman et al., 1992). It is also reported that vascular dementia patients have dramatically lower serum uric acid (UA) levels in comparison to nondemented controls. Lower serum UA levels are linked to cognitive dysfunction and can serve as a potential predictor for vascular dementia (Xu et al., 2016). Another recent study has indicated that plasma microRNA can be used as a biomarker for vascular dementia (Prabhakar et al., 2017). It is reported that plasma miR-409-3p, miR-502-3p, miR-486-5p, and miR-451a can be used to differentiate small vessel vascular dementia patients from healthy controls (Fig. 5.3).

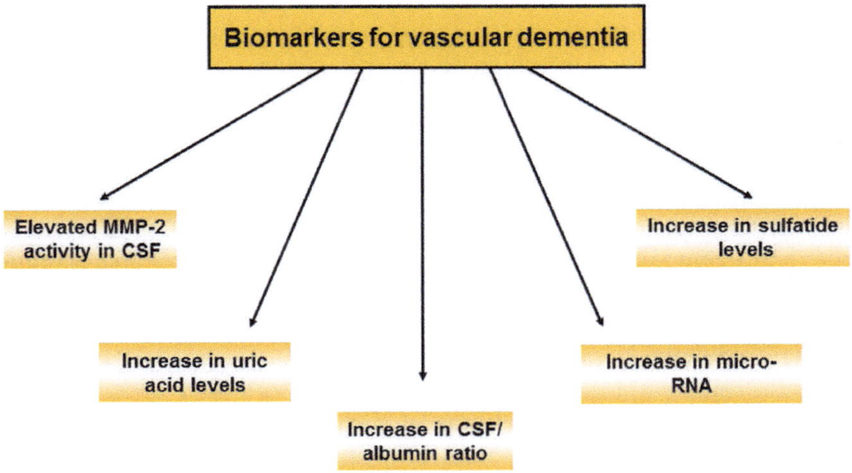

FIGURE 5.3 Hypothetical diagram showing neuropathological mechanisms contributing to vascular dementia.
ApoEε4, apolipoprotein Eε4; *BBB*, blood–brain barrier; *BDNF*, brain-derived neurotrophic factor; *LTP*, long-term potentiation.

Involvement of genes in the pathogenesis of vascular dementia has also been described (Forti et al., 2010; Marchesi, 2014; Gridley, 2007; Haritunians et al., 2005). These genes are classified into two classes: (1) genes that predispose individuals to cerebrovascular disease; and (2) genes that determine tissue responses to cerebrovascular disease (e.g., genes conveying ischemic tolerance or susceptibility, or the ability to recover from ischemic insult) (Forti et al., 2010). In the first category, genes that confer susceptibility to hypertension and atherosclerosis have been associated with some monogenic forms of disease such as CADASIL induced by mutations in NOTCH 3 gene. From the second category, genes that modify tissue responses to injury have also been identified including at least three sets of genes related to pathways involved in the pathogenesis of AD (presenilins, APP, and APOE). These genes are known to modulate the VCI disease pathway. The presenilin mutations associated with AD have been shown to interact directly with Notch proteins, including Notch 3 (mutations of which cause CADASIL) (Marchesi, 2014; Gridley, 2007; Haritunians et al., 2005).

DIAGNOSIS OF VASCULAR DEMENTIA

The diagnosis of vascular dementia is difficult due to the presence of various types and the number of lesions and their locations in the brain. Factors that increase the risk of vascular diseases, such as stroke, hypertension, high cholesterol, and smoking, also raise the risk of vascular dementia. Therefore, controlling these risk factors can help lower the chances of developing vascular dementia. At present there are two major issues regarding the assessment and diagnosis of vascular dementia. First, there are no currently accepted neuropathological criteria regarding the assessment of vascular dementia, VCI, and cerebrovascular pathology or related lesions (Grinberg and Heinsen, 2010). Thus, there is no generally accepted neuropathological criteria for the diagnosis of vascular dementia available. Second, general assumptions regarding the underlying pathology of frequently observed in vivo MRI findings may not always be accurate. Still, the diagnosis of vascular dementia relies on a good clinical history, supported by formal cognitive testing, to identify the subtype of dementia. As per the California criteria, diagnosis of vascular dementia not only requires neuropathological assessment, CT, positron-emission tomography (PET), MRI, and magnetic resonance spectroscopy, but also neuropsychiatric evaluation (Kirshner, 2009; Erkinjuntti and Gauthier, 2009). Among the abovementioned neuroimaging techniques, functional imaging can help to confirm neurodegeneration and to distinguish dementia subtypes when structural imaging has been inconclusive. Amyloid-PET scans reflect neuritic plaque burden and identify the earliest pathological changes in AD, but their value outside research settings is still uncertain. Structural neuroimaging has been used in most patients, not just to identify potentially reversible surgical pathology, but also to detect vascular changes and patterns of cerebral atrophy. Based on diffusion tensor imaging (DTI) studies, Palesi et al. have reported that AD can be distinguished from vascular dementia on the basis of changes in parahippocampal tracts (Palesi et al., 2018). It is reported that AD is accompanied by alterations in parahippocampal tracts, while vascular dementia patients show more widespread white matter damage associated with the involvement of thalamic radiations (Palesi et al., 2018). The genu of corpus callosum (cc) is predominantly affected in vascular dementia, while the splenium is predominantly affected in AD, revealing the existence of specific patterns of alteration that are useful in distinguishing between vascular dementia and AD. It is proposed that DTI parameters of these regions can be informative to understand the pathogenesis and support the etiological diagnosis of dementia (Palesi et al., 2018). Collective evidence suggests that accurate diagnosis of vascular dementia relies on wide-ranging clinical, neuropsychometric, and neuroimaging measures with subsequent pathological confirmation and more studies are needed on the molecular mechanisms and neuropathological to clearly define microvascular disease and vascular substrates of various subtypes of vascular dementia (Kalaria, 2017).

BIOCHEMICAL AND NEUROPATHOLOGICAL CHANGES IN VASCULAR DEMENTIA

Many biochemical mechanisms contribute to the development of vascular dementia. Among them lipid metabolism plays a vital role in the pathogenesis of vascular dementia. Thus, both high levels of low-density lipoprotein (LDL) cholesterol and low levels of high-density lipoprotein (HDL) cholesterol are major risk factors for development of carotid atherosclerosis coronary artery disease, SVD, and CAA (Reitz et al., 2004). These conditions not only promote cognitive impairment, but also cerebral hypoperfusion or embolism, leading to oxidative stress and insulin resistance (Fig. 5.4) (Breteler et al., 1994). HDL cholesterol may be involved in the removal of excess cholesterol from the brain mediated by APOE and heparin sulfate proteoglycans in the subendothelial space of cerebral microvessels (Mulder and Terwel, 1998). In addition, HDL particles reverse the inhibitory action of oxidized LDL

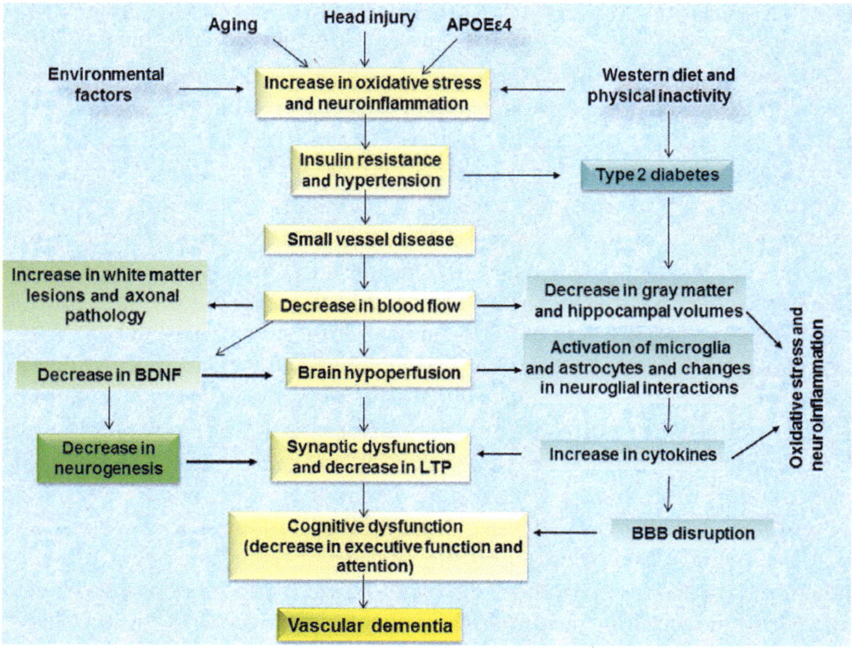

FIGURE 5.4 Factors and processes contributing to the pathogenesis of vascular dementia.

particles on endothelium-dependent arterial relaxation (Matsuda et al., 1993) and also inhibit cytokine-induced expression of endothelial cell adhesion molecules (Cockerill et al., 1995); both of which may be potential mechanisms in the development of vascular dementia. It is becoming increasingly evident that the NVU not only controls cerebral blood flow and BBB permeability, but also maintains the chemical composition of the neuronal "milieu," which modulates functioning of neuronal circuits. The uncoupling of the NVU can ultimately lead to mitochondrial dysfunction and oxidative stress, neuronal death, and brain tissue atrophy (Marlatt et al., 2008; Chen et al., 2009). The NVU also controls neuronal activity by modulating glucose uptake. A decreased cerebral glucose metabolism is an early event in the pathogenesis of vascular dementia and may precede the neuropathological changes associated with the pathogenesis of vascular dementia. Mild hypoperfusion during vascular dementia has been reported to decrease synaptic plasticity by regulating protein synthesis in the NVU (Fig. 5.5) (Iadecola, 2004). Thus, moderate to severe reduction in cerebral blood flow not only reduces ATP synthesis and diminishes (Na^+, K^+) ATPase activity, but also reduces synaptic plasticity by decreasing protein synthesis. These alterations inhibit the ability of neurons to generate action potentials

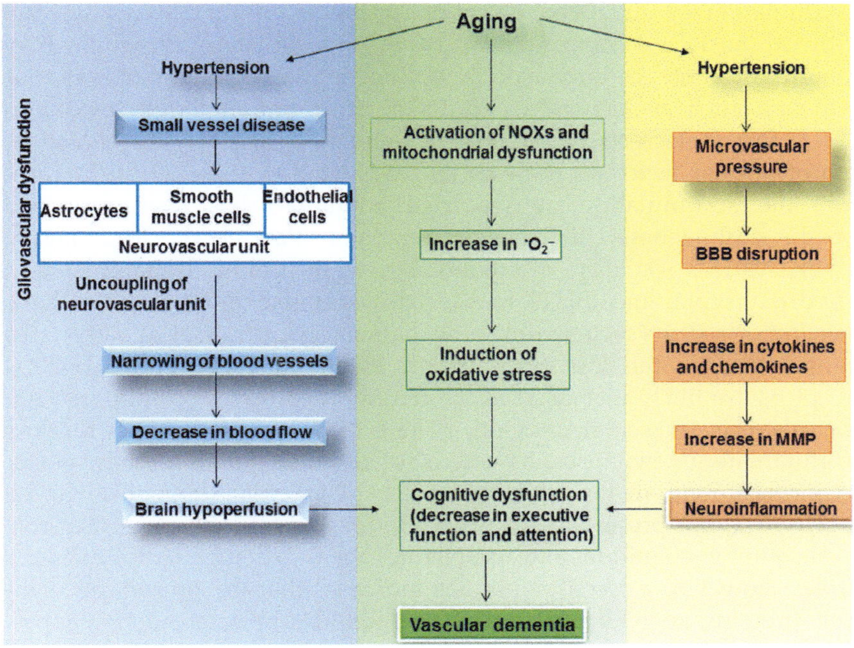

FIGURE 5.5 Involvement of the neurovascular unit, oxidative stress, and BBB disruption in the pathogenesis of vascular dementia.
BBB, blood–brain barrier; *MMP*, matrix metalloproteinase; *NOXs*, NADPH oxidases; O_2^-, superoxide.

producing alterations in ion homeostasis (Kalaria, 2010). In addition, a reduction in cerebral blood flow lowers the pH not only by altering electrolyte balances and water gradients, but also by promoting the development of cerebral edema, white matter lesions, and the accumulation of glutamate and proteinaceous toxins (amyloid-β and hyperphosphorylated tau). These processes contribute to excitotoxicity and oxidative stress in the brain. Furthermore, a decrease in cerebral blood flow also impairs the clearance of neurotoxic molecules that accumulate and/or are deposited in the interstitial fluid, nonneuronal cells, and neurons. These processes promote vascular dementia and cognitive impairments. In addition, progressive reductions in cerebral blood flow also cause serious consequences in neuronal function. In vascular dementia, cholinergic reductions correlate with cognitive impairment, and cholinesterase inhibitors have been reported to produce some beneficial effects, supporting the view that cerebral blood flow modulates neurotransmission (Kalaria, 2010; O'Brien and Thomas, 2015; Iadecola, 2013; Jellinger, 2013).

Other biochemical changes in vascular dementia involve mitochondrial oxidative stress, hypoxic/ischemia injury, neuroinflammation, and accumulation of advanced glycation products (AGEs). AGEs interact with cell surface receptors and downstream signal transduction, may help to promote certain aspects of the etiopathogenesis of vascular dementia such as increased expression of proinflammatory cytokines (IL-1β, tumor necrosis factor (TNF-α), and IL-6, along with the activation of nuclear factor-κB (NF-κB) (Fig. 5.5) (Jagtap et al., 2015). In addition, the accumulation of abnormal amyloid-β has been proposed for the pathogenesis of vascular dementia (Ray et al., 2013; Jagtap et al., 2015). Deficits in monoamines including dopamine and 5-hydroxytryptamine (5HT) in the basal ganglia and neocortex have been reported in vascular dementia patients (Gottfries et al., 1994). To compensate for the loss (Ellis et al., 1996), 5-HT(1A) and 5-HT(2A) receptors are likely increased in the temporal cortex in multiinfarct, but not subcortical vascular dementia. There have been reports on the loss of glutamatergic synapses, assessed by vesicular glutamate transporter 1 concentrations, in the temporal cortex of vascular dementia (Kirvell et al., 2011), but preservation of these in the frontal lobe suggests a role in sustaining cognition and protecting against dementia following a stroke. However, a recent study has indicated that the presynaptic synaptic proteins, such as a 313-amino acid, 38-kDa protein called synaptophysin and synaptosomal-associated protein 25, a protein associated with synaptic vesicle membrane docking and fusion, are reduced, whereas drebrin, a protein contributing to increase in dendritic length, size, and density (Ivanov et al., 2009) and in regulation of NMDR receptor (Lee and Aoki, 2012) is increased. This observation suggests that drebrin may be involved in a compensatory response to the ischemia caused by the vascular dementia.

OXIDATIVE STRESS IN VASCULAR DEMENTIA

In the brain, oxidative stress due to chronic cerebral hypoperfusion is considered to be the major risk factor in the pathogenesis of vascular dementia. As stated in earlier chapters, oxidative stress is caused by an environment where imbalance between the production of reactive oxygen species (ROS) and removal of ROS by antioxidant species is linked to the pathogenesis of dementia (Bennett et al., 2009). Major ROS-generating systems in the cardio- and cerebrovascular walls include NADPH oxidase, xanthine oxidase, the mitochondrial electron transport chain, and uncoupled endothelial nitric oxide (NO) synthase (Fig. 5.6) (Bennett et al., 2009). Oxidation of arachidonic acid (ARA) also generates highly electrophilic α,β-unsaturated carbonyl derivatives, including acrolein, 4-hydroxy-2-nonenal, and 4-oxononenal (LoPachin et al., 2009).

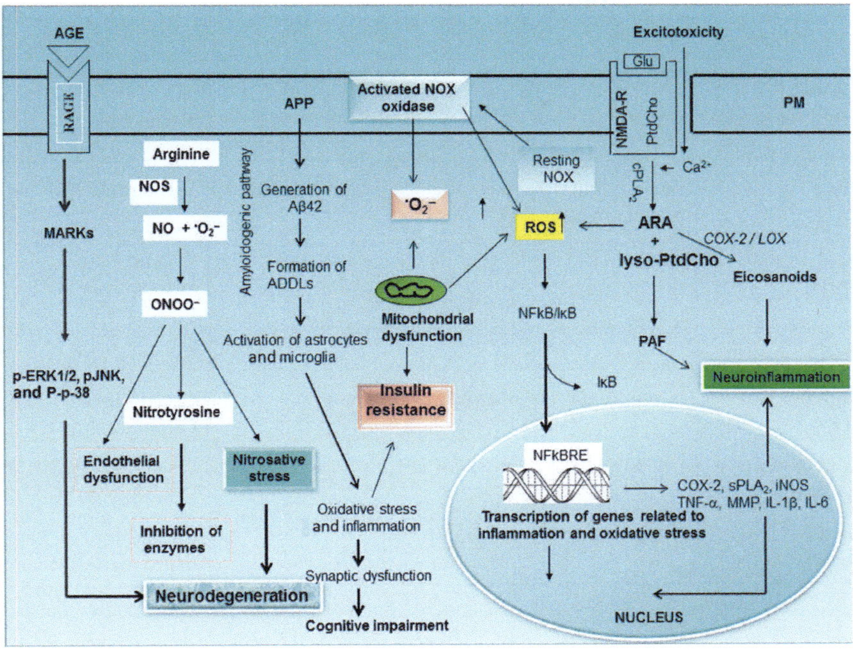

FIGURE 5.6 Diagram showing the induction of oxidative stress and neuroinflammation in vascular dementia.
Aβ, beta-amyloid; *ADDLs*, Aβ-derived diffusible ligands; *AGE*, advanced glycation end-product; *APP*, amyloid precursor protein; *ARA*, arachidonic acid; *COX*, cyclooxygenase; *cPLA₂*, cytosolic phospholipase A_2; *ERK*, extracellular signal regulated kinase; *IL-1β*, interleukin-1beta; *IL-6*, interleukin-6; *JNK*, Jun amino-terminal kinases; *LOX*, lipoxygenase; *lyso-PtdCho*, lysophosphatidylcholine; *MARK*, mitogen-activated protein kinase; *NF-κB*, nuclear factor-kappa B; *NF-κB-RE*, nuclear factor-kappa B response element; *NMDA-R*, N-methyl-D-aspartate receptors; *NOX*, nicotinamide adenine dinucleotide phosphate (NADPH)-oxidase; *ONOO⁻*, peroxynitrite; *PAF*, platelet activating factor; *PM*, plasma membrane; *PtdCho*, phosphatidylcholine; *ROS*, reactive oxygen species; *TNF-α*, tumor necrosis factor-alpha; *NO*, nitric oxide; *RAGE*, receptors for advanced glycation end-product.

In vascular dementia patients, levels of acrolein are significantly higher in several brain regions such as in the hippocampus, amygdala, middle temporal gyrus, and cerebellum (Williams et al., 2006; Bradley et al., 2010). In vitro studies have indicated that the toxicity of acrolein is higher than 4-hydroxy-2-nonenal in primary neuronal cultures from the hippocampus (Lovell et al., 2001). It is also shown that acrolein produces its toxicity not only through oxidative damage, and activation of several redox-sensitive pathways, but also via the induction of endoplasmic reticulum stress and disruption of tight junction protein (Chen et al., 2017). Although studies measuring markers of oxidative stress (increase in lipid peroxidation products, elevation in DNA oxidation products in

TABLE 5.1 Effects of High Oxidative Stress on Endothelial, Neurons, and Glial Cells

Extent of oxidative stress	Effects	Reference
High oxidative stress	Macromolecular damage to mitochondria, nucleus, endoplasmic reticulum, cytoplasm, and plasma membranes	Go and Jones (2014), Farooqui (2014)
High oxidative stress	Uncontrollable kinase-phosphates activation/deactivation	Go and Jones (2014), Farooqui (2014)
High oxidative stress	Inactivation of redox regulatory enzymes	Go and Jones (2014), Farooqui (2014)
High oxidative stress	Accumulation of misfolded protein	Go and Jones (2014), Farooqui (2014)
High oxidative stress	Uncoupling of neurovascular unit	Go and Jones (2014), Farooqui (2014)
High oxidative stress	Decrease in telomere length	Go and Jones (2014), Farooqui (2014)

CSF) specifically in vascular dementia are limited, there has been considerable evidence supporting the involvement of oxidative stress in vascular brain injury (Bennett et al., 2009). Oxidative stress-mediated injury to vascular endothelial cell, glia, and neuron not only impairs vascular function and NVU coupling, but also damages subcellular structures through several mechanisms, such as inactivation of redox regulatory enzymes, accumulation of misfolded proteins, and decrease in telomere length (Table 5.1). In addition to oxidative stress, the brain also undergoes nitrosative stress. During this process, superoxide radicals react with nitric oxide ($NO^{\bullet}$) to produce the peroxynitrite anion ($ONOO^{-}$), a nonradical product, which is as toxic as the $^{\bullet}OH$ in terms of its ability to oxidize and destroy bystander molecules. $NO^{\bullet}$ and $ONOO^{-}$ are often referred to as reactive nitrogen species (RNS) (Farooqui, 2014). Like ROS, RNS oxidize lipids and proteins components. Studies on ROS and RNS indicate that ROS/RNS are highly reactive and short-lived species that do not accumulate to significant levels and it is not possible to measure them directly; rather, one must measure either the accumulation of biomolecules or the exogenously added indicators that are modified by ROS and RNS (Farooqui, 2014). In other words, the generation of ROS and RNS may leave its footprint in the cell in the form of different oxidatively modified components. In clinical cases cerebral hypofunction may not only lead to induction of severe oxidative stress, but also result in cognitive dysfunction. Collectively, these studies suggest that cerebral hypoperfusion, oxidative stress, and neuroinflammation are key factors in the development of cognitive impairment (Liu and Zhang, 2012).

NEUROINFLAMMATION IN VASCULAR DEMENTIA

Neuroinflammation is a complex host defence mechanism that isolates the damaged brain tissue from the uninjured area, destroys injured cells, and repairs the ECM (Minghetti et al., 2005). Neuroinflammation is orchestrated by microglia and astrocytes reestablishing homeostasis in the brain after injury-mediated disequilibrium of normal physiology. Recently the role of inflammation in brain health has become a major focal point of studies related to aging and age-related neurological disorders (Farooqui, 2014). Activation of inflammatory pathways in the brain has been increasingly emphasized as a major risk factor for the initiation, development, and progression of the pathogenesis of various types of dementia (Farooqui, 2014). As stated earlier there are two types of neuroinflammation. Acute neuroinflammation develops rapidly with the experience of pain, whereas chronic inflammation develops slowly. Acute neuroinflammation is accompanied by rapid activation of microglia, damage to the BBB, and acute upregulation of proinflammatory cytokines such as IL-1β, TNF-α, and IL-6 (Schmidt et al., 2005; Farooqui, 2014). Chronic neuroinflammation differs from acute inflammation in that it is below the threshold of pain perception. As a result, the immune system continues to attack at the cellular level. Chronic inflammation lingers for years causing continued insult to the brain tissue before reaching the threshold of detection (Wood, 1998) and initiating the pathogenesis of chronic disease. Chronic inflammation disrupts hormonal signaling networks not only in the brain, but also in the visceral organs. In vascular dementia, the development of SVD in the brain produces a narrowing of the cerebral blood vessel leading to hypoperfusion and chronic hypertension. The onset of hypoperfusion causes the activation and degeneration of astrocytes inducing fibrosis of the ECM (Rosenberg, 2017). Elasticity is lost in fibrotic cerebral vessels, reducing the response of stiffened blood vessels in times of increased metabolic need (Rosenberg, 2017). In vascular dementia, intermittent hypoxic/ischemic injury activates a molecular injury cascade, producing an incomplete infarction that is most damaging to the deep white matter, which is a watershed region for cerebral blood flow. Neuroinflammation induced by hypoxic injury activates microglial cells to release proteases and free radicals that perpetuate the damage over time to molecules in the ECM and the NVU. It is proposed that MMPs, secreted in an attempt to remodel the blood vessel wall, have the undesired consequences of opening the BBB and attacking myelinated fibers.

This dual effect of the MMPs causes vasogenic edema in white matter and vascular demyelination, which are the hallmarks of vascular dementia and related diseases (Rosenberg, 2017),

Oxidative stress and neuroinflammation are closely interlinked processes in the brain.

As stated above, high levels of ROS/RNS may lead to the activation of the transcription factor NF-κB which induces the overexpression of NO synthases in astrocytes and microglia, in particular NADPH oxidase and iNOS, resulting in peroxynitrite production by superoxide and NO reaction producing neuronal damage (Morgan and Liu, 2011). Moreover, NF-κB activation induces the expression of COX-2 and cytosolic phospholipase A_2, which in turn stimulate the generation of prostaglandins, promoting inflammation and oxidative stress (Hsieh and Yang, 2013). Formation of peroxynitrite $ONOO^-$ also leads to protein nitration in enzymes, such as alpha and gamma enolases, which are implicated in brain glucose metabolism (Castegna et al., 2003). Thus, the signaling pathway NF-κB, which is also heavily involved in inflammatory reactions, has been proposed to be involved in oxidative stress, since a direct cross-talk between ROS and NF-κB has been reported (Turillazzi et al., 2016). It is difficult to establish the temporal sequence of their relationship between oxidative stress and neuroinflammation (Bryan et al., 2013). However, cross-talk between oxidative stress and neuroinflammation establishes a vicious circle between ROS production and cytokines and chemokines expression. Onset of chronic inflammation and oxidative stress not only leads to vascular dysfunction and heart disease, but also promotes the pathogenesis of type II diabetes and MetS, which are risk factors for stroke and chronic age-related neurodegenerative disorders, such as AD, Parkinson disease, and their related dementias (Farooqui, 2014).

ANIMAL MODELS FOR VASCULAR DEMENTIA

Several animal models have been developed to mimic the neuropathological and behavioral changes of vascular dementia (Venkat et al., 2015; Jiwa et al., 2010). Of these, the bilateral common carotid artery occlusion (BCCAO) model in rats is the most commonly used. In this model, white matter lesions manifest axonal and myelin injury, vacuolization, and glial cell activation (Venkat et al., 2015; Jiwa et al., 2010). Thus, the rat BCCAO model mimics hypoperfusion dementia and subcortical vascular dementia in that white matter injury is consistently observed in both subtypes (Iadecola, 2013; Jiwa et al., 2010). In one model of hypoperfusion, dementia is developed by partial or complete blockage of the carotid arteries (Iadecola, 2013; Jiwa et al., 2010). In humans, carotid artery stenosis or occlusion is found to be associated with white matter injury (Iadecola, 2013), which is observed in the regions including the cc (Yamauchi et al., 1993; Lin et al., 2016) and optic nerve (Terelak-Borys et al., 2012), a bundle of which forms the optic tract in the visual pathway. In the rat BCCAO model, optic nerve injury correlates with loss of the pupillary light reflex, a reflex that controls the pupil diameter in response to light intensity (Stevens et al., 2002). Conversely, in the subcortical vascular dementia, the most common form of vascular dementia, one etiology is caused by partial blockage of small vessels, which also leads to white matter injury (Iadecola, 2013; Román et al., 2002).

IMMUNE RESPONSES IN VASCULAR DEMENTIA

Immune responses have recently emerged as important elements contributing to the progression of many neurological disorders including vascular dementia. Brain lesions in vascular dementia are not only associated with the release of inflammatory mediators (TNF-α, IL-1β, IL-6) and lymphocytes entry, but also contribute to neurological deficits in acute cerebral hemorrhage (de Laat et al., 2011; van der Holst et al., 2017). Hypoxic/ischemic injury results in SVD, which is accompanied by BBB leakage (Grau-Olivares et al., 2010; Duering et al., 2012), central nervous system antigen release into the peripheral circulation, and lymphocyte infiltration into brain tissue, which allow for the possibility of novel antigens deprived from the brain to encounter the lymphocytes (Bene et al., 2013; Topol et al., 2003). In addition to BBB disruption, blood proteins at the NVU activate microglia to produce cytokines and chemokines, which cause peripheral inflammatory cells to migrate to the brain, creating a chronic inflammatory microenvironment and encouraging activated lymphocytes to encounter brain antigens (Lavallee et al., 2013; Allen and Bayraktutan, 2009). Immune responses in SVD have not been well characterized. However, it is proposed that immune responses may contribute to the pathogenesis of SVD-mediated injury in diseases such as multiple sclerosis and neuromyelitis optica, classic autoimmune disorders.

VASCULAR DEMENTIA AND COGNITIVE DYSFUNCTION

Vascular dementia is a type of cognitive disorder mediated by vascular abnormalities. The major hemodynamic alteration in this condition is a chronic and significant decrease in cerebral blood flow (Sabayan et al., 2012; O'Brien and Thomas, 2015), which is caused by diverse types of pathology such as atherosclerosis, arteriolosclerosis, infarcts, white matter changes, and microbleeds. The human heart uses the left ventricle to perfuse blood to the brain via central elastic arteries, which stiffen with advancing age and may increase the risk of vascular dementia (Henskens et al., 2008). The brain is a high flow, low resistance organ that is continuously exposed to the mechanical forces of cardiac pulsations (O'Rourke and Safar, 2005). In healthy young individuals, central elastic arteries (e.g., aorta and carotid artery) expand and recoil effectively within each cardiac cycle, providing a Windkessel effect to dampen hemodynamic pulsatility and facilitate a continuous blood flow in the capillaries (Nichols and O'Rourke, 2005). In contrast, age and vascular dementia-mediated increases in central arterial stiffness may lead to a less effective Windkessel function and enhanced cerebral hemodynamic pulsatility. These processes elevate the risk of heart disease (Nichols and O'Rourke, 2005; Scuteri et al., 2011). Indeed, population-based epidemiologic studies have shown that an age-related increase in central arterial stiffness is an important risk factor for white matter damage and cognitive decline in older adults (Tsao et al., 2013). Thus, vasculature changes in vascular dementia patients produce cognitive impairments due to the induction of SVD for supplying blood to the brain. The mechanisms by which vascular disease and its risk factors cause pathological changes in vascular dementia and how such changes impact cognitive function are not fully understood. However, it is proposed that changes in vasculature result in the induction of hypoperfusion (Marshall, 2012; Jellinger, 2013), a process which not only promotes brain atrophy in both the temporal and frontal lobes, but also decreases long-term potentiation and synaptic plasticity (Fig. 5.5), processes important for maintaining normal neuronal activity and proper neuronal functioning in the nervous system. It is crucial for regulating synaptic transmission or electrical signal transduction to neuronal networks, for sharing essential information among neurons, and for maintaining homeostasis in the body. Moreover, changes in synaptic or neural plasticity are associated with AD type of dementia. Maintenance of adequate tissue perfusion through a dense cerebromicrovascular network is vital for the preservation of normal brain function (Iadecola, 2013). Astrocytes contribute to the regulation of cerebral blood flow through the involvement of glutamate. Glutamate-mediated signaling not only facilitates the release of nitric oxide from arginine in neurons, but also promotes the generation and release of ARA and its downstream metabolites such as prostaglandin E2 (PGE2) and 20-hydroxyeicosatetraenoic acid (20-HETE) (Farooqui et al., 2008). These metabolites along with nitric oxide can either increase or decrease cerebral blood flow, depending on the local O_2 concentration. Earlier studies, which have indicated that cerebral blood flow is controlled solely by arterioles have been challenged, with the finding that contractile cells called pericytes can control the diameter of capillaries, and that damage to these cells contributes to the long-lasting decrease in cerebral blood flow that occurs at the injury site after neuronal injury in neurological disorders. Cerebral hypoperfusion is one of the key factors in the development of hypertension, vascular dementia, and cognitive impairment. At the molecular level, cerebral hypoperfusion leads to gait disturbances and a decline in cognitive performance, executive function, and processing speed (Kim et al., 2016; Pinter et al., 2015; Webb et al., 2012), not only through the involvement of neuroinflammation and oxidative stress, but also due to alterations in synaptic plasticity and connectivity, as well as epigenetic and other environmental/psychosocial factors (Kremen et al., 2012; DeCarli et al., 2012; Kosik et al., 2012). Furthermore, cerebral hypoperfusion also contributes to the pathogenesis of diffuse white matter disease, which involves microvascular injury, BBB disruption, and consequential demyelination. There is growing evidence suggesting a causal relationship between cerebral autoregulatory dysfunction and brain WMH in older adults (Purkayastha et al., 2014). An important question is whether different vascular disease factors (arteriosclerosis, BBB disruption, lifestyle, and genetic susceptibility and predisposition) can be separated from each other and from the effect of aging itself in order to identify their unique individual impacts on cognitive function (Stephan et al., 2009). A more complete understanding of the relationship between vascular disease, cognitive decline, and dementia risk will have important implications in identifying vulnerable population subgroups and a potential treatment target. There is growing evidence the NVU is compromised under pathological conditions such as stroke, diabetes, hypertension, dementias, and with aging. All these conditions trigger a cascade of inflammatory and oxidative stress processes that exacerbate brain damage. Hence, tight regulation and maintenance of neurovascular coupling is central for brain homeostasis. The functional uncoupling of the NVU develops in early stages of vascular dementia (Balbi et al., 2015). This uncoupling may contribute to cognitive impairment (Tarantini et al., 2017; Toth et al., 2017). This process may result in dysregulated release and/or increased degradation of nitric oxide, epoxyeicosatrienoic acids, and prostaglandins (Stefanova et al., 2013). The energy demands of active neurons are high and their proper function depends on constant, tightly controlled delivery of oxygen and nutrients via the microcirculatory network. With increased neuronal activity there is a requirement for rapid compensatory increases in oxygen and glucose delivery to the active brain regions. Thus, neuronal activation triggers hemodynamic responses resulting in vasodilation and increased CBF. The chemical mediators of neurovascular coupling are mediators of metabolic degradation of neuronal and glial cell metabolism with vasodilatory properties such as adenosine, nitric oxide, ions like hydrogen (H^+), potassium (K^+), calcium (Ca^{2+}), and lactate. NO is a powerful vasodilator, which is synthesized by neurons, glial cells, vascular cells, and endothelial cells lining the cerebral vessels (Faraci and Heistad, 1998). In the hippocampus, direct and simultaneous in vivo measurements of NO and CBF changes has revealed that neurovascular coupling is mediated by diffusion of NO between active glutamatergic neurons and blood vessels (Lourenco et al., 2014). It is possible that brain hypoperfusion may lead to the uncoupling of the NVU resulting in cognitive impairment (Attwell et al., 2010; Tarantini et al., 2017; Toth et al., 2017). The mechanisms and consequences of astrocyte dysfunction (including potential alteration of astrocytic end-feet calcium signaling, dysregulation of eicosanoid gliotransmitters, and astrocyte energetics) and functional impairment of the microvascular endothelium are closely associated with hypoperfusion. Age-related mechanisms (cellular oxidative stress, senescence, circulating IGF-1 deficiency) play an important role in the functioning of cells of the NVU (Tarantini et al., 2017; Toth et al., 2017). The presence of lacunar infarcts and leukoaraiosis is another feature of vascular dementia. Lifestyle, lacunar infarcts, and leukoaraiosis may contribute to hypoperfusion and arterial stiffness. Patients with leukoaraiosis have higher pulse wave velocity that transmits increased pulse pressure into the brain through the middle cerebral artery (Webb et al., 2012). Collective evidence suggests that more understanding of the neurochemical and molecular mechanisms is needed to better define microvascular disease and vascular substrates of vascular dementia. The investigation of relevant animal models can be a valuable tool in exploring the pathogenesis as well as prevention of the vascular causes of cognitive impairment in vascular dementia.

CONCLUSION

Vascular dementia is a progressive neurocognitive clinical syndrome, which is characterized by damaged brain tissue due to the decrease in cerebral blood flow and neuronal death.

The mechanisms of vascular dementia are complex. In general, the pathogenesis of vascular dementia involves chronic or acute global or local hypoperfusion and thromboembolic events, oxidative stress, and the inflammatory responses. Vascular dementia is accompanied not only by SVD, but also by vascular lesions such as lacunar, cortical or subcortical infarcts, cerebral hemorrhage, and cardiogenic embolism, contributing to cognitive decline. Causes of vascular dementia include

SVD, blood clots, ruptured blood vessels, or narrowing or hardening of blood vessels that supply blood to the brain. The heterogeneity of cerebrovascular disease makes it challenging to elucidate the neuropathological substrates and mechanisms of vascular dementia, as well as VCI. Hypertension has been identified as an important risk factor for vascular dementia. It not only produces changes in cerebral vessel structure and function, but also predisposes to lacuna infarcts and small vessel hemorrhages in the frontostriatal loop leading to executive dysfunction and other cognitive impairments. Consensus and accurate diagnosis of vascular dementia relies on wide-ranging clinical, neuropsychometric, and neuroimaging measures. Atherosclerotic and cardioembolic diseases appear the most common substrates of vascular brain injury or infarction. SVD characterized by arteriolosclerosis and lacunar infarcts also causes cortical and subcortical microinfarcts, which appear to be the most robust substrates of cognitive impairment. Symptoms of vascular dementia include problems with memory and concentration, confusion, changes in personality and behavior, loss of speech and language skills, and sometimes physical symptoms such as weakness or tremors.

References

Abraham, H.M., Wolfson, L., Moscufo, N., Guttmann, C.R., Kaplan, R.F., et al., 2016. Cardiovascular risk factors and small vessel disease of the brain: blood pressure, white matter lesions, and functional decline in older persons. J. Cereb. Blood Flow Metab. 36, 132–142.

Adluru, N., Destiche, D.J., Lu, S.Y., Doran, S.T., Birdsill, A.C., et al., 2014. White matter microstructure in late middle-age: effects of apolipoprotein E4 and parental family history of Alzheimer's disease. Neuroimage Clin. 4, 730–742.

Allen, C.L., Bayraktutan, U., 2009. Oxidative stress and its role in the pathogenesis of ischaemic stroke. Int. J. Stroke. 4, 461–470.

Amberla, K., Wäljas, M., Tuominen, S., Almkvist, O., Pöyhönen, M., Tuisku, S., et al., 2004. Insidious cognitive decline in CADASIL. Stroke 35, 1598–1602.

Attwell, D., Buchan, A.M., Charpak, S., Lauritzen, M., Macvicar, B.A., et al., 2010. Glial and neuronal control of brain blood flow. Nature 468, 232–243.

Balbi, M., Ghosh, M., Longden, T.A., Jativa Vega, M., Gesierich, B., et al., 2015. Dysfunction of mouse cerebral arteries during early aging. J. Cereb. Blood Flow Metab. 35, 1445–1453.

Baumgart, M., et al., 2015. Summary of the evidence on modifiable risk factors for cognitive decline and dementia: a population-based perspective. Alzheimer Dement. 2015 (11), 718–726.

Bene, A., Makin, S., Doubal, F., Inzitari, D., Wardlaw, J., 2013. Variation in risk factors for recent small subcortical infarcts with infarct size, shape, and location. Stroke 44, 3000–3006.

Bennett, S., Grant, M.M., Aldred, S., 2009. Oxidative stress in vascular dementia and Alzheimer's disease: a common pathology. J. Alzheimers Dis. 17, 245–257.

Bradley, M.A., Markesbery, W.R., Lovell, M.A., 2010. Increased levels of 4-hydroxynonenal and acrolein in the brain in preclinical Alzheimer disease. Free Radic. Biol. Med. 48, 1570–1576.

Breteler, M.M., Claus, J.J., Grobbee, D.E., Hofman, A., 1994. Cardiovascular disease and distribution of cognitive function in elderly people: the rotterdam study. BMJ 308, 1604–1608.

Bryan, S., Baregzay, B., Spicer, D., Singal, P.K., Khaper, N., 2013. Redox-inflammatory synergy in the metabolic syndrome. Can. J. Physiol. Pharmacol. 91, 22–30.

Buffon, F., Porcher, R., Hernandez, K., Kurtz, A., Pointeau, S., Vahedi, K., et al., 2006. Cognitive profile in CADASIL. J. Neurol. Neurosurg. Psychiatry 77, 175–180.

Castegna, A., Thongboonkerd, V., Klein, J.B., Lynn, B., Markesbery, W.R., Butterfield, D. A., 2003. Proteomic identification of nitrated proteins in Alzheimer's disease brain. J. Neurochem. 85, 1394–1401.

Chabriat, H., Joutel, A., Dichgans, M., Tournier-Lasserve, E., Bousser, M.G., 2009. Cadasil. Lancet Neurol. 8, 643–653.

Chan, S.L., Sweet, J.G., Cipolla, M.J., 2013. Treatment for cerebral small vessel disease: effect of relaxin on the function and structure of cerebral parenchymal arterioles during hypertension. FASEB J. 27, 3917–3927.

Chen, B., Friedman, B., Cheng, Q., Tsai, P., Schim, E., Kleinfeld, D., et al., 2009. Severe blood–brain barrier disruption and surrounding tissue injury. Stroke 40, e666–e674.

Chen, W.Y., Wang, M., Zhang, J., Barve, S.S., McClain, C.J., et al., 2017. Acrolein disrupts tight junction proteins and causes endoplasmic reticulum stress-mediated epithelial cell death leading to intestinal barrier dysfunction and permeability. Am. J. Pathol. 187, 2686–2697.

Chui, H., 2001. Dementia due to subcortical ischemic vascular disease. Clin. Cornerstone 3, 40–51.

Cockerill, G.W., Rye, K.A., Gamble, J.R., Vadas, M.A., Barter, P.J., 1995. High-density lipoproteins inhibit cytokine-induced expression of endothelial cell adhesion molecules. Arterioscler. Thromb. Vasc. Biol. 15, 1987–1994.

Craft, S., 2009. The role of metabolic disorders in alzheimer disease and vascular dementia: two roads converged. Arch. Neurol. 66, 300–305.

Dearborn, J.L., Schneider, A.L., Sharrett, A.R., Mosley, T.H., Bezerra, D.C., et al., 2015. Obesity, insulin resistance, and incident small vessel disease on magnetic resonance imaging: atherosclerosis risk in communities study. Stroke 46, 3131–3136.

DeCarli, C., Kawas, C., Morrison, J.H., Reuter-Lorenz, P.A., Sperling, R.A., et al., 2012. Session II: mechanisms of age-related cognitive change and targets for intervention: neural circuits, networks, and plasticity. J. Gerontol. A Biol. Sci. Med. Sci. 67, 747–753.

de la Torre, J.C., 2012. Cardiovascular risk factors promote brain hypoperfusion leading to cognitive decline and dementia. Cardiovasc. Psychiatry Neurol. 2012, 367516.

de Laat, K.F., Tuladhar, A.M., van Norden, A.G., Norris, D.G., Zwiers, M.P., et al., 2011. Loss of white matter integrity is associated with gait disorders in cerebral small vessel disease. Brain 134 (Pt 1), 73–83.

Duering, M., Righart, R., Csanadi, E., et al., 2012. Incident subcortical infarcts induce focal thinning in connected cortical regions. Neurology 79, 2025–2028.

Duschek, S., Schandry, R., 2007. Reduced brain perfusion and cognitive performance due to constitutional hypotension. Clin. Auton. Res. 17, 69–76.

Elias, M.F., Goodell, A.L., Dore, G.A., 2012. Hypertension and cognitive functioning: a perspective in historical context. Hypertension 60, 260–268.

Ellis, R.J., Olichney, J.M., Thal, L.J., Mirra, S.S., Morris, J.C., et al., 1996. Cerebral amyloid angiopathy in the brains of patients with Alzheimer's disease: the CERAD experience, Part XV. Neurology 46, 1592–1596.

Erkinjuntti, T., Gauthier, S., 2009. The concept of vascular cognitive impairment. Front. Neurol. Neurosci. 24, 79–85.

Faraci, F.M., Heistad, D.D., 1998. Regulation of the cerebral circulation: role of endothelium and potassium channels. Physiol. Rev. 78, 53–97.

Farooqui, A.A., 2013. Metabolic Syndrome: An Important Risk Factor for Stroke, Alzheimer Disease, and Depression. Springer Science-Business Media, New York.

Farooqui, A.A., 2014. Inflammation and Oxidative Stress in Neurological Disorders. Springer International Publishing, Switzerland.

Farooqui, A.A., 2015. High Calorie Diet and the Human Brain. Springer International Publishing, Switzerland.

Farooqui, A.A., 2017. Neurochemical Aspects of Alzheimer's Disease. Academic Press, an Imprint of Elsevier, San Diego.

Farooqui, A.A., Farooqui, T., 2018. Effects of Mediterranean diet components on neurodegenerative diseases. In: Farooqui, T., Farooqui, A.A. (Eds.), Role of the Mediterranean Diet in the Brain andNeurodegenerative Diseases. Academic Press, an Imprint of Elsevier, San Diego.

Farooqui, A.A., Ong, W.Y., Horrocks, L.A., 2008. Neurochemical Aspects of Excitotoxicity. Springer, New York.

Farrer, L.A., Cupples, L.A., Haines, J.L., Hyman, B., Kukull, W.A., et al., 1997. Effects of age, sex, and ethnicity on the association between apolipoprotein E genotype and Alzheimer disease: a meta-analysis. JAMA 278, 1349–1356.

Ferrari, C., Nacmias, B., Sorbi, S., 2017. The diagnosis of dementias: a practical tool not to miss rare causes. Neurol. Sci. Available from: https://doi.org/10.1007/s10072-017-3206-0.

Fillit, H., Nash, D.T., Rundek, T., Zuckerman, A., 2008. Cardiovascular risk factors and dementia. Am. J. Geriatr. Pharmacother. 6, 100–118.

Förstermann, U., Xia, N., Li, H., 2017. Roles of vascular oxidative stress and nitric oxide in the pathogenesis of atherosclerosis. Circ. Res. 120, 713–735.

Forti, P., Pisacane, N., Rietti, E., et al., 2010. Metabolic syndrome and risk of dementia in older adults. J. Am. Geriatr. Soc. 58, 487–492.

Fredman, P., Wallin, A., Blennow, K., et al., 1992. Sulfatide as a biochemical marker in cerebrospinal fluid of patients with vascular dementia. Acta Neurol. Scand. 85, 103–106.

Friedman, J.I., Tang, C.Y., de Haas, H.J., Changchien, L., Goliasch, G., et al., 2014. Brain imaging changes associated with risk factors for cardiovascular and cerebrovascular disease in asymptomatic patients. JACC Cardiovasc. Imaging 7, 1039–1053.

Gąsecki, D., Kwarciany, M., Nyka, W., Narkiewicz, K., 2013. Hypertension, brain damage and cognitive decline. Curr. Hypertens. Rep. 15, 547–558.

Go, Y.M., Jones, D.P., 2014. Redox biology: interface of the exposome with the proteome, epigenome and genome. Redox Biol. 2, 358–360.

Gorelick, P.B., Scuteri, A., Black, S.E., DeCarli, C., Greenberg, S.M., et al., 2011. Vascular contributions to cognitive impairment and dementia: a statement for healthcare professionals from theamerican heart association/american stroke association. Stroke 42, 2672–2713.

Gottfries, C.G., Blennow, K., Karlsson, I., Wallin, A., 1994. The neurochemistry of vascular dementia. Dementia 5, 163–167.

Grau-Olivares, M., Arboix, A., Junque, C., Arenaza-Urquijo, E., Rovira, M., et al., 2010. Progressive gray matter atrophy in lacunar patients with vascular mild cognitive impairment. Cerebrovasc. Dis. 30, 157–166.

Gridley, T., 2007. Notch signaling in vascular development and physiology. Development 134, 2709–2718.

Grinberg, L.T., Heinsen, H., 2010. Toward a pathological definition of vascular dementia. J. Neurol. Sci. 299, 136–138.

Haritunians, T., Chow, T., De Lange, R.P.J., et al., 2005. Functional analysis of a recurrent missense mutation in Notch3 in CADASIL. J. Neurol. Neurosurg. Psychiatry 76, 1242–1248.

Henskens, L.H., Kroon, A.A., van Oostenbrugge, R.J., Gronenschild, E.H., Fuss-Lejeune, M.M., et al., 2008. Increased aortic pulse wave velocity is associated with silent cerebral small-vessel disease in hypertensive patients. Hypertension 52, 1120–1126.

Hsieh, H.-L., Yang, C.-M., 2013. Role of redox signaling in neuroinflammation and neurodegenerative diseases. Biomed. Res. Int. 2013, e484613.

Iadecola, C., 2004. Neurovascular regulation in the normal brain and in Alzheimer's disease. Nat. Rev. Neurosci. 5, 347–360.

Iadecola, C., 2010. The overlap between neurodegenerative and vascular factors in the pathogenesis of dementia. Acta Neuropathol. 120, 287–296.

Iadecola, C., 2013. The pathobiology of vascular dementia. Neuron 80, 844–866.

Iadecola, C., 2014. Hypertension and dementia. Hypertension 64, 3–5.

Imamine, R., Kawamura, T., Umemura, T., Umegaki, H., Kawano, N., Hotta, M., et al., 2011. Does cerebral small vessel disease predict future decline of cognitive function in elderly people with type 2 diabetes? Diabetes Res. Clin. Pract. 94, 91–99.

Institute of Medicine, 2015. Cognitive Aging: Progress in Understanding and Opportunity for Action. The National Academies Press, Washington, DC.

Ivanov, A., Esclapez, M., Ferhat, L., 2009. Role of drebrin A in dendritic spine plasticity and synaptic function: implications in neurological disorders. Commun. Integr. Biol. 2, 268–270.

Jagtap, A., Gawande, S., Sharma, S., 2015. Biomarkers in vascular dementia: a recent update. Biomarkers and Genomic. Medicine 7, 43–56.

Jellinger, K.A., 2013. Pathology and pathogenesis of vascular cognitive impairment-a critical update. Front. Aging Neurosci. 5, 17.

Jellinger, K.A., Attems, J., 2012. Neuropathology and general autopsy findings in nondemented aged subjects. Clin. Neuropathol. 31, 87–98.

Jiwa, N.S., Garrard, P., Hainsworth, A.H., 2010. Experimental models of vascular dementia and vascular cognitive impairment: a systematic review. J. Neurochem. 115, 814–828.

Kalaria, R.N., 2010. Vascular basis for brain degeneration: faltering controls and risk factors for dementia. Nutr. Rev. 68 (Suppl. 2), S74–S87.

Kalaria, R.N., 2016. Neuropathological diagnosis of vascular cognitive impairment and vascular dementia with implications for Alzheimer's disease. Acta Neuropathol. 131, 659–685.

Kalaria, R.N., 2017. The pathology and pathophysiology of vascular dementia. Neuropharmacology 19. pii: S0028-3908(17)30627-30635.

Kennelly, S., Collins, O., 2012. Walking the cognitive "minefield" between high and low blood pressure. J. Alzheimer's Dis. 32, 609–621.

Kim, Y.J., Kwon, H.K., Lee, J.M., Cho, H., Kim, H.J., et al., 2016. Gray and white matter changes linking cerebral small vessel disease to gait disturbances. Neurology 86, 1199–1207.

Kirvell, S.L., Elliott, M.S., Kalaria, R.N., Hortobagyi, T., Ballard, C.G., et al., 2011. Vesicular glutamate transporter and cognition in stroke: a case-control autopsy study. Neurology 75, 1803–1809.

Kirshner, H.S., 2009. Vascular dementia: a review of recent evidence for prevention and treatment. Curr. Neurol. Neurosci. Rep. 9, 437–442.

Kosik, K.S., Rapp, P.R., Raz, N., Small, S.A., Sweatt, J.D., et al., 2012. Mechanisms of age-related cognitive change and targets for intervention: epigenetics. J. Gerontol. A Biol. Sci. Med. Sci. 67, 741–746.

Kremen, W.S., Lachman, M.E., Pruessner, J.C., Sliwinski, M., Wilson, R.S., 2012. Mechanisms of age-related cognitive change and targets for intervention: social interactions and stress. J. Gerontol. A Biol. Sci. Med. Sci. 67, 760–765.

Langa, K.M., Foster, N.L., Larson, E.B., 2004. Mixed dementia: emerging concepts and therapeutic implications. JAMA 292, 2901–2908.

Lavallee, P.C., Labreuche, J., Faille, D., et al., 2013. Circulating markers of endothelial dysfunction and platelet activation in patients with severe symptomatic cerebral small vessel disease. Cerebrovasc. Dis. 36, 131–138.

Lawrence, A.J., Chung, A.W., Morris, R.G., Markus, H.S., Barrick, T.R., 2014. Structural network efficiency is associated with cognitive impairment in small-vessel disease. Neurology 83, 304–311.

Leblanc, G.G., Meschia, J.F., Stuss, D.T., et al., 2006. Genetics of vascular cognitive impairment. the opportunity and the challenges. Stroke 37, 248–255.

Lee, D., Aoki, C., 2012. Presenilin conditional double knockout mice exhibit decreases in drebrin a at hippocampal CA1 synapses. Synapse 66, 870–879.

Leritz, E.C., McGlinchey, R.E., Kellison, I., Rudolph, J.L., Milberg, W.P., 2011. Cardiovascular disease risk factors and cognition in the elderly. Curr. Cardiovasc. Risk Rep. 5, 407–412.

Li, H., Horke, S., Forstermann, U., 2014. Vascular oxidative stress, nitric oxide and atherosclerosis. Atherosclerosis 237, 208–219.

Li, L., Yiin, G.S., Geraghty, O.C., Schulz, U.G., Kuker, W., et al., 2015. Incidence, outcome, risk factors, and long-term prognosis of cryptogenic transient ischaemic attack and ischaemic stroke: a population-based study. Lancet Neurol. 14, 903–913.

Lim, S.S., Vos, T., Flaxman, A.D., Danaei, G., Shibuya, K., et al., 2012. A comparative risk assessment of burden of disease and injury attributable to 67 risk factors and risk factor clusters in 21 regions, 1990–2010: a systematic analysis for the Global Burden of Disease Study 2010 [published corrections appear in *Lancet.* 2013; 381:628 and *Lancet.* 2013; 381:1276]. Lancet 380, 2224–2260.

Lin, C.J., Chang, F.C., Chou, K.H., Tu, P.C., Lee, Y.H., Lin, C.P., et al., 2016. Intervention versus aggressive medical therapy for cognition in severe asymptomatic carotid stenosis. AJNR Am. J. Neuroradiol. 37, 1889–1897.

Liu, H., Zhang, J., 2012. Cerebral hypoperfusion and cognitive impairment: the pathogenic role of vascular oxidative stress. Int. J. Neurosci. 122, 494–499.

LoPachin, R.M., Gavin, T., Petersen, D.R., et al., 2009. Molecular mechanisms of 4 hydroxy 2-nonenal and acrolein toxicity: nucleophilic targets and adduct formation. Chem. Res. Toxicol. 22, 1499–1508.

Lopez, O.L., Jagust, W.J., Dulberg, C., Becker, J.T., DeKosky, S.T., Fitzpatrick, A., et al., 2003. Risk factors for mild cognitive impairment in the cardiovascular health study cognition study: part 2. Arch. Neurol. 60, 1394–1399.

Lopez, O.L., Kuller, L.H., Becker, J.T., Jagust, W.J., DeKosky, S.T., et al., 2005. Classification of vascular dementia in the Cardiovascular Health Study Cognition Study. Neurology 64, 1539–1547.

Lourenco, C.F., Santos, R.M., Barbosa, R.M., Cadenas, E., Radi, R., Laranjinha, J., 2014. Neurovascular coupling in hippocampus is mediated via diffusion by neuronal-derived nitric oxide. Free Radic. Biol. Med. 73, 421–429.

Lovell, M.A., Xie, C., Markesbery, W.R., 2001. Acrolein is increased in Alzheimer's disease brain and is toxic to primary hippocampal cultures. Neurobiol. Aging 2001 (22), 187–194.

Marchesi, V.T., 2014. Alzheimer's disease and CADASIL are heritable, adult-onset dementias that both involve damaged small blood vessels. Cell Mol. Life Sci. 71, 949–955.

Marlatt, M.W., Lucassen, P.J., Perry, G., Smith, M.A., Zhu, X., 2008. Alzheimer's disease: cerebrovascular dysfunction, oxidative stress, and advanced clinical therapies. J. Alzheimers Dis. 15, 199–210.

Marshall, R.S., 2012. Effects of altered cerebral hemodynamics on cognitive function. J. Alzheimers Dis. 32, 633–642.

Marshall, R.S., Lazar, R.M., 2011. Pumps, aqueducts, and drought management: vascular physiology in vascular cognitive impairment. Stroke 32, 221–226.

Matsuda, Y., Hirata, K., Inoue, N., Suematsu, M., Kawashima, S., et al., 1993. High-density lipoprotein reverses inhibitory effect of oxidized low-density lipoprotein on endothelium-dependent arterial relaxation. Circ. Res. 72, 1103–1109.

Minghetti, L., Ajmone-Cat, M.A., De Berardinis, M.A., De Simone, R., 2005. Microglial activation in chronic neurodegenerative diseases: roles of apoptotic neurons and chronic stimulation. Brain Res. Rev. 48, 251–256.

Moorhouse, P., Rockwood, K., 2008. Vascular cognitive impairment: current concepts and clinical developments. Lancet Neurol. 7, 246–255.

Morgan, M.J., Liu, Z., 2011. Crosstalk of reactive oxygen species and NF-κB signaling. Cell Res. 21, 103–115.

Moskowitz, M.A., Lo, E.H., Iadecola, C., 2010. The science of stroke: mechanisms in search of treatments. Neuron 67, 181–198.

Mozaffarian, D., Benjamin, E.J., Go, A.S., Arnett, D.K., Blaha, M.J., et al., American Heart Association Statistics Committee and Stroke Statistics Subcommittee 2015. Heart disease and stroke statistics–2015 update: a report from the American Heart Association. Circulation 131, e535.

Mulder, M., Terwel, D., 1998. Possible link between lipid metabolism and cerebral amyloid angiopathy in alzheimer's disease: a role for high-density lipoproteins? Haemostasis 28, 174–194.

Mulugeta, E., Molina-Holgado, F., Elliott, M.S., Hortobagyi, T., Perry, R., et al., 2008. Inflammatory mediators in the frontal lobe of patients with mixed and vascular dementia. Dement. Geriatr. Cogn. Disord. 25, 278–286.

Nichols, W.W., O'Rourke, M.E., 2005. Mcdonald's Blood Flow in Arteries. Oxford University Press, New York.

NIH, 2010. Consensus Development Conference Statement on Preventing Alzheimer's Disease and Cognitive Decline. National Institutes of Health, Bethesda, MD.

O'Brien, J.T., Thomas, A., 2015. Vascular dementia. Lancet 386, 1698–1706.

Okonkwo, O.C., Cohen, R.A., Gunstad, J., Tremont, G., Alosco, M.L., Poppas, A., 2010. Longitudinal trajectories of cognitive decline among older adults with cardiovascular disease. Cerebrovasc. Dis. 32, 362–373.

O'Rourke, M.F., Safar, M.E., 2005. Relationship between aortic stiffening and microvascular disease in brain and kidney: cause and logic of therapy. Hypertension 46, 200–204.

Palesi, F., De Rinaldis, A., Vitali, P., Castellazzi, G., Casiraghi, L., Germani, G., et al., 2018. Specific patterns of white matter alterations help distinguishing Alzheimer's and vascular dementia. Front. Neurosci. 12, 274.

Pantoni, L., 2010. Cerebral small vessel disease: from pathogenesis and clinical characteristics to therapeutic challenges. Lancet Neurol. 9, 689–701.

Patel, B., Markus, H.S., 2011. Magnetic resonance imaging in cerebral small vessel disease and its use as a surrogate disease marker. Int. J. Stroke 6, 47–59.

Pinter, D., Enzinger, C., Fazekas, F., 2015. Cerebral small vessel disease, cognitive reserve and cognitive dysfunction. J. Neurol. 262, 2411–2419.

Prabhakar, P., Chandra, S.R., Christopher, R., 2017. Circulating microRNAs as potential biomarkers for the identification of vascular dementia due to cerebral small vessel disease. Age Ageing 46, 861–864.

Purkayastha, S., Fadar, O., Mehregan, A., Salat, D.H., Moscufo, N., et al., 2014. Impaired cerebrovascular hemodynamics are associated with cerebral white matter damage. J. Cereb. Blood Flow Metab. 34, 228–234.

Quaegebeur, A., Lange, C., Carmeliet, P., 2011. The neurovascular link in health and disease: molecular mechanisms and therapeutic implications. Neuron 71, 406–424.

Ray, K.V., Behera, P., et al., 2013. Serum homocysteine, dehydroepiandrosterone sulphate and lipoprotein (a) in Alzheimer's disease and vascular dementia. Aging Dis. 4, 57–64.

Reijmer, Y.D., Fotiadis, P., Martinez-Ramirez, S., Salat, D.H., Schultz, A., Shoamanesh, A., 2015. Structural network alterations and neurological dysfunction in cerebral amyloid angiopathy. Brain 138, 179–188.

Reitz, C., Tang, M.X., Luchsinger, J., Mayeux, R., 2004. Relation of plasma lipids to Alzheimer disease and vascular dementia. Arch. Neurol. 61, 705–714.

Richardson, K., Stephan, B.C., Ince, P.G., Brayne, C., Matthews, F.E., et al., 2012. The neuropathology of vascular disease in the Medical Research Council Cognitive Function and Ageing Study (MRC CFAS). Curr. Alzheimer Res. 9, 687–696.

Rincon, F., Wright, C.B., 2014. Current pathophysiological concepts in cerebral small vessel disease. Front. Aging Neurosci. 6, 24.

Román, G.C., Erkinjuntti, T., Wallin, A., Pantoni, L., Chui, H.C., 2002. Subcortical ischaemic vascular dementia. Lancet Neurol. 1, 426–436.

Rosenberg, G.A., 2016. Matrix metalloproteinase-mediated neuroinflammation in vascular cognitive impairment of the binswanger type. Cell Mol. Neurobiol. 36, 195–202.

Rosenberg, G.A., 2017. Extracellular matrix inflammation in vascular cognitive impairment and dementia. Clin. Sci. (Lond.) 131, 425–437.

Sabayan, B., Oleksik, A.M., Maier, A.B., van Buchem, M.A., Poortvliet, R.K., de Ruijter, W., et al., 2012. High blood pressure and resilience to physical and cognitive decline in the oldest old: the Leiden 85-plus Study. Am. Geriatr. Soc 60, 2014–2019.

Saunders, A.M., Strittmatter, W.J., Schmechel, D., George-Hyslop, P.H., Pericak-Vance, M. A., et al., 1993. Association of apolipoprotein E allele epsilon 4 with late-onset familial and sporadic Alzheimer's disease. Neurology 43, 1467–1472.

Schmidt, O., Heyde, C., Ertel, W., Stahel, P., 2005. Closed head injury – an inflammatory disease? Brain Res. Rev. 48, 388–399.

Scuteri, A., Nilsson, P.M., Tzourio, C., Redon, J., Laurent, S., 2011. Microvascular brain damage with aging and hypertension: pathophysiological consideration and clinical implications. J. Hypertens. 29, 1469–1477.

Snyder, H.M., Corriveau, R.A., Craft, S., Faber, J.E., Greenberg, S.M., et al., 2015. Vascular contributions to cognitive impairment and dementia including Alzheimer's disease. Alzheimer's Dement. 11, 710–717.

Stefanova, I., Stephan, T., Becker-Bense, S., Dera, T., Brandt, T., Dieterich, M., 2013. Age-related changes of blood-oxygen-level-dependent signal dynamics during optokinetic stimulation. Neurobiol. Aging 34, 2277–2286.

Stephan, B.C., Matthews, F.E., Khaw, K.T., Dufouil, C., Brayne, C., 2009. Beyond mild cognitive impairment: vascular cognitive impairment, no dementia (VCIND). Alzheimers Res. Ther. 1, 4.

Stevens, W.D., Fortin, T., Pappas, B.A., 2002. Retinal and optic nerve degeneration after chronic carotid ligation: time course and role of light exposure. Stroke 33, 1107–1112.

Sun, X., Salat, D., Upchurch, K., Deason, R., Kowall, N., Budson, A., 2014. Destruction of white matter integrity in patients with mild cognitive impairment and Alzheimer disease. J. Invest. Med. 62, 927–933.

Tarantini, S., Tran, C.H., Gordon, G.R., Ungvari, Z., Csiszar, A., 2017. Impaired neurovascular coupling in aging and Alzheimer's disease: contribution of astrocyte dysfunction and endothelial impairment to cognitive decline. Exp. Gerontol. 94, 52–58.

Terelak-Borys, B., Skonieczna, K., Grabska-Liberek, I., 2012. Ocular ischemic syndrome – a systematic review. Med. Sci. Monit. 18, RA138–RA144.

Toledo, J.B., Arnold, S.E., Raible, K., Brettschneider, J., Xie, S.X., et al., 2013. Contribution of cerebrovascular disease in autopsy confirmed neurodegenerative disease cases in the National Alzheimer's Coordinating Centre. Brain 136, 2697–2706.

Topol, E., Easton, D., Harrington, R., et al., 2003. Randomised, double-blind, placebo-controlled, international trial of the oral IIb/IIIa antagonist lotrafiban in coronary and cerebrovascular disease. Am. Heart Assoc. 108, 399–406.

Toth, P., Tarantini, S., Csiszar, A., Ungvari, Z., 2017. Functional vascular contributions to cognitive impairment and dementia: mechanisms and consequences of cerebral autoregulatory dysfunction, endothelial impairment, and neurovascular uncoupling in aging. Am. J. Physiol. Heart Circ. Physiol. 312, H1–H20.

Tsao, C.W., Seshadri, S., Beiser, A.S., Westwood, A.J., DeCarli, C., Au, R., et al., 2013. Relations of arterial stiffness and endothelial function to brain aging in the community. Neurology 81, 1–8.

Tucsek, Z., Toth, P., Tarantini, S., Sosnowska, D., Gautam, T., Warrington, J.P., et al., 2014. Aging exacerbates obesity-induced cerebromicrovascular rarefaction, neurovascular uncoupling, and cognitive decline in mice. J. Geront. A Biomed. Sci. Med. Sci. 69, 1339–1352.

Tuladhar, A.M., van Dijk, E., Zwiers, M.P., van Norden, A.G.W., de Laat, K., Shumskaya, E., 2016. Structural network connectivity and cognition in cerebral small vessel disease. Hum. Brain Mapp. 37, 300–310.

Turillazzi, E., Neri, M., Cerretani, D., Cantatore, S., Frati, P., Moltoni, L., et al., 2016. Lipid peroxidation and apoptotic response in rat brain areas induced by long-term administration of nandrolone: the mutual crosstalk between ROS and NF-kB. J. Cell Mol. Med. 2016 (20), 601–612.

van der Holst, H.M., Tuladhar, A.M., Zerbi, V., van Uden, I.W.M., de Laat, K.F., et al., 2017. White matter changes and gait decline in cerebral small vessel disease. Neuroimage Clin. 17, 731–738.

van Dijk, E.J., Breteler, M.M., Schmidt, R., Berger, K., Nilsson, L.G., et al., 2004. The association between blood pressure, hypertension, and cerebral white matter lesions: cardiovascular determinants of dementia study. Hypertension 44, 625–630.

Venkat, P., Chopp, M., Chen, J., 2015. Models and mechanisms of vascular dementia. Exp. Neurol. 272, 97–108.

Wallin, A., Sjögren, M., 2001. Cerebrospinal fluid cytoskeleton proteins in patients with subcortical white-matter dementia. Mech. Ageing Dev. 122, 1937–1949.

Wallin, A., Kapaki, E., Boban, M., Engelborghs, S., Hermann, D.M., et al., 2017. Biochemical markers in vascular cognitive impairment associated with subcortical small vessel disease – a consensus report. BMC Neurol. 17, 102.

Wardlaw, J.M., Smith, C., Dichgans, M., 2013a. Mechanisms of sporadic cerebral small vessel disease: insights from neuroimaging. Lancet Neurol. 12, 483–497.

Wardlaw, J.M., Smith, E.E., Biessels, G.J., Cordonnier, C., Fazekas, et al., 2013b. Neuroimaging standards for research into small vessel disease and its contribution to ageing and neurodegeneration. Lancet Neurol. 12, 822–838.

Webb, A.J., Simoni, M., Mazzucco, S., Kuker, W., Schulz, U., et al., 2012. Increased cerebral arterial pulsatility in patients with leukoaraiosis: arterial stiffness enhances transmission of aortic pulsatility. Stroke 43, 2631–2636.

Williams, T.I., Lynn, B.C., Markesbery, W.R., et al., 2006. Increased levels of 4-hydroxynonenal and acrolein, neurotoxic markers of lipid peroxidation, in the brain in mild cognitive impairment and early Alzheimer's disease. Neurobiol. Aging 27, 1094–1099.

Wood, P.L., 1998. Neuroinflammation: Mechanisms and Management. Humana Press, Totowa, NJ.

Xu, Y., Wang, Q., Cui, R., Lu, K., Liu, Y., Zhao, Y., 2016. Uric acid is associated with vascular dementia in Chinese population. Brain Behav. 7, e00617.

Yamauchi, H., Fukuyama, H., Harada, K., Nabatame, H., Ogawa, M., Ouchi, Y., et al., 1993. Callosal atrophy parallels decreased cortical oxygen metabolism and neuropsychological impairment in Alzheimer's disease. Arch. Neurol. 50, 1070–1074.

Zlokovic, B.V., 2008. The blood-brain barrier in health and chronic neurodegenerative disorders. Neuron 57, 178–201.
Zlokovic, B.V., 2011. Neurovascular pathways to neurodegeneration in Alzheimer's disease and other disorders. Nat. Rev. Neurosci. 12, 723–738.
Zuccala, G., Onder, G., Marzetti, E., Monaco, M.R., Cesari, M., Cocchi, A., et al., 2005. Use of angiotensin-converting enzyme inhibitors and variations in cognitive performance among patients with heart failure. Eur. Heart J. 32, 226–233.

Further Reading

Allen, C., Srivastava, K., Bayraktutan, U., 2010. Small GTPase RhoA and its effector rho kinase mediate oxygen glucose deprivation-evoked in vitro cerebral barrier dysfunction. Stroke 41, 2056–2063.
Bornstein, R.A., Starling, R.C., Myerowitz, P.D., Haas, G.J., 1995. Neuropsychological function in patients with end-stage heart failure before and after cardiac transplantation. Acta Neurol. Scand. 32, 260–265.
Sinclair, L.I., Tayler, H.M., Love, S., 2015. Synaptic protein levels altered in vascular dementia. Neuropathol. Appl. Neurobiol. 41, 533–543.

CHAPTER

6

Neurochemical Aspects of Frontotemporal Dementia

INTRODUCTION

Frontotemporal dementia (FTD) is a clinically and pathologically heterogeneous group of non-Alzheimer presenile dementias. Anatomically, FTD is characterized by relatively selective, progressive atrophy involving the frontal or temporal lobes, or both. Diffusion tensor imaging (DTI) studies have indicated that FTD is associated with a reduction in fractional anisotropy (FA) in the anterior corpus callosum, bilateral anterior, and descending cingulum and uncinate fiber tracts (Zhang et al., 2009). A voxel-by-voxel analysis shows even more widespread FA reduction in FTD which involves frontal and temporal white matter regions, expanded to parietal, and spared occipital white matter (Zhang et al., 2009; Avants et al., 2010). In contrast, Alzheimer's disease (AD) is associated with FA reduction in bilateral descending cingulum, left posterior and anterior cingulum, and left uncinate fiber tracts, which is in agreement with previous DTI studies. Furthermore, despite differences in memory profiles, AD and FTD both cause severe hippocampal hypometabolism and atrophy but differ in the degree of involvement of other memory related structures. FTD is characterized by progressive behavioral change, executive dysfunction, and language difficulties (Mackenzie et al., 2011; Rabinovici and Miller, 2010; Warren et al., 2013). At the molecular level, FTD is characterized by a disorder of tau metabolism (Lee et al., 2001) or the accumulation of a ubiquitinated protein known as TDP-43 (Neumann et al., 2006). TDP-43 is a DNA- and RNA-binding protein, which is normally found in the nucleus. It is involved in the regulation of gene expression by controlling several processes, including gene transcription, RNA splicing, mRNA stabilization and transport, miRNA binding, and regulation (Geser et al., 2009; Buratti et al., 2010; Igaz et al., 2009). Under physiological conditions,

Molecular Mechanisms of Dementia
DOI: https://doi.org/10.1016/B978-0-12-816347-4.00006-4

TDP-43 actively shuttles between the nucleus and the cytoplasm (Fallini et al., 2012). In contrast, under pathological conditions TDP-43 migrates from the nucleus to the cytoplasm, where it accumulates in both neurons and glia (Van Deerlin et al., 2008; Igaz et al., 2009; Cairns et al., 2007; Hasegawa et al., 2008). It is suggested that TDP-43 proteinopathies (amyotrophic lateral sclerosis (ALS) and frontotemporal lobe dementia) can be caused by a loss of function due to nuclear depletion, by a gain of function due to cytoplasmic aggregation, or by a combination of both (Wegorzewska et al., 2009; Neumann, 2009). Under pathological conditions, TDP-43 is abnormally ubiquitinated, hyperphosphorylated and N-terminally cleaved to generate C-terminal fragments (20–25 kDa) (Arai et al., 2006). Degradation of TDP-43 is closely associated with neurodegenerative process in TDP-43 proteinopathies. The fundamental question as to whether TDP-43 mediates neurodegeneration via a gain of function or a loss of function remains unanswered.

FTD is a leading cause of early onset dementia found in 4% of the general dementia population and is present in 20%–30% of dementia patients younger than 65 years (Rabinovici and Miller, 2010; Warren et al., 2013). The majority of FTD cases are sporadic and likely caused by the interaction between genetic and environmental factors. A number of cases, however, present familial aggregation and are inherited in an autosomal dominant fashion, suggesting a genetic cause (Ratnavalli et al., 2002; Bird et al., 2003). Up to 40% of patients have a positive family history, with a diagnosis of dementia in at least one extra family member (Ratnavalli et al., 2002; Goldman et al., 2005). Based on anatomic, genetic, and neuropathologic categorizations, the six clinical subtypes of FTD are: (1) behavioral variant of FTD (bvFTD); (2) semantic variant primary progressive aphasia (SD); (3) nonfluent agrammatic variant primary progressive aphasia (PA); (4) corticobasal syndrome; (5) progressive supranuclear palsy (PSP); and (6) FTD associated with motor neuron disease (MND). Variants of FTD differ from each other in their clinical presentation, cognitive deficits, and affected brain regions (Rabinovici and Miller, 2010; Warren et al., 2013). Thus, patients with bvFTD have profound alterations in personality, social conduct, and behavior, which have been related to atrophy of prefrontal brain areas, particularly the ventromedial prefrontal cortex but also anterior temporal atrophy (Halabi et al., 2013). In addition, bvFTD patients show increased food consumption with a craving for sweets. This is one of the characteristic and discriminating features of bvFTD (Rascovsky et al., 2011). It is reported that there is an early involvement of the hypothalamus (Piguet et al., 2011) as well as alterations in a complex network (cingulate and orbitofrontal cortices and cerebellum) that controls food intake (Ahmed et al., 2016). Alterations in food consumption are accompanied by changes in cholesterol, insulin, neuroendocrine levels, and metabolic rate which appear to have significant effect on survival (Ahmed et al., 2014) supporting the view that bvFTD should not be regarded as a neural disease, but a pathological condition, which has a global impact on body structure and functions. Patients with SD are characterized by anomia and impaired word comprehension with concomitant asymmetrical rostral temporal lobe atrophy (Nestor, 2007). Finally, progressive nonfluent aphasia (PNFA) presents with expressive language deficits, characterized by effortful speech production, phonologic and grammatical errors (Gorno-Tempini et al., 2011) and atrophy in the left perisylvian region (Gorno-Tempini et al., 2008). Up to 40% of FTD patients report a family history of neurodegenerative illness, and one-third to one-half of familial cases of FTD follow an autosomal dominant inheritance pattern. The mean disease duration from onset of symptoms to death is 6–8 years (Neary et al., 2005) The various neuropsychiatric symptoms associated with FTD are apathy, disinhibition, agitation and aggression, eating disturbances, and other behavioral abnormalities include repetitive stereotypical behaviors such as verbal perseveration, hoarding, and rituals (Rabinovici and Miller, 2010; Warren et al., 2013; Onyike and Diehl-Schmid, 2013). Pathologically, variants of FTD are characterized by mild gliosis, neuronal loss, and superficial spongiform degeneration in the frontal and/or temporal cortexes. Ballooned neurons (Pick cells) occur with variable frequency in all subtypes of FTD (Kertesz and Munoz, 2002). Furthermore, based on the presence of tau-inclusions in FTD, corticobasal degeneration (CBD), and progressive supranuclear palsy (PSP), some authors have proposed that the pathology of FTD can be divided into tau-positive and tau-negative variants. Clinically, these variants differ from each other on the localization of tau-inclusions in the affected brain regions (Sha et al., 2006; Kertesz, 2005). FTD is a highly heritable disorder despite varying heritability among different clinical syndromes and subtypes due to a range of gene mutations (Rohrer et al., 2009). There is little or no Aβ pathology in FTD. The p-tau pathology is usually confined to the cerebral cortex gray matter and white matter. Atrophy of the frontal and temporal lobes is severe. Up to half of FTD cases with autosomal-dominant inheritance report a family history of FTD (Table 6.1) (Rademakers et al., 2012). Familial FTD also accounts for approximately one-third to one-half of all FTD cases and presents more commonly as bvFTD than other FTD subtypes (Capozzo et al., 2017). At present, three major causal genes have been identified: Microtubule Associated Protein Tau (*MAPT*), Progranulin (*GRN*), and Chromosome 9 Open Reading Frame 72 (*C9ORF72*) (Sudre et al., 2017; Lashley et al., 2014; Snowden et al., 2006; Mahoney et al., 2012; Khan et al., 2012; Ferrari et al., 2014). Although FTD presentations are relatively

TABLE 6.1 Genes and Proteins Associated With Various Forms of FTD

Gene	Chromosome	Protein	FTD phenotype	Mode of inheritance
C90RF72	9p21.2	Unknown	bvFTD, ALS, FTLD-ALS	Autosomal dominant
CHMP2B	3p11.2	Chromatin-modifying protein B	bvFTD, FTLD-ALS	Autosomal dominant
FUS	16p11.2	Fused in sarcoma protein	ALS, bvFTD, FTLD-ALS	Autosomal dominant
GRN	17q21.32	Progranulin	bvFTD, PPA, CBS	Autosomal dominant
MAPT	17q21.32	Microtubule associated tau protein	bvFTD, PSP, CBS	Autosomal dominant
TDP-43		Transactive response DNA-binding protein 43 kDa	ALS, FTD	Autosomal dominant
TMEM 106B	6p21.1	Transmembrane protein 106B	PLOSL, bvFTD	Autosomal dominant
VCP	9p13.3	Valosin containing protein	Multisystem proteinopathy	Autosomal dominant

homogenous early in the disease course, different biological correlates and varying genetic mutations ultimately result in diverging clinical courses (Piguet et al., 2004). *MAPT*-associated familial FTD typically presents younger than FTD associated with other mutations. Although FTD is traditionally associated with cortical atrophy, which is thought to be the major determinant of their symptoms, there is growing evidence for concomitant involvement of subcortical brain regions that may contribute to some symptoms of FTD. For example, a postmortem study suggested that basal ganglia structures are affected from an early disease stage (Broe et al., 2003). At more advanced stages, basal ganglia degeneration becomes very evident as indicated by grossly dilated frontal horns of the lateral ventricles. These pathological findings can be confirmed in vivo (Chow et al., 2008; Looi et al., 2008) using MRI volumetrics demonstrating striatal atrophy, especially in those with bvFTD and PNFA (Garibotto et al., 2011; Halabi et al., 2013). There is significant overlap in pathogenic processes between FTD and other neurodegenerative diseases, such as AD and ALS;

attempts are underway to treat FTD with drugs used for the treatment of AD and ALS. However, on the basis of neuroimaging studies, it is hypothesized that patients with bvFTD demonstrate significantly greater white matter breakdown compared to AD in the late-myelinating regions of frontal white matter and the genu of corpus callosum but not in the splenium of the corpus callosum, an early-myelinating association area. These changes are associated with expression of emotional and behavioral symptoms (Lu et al., 2014). Collective evidence suggests that FTD is a heterogeneous disorder with a common feature of relatively selective neurodegeneration in the frontal and temporal lobes. In addition, most cases of FTD are accompanied by the abnormal intracellular accumulation of some disease-specific protein (Table 6.1).

DIAGNOSIS OF FRONTOTEMPORAL DEMENTIA

FTD is diagnosed by neuroimaging techniques including magnetic resonance imaging (MRI) or computed tomography (CT), positron-emission tomography (PET), or single-photon emission computed tomography (SPECT) imaging. Thus, 18F FDG-PET imaging studies have indicated patterns of hypometabolism that correlate with areas of atrophy. AD can be differentiated from FTD on the basis of 18F FDG-PET imaging studies. In FTD, PET studies typically demonstrate hypometabolism in the anterior frontotemporal regions including the cingulate gyri, uncus, insula, subcortical areas, basal ganglia, and medial thalamic regions. Hypometabolism is limited to frontal, parietal, and temporal cortices during the early stages of FTD but spreads outwards as the disease progresses. PET scanning has indicated the presence of hypometabolism in orbitofrontal, dorsolateral, medial prefrontal cortex, and anterior in bvFTD patients. Likewise, svPPA patients exhibit more asymmetrical presentations with exclusively left temporal lobe hypometabolism, nfvPPA imaging studies show hypometabolism in the left frontal and superior temporal regions, and logopenic variants demonstrate a left parietotemporal hypometabolism extending to anterior temporal and frontal regions (Gordon et al., 2016; Day et al., 2013; Ranasinghe et al., 2016; Kato et al., 2016). However, it is important to note that there is considerable individual variability when using imaging to classify FTD related disorders as a consequence of heterogeneity in genetic associations and the underlying pathological causes that are yet to be fully understood. Consequently, there are no specific neurobiological and imaging biomarkers for the diagnosis and classification of FTD subtypes.

COMMONALITIES BETWEEN FRONTOTEMPORAL DEMENTIA AND AMYOTROPHIC LATERAL SCLEROSIS

FTD shares some genetic risk factors, pathological hallmarks, and neurochemical processes with Amyotrophic lateral sclerosis (ALS), a neurological disorder that causes death of the motor neurons in the brain and spinal cord. These include abnormalities in several key proteins, such as SOD1 (superoxide dismutase 1), TARDBP/TDP-43, FUS, C9orf72, and dipeptide repeat proteins. Among these proteins, TDP-43 (encoded by the *TARDBP* gene) and FUS (encoded by the *FUS* gene) are the major components of pathological inclusions in over 90% of all ALS and 55% of FTD cases regardless of the cause (Ling et al., 2013; Mackenzie et al., 2010). TDP-43 and FUS are nucleic acid-binding proteins involved in the biogenesis and processing of coding and noncoding RNAs (Fig. 6.1). In contrast, FUS is a 526-amino acid protein containing a prion-like, low-complexity domain, which is enriched with glutamine, glycine, serine, and tyrosine (Q/G/S/Y) residues (Kato et al., 2012; Cushman et al., 2010), followed by a nuclear export signal, a RNA recognition motif domain, arginine/glycine (R/G)-rich domains, and a zinc-finger motif and nuclear localization signal. FUS not only associates itself with RNA polymerase II at the promoter region (Kwon et al., 2013), but also contributes to the directionality of transcription (Masuda et al., 2015). Furthermore, the interactions of FUS with U1-snRNP ensures transcription-splicing coupling (Lagier-Tourenne et al., 2012; Yu and Reed, 2015). FUS is also

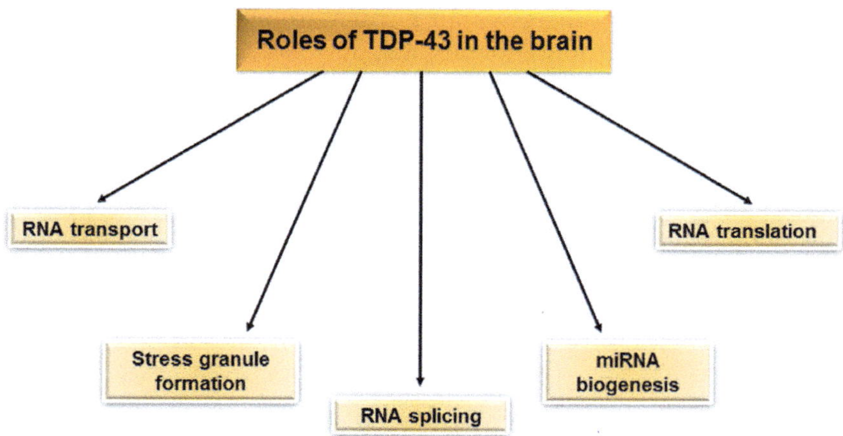

FIGURE 6.1 Roles of TDP-43 in the brain.

involved in alternative splicing and polyadenylation site selection (Lagier-Tourenne et al., 2012; Masuda et al., 2015; Ishigaki et al., 2012). FUS shuttles between the nucleus and the cytosol (Zinszner et al., 1997) and is implicated in RNA localization and translation (Fujii and Takumi, 2005; Yasuda et al., 2013). Among the pleiotropic effects induced by TDP-43 and FUS dysfunctions, neurons that are depleted of TDP-43 and FUS, or express dominant mutations in TDP-43 and FUS, show morphological and molecular changes that indicate potential neuronal and synaptic dysfunctions. It should be noted that TDP-43 and FUS shuttle between the nucleus and the cytosol, where they may form cytoplasmic RNP granules (Bowden and Dormann, 2016) that transport within dendrites and axons. These transporting RNA granules provide a pathway to regulate synaptic strength through localized translation (Holt and Schuman, 2013). Pathological conditions inducing mutations in genes that encode pathological hallmark proteins are commonly observed in the major adult-onset neurodegenerative diseases, underscoring the critical role of TDP-43 and FUS in driving ALS and FTD pathogenesis. Curiously, a common characteristic of TDP-43 pathology is the loss of nuclear TDP-43 with concomitant cytoplasmic TDP-43 accumulation in neurons and glia (Neumann et al., 2006; Mackenzie et al., 2010). This nuclear clearing supports a mechanism of disease that is at least partially driven by the loss of normal TDP-43 function in the nucleus, whereas the presence of cytoplasmic protein inclusions suggests a gain of one or more toxic properties (Ling et al., 2013; Mackenzie et al., 2010). This gene–pathology–phenotype relationship indicates that (1) alterations in TDP-43 and FUS functions may be responsible for triggering disease cascades as mutations in the *TARDBP*and *FUS* genes are closely associated with the pathogenesis of ALS and FTD; (2) regardless of the causes, the pathogenic process converges on TDP-43 as pathological TDP-43 inclusions are present in the majority of ALS and FTD patients (to a much lesser extent for FUS); and (3) the pathogenic mechanisms for TDP-43 and FUS are likely to be a combination of both loss-of-function and gain-of-function properties. Thus, it is critical to first understand the physiological and pathophysiological roles of TDP-43 and FUS in ALS and FTD. Although these proteins are structurally and functionally different and have rather different pathological functions, they all possess some levels of intrinsic disorder and are either directly engaged in or are at least related to the physiological liquid–liquid phase transitions leading to the formation of various proteinaceous membraneless organelles, both normal and pathological (Uversky, 2017). Furthermore, the accumulation of intracellular protein aggregates is a common pathological hallmark of both these conditions. Emerging evidence suggests that impaired RNA processing, disrupted protein homeostasis and cytoplasmic protein aggregation and induction of oxidative stress are major pathogenic pathways that are common in these diseases (Deng et al., 2017). In eukaryotic cells, the clearance of toxic-aggregated proteins is critical for cell survival. It is performed mainly by two protein degradation systems: the ubiquitin–proteasome system (UPS) and autophagy (Ravid and Hochstrasser, 2008). The UPS is mainly used for the degradation of short-lived proteins, while autophagy is a conserved intracellular mechanism for maintaining cellular homeostasis in which damaged or dysfunctional proteins, lipids, and organelles are degraded by the lysosome (Levine and Klionsky, 2004). Autophagy is preferentially used for the selective degradation of long-lived proteins and damaged organelles (Rubinsztein, 2006). There are three distinct autophagic pathways (Cuervo, 2004): (1) macroautophagy; (2) microautophagy; and (3) chaperone-mediated autophagy. Autophagy is not only linked with neuronal cell survival and neurodegeneration (Chu, 2006), but also with transformation. Macroautophagy is constitutively active and highly efficient in neurons under physiological and disease conditions. The disruption of endoplasmic reticulum (ER) mitochondrial signaling results in disruption of Ca^{2+} homeostasis, axonal transport defect, and induction of autophagy. ER and mitochondrial signaling involves tight functional contact between ER and mitochondria. The formation of these contacts involves "tethering proteins" that function to recruit regions of ER to mitochondria (Gomez-Suaga et al., 2017). The integral ER protein VAPB (VAMP associated protein B and C) binds to the outer mitochondrial membrane protein, RMDN3/PTPIP51 (regulator of microtubule dynamics 3) to form one such set of tethers. The tethering of VAPB-RMDN3 regulates autophagy (Gomez-Suaga et al., 2017). Treatment with small interfering RNA (siRNA) knocks down the VAPB or RMDN3 and loosens contact between ER and mitochondria leading to the stimulation of autophagosome formation. In contrast, the overexpression of VAPB or RMDN3 tightens contacts and prevents the formation of autophagosomes. Artificial tethering of ER with mitochondria via expression of a synthetic linker protein also reduces autophagy and this artificial tether rescues the effects of VAPB- or RMDN3-targeted siRNA loss on autophagosome formation. It is also reported that the modulatory effects of ER–mitochondria contacts on autophagy may involve the delivery of Ca^{2+} from ER stores to mitochondria via ITPR (inositol 1,4,5-trisphosphate receptor) signaling (Gomez-Suaga et al., 2017). While the dysfunction of either the UPS or autophagy has been implicated in the formation of ALS-FTD-linked protein aggregates, accumulating evidence suggests that proper functioning of autophagy is the major determinant of motor neuron survival in ALS (Ferrucci et al., 2011; Madeo et al., 2009). These processes may be common to FTD and ALS (Deng et al., 2017; Paillusson et al., 2016). Axon degeneration is characterized by several morphological features including the accumulation of axoplasmic organelles, disassembly of microtubules, and fragmentation of the axonal cytoskeleton. Furthermore, both FTD and ALS can be caused by many mutations in the same set of genes; the most prevalent of these mutations is a GGGGCC repeat expansion in the first intron of *C9ORF72*. Although, many attempts have been made to elucidate the molecular mechanisms underlying the role of this repeat in disease, the exact pathomechanisms are still unclear (Todd and Petrucelli, 2016). A reduction in the expression of the *C9orf72* gene is observed in patients, but a gain-of-function model is now the preferred hypothesis. The hexanucleotide repeat expansion forms RNA foci in the brain of repeat-positive FTD and ALS patients, and these foci are believed to sequester RNA-binding proteins and impair their function in RNA processing. At the same time, the repeat undergoes repeat-associated non-ATG translation to produce dipeptide repeat proteins that also form inclusions in the patient's brain (Todd and Petrucelli, 2016). Collective evidence suggests that many signaling pathways are dysregulated in the ALS and FTD. These include the disruption of ER mitochondrial dysfunction, nucleocytoplasmic transport, mutations in GGGGCC repeat expansion, abnormalities in DNA damage repair, pre-mRNA splicing, and stress granule dynamics pathway.

DIAGNOSIS OF FRONTOTEMPORAL DEMENTIA

Differential diagnosis of various forms of FTD is made on the basis of careful history that examines the progression of behavioral changes, family history, behavior in face-to-face interviews, performance on neuropsychological testing, laboratory studies, and neuroimaging. Blood work can be included with a comprehensive metabolic panel including liver and kidney function tests, complete blood count, vitamin B12 concentration, and thyroid studies. CSF is examined for low levels of Aβ and very high levels of tau protein and onset of rapidly progressive dementia. Among clinical syndromes mentioned above, bvFTD variants of FTD are diagnosed by the presence of three or more following features: (1) early behavioral disinhibition described as socially inappropriate behavior, loss of manners/decorum, impulsivity, and rash actions; (2) early apathy or inertia; (3) early loss of empathy or sympathy, including diminished response to other people's need or sympathies, and diminished social interest; (4) early perseverative or compulsive/ritualistic behavior (i.e., simple repetitive movements, complex, compulsive, or ritualistic behavior, stereotypy of speech); (5) hyperorality and dietary changes, including altered food preferences, binge eating, increased consumption of alcohol and cigarettes, oral exploration or consumption of inedible objects; and (6) neuropsychological profile demonstrates deficits in executive functioning and relative sparing of episodic memory and visuospatial functioning (Fig. 6.2). The diagnosis of the earlier features is aided by magnetic MRI or CT. Alternatively, PET or SPECT imaging has been used to demonstrate hypoperfusion or hypometabolism in the frontal or anterior temporal regions of the brain. It should be noted that there is variable overlap not only among various variants of FTD, but also with AD and atypical parkinsonism and MND. New consensus diagnostic criteria for FTD (Gorno-Tempini et al., 2011) and the progressive aphasias (Rascovsky et al., 2011) have recently been formulated, but they are likely to be refined as more specific information about disease pathophysiology arises and neuroimaging by SPECT demonstrating hypoperfusion or hypometabolism in the frontal or anterior temporal regions can be used to diagnose various subtypes of FTD in general and bvFTD in particular. Definitive bvFTD with definite FTD pathology is diagnosed when a patient meets criteria for possible bvFTD and has one or both of the following: histopathological evidence of FTD or evidence of a known pathogenic mutation.

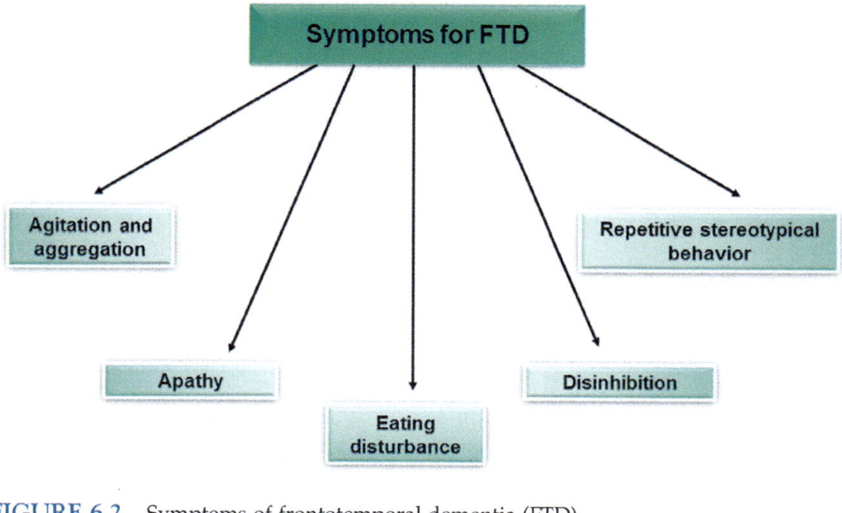

FIGURE 6.2 Symptoms of frontotemporal dementia (FTD).

BIOMARKERS FOR FRONTOTEMPORAL DEMENTIA

FTD poses several diagnostic challenges to clinicians because symptoms of FTD are often mistaken for psychiatric or neurological diseases causing a delay in correct diagnosis, and the majority of patients with FTD present with symptoms at ages between 45 and 65. There is no specific biomarker of FTD. However, there is an increase in ratio of p-tau to total tau in serum and CSF. There is also an increase in levels of neurofilament light-chain protein and the presence of TDP-43 in serum and CSF. These parameters have been used to identify FTD from other types of dementias (Fig. 6.3). As stated previously, binge eating (increased consumption of sweet foods (sugar) and alcohol) is a core criterion for the diagnosis of bvFTD. This results in increased body mass index (BMI). In contrast to earlier studies, which have indicated considerable early loss of significant amounts of brain tissue from bvFTD patients (Broe et al., 2003; Kril and Halliday, 2004; Kril et al., 2005), several recent studies have indicated high levels of triacylglycerol and decreased levels of high-density lipoprotein-cholesterol in a cohort of bvFTD compared to controls, indicating the presence of lipid metabolic abnormality (Ahmed et al., 2014, 2016). A comprehensive lipidomics analysis using liquid chromatography-tandem mass spectrometry of blood samples from patients with bvFTD, AD, and age-matched control subjects has indicated the presence of glycerolipids, phospholipids, sphingolipids, and sterols in the plasma (Kim et al., 2018). Seventeen subclasses of lipids and 3225 putative individual lipid species have been reported to occur in the brain.

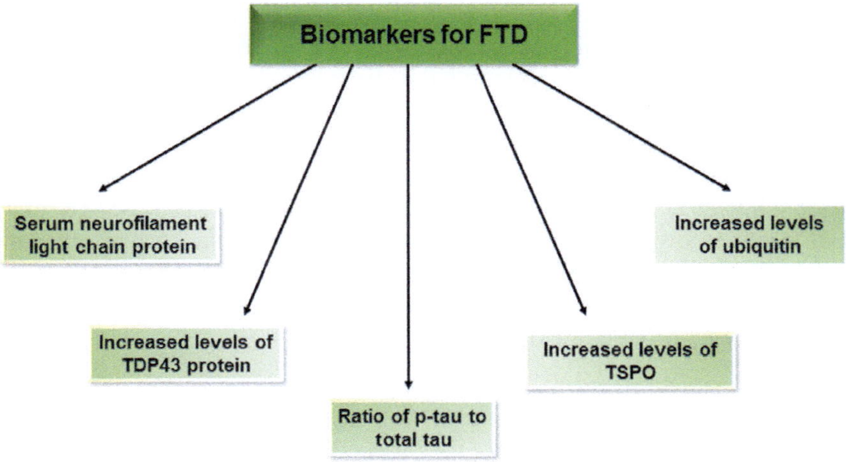

FIGURE 6.3 Biomarkers of frontotemporal dementia (FTD).

Levels of numerous lipid species are significantly altered in bvFTD compared to AD and control. Thus, the total abundance of triacylglycerol is increased significantly in bvFTD, whereas levels of phosphatidylserine and phosphatidylglycerol are decreased significantly in bvFTD. Consumption of a diet high in refined carbohydrates (sugar) by bvFTD patients is known to increase TG levels. This correlation is stronger for those with high BMI (≥28) (Parks, 2002). These results not only suggest manifestation of insulin resistance, but also hypertriglyceridemia and hypoalphalipoproteinemia in bvFTD patients (Kim et al., 2018).

Moreover, levels of monogalactosyldiacylglycerol and sitosteryl ester are significant decreased in bvFTD indicating alterations in eating behavior in bvFTD. It is proposed that occurrence of lipid abnormalities can be used to identify biomarkers for bvFTD (Kim et al., 2018).

In recent years investigators have turned their attention from clinical diagnosis to a biomarker-supported diagnosis, and molecular neuroimaging techniques such as PET have played a leading role in the dementia diagnostic work-up (Gorno-Tempini et al., 2011; Rascovsky et al., 2011). PET techniques have provided major advances, promoting novel approaches to support an early and differential dementia diagnosis (Iaccarino et al., 2017). Thus, ^{18}F-FDG PET, "Pittsburgh compound B" (^{11}C-PiB), ^{18}F-florbetapir, ^{18}F-florbetaben, ^{18}F-flutemetamol, ^{18}F-THK5351, ^{18}F-THK5117, ^{18}F-THK5105, ^{18}F-T807 (also known as ^{18}F-AV1451 or ^{18}F-Flortaucipir), and the ^{11}C-PBB3 have been used to distinguish among AD, PD, CBD, and PSD (Iaccarino et al., 2017). Neuroinflammation in the above pathological conditions is detected by neuroinflammation in PET using ^{11}C or ^{18}F isotopes, such as ^{11}C-PBR28, ^{18}F-DPA714, and ^{11}C-(R)-PK11195 (Iaccarino et al., 2017). The definitive diagnoses of FTD can only be made at autopsy, interim diagnoses usually take into account the base rates of the disorder, clinical criteria, medical history, physical examination, brain imaging, and neuropsychological assessments (Yeaworth and Burke, 2000; Moss et al., 1992). Unfortunately, the diagnosis of FTD can be difficult because of its insidious and gradual onset (Hou et al., 2005) and it can also be misdiagnosed as AD (Rankin et al., 2005). However, accurate differential diagnosis has become increasingly important because of the recent availability of pharmacological treatments that temporarily improve the cognitive and functional abilities of people with AD (Standridge, 2004; Cummings, 2004).

RISK FACTORS FOR FRONTOTEMPORAL DEMENTIA

Like other types of dementias, FTD is characterized by the abnormal protein inclusions (or proteinopathies) in neurons and/or glial cells.

These aggregates are made up of microtubule-associated protein tau (MAPT), the transactive response DNA-binding protein with molecular weight 43 kDa (TDP-43), the fused in sarcoma protein (FUS), and dipeptide proteins generated from mutant forms of the *C9ORF72* gene (Riedl et al., 2014). In FTD and AD, tau becomes increasingly hyperphosphorylated, that is, more phosphorylated at physiological sites and, in addition, de novo at pathological sites (Alonso et al., 1996). Hyperphosphorylation detaches tau from microtubules, and makes it prone to form filamentous inclusions, including neurofibrillary tangles (NFTs) in AD and FTD, and Pick bodies in Pick disease (Lee et al., 2001). However, it is only partly understood how aggregated tau interferes with cellular functions. Of the FTD cases, 25%–50% are inherited (Rademakers et al., 2012), and the mutations are in the genes for MAPT, progranulin (GRN), valosin containing protein (VCP) genes in the chromosome 9 open reading frame 72 (C9orf72), and the charged multivesicular body protein 2B (CHMP2B) (Fig. 6.4) (Riedl et al., 2014; Dejesus-Hernandez et al., 2011). In addition, there are many nonmodifiable risk factors for FTD. They include age, family history, sex, and traumatic brain injury (Dejesus-Hernandez et al., 2011). Furthermore, a link has been proposed recently between neuroinflammation and specific forms of FTD, suggesting that neuroinflammation contribution is an important component of FTD (Bai et al., 2007; Zhang, 2015; Miller et al., 2013). Recent studies on peripheral levels of tumor necrosis factor-α (TNF-α) suggest that early dysregulation of this inflammation mediators is associated with the neurodegenerative process in bvFTD (Paquet et al., 2009; Noble et al., 2010; Zhang, 2015). Detailed investigations have been performed on levels of several

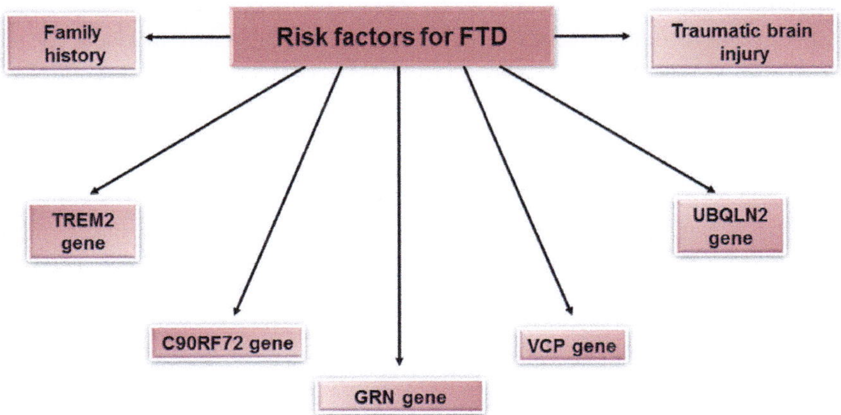

FIGURE 6.4 Risk factors for frontotemporal dementia (FTD).

inflammatory cytokines and chemokines in the CSF from sporadic frontotemporal lobar degeneration (FTLD) patients and it is reported that CSF levels of MCP-1 are unchanged, whereas Interferon-γ-inducible protein-10 (IP-10) levels are increased. In the same group, TNF-α and Interleukin (IL)-15 levels are decreased. These observations support the view that alterations in neuroinflammatory processes is a characteristic feature of both sporadic FTLD and GRN carriers compared to controls (Galimberti et al., 2015). It is also reported that neuroinflammatory responses in FTD begin before patients start experiencing FTD symptoms (Heneka et al., 2014); thus, it is very likely that neuroinflammation can be an early marker for FTD. This has been recently confirmed by PET imaging, a technique which utilizes radioligands to label activated microglia, a key cellular component of the neuroinflammatory response. This procedure offers a potential means to characterize neuroinflammation in vivo. Increases in the translocator protein (TSPO, 18 kDa) expression detected by PET imaging with radioligands of TSPO or peripheral benzodiazepine receptor (PBR) are recognized as a biomarker of activated microglia (Venneti et al., 2009). This procedure may aid in the diagnosis of early FTD. Furthermore, another recent study indicates that autoimmune abnormalities also contribute to increased vulnerability of neurons in FTD syndromes. It is reported that rates of nonthyroid-spectrum autoimmune disorders are twice as common in patients with svPPA and in individuals with a mutation in the GRN gene than in the general population (Yoshiyama et al., 2007). Other factors such as language abnormalities, neurological/psychiatric symptoms, and family history in patient's first-degree relatives to diagnosis FTD indicate the induction of cognitive decline, hypersexuality, and bizarre compulsions (Santacruz et al., 2005; Zahs and Ashe, 2010). Miller et al. also suggest the possibility of a relationship between atypical brain hemispheric lateralization and FTLD-TAU, with an increased level of nonright-handedness in svPPA patients compared with the general population (Santacruz et al., 2005).

NEUROCHEMICAL CHANGES IN FRONTOTEMPORAL DEMENTIA

Very little information is available on neurochemical changes in FTD. A few studies have indicated that neurochemical changes in FTD involve pre- and postsynaptic alterations in neurotransmitters (serotonin and dopamine) along with a decrease in serotonin receptors in frontal and temporal cortex of FTD patients. The disruption in the serotonergic and cholinergic systems (5HT dysfunction) are linked to behavioral changes in FTD (Sparks and Markesbery, 1991; Bowen et al., 2008; Murley and Rowe, 2018). In FTD, pre- and postsynaptic changes in serotonin may play a role in the behavioral disorders of this disease (Sparks and Markesbery, 1991). Although many symptoms may be anatomically specific, the disruption of circuits and networks in the brains of affected patients may produce behavioral symptoms associated with regions far from the areas of tissue loss (Geda et al., 2013). These circuits include the dorsolateral circuit (which mediates aspects of executive function), the preprefrontal basal ganglia (responsible for motivation), and the orbitofrontal circuit (inhibition and social appropriateness) (Kales et al., 2015). Collective evidence suggests that neurochemical changes in FTD include (1) progressive deterioration in social function and personality and (2) insidious decline in language skills, known as primary progressive aphasia which can, in turn, be subdivided according to the predominant pattern of language breakdown into progressive nonfluent aphasia and semantic dementia (Grossman, 2010). Circuits involved in the above changes include the dorsolateral circuit (which mediates aspects of executive function), the preprefrontal basal ganglia (responsible for motivation), and the orbitofrontal circuit (inhibition and social appropriateness) (Kales et al., 2015). In some cases, FTD overlaps with MND both clinically and pathologically, and with a number of the extrapyramidal motor disorders. Around 10% of patients with FTD develop clinical and neurophysiological evidence of MND (Lillo et al., 2010) and likewise patients with MND show behavioral and/or language changes which, in some instances, are severe enough to qualify for a diagnosis of FTD (Lillo et al., 2011) Of the extrapyramidal disorders, CBD and progressive PSP show substantial overlap with FTD and share the finding of abnormal tau pathology (Kertesz et al., 2005). Although, there is no cure for FTD, symptom management with selective serotonin reuptake inhibitors, antipsychotics, and galantamine has been shown to be beneficial. Primary care physicians may play a critical role in identifying patients with FTD and assembling an interdisciplinary team to treat FTD patients.

OXIDATIVE STRESS IN FRONTOTEMPORAL DEMENTIA

Accumulation of misfolded proteins (MAPT, TDP-43, FUS, and tau) in FTD in synapses promote neurodegeneration by altering functioning of a variety of subcellular organelles (Fig. 6.5). In the brain, misfolded

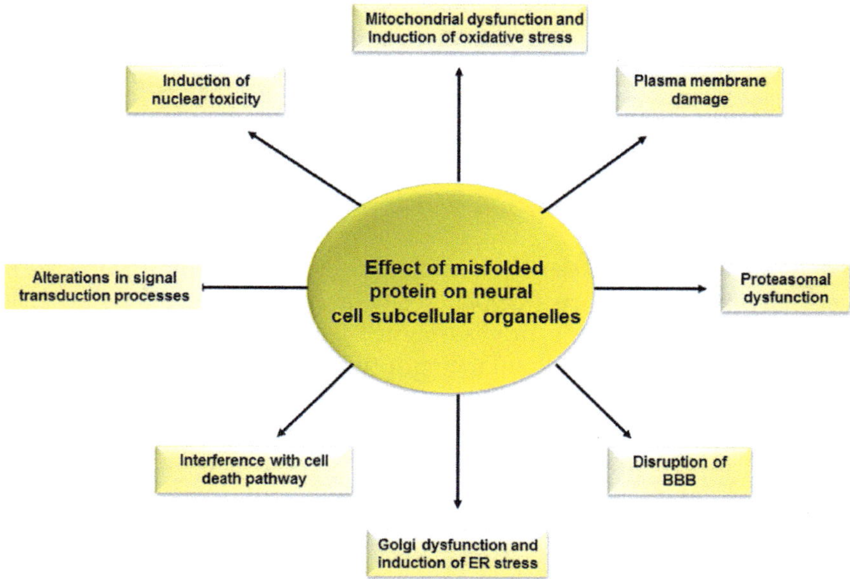

FIGURE 6.5 Effect of protein misfolding on neural cell function.

proteins are degraded via two pathways (Fig. 6.6). The UPS is the main degradation pathway for the majority of intracellular proteins. This system is linked to ER stress responses since ER-associated degradation feeds proteins into the UPS for degradation. The second pathway associated with metabolism of misfolded protein metabolism is called autophagy. This process is used by neural cells not only to degrade misfolded proteins, but also to eliminate unwanted or damaged organelles via lysosomal degradation (Zheng et al., 2014; Son et al., 2012; Martinez-Vicente and Cuervo, 2007). It is becoming increasingly evident that in model systems neurodegeneration can be induced acutely by excessive generation of reactive oxygen species (ROS), induction of ER stress, inhibition of the UPS, inhibition of autophagy, or excessive autophagy (Zheng et al., 2014; Martinez-Vicente and Cuervo, 2007). Neurons have ability to protect themselves from a certain amount of damage before there is severe disruption to function and viability. This suggests that the life or death of an individual neuron is dependent upon the overall burden of accumulated misfolded or aggregated proteins, balanced against the overall capacity of the cell to deal with this effectively and safely (Zheng et al., 2014; Farooqui, 2014). Accumulation of misfolded proteins not only produces more oxidative stress, but also disrupts the blood–brain barrier (Farooqui, 2018). Some risk factors for FTD, such as TBI, contribute to neuronal injury by promoting diverse

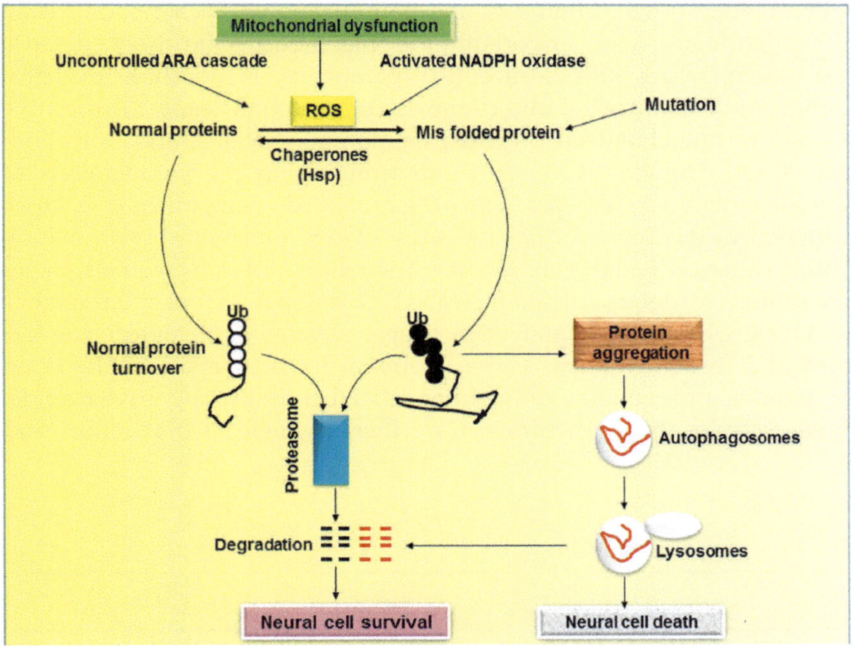

FIGURE 6.6 Metabolism of normal and misfolded proteins in neural cells. Misfolded proteins can be refolded either by chaperones (Hsps) or processed through proteasomal degradation by the addition of ubiquitin (Ub). The accumulation of misfolded proteins or protein aggregates leads to neuronal cell death by an unknown mechanism.

pathological mechanisms including cerebral hypoperfusion, glucose hypometabolism, and mitochondrial dysfunction, which produce ROS. As stated in earlier chapters, ROS comprise hydrogen peroxide (H_2O_2), nitric oxide (NO), superoxide anions, and the highly reactive hydroxyl (OH) and NO. Low or moderate concentrations of ROS are involved in physiological functions in cellular signaling systems. ROS signaling affects cellular energetics by acutely regulating adenosine triphosphate production via activation of uncoupling proteins (Echtay et al., 2002). Moreover, ROS are required for transduction growth signals through tyrosine kinases (Sundaresan et al., 1995). High levels of ROS along with Ca^{2+} overload induces mitochondrial depolarization through activation of the DNA repairing enzyme poly(ADP-ribose) polymerase-1 (PARP-1) and the opening of mitochondrial permeability transition pores. ROS significantly reduce the level of GSH in both astrocytes and neurons, an effect which is dependent on external calcium. Thus, the presence of high levels of Ca^{2+} signal in astrocytes can downregulate the GSH level triggering for neurotoxicity (Angelova and Abramov, 2014). Nitric oxide is an intercellular messenger which modulates cerebral blood flow, thrombosis, and neural activity (Pacher et al., 2007). Nitric oxide is also important for the nonspecific host defense mechanism, which kills intracellular pathogens. In addition, overproduction of ROS through mitochondrial dysfunction leads to oxidative damage to mitochondrial proteins, membrane proteins, and mitochondrial DNA (Higgins et al., 2010) contributing to the pathogenesis of neurodegenerative diseases and their related dementias (Fig. 6.7). Induction of oxidative-nitrosative stress (ONS) and neuroinflammation in turn decrease the availability of nitric oxide and enhance endothelin generation (Farooqui, 2018). Increased expression of proinflammatory cytokines, endothelin-1, and ONS trigger several pathological feedforward and feedback loops. These upstream factors persist in the brain for decades, upregulating amyloid and tau, before the cognitive decline. These cascades lead to neuronal Ca^{2+} increase, neurodegeneration, and

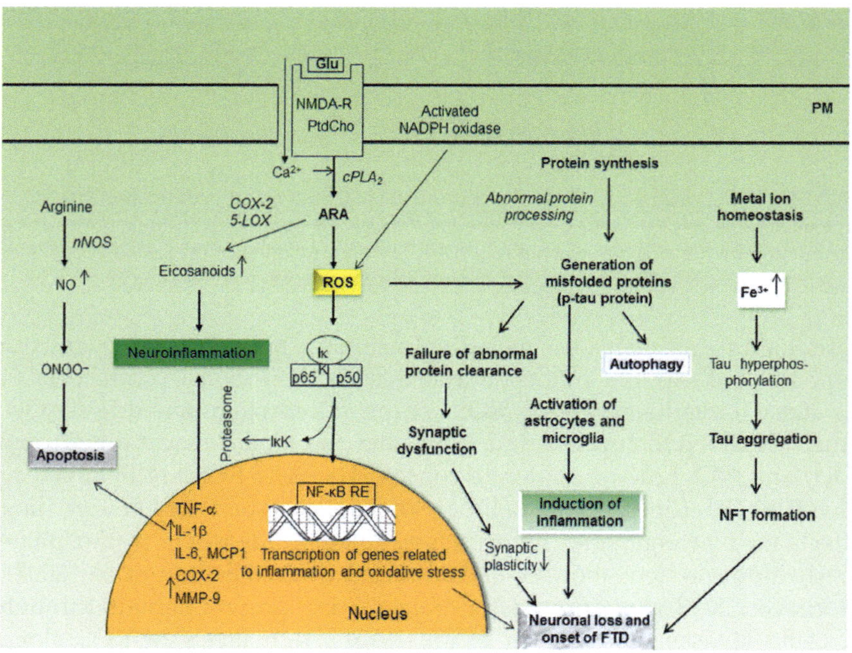

FIGURE 6.7 Hypothetical diagram showing neurochemical processes contributing to the pathogenesis of frontotemporal dementia (FTD).
ARA, arachidonic acid; *cPLA*$_2$, cytosolic phospholipase A_2; *COX-2*, cyclooxygenase-2; *Glu*, Glutamate; *I-κB*, inhibitory subunit of NF-κB; *IL-1β*, interleukin-1β; *IL-6*, interleukin-6; *5-LOX*, 5-lipoxygenase; *MCP-1*, monocyte chemoattractant protein-1; *NF-κB*, nuclear factor-κB; *NF-κB-RE*, nuclear factor-κB-response element; *NMDA-R*, NMDA receptor; *NO*, nitric oxide; *ONOO*$^-$, peroxynitrite; *PtdCho*, phosphatidylcholine; *ROS*, reactive oxygen species; *TNF-α*, tumor necrosis factor-α.

cognitive/memory decline in FTD. There is significant evidence that pathways involving inflammation and ONS play a key pathophysiological role in promoting cognitive dysfunction in various types of dementia including FTD.

NEUROINFLAMMATION IN FRONTOTEMPORAL DEMENTIA

Neuroinflammation represents a normal response not only to pathogen invasion, but also to acute neural trauma and is critical for initiating brain damage repair, upholding basal cognitive functions, and maintaining homeostatic function (Farooqui et al., 2007; Farooqui, 2014). Accumulation of abnormal protein aggregates in AD, FTD, and Lewy body dementia (LDB) along with inflammatory response, which involves the activation of microglia, astrocytes, and increased expression of proinflammatory cytokines (TNF-α, IL-6, IL-1β), and chemokines (MCH-1), contribute to the neurodegeneration (Fig. 6.5) (Wyss-Coray and Mucke, 2002; Farooqui, 2014). In addition to the above parameters, 18 kDa translocator protein (TSPO) is a key biomarker for measuring neuroinflammation in the brain via PET (Kreisl et al., 2018). Increased TSPO density has been observed in brain tissue from patients with AD, FTD, and LDB, which colocalizes with activated microglia and reactive astrocytes. Several radioligands have been developed to measure TSPO density in vivo with PET, and these have been used in clinical studies of different dementia syndromes. However, TSPO radioligands have limitations, including low specific-to-nonspecific signal and differential affinity to a polymorphism on the TSPO gene, which must be taken into consideration in designing and interpreting human PET studies (Kreisl et al., 2018). Nonetheless, most PET studies have shown that increased TSPO binding is associated with various dementias, suggesting that TSPO has potential as a biomarker to further explore the role of neuroinflammation in dementia pathogenesis and may prove useful in monitoring disease progression (Kreisl et al., 2018).

Neuroinflammation in FTD upregulates cerebrovascular pathology through proinflammatory cytokines, endothelin-1, and NO. Neuroinflammation-mediated inflammation and ONS promotes long-term damage involving fatty acids, proteins, DNA, and mitochondria; these amplify and perpetuate several feedforward and feedback pathological loops. The latter includes dysfunctional energy metabolism (compromised mitochondrial ATP production), generation of misfolded proteins, endothelial dysfunction, and blood–brain barrier disruption.

These lead to decreased cerebral blood flow and chronic cerebral hypoperfusion, which modulates metabolic dysfunction and neurodegeneration. In essence, hypoperfusion deprives the brain of its two paramount trophic substances, that is, oxygen and nutrients. Consequently, the brain suffers from synaptic dysfunction and neuronal degeneration/loss, leading to both gray and white matter atrophy, cognitive dysfunction, and AD.

IMMUNE RESPONSES IN FRONTOTEMPORAL DEMENTIA

Microglial cells are known to secrete and use a diverse repertoire of proteins in the innate immune system to regulate synapse formation and maintenance. For example, in early postnatal life, microglia use the classical complement pathway to regulate synapse development in the lateral geniculate nucleus (Ransohoff and Perry, 2009; Schafer et al., 2012; Stevens et al., 2007). During aging, progressive accumulation of complement C1qa in the dentate gyrus of hippocampus not only promotes the induction of cognitive decline, but also impairs memory formation (Stephan et al., 2013). In contrast, loss of complement C3 protein protects against age-dependent declines in synaptic and dendritic spine density in the CA3 region of the hippocampus, and rescues attenuation of long-term potentiation (LTP) (Shi et al., 2015). In addition, microglial cells also use fractalkine receptor CX3CR1 to regulate the growth and maintenance of dendritic spines on hippocampal neurons, which in turn serve as the structural basis of synapse formation (Paolicelli et al., 2011). Finally, genetic ablation of microglia in the adult brain further reveals the essential role of microglia in the maintenance of synaptic functions and motor learning (Parkhurst et al., 2013). It is well known that drastic reduction in progranulin (PGRN) levels and mutation in the human *progranulin* (*GRN*) gene contribute to the pathogenesis of familial FTLD, a proteinopathy characterized by the appearance of neuronal inclusions containing ubiquitinated and fragmented TDP-43 (encoded by TARDBP). The neurotrophic and neuro-immunomodulatory properties of progranulin have recently been reported but are still not well understood (Ghidoni et al., 2008; Sleegers et al., 2009; Kumar-Singh, 2011). Studies on *Grn* knockout ($Grn^{-/-}$) and microglia-specific *Grn* knockout ($Cd11b\text{-}Cre;Grn^{fl/fl}$) mutant mice indicate that deficiency of progranulin (PGRN) results in an age-dependent upregulation of lysosomal and innate immunity genes in microglia. This increases complement production and synaptic pruning activity by microglial cells to preferentially eliminate inhibitory synapses in the ventral thalamus. These defects promote hyperexcitability in the thalamocortical circuits and obsessive-compulsive disorder-like grooming behaviors. Inhibition or blocking of complement activation significantly decreases synaptic pruning by *Grn*$^{-/-}$ microglia, but also mitigates neurodegeneration, behavioral phenotypes, and premature mortality in *Grn*$^{-/-}$ mice. These results suggest that PGRN not only suppresses microglia activation, but also support the view that complement activation and microglia activation are major drivers, rather than consequences, of neurodegeneration caused by PGRN deficiency.

FRONTOTEMPORAL DEMENTIA AND COGNITIVE DYSFUNCTION

Age-related cognitive dysfunction in elderly subjects represents a complex phenomenon that has a heterogeneous etiology. Multiple factors and mechanisms may contribute to the erosion of cognitive function. Several mechanisms may contribute to cognitive loss during aging. These include oxidative stress, inflammation (Craft et al., 2012), alterations in brain neuroplasticity and connectivity (DeCarli et al., 2012), epigenetics (Kosik et al., 2012), and environmental/psychosocial factors (Kremen et al., 2012). These processes are connected with each with each other through multiple metabolic signal transduction pathways, such as increased production of ROS, release of proinflammatory cytokines from microglia, induction of insulin resistance, hypertension, and decrease in misfolded protein clearance. Alterations in the abovementioned pathways along with a decrease in cerebral blood flow may contribute to executive dysfunction, the slowing of attention and mental processing speed, and later to memory deficits, which play an important role in cognitive aging and affect social and occupational activities in elderly humans. The main pathological hallmarks of FTD are the presence of intraneuronal NFTs primarily composed of hyperphosphorylated MAPT, and brain atrophy, together with increased brain oxidative stress and neuroinflammation (Bai et al., 2007; Zhang, 2015; Miller et al., 2013). It is well known that the rate of cross-talk between neurons and glial cells actively controls the generation of dendrites, the increase in synaptic plasticity, and the generation of neurotransmitters (Perea et al., 2009; Fields et al., 2014). The rate of cross-talk between neurons and glial cells is higher in the young human brain then in the elderly because the rate of neurogenesis is higher in younger adults than the elderly. Neurogenesis is a crucial factor in preserving the cognitive function and repair of damaged brain cells affected by aging and brain disorders. Intrinsic factors such as aging, neuroinflammation, oxidative stress, and brain injury, as well as unhealthy lifestyle factors (diet, exercise, and sleep) negatively affect adult neurogenesis (Farooqui, 2015). Proinflammatory cytokines play a key role in this cross-talk. Proinflammatory cytokines are regulators of immune responses, inflammation, and reactions to trauma. Interleukin-1 beta (IL-1β) plays a principal role in immune-to-brain communication. It is released by neurons astrocytes, microglia, and endothelial cells in response to aging in the elderly. IL-1β can also induce the production of other cytokines such as IL-6, and TNFα, which in turn have secondary effects on neural cells. The interruption of astrocytes' functions and hence in glia transmission, may contribute to the pathogenesis of different neuropsychiatric disorders (Webster et al., 2005), as well as neurodegenerative diseases, such as AD, PD, and FTD (Forman et al., 2005; Halassa et al., 2007).

The concept of "tripartite synapse" refers to a cellular network involving both presynaptic and postsynaptic neurons, as well as astrocytes (Perea et al., 2009). Numerous gliotransmitters such as hydrogen sulfide (H_2S), nitric oxide (NO), and carbon monoxide (CO) are released from astrocytes. These gliotransmitters are not only necessary for maintenance of synaptic plasticity in different brain structures (Pascual et al., 2005; Panatier et al., 2006), such as the cortex (Ding et al., 2007) and hippocampus (Jourdain et al., 2007), but are also involved in the modulation of memory and learning processes. The molecular mechanisms contributing to synaptic plasticity are broadly linked to long-term memory. Synapse modifications involve two important processes: LTP and long-term depression (LTD), which cause an increase or a reduction in synaptic strength, respectively. LTP and LTD also play important roles in memory and learning (Kumar, 2011). Neurotransmitters are the chemical messengers that activate, amplify, and harmonize signals between neurons and other cells in the body. Neuronal functions rely on a balance between the number of relevant excitatory and inhibitory processes, which may happen individually or concomitantly (Rico et al., 2011). Regular physical exercise, which includes both aerobic exercises (e.g., walking and cycling) and nonaerobic exercises (e.g., strength and resistance training; flexibility and balance exercises) stimulate synaptic plasticity and neurogenesis in the young's as well the elderly's brains. In addition, neurogenesis can also be aided by hobbies (e.g., reading, word puzzles, and card games) and cognitive training (e.g., computer training games/paradigms that target specific cognitive domains such as memory and attention). Social interactions such as the participation in AD patient support group-related activities, such as mealtime conversations and other forms of social engagement may also have positive effects on neurogenesis (Hughes et al., 2010; Stern, 2002; Scarmeas and Stern, 2003, 2004; Mowszowski et al., 2010). In addition to improving cerebral blood flow aerobic exercise increases the expression of synaptic plasticity genes, gene products (synapsin I and synaptophysin), and various neuroplasticity-related transcription factors such as cyclic adenosine monophosphate response element binding and intracellular kinases (Stranahan et al., 2010; Vaynman et al., 2006). Exercise also modulates genes involved in insulin-like signaling, energy metabolism, and synaptic plasticity along with learning and memory (Reagan, 2007; van Praag et al., 2005). The molecular mechanism by which exercise modulates insulin signaling in brain cells is not fully understood, but based on lifelong running studies in rats, it is proposed that MAP kinase and Wnt signaling may contribute to hippocampal plasticity, neurogenesis, and learning and memory (Reichardt, 2006; Sweatt, 2004; Stranahan et al., 2010).

CONCLUSION

FTD is a devastating neurodegenerative disorder, primarily affecting the frontal and/or temporal lobes of the brain. It is the second most frequent cause of presenile neurodegenerative dementia in those less than 65 years of age. FTD encompasses several disorders, including bvFTD, CBD, and primary progressive aphasia, with symptoms including behavioral and language changes, and executive dysfunction. Typical symptoms of FTD include apathy, agitation and aggression, eating disturbance, and repetitive stereotypical behavior. In bvFTD, eating behavioral changes are common, including hyperphagia, increased sweet preference, and changes in food preference that may be associated with increased BMI, dyslipidemia, and insulin resistance.

While some cases involve tau tangles, others present with aggregations of TDP-43. This protein is also a hallmark of ALS. Other common behavioral features include loss of insight, social inappropriateness, and emotional blunting. MRI studies demonstrate focal atrophy. A careful history and physical examination and use of MRI can help in distinguishing FTD from other common forms of dementia, including AD, dementia with Lewy bodies, and vascular dementia. Despite a wealth of information on FTD, a lot remains unknown about FTD, including the causes of the sporadic forms of FTD. This is not only because of the heterogeneity of clinical presentation and age at disease onset, but also due to the rate of progression of the disease. The overlap with AD makes FTD an even more complex neurological disorder.

References

Ahmed, R.M., MacMillan, M., Bartley, L., Halliday, G.M., Kiernan, M.C., et al., 2014. Systemic metabolism in frontotemporal dementia. Neurology 83, 1812–1818.

Ahmed, R.M., Irish, M., Henning, E., Dermody, N., Bartley, L., 2016. Assessment of eating behavior disturbance and associated neural networks in frontotemporal dementia. JAMA Neurol. 73, 282–290.

Alonso, A.C., Grundke-Iqbal, I., Iqbal, K., 1996. Alzheimer's disease hyperphosphorylated tau sequesters normal tau into tangles of filaments and disassembles microtubules. Nat. Med. 2, 783–787.

Angelova, P.R., Abramov, A.Y., 2014. Interaction of neurons and astrocytes underlies the mechanism of Aβ-induced neurotoxicity. Biochem. Soc. Trans. 42, 1286–1290.

Arai, T., Hasegawa, M., Akiyama, H., Ikeda, K., Nonaka, T., Mori, H., et al., 2006. TDP-43 is a component of ubiquitin-positive tau-negative inclusions in frontotemporal lobar degeneration and amyotrophic lateral sclerosis. Biochem. Biophys. Res. Commun. 351, 602–611.

Avants, B.B., Cook, P.A., Ungar, L., Gee, J.C., Grossman, M., 2010. Dementia induces correlated reductions in white matter integrity and cortical thickness: a multivariate neuroimaging study with sparse canonical correlation analysis. Neuroimage 50, 1004–1016.

Bai, Q., Garver, J.A., Hukriede, N.A., Burton, E.A., 2007. Generation of a transgenic zebrafish model of Tauopathy using a novel promoter element derived from the zebrafish eno2 gene. Nucleic Acids Res. 35, 6501–6516.

Bird, T., Knopman, D., VanSwieten, J., Rosso, S., Feldman, H., et al., 2003. Epidemiology and genetics of frontotemporal dementia/Pick's disease. Ann. Neurol. 54, S29–S31.

Bowden, H.A., Dormann, D., 2016. Altered mRNP granule dynamics in FTLD pathogenesis. J. Neurochem. 138, 112–133.

Bowen, D.M., Procter, A.W., Mann, D.M., Snowden, J.S., Esiri, M.M., et al., 2008. Imbalance of a serotonergic system in frontotemporal dementia: implication for pharmacotherapy. Psychopharmacology (Berl.) 196, 603–610.

Broe, M., Hodges, J.R., Schofield, E., Shepherd, C.E., Kril, J.J., Halliday, G.M., 2003. Staging disease severity in pathologically confirmed cases of frontotemporal dementia. Neurology 60, 1005–1011.

Buratti, E., De Conti, L., Stuani, C., Romano, M., Baralle, M., Baralle, F., et al., 2010. Nuclear factor TDP-43 can affect selected microRNA levels. FEBS J. 277, 2268–2281.

Cairns, N.J., Neumann, M., Bigio, E.H., Holm, I.E., Troost, D., Hatanpaa, K.J., et al., 2007. TDP-43 in familial and sporadic frontotemporal lobar degeneration with ubiquitin inclusions. Am. J. Pathol. 171, 227–240.

Capozzo, R., Sassi, C., Hammer, M.B., et al., 2017. Clinical and genetic analyses of familial and sporadic frontotemporal dementia patients in Southern Italy. Alzheimers Dement. 13, 858–869.

Chow, T.W., Izenberg, A., Binnset, M.A., et al., 2008. Magnetic resonance imaging in frontotemporal dementia shows subcortical atrophy. Dement. Geriatr. Cogn. Disord. 26, 79–88.

Chu, C.T., 2006. Autophagic stress in neuronal injury and disease. J. Neuropathol. Exp. Neurol. 65, 423–432.

Craft, S., Foster, T.C., Landfield, P.W., Maier, S.F., Resnick, S.M., Yaffe, K., 2012. Session III: mechanisms of age-related cognitive change and targets for intervention: inflammatory, oxidative, and metabolic processes. J. Gerontol. A Biol. Sci. Med. Sci. 67, 754–759.

Cuervo, A.M., 2004. Autophagy: in sickness and in health. Trends Cell Biol. 14, 70–77.

Cummings, J.L., 2004. Alzheimer's disease. N. Engl. J. Med. 351, 56–67.

Cushman, M., Johnson, B.S., King, O.D., Gitler, A.D., Shorter, J., 2010. Prion-like disorders: blurring the divide between transmissibility and infectivity. J. Cell Sci. 123, 1191–1201.

Day, G.S., Farb, N.A., Tang-Wai, D.F., Masellis, M., Black, S.E., Freedman, M., et al., 2013. Salience network resting-state activity: prediction of frontotemporal dementia progression. JAMA Neurol. 70, 1249–1253.

DeCarli, C., Kawas, C., Morrison, J.H., Reuter-Lorenz, P.A., Sperling, R.A., et al., 2012. Session II: mechanisms of age-related cognitive change and targets for intervention: neural circuits, networks, and plasticity. J. Gerontol. A Biol. Sci. Med. Sci. 67, 747–753.

Dejesus-Hernandez, M., Mackenzie, I.R., Boeve, B.F., et al., 2011. Expanded GGGGCC hexanucleotide repeat in non-coding region of C9ORF72 causes chromosome 9p-linked frontotemporal dementia and amyotrophic lateral sclerosis. Neuron 72, 245–256.

Deng, Z., Sheehan, P., Chen, S., Yue, Z., 2017. Is amyotrophic lateral sclerosis/frontotemporal dementia an autophagy disease? Mol. Neurodegener. 12, 90.

Ding, S., Fellin, T., Zhu, Y., Lee, S.-Y., Auberson, Y.P., et al., 2007. Enhanced astrocytic Ca^{2+} signals contribute to neuronal excitotoxicity after status epilepticus. J. Neurosci. 27, 10674–10684.

Echtay, K.S., Roussel, D., St-Pierre, J., et al., 2002. Superoxide activates mitochondrial uncoupling proteins. Nature 415, 96–99.

Fallini, C., Bassell, G.J., Rossoll, W., 2012. The ALS disease protein TDP-43 is actively transported in motor neuron axons and regulates axon outgrowth. Hum. Mol. Genet. 21, 3703–3718.

Farooqui, A.A., 2014. Inflammation and Oxidative stress in Neurological Disorders. Springer International Publishing, Switzerland.

Farooqui, A.A., 2015. High Calorie Diet and the Human Brain. Springer International Publishing, Switzerland.

Farooqui, A.A., 2018. Ischemic and Traumatic Brain and Spinal cord Injuries. Academic Press, an Imprint of Elsevier, San Diego.

Farooqui, A.A., Horrocks, L.A., Farooqui, T., 2007. Modulation of inflammation in brain: a matter of fat. J. Neurochem. 577–599.

Ferrari, R., Hernandez, D.G., Nalls, M.A., et al., 2014. Frontotemporal dementia and its subtypes: a genome-wide association study. Lancet Neurol. 13, 686–699.

Ferrucci, M., Fulceri, F., Toti, L., Soldani, P., Siciliano, G., et al., 2011. Protein clearing pathways in ALS. Arch. Italiennesde Biol. 149, 121–149.

Fields, R.D., Araque, A., Johansen-Berg, H., Lim, S.-S., Lynch, G., et al., 2014. Glial biology in learning and cognition. Neuroscientist 2014 (20), 426–431.

Forman, M.S., Lal, D., Zhang, B., Dabir, D.V., Swanson, E., et al., 2005. Transgenic mouse model of tau pathology in astrocytes leading to nervous system degeneration. J. Neurosci. 253539–253550.

Fujii, R., Takumi, T., 2005. TLS facilitates transport of mRNA encoding an actin-stabilizing protein to dendritic spines. J. Cell Sci. 118, 5755–5765.

Galimberti, D., Bonsi, R., Fenoglio, C., Serpente, M., Cioffi, S.M., et al., 2015. Inflammatory molecules in Frontotemporal Dementia: cerebrospinal fluid signature of progranulin mutation carriers. Brain Behav. Immun. 49, 182–187.

Garibotto, V., Borroni, B., Agosti, C., et al., 2011. Subcortical and deep cortical atrophy in frontotemporal lobar degeneration. Neurobiol. Aging 32, 875–884.

Geda, Y.E., Schneider, L.S., Gitlin, L.N., Miller, D.S., Smith, G.S., Bell, J., et al., 2013. Neuropsychiatric symptoms in Alzheimer's disease: past progress and anticipation of the future. Alzheimers Dement. 9, 602–608.

Geser, F., Martinez-Lage, M., Kwong, L.K., Lee, V.M., Trojanowski, J.Q., et al., 2009. Amyotrophic lateral sclerosis, frontotemporal dementia and beyond: the TDP-43 diseases. J. Neurol. 256, 1205–1214.

Ghidoni, R., Benussi, L., Glionna, M., Franzoni, M., Binetti, G., 2008. Low plasma progranulin levels predict progranulin mutations in frontotemporal lobar degeneration. Neurology 71, 1235–1239.

Goldman, J.S., Farmer, J.S., Wood, E.M., Johnson, J.K., Boxer, A., et al., 2005. Comparison of family histories in FTLD subtypes and related tauopathies. Neurology 65, 1817–1819.

Gomez-Suaga, P., Paillusson, S., Miller, C.C.J., 2017. ER-mitochondria signaling regulates autophagy. Autophagy 13, 1250–1251.

Gordon, E., Rohrer, J.D., Fox, N.C., 2016. Advances in neuroimaging in frontotemporal dementia. J. Neurochem. 138 (Suppl. 1), 193–210.

Gorno-Tempini, M.L., Brambati, S.M., Ginexet, V., et al., 2008. The logopenic/phonological variant of primary progressive aphasia. Neurology 71, 1227–1234.

Gorno-Tempini, M.L., Hillis, A.E., Weintraub, S., Kertesz, A., Mendez, M., Cappa, S.F., et al., 2011. Classification of primary progressive aphasia and its variants. Neurology 76, 1006–1014.

Grossman, M., 2010. Primary progressive aphasia: clinicopathological correlations. Nat. Rev. Neurol. 6, 88–97.

Halabi, C., Halabi, A., Dean, D.L., et al., 2013. Patterns of striatal degeneration in frontotemporal dementia. Alzheimer Dis. Assoc. Disord. 27, 74–83.

Halassa, M.M., Fellin, T., Haydon, P.G., 2007. The tripartite synapse: roles for gliotransmission in health and disease. Trends Mol. Med. 13, 54–63.

Hasegawa, M., Arai, T., Nonaka, T., Kametani, F., Yoshida, M., Hashizume, Y., et al., 2008. Phosphorylated TDP-43 in frontotemporal lobar degeneration and amyotrophic lateral sclerosis. Ann. Neurol. 64, 60–70.

Heneka, M.T., Kummer, M.P., Latz, E., 2014. Innate immune activation in neurodegenerative disease. Nat. Rev. Immunol. 14, 463–477.

Higgins, G.C., Beart, P.M., Shin, Y.S., et al., 2010. Oxidative stress: emerging mitochondrial and cellular themes and variations in neuronal injury. J. Alzheimers Dis. 20 (Suppl. 2), S453–S473.

Holt, C.E., Schuman, E.M., 2013. The central dogma decentralized: new perspectives on RNA function and local translation in neurons. Neuron 80, 648–657.

Hou, C.E., Miller, B.L., Kramer, J.H., 2005. Patterns of autobiographical memory loss in dementia. Int. J. Geriatr. Psychiatry. 20, 809–815.

Hughes, T.F., Chang, C.C.H., Vander Bilt, J., Ganguli, M., 2010. Engagement in reading and hobbies and risk of incident dementia: the MoVIES project. Am. J. Alzheimer's Dis. Other Demen. 25, 432–438.

Iaccarino, L., Sala, A., Caminiti, S.P., Perani, D., 2017. The emerging role of PET imaging in dementia. F1000Res. 6, 1830.

Igaz, L.M., Kwong, L.K., Chen-Plotkin, A., Winton, M.J., Unger, T.L., Xu, Y., et al., 2009. Expression of TDP-43 C-terminal fragments in vitro recapitulates pathological features of TDP-43 proteinopathies. J. Biol. Chem. 284, 8516–8524.

Ishigaki, S., Masuda, A., Fujioka, Y., et al., 2012. Position-dependent FUS-RNA interactions regulate alternative splicing events and transcriptions. Sci. Rep. 2, 529.

Jourdain, P., Bergersen, L.H., Bhaukaurally, K., Bezzi, P., Santello, M., et al., 2007. Glutamate exocytosis from astrocytes controls synaptic strength. Nat. Neurosci. 10, 331–339.

Kales, H.C., Gitlin, L.N., Lyketsos, C.G., 2015. Assessment and management of behavioral and psychological symptoms of dementia. BMJ 350, h369.

Kato, M., Han, T.W., Xie, S., et al., 2012. Cell-free formation of RNA granules: low complexity sequence domains form dynamic fibers within hydrogels. Cell 149, 753–767.

Kato, T., Inui, Y., Nakamura, A., Ito, K., 2016. Brain fluorodeoxyglucose (FDG) PET in dementia. Ageing Res. Rev. 30, 73–84.

Kertesz, A., 2005. Frontotemporal dementia: one disease, or many? Probably one, possibly two. Alzheimer Dis. Assoc. Disord. 19, S19–S24.

Kertesz, A., Munoz, D.G., 2002. Frontotemporal dementia. Med. Clin. North Am. 86, 501–518.

Kertesz, A., McMonagle, P., Blair, M., Davidson, W., Munoz, D.G., 2005. The evolution and pathology of frontotemporal dementia. Brain 128, 1996–2005.

Khan, B.K., Yokoyama, J.S., Takada, L.T., Sha, S.J., Rutherford, N.J., Fong, J.C., et al., 2012. Atypical, slowly progressive behavioural variant frontotemporal dementia associated with C9ORF72 hexanucleotide expansion. J. Neurol. Neurosurg. Psychiatry. 83, 358–364.

Kim, W.S., Jary, E., Pickford, R., He, Y., Ahmed, R.M., Piguet, O., et al., 2018. Lipidomics analysis of behavioral variant frontotemporal dementia: a scope for biomarker development. Front. Neurol. 9, 104.

Kosik, K.S., Rapp, P.R., Raz, N., Small, S.A., Sweatt, J.D., et al., 2012. Mechanisms of age-related cognitive change and targets for intervention: epigenetics. J. Gerontol. A Biol. Sci. Med. Sci. 67, 741–746.

Kreisl, W.C., Henter, I.D., Innis, R.B., 2018. Imaging translocator protein as a biomarker of neuroinflammation in dementia. Adv. Pharmacol. 82, 163–185.

Kremen, W.S., Lachman, M.E., Pruessner, J.C., Sliwinski, M., Wilson, R.S., 2012. Mechanisms of age-related cognitive change and targets for intervention: social interactions and stress. J. Gerontol. A Biol. Sci. Med. Sci. 67, 760–765.

Kril, J.J., Halliday, G.M., 2004. Clinicopathological staging of frontotemporal dementia severity: correlation with regional atrophy. Dement. Geriatr. Cogn. Disord. 17, 311–315.

Kril, J.J., Macdonald, V., Patel, S., Png, F., Halliday, G.M., 2005. Distribution of brain atrophy in behavioral variant frontotemporal dementia. J. Neurol. Sci. 232, 83–90.

Kumar, A., 2011. Long-term potentiation at CA3–CA1 hippocampal synapses with special emphasis on aging, disease, and stress. Front. Aging Neurosci. 3, 1–20.

Kumar-Singh, S., 2011. Progranulin and TDP-43: mechanistic links and future directions. J. Mol. Neurosci. 45, 561–573.

Kwon, I., Kato, M., Xiang, S., et al., 2013. Phosphorylation-regulated binding of RNA polymerase II to fibrous polymers of low-complexity domains. Cell 155, 1049–1060.

Lagier-Tourenne, C., Polymenidou, M., Hutt, K.R., et al., 2012. Divergent roles of ALS-linked proteins FUS/TLS and TDP-43 intersect in processing long pre-mRNAs. Nat. Neurosci. 15, 1488–1497.

Lashley, T., Rohrer, J.D., Mahoney, C., et al., 2014. A pathogenic progranulin mutation and C9orf72 repeat expansion in a family with frontotemporal dementia. Neuropathol. Appl. Neurobiol. 40, 502–513.

Lee, V.M., Goedert, M., Trojanowski, J.Q., 2001. Neurodegenerative tauopathies. Annu. Rev. Neurosci. 24, 1121–1159.

Levine, B., Klionsky, D.J., 2004. Development by self-digestion: molecular mechanisms and biological functions of autophagy. Dev. Cell. 6, 463–477.

Lillo, P., Garcin, B., Hornberger, M., Bak, T.H., Hodges, J.R., 2010. Neurobehavioral features in frontotemporal dementia with amyotrophic lateral sclerosis. Arch. Neurol. 67, 826–830.

Lillo, P., Mioshi, E., Zoing, M.C., Kiernan, M.C., Hodges, J.R., 2011. How common are behavioural changes in amyotrophic lateral sclerosis? Amyotroph. Lateral Scler. 12, 45–51.

Ling, S.-C., Polymenidou, M., Cleveland, D.W., 2013. Converging mechanisms in ALS and FTD: disrupted RNA and protein homeostasis. Neuron 79, 416–438.

Looi, J.C., Lindberg, O., Zandbelt, B.B., et al., 2008. Caudate nucleus volumes in frontotemporal lobar degeneration: differential atrophy in subtypes. AJNR Am. J. Neuroradiol. 29, 1537–1543.

Lu, P.H., Lee, G.J., Shapira, J., Jimenez, E., Mather, M.J., Thompson, P.M., et al., 2014. Regional differences in white matter breakdown between frontotemporal dementia and early-onset Alzheimer's Disease. J. Alzheimers Dis. 39, 261–269.

Mackenzie, I.R., Rademakers, R., Neumann, M., 2010. TDP-43 and FUS in amyotrophic lateral sclerosis and frontotemporal dementia. Lancet Neurol. 9, 995–1007.

Mackenzie, I.R., Neumann, M., Baborie, A., 2011. A harmonized classification system for FTD-TDP pathology. Acta Neuropathol. 122, 111–113.

Madeo, F., Eisenberg, T., Kroemer, G., 2009. Autophagy for the avoidance of neurodegeneration. Genes Dev. 23, 2253–2259.

Mahoney, C.J., Beck, J., Rohrer, J.D., et al., 2012. Frontotemporal dementia with the C9ORF72 hexanucleotide repeat expansion: clinical, neuroanatomical and neuropathological features. Brain 135, 736–750.

Martinez-Vicente, M., Cuervo, A.M., 2007. Autophagy and neurodegeneration: when the cleaning crew goes on strike. Lancet Neurol. 6, 352–361.

Masuda, A., Takeda, J.-I., Okuno, T., et al., 2015. Position-specific binding of FUS to nascent RNA regulates mRNA length. Genes Dev. 29, 1045–1057.

Miller, Z.A., Rankin, K.P., Graff-Radford, N.R., Takada, L.T., Sturm, V.E., Cleveland, C.M., et al., 2013. TDP-43 frontotemporal lobar degeneration and autoimmune disease. J. Neurol. Neurosurg. Psychiatry. 84, 956–962.

Moss, M.B., Albert, M.S., Kemper, T.L., 1992. Neuropsychology of frontal lobe dementia. In: White, R.F. (Ed.), Clinical Syndromes in Adult Neuropsychology: The Practitioner's Handbook. Elsevier Science, New York, NY, pp. 287–303.

Mowszowski, L., Batchelor, J., Naismith, S.L., 2010. Early intervention for cognitive decline: can cognitive training be used as a selective prevention technique? Int. Psychogeriatr. 22, 537–548.

Murley, A.G., Rowe, J.B., 2018. Neurotransmitter deficits from frontotemporal lobar degeneration. Brain 24, 1–23.

Neary, D., Snowden, J., Mann, D., 2005. Frontotemporal dementia. Lancet Neurol. 4, 771–780.

Nestor, P.J., 2007. Neuroimaging and Other Investigative Findings. Cambridge University Press.

Neumann, M., 2009. Molecular neuropathology of TDP-43 proteinopathies. Int. J. Mol. Sci. 10, 232–246.

Neumann, M., Sampathu, D.M., Kwong, L.K., Truax, A.C., Micsenyi, M.C., et al., 2006. Ubiquitinated TDP-43 in frontotemporal lobar degeneration and amyotrophic lateral sclerosis. Science 314, 130–133.

Noble, W., Hanger, D.P., Gallo, J.M., 2010. Transgenic mouse models of tauopathy in drug discovery. CNS Neurol. Disord. Drug Targets 9, 403–428.

Onyike, C.U., Diehl-Schmid, J., 2013. The epidemiology of frontotemporal dementia. Int. Rev. Psychiatry 25, 130–137.

Pacher, P., Beckman, J.S., Liaudet, L., 2007. Nitric oxide and peroxynitrite in health and disease. Physiol. Rev. 87, 315–424.

Paillusson, S., Stoica, R., Gomez-Suaga, P., Lau, D.H.W., Mueller, S., et al., 2016. There's something wrong with my MAM; the ER–mitochondria axis and neurodegenerative diseases. Trends Neurosci. 39, 146–157.

Panatier, A., Theodosis, D.T., Mothet, J.-P., Touquet, B., Pollegioni, L., Poulain, D., et al., 2006. Glia-derived D-serine controls NMDA receptor activity and synaptic memory. Cell 125, 775–784.

Paolicelli, R.C., Bolasco, G., Pagani, F., Maggi, L., Scianni, M., et al., 2011. Synaptic pruning by microglia is necessary for normal brain development. Science 333, 1456–1458.

Paquet, D., Bhat, R., Sydow, A., et al., 2009. A zebrafish model of tauopathy allows in vivo imaging of neuronal cell death and drug evaluation. J. Clin. Invest. 119, 1382–1395.

Parkhurst, C.N., Yang, G., Ninan, I., Savas, J.N., Yates 3rd, J.R., et al., 2013. Microglia promote learning-dependent synapse formation through brain-derived neurotrophic factor. Cell 155, 1596–1609.

Parks, E.J., 2002. Dietary carbohydrate's effects on lipogenesis and the relationship of lipogenesis to blood insulin and glucose concentrations. Br. J. Nutr. 87 (Suppl. 2), S247–S253.

Pascual, O., Casper, K.B., Kubera, C., Zhang, J., Revilla-Sanchez, R., et al., 2005. Astrocytic purinergic signaling coordinates synaptic networks. Science 310, 113–116.

Perea, G., Navarrete, M., Araque, A., 2009. Tripartite synapses: astrocytes process and control synaptic information. Trends Neurosci. 32, 421–431.

Piguet, O., Brooks, W.S., Halliday, G.M., et al., 2004. Similar early clinical presentations in familial and non-familial frontotemporal dementia. J. Neurol. Neurosurg. Psychiatry 75, 1743–1745.

Piguet, O., Petersen, A., Yin Ka Lam, B., Gabery, S., Murphy, K., et al., 2011. Eating and hypothalamus changes in behavioral-variant frontotemporal dementia. Ann. Neurol. 69, 312–319.

Rabinovici, G.D., Miller, B.L., 2010. Frontotemporal lobar degeneration: epidemiology, pathophysiology, diagnosis and management. CNS Drugs 24, 375–398.

Rademakers, R., Neumann, M., Mackenzie, I.R., 2012. Advances in understanding the molecular basis of frontotemporal dementia. Nat. Rev. Neurol. 8, 423–434.

Ranasinghe, K.G., Rankin, K.P., Pressman, P.S., Perry, D.C., Lobach, I.V., Seeley, W.W., et al., 2016. Distinct subtypes of behavioral variant frontotemporal dementia based on patterns of network degeneration. JAMA Neurol. 73, 1078–1088.

Rankin, K.P., Baldwin, E., Pace-Savitsky, C., et al., 2005. Self-awareness and personality change in dementia. J. Neurol. Neurosurg. Psychiatry 76, 632–639.

Ransohoff, R.M., Perry, V.H., 2009. Microglial physiology: unique stimuli, specialized responses. Annu. Rev. Immunol. 27, 119–145.

Rascovsky, K., Hodges, J.R., Knopman, D., Mendez, M.F., Kramer, J.H., et al., 2011. Sensitivity of revised diagnostic criteria for the behavioural variant of frontotemporal dementia. Brain 134, 2456–2477.

Ratnavalli, E., Brayne, C., Dawson, K., Hodges, J.R., 2002. The prevalence of frontotemporal dementia. Neurology 58, 1615–1621.

Ravid, T., Hochstrasser, M., 2008. Diversity of degradation signals in the ubiquitin-proteasome system. Nat. Rev. Mol. Cell Biol. 9, 679–690.

Reagan, L.P., 2007. Insulin signaling effects on memory and mood. Curr. Opin. Pharmacol. 7, 633–637.

Reichardt, L.F., 2006. Neurotrophin-regulated signalling pathways. Philos. Trans. R. Soc. Lond. B Biol. Sci. 361, 1545–1564.

Rico, E.P., Rosemberg, D.B., Seibt, K.J., Capiotti, K.M., Da Silva, R.S., Bonan, C.D., 2011. Zebrafish neurotransmitter systems as potential pharmacological and toxicological targets. Neurotoxicol. Teratol. 33, 608–617.

Riedl, L., Mackenzie, I.R., Förstl, H., Kurz, A., Diehl-Schmid, J., 2014. Frontotemporal lobar degeneration: current perspectives. Neuropsychiatr. Dis. Treat. 10, 297–310.

Rohrer, J.D., Guerreiro, R., Vandrovcova, J., et al., 2009. The heritability and genetics of frontotemporal lobar degeneration. Neurology 73, 1451–1456.

Rubinsztein, D.C., 2006. The roles of intracellular protein-degradation pathways in neurodegeneration. Nature 443, 780–786.

Santacruz, K., Lewis, J., Spires, T., et al., 2005. Tau suppression in a neurodegenerative mouse model improves memory function. Science 309, 476–481.

Scarmeas, N., Stern, Y., 2003. Cognitive reserve and lifestyle. J. Clin. Exp. Neuropsychol. 25, 625–633.

Scarmeas, N., Stern, Y., 2004. Cognitive reserve: implications for diagnosis and prevention of Alzheimer's disease. Curr. Neurol. Neurosci. Rep. 4, 374–380.

Schafer, D.P., Lehrman, E.K., Kautzman, A.G., Koyama, R., Mardinly, A.R., et al., 2012. Microglia sculpt postnatal neural circuits in an activity and complement-dependent manner. Neuron 74, 691–705.

Sha, S., Hou, C., Viskontas, I.V., Miller, B.L., 2006. Are frontotemporal lobar degeneration, progressive supranuclear palsy and corticobasal degeneration distinct diseases? Nat. Clin. Pract. Neurol. 2, 658–665.

Shi, Q., Colodner, K.J., Matousek, S.B., Merry, K., Hong, S., et al., 2015. Complement C3-deficient mice fail to display age-related hippocampal decline. J. Neurosci. 35, 13029–13042.

Sleegers, K., Brouwers, N., Van Damme, P., Engelborghs, S., Gijselinck, I., van der Zee, J., et al., 2009. Serum biomarker for progranulin-associated frontotemporal lobar degeneration. Ann. Neurol. 65, 603–609.

Snowden, J.S., Pickering-Brown, S.M., Mackenzie, I.R., et al., 2006. Progranulin gene mutations associated with frontotemporal dementia and progressive non-fluent aphasia. Brain 129, 3091–3102.

Son, J.H., Shim, J.H., Kim, K.H., Ha, J.Y., Han, J.Y., 2012. Neuronal autophagy and neurodegenerative diseases. Exp. Mol. Med. 44, 89–98.

Sparks, D.L., Markesbery, W.R., 1991. Altered serotonergic and cholinergic synaptic markers in Pick's disease. Arch. Neurol. 48, 796–799.

Standridge, J.B., 2004. Pharmacotherapeutic approaches to the treatment of Alzheimer's disease. Clin. Ther. 26, 615–630.

Stephan, A.H., Madison, D.V., Mateos, J.M., Fraser, D.A., Lovelett, E.A., Coutellier, L., et al., 2013. A dramatic increase of C1q protein in the CNS during normal aging. J. Neurosci. 33, 13460–13474.

Stern, Y., 2002. What is cognitive reserve? Theory and research application of the reserve concept. J. Int. Neuropsychol. Soc. 8, 448–460.

Stevens, B., Allen, N.J., Vazquez, L.E., Howell, G.R., Christopherson, K.S., Nouri, N., et al., 2007. The classical complement cascade mediates CNS synapse elimination. Cell 131, 1164–1178.

Stranahan, A.M., Lee, K., Becker, K.G., Zhang, Y., Maudeley, S., et al., 2010. Hippocampal gene expression patterns underlying the enhancement of memory by running in aged mice. Neurobiol. Aging 31, 1937–1949.

Sudre, C.H., Bocchetta, M., Cash, D., et al., 2017. White matter hyperintensities are seen only in GRN mutation carriers in the GENFI cohort. Neuroimage Clin. 15, 171–180.

Sundaresan, M., Yu, Z.X., Ferrans, V.J., et al., 1995. Requirement for generation of H_2O_2 for platelet-derived growth factor signal transduction. Science 270, 296–299.

Sweatt, J.D., 2004. Mitogen-activated protein kinases in synaptic plasticity and memory. Curr. Opin. Neurobiol. 14, 311–317.

Todd, T.W., Petrucelli, L., 2016. Insights into the pathogenic mechanisms of Chromosome 9 open reading frame 72 (C9orf72) repeat expansions. J. Neurochem. 138 (Suppl. 1), 145–162.

Uversky, V.N., 2017. The roles of intrinsic disorder-based liquid-liquid phase transitions in the "Dr. Jekyll-Mr. Hyde" behavior of proteins involved in amyotrophic lateral sclerosis and frontotemporal lobar degeneration. Autophagy 13, 2115–2162.

Van Deerlin, V.M., Leverenz, J.B., Bekris, L.M., Bird, T.D., Yuan, W., Elman, L.B., et al., 2008. TARDBP mutations in amyotrophic lateral sclerosis with TDP-43 neuropathology: a genetic and histopathological analysis. Lancet Neurol. 7, 409–416.

van Praag, H., Shubert, T., Zhao, C., Gage, F.H., 2005. Exercise enhances learning and hippocampal neurogenesis in aged mice. J. Neurosci. 25, 8680–8685.

Vaynman, S., Ying, Z., Wu, A., Gomez-Pinilla, F., 2006. Coupling energy metabolism with a mechanism to support brain-derived neurotrophic factor-mediated synaptic plasticity. Neuroscience 139, 1221–1234.

Venneti, S., Wiley, C.A., Kofler, J., 2009. Imaging microglial activation during neuroinflammation and Alzheimer's disease. J. Neuroimmune Pharmacol. 4, 227–243.

Warren, J.D., Rohrer, D., Rossor, M.N., 2013. Frontotemporal dementia. BMJ 347, f4827.

Webster, M.J., O'Grady, J., Kleinman, J.E., Weickert, C.S., 2005. Glial fibrillary acidic protein mRNA levels in the cingulate cortex of individuals with depression, bipolar disorder and schizophrenia. Neuroscience 133, 453–461.

Wegorzewska, I., Bell, S., Cairns, N.J., Miller, T.M., Baloh, R.H., 2009. TDP-43 mutant transgenic mice develop features of ALS and frontotemporal lobar degeneration. Proc. Natl. Acad. Sci. U.S.A. 106, 18809–18814.

Wyss-Coray, T., Mucke, L., 2002. Inflammation in neurodegenerative disease – a double-edged sword. Neuron 35, 419–432.

Yasuda, K., Zhang, H., Loiselle, D., Haystead, T., Macara, I.G., Mili, S., 2013. The RNA-binding protein Fus directs translation of localized mRNAs in APC-RNP granules. J. Cell Biol. 203, 737–746.

Yeaworth, R.C., Burke, W.J., 2000. Frontotemporal dementia: a different kind of dementia. Arch. Psychiatr. Nurs. 14, 249–253.

Yoshiyama, Y., Higuchi, M., Zhang, B., et al., 2007. Synapse loss and microglial activation precede tangles in a P301S tauopathy mouse model. Neuron 53, 337–351.

Yu, Y., Reed, R., 2015. FUS functions in coupling transcription to splicing by mediating an interaction between RNAP II and U1 snRNP. Proc. Natl. Acad. Sci. U.S.A. 112, 8608–8613.

Zahs, K.R., Ashe, K.H., 2010. 'Too much good news' - are Alzheimer mouse models trying to tell us how to prevent, not cure, Alzheimer's disease? Trends Neurosci. 33, 381–389.

Zhang, J., 2015. Mapping neuroinflammation in frontotemporal dementia with molecular PET imaging. J. Neuroinflammation 12, 108.

Zhang, Y., Schuff, N., Du, A.T., Rosen, H.J., Kramer, J.H., et al., 2009. White matter damage in frontotemporal dementia and Alzheimer's disease measured by diffusion MRI. Brain 132, 2579–2592.

Zheng, C., Geetha, T., Babu, J.R., 2014. Failure of ubiquitin proteasome system: risk for neurodegenerative diseases. Neurodegener. Dis. 14, 161–175.

Zinszner, H., Sok, J., Immanuel, D., Yin, Y., Ron, D., 1997. TLS (FUS) binds RNA in vivo and engages in nucleo-cytoplasmic shuttling. J. Cell Sci. 110, 1741–1750.

Further Reading

Boeve, B.F., Hutton, M., 2008. Refining frontotemporal dementia with parkinsonism linked to chromosome 17: introducing FTDP-17 (MAPT) and FTDP-17 (PGRN). Arch. Neurol. 65, 460–464.

Ghetti, B., Oblak, A.L., Boeve, B.F., et al., 2015. Invited review: frontotemporal dementia caused by microtubule-associated protein tau gene (MAPT) mutations: a chameleon for neuropathology and neuroimaging. Neuropathol. Appl. Neurobiol. 41, 24–46.

Lagier-Tourenne, C., Cleveland, D.W., 2009. Rethinking ALS: the FUS about TDP-43. Cell 136, 1001–1004.

Lau, D.H.W., Hartopp, N., Welsh, N.J., Mueller, S., Glennon, E.B., et al., 2018. Disruption of ER-mitochondria signalling in fronto-temporal dementia and related amyotrophic lateral sclerosis. Cell Death Dis. 9, 327.

Moreno, F., Indakoetxea, B., Barandiaran, M., et al., 2017. The unexpected co-occurrence of GRN and MAPT p.A152T in Basque families: clinical and pathological characteristics. PLoS One 12, e0178093.

O'Brien, J.S., Sampson, E.L., 1965. Lipid composition of the normal human brain: gray matter, white matter, and myelin. J. Lipid Res. 6, 537–544.

Walker, A.J., Meares, S., Sachdev, P.S., et al., 2005. The differentiation of mild frontotemporal dementia from Alzheimer's disease and healthy aging by neuropsychological tests. Int. Psychogeriatr. 17, 57–68.

CHAPTER

7

Potential Treatment Strategies for Dementia With Pharmacological and Nonpharmacological Interventions

INTRODUCTION

Dementia is an irreversible, progressive, and multifactorial neurodegenerative disorder associated with deterioration of memory, disturbances in language, psychological and psychiatric changes, and impairments in activities of daily living. Dementia is accompanied by neuronal death that leads to brain atrophy years before its symptoms are manifested. Currently, there is no FDA-approved pharmacological treatment for dementia and extensive investigations are underway to slow symptoms and reduce cognitive impairment caused by dementia (Ijaopo, 2017). As stated in Chapter 1, Neurochemical Aspects of Dementia, the most common types of dementia are Alzheimer's disease (AD), vascular dementia, and Lewy body dementia (LBD). These dementias coexist in the brain and share common symptoms (agitation, delusions, hallucinations, dysphoria, anxiety, aggression, euphoria, disinhibition, irritability/lability, and apathy) and modifiable risk factors, which have been used as treat targets to treat dementia in numerous prevention trials. As stated in earlier chapters, dementia is not only accompanied by blood–brain barrier (BBB) disruption, induction of oxidative stress, mitochondrial impairment, neuroinflammation, hypoperfusion, hypometabolism, and aberrant cell-cycle reentry, but is

Molecular Mechanisms of Dementia
DOI: https://doi.org/10.1016/B978-0-12-816347-4.00007-6

also mediated by the accumulation of misfolded proteins (Aβ, hyperphosphorylated tau, and α-synuclein). In addition, AD and Parkinson's disease (PD) types of dementias not only involve a decrease in acetylcholine levels and a reduction of cerebral blood flow (Mondragon-Rodriguez et al., 2010; Farooqui, 2013), but are also linked with insulin resistance, type 2 diabetes, and metabolic syndrome (Farooqui, 2013). Involvement of cholinergic dysfunction in the pathogenesis of dementia is supported by molecular neuroimaging [single photon emission computed tomography (SPECT) and positron emission tomography (PET)] studies (Roy et al., 2016; Pagano et al., 2017). SPECT and PET studies using selective radioligands for cholinergic markers, such as [^{11}C]MP4A and [^{11}C]PMP PET for acetylcholinesterase (AChE), [^{123}I]5IA SPECT for the α4β2 nicotinic acetylcholine receptor, and [^{123}I]IBVM SPECT for the vesicular acetylcholine transporter, have indicated that cortical AChE activity is significantly decreased in AD, LBD, and Parkinson's disease dementia and this decrease is correlated with certain aspects of cognitive function (attention and working memory) (Roy et al., 2016; Pagano et al., 2017).

Key to treating dementia is the complete understanding of the processes and molecular mechanisms that trigger the neurodegenerative process. Treatments for dementia are divided into two categories: pharmacological treatments and nonpharmacological treatments. The pharmacological treatment of dementia involves important challenges such as complexities in the clinical presentation and diagnosis of dementia. Present pharmacological treatments of dementias are based on the treatment of symptoms and not the neurochemical mechanisms that contribute to dementia syndrome. Neuroleptics, antidepressants, sedatives/hypnotics, and anxiolytics are frequently prescribed. These medications have many issues related to tolerability and side effects and are known to produce increased risk of stroke and mortality in the elderly patients with dementia (Porsteinsson et al., 2014). The use of benzodiazepines to treat agitation in dementia patients may increase cognitive decline (Bierman et al., 2007) and expose patients to an immediate risk of injurious falls (Berry et al., 2016). It is suggested that the pharmacological treatment of dementia should be planned after the comprehensive diagnosis of the subtype of dementia along with analysis of behavioral and psychological changes. The failure of a correct diagnosis may have significant and deleterious impacts on quality of life for a person suffering from dementia.

In animal models, AD type of dementia is treated by using acetylcholinesterase inhibitors and *N*-methyl-D-aspartate (NMDA) receptor antagonists, secretase inhibitors, amyloid binders, and Tau therapies. Ongoing clinical trials with Aβ antibodies (solanezumab, gantenerumab, crenezumab) seem to be promising, while vaccines against the tau protein (AADvac1 and ACI-35) are now in early-stage trials (Godyń et al., 2016). Furthermore, interesting results have also been obtained in trials using small molecules such as inhibitors of β-secretase (MK-8931, E2609), a combination of 5-HT6 antagonist (idalopirdine) with donepezil, inhibition of advanced glycation end product receptors by azeliragon, or modulation of the acetylcholine response of α-7 nicotinic acetylcholine receptors by encenicline (Godyń et al., 2016). Until more evidence is available in human subjects, the abovementioned drugs cannot be used for the treatment of dementia. It must be mentioned that the discovery of new and effective pharmacological drugs for the treatment of AD, PD, and their related dementias is a difficult and time-consuming process because molecular mechanisms contributing to these neurodegenerative diseases and related dementia are not known and clinical trials of the abovementioned pharmacological drugs are complicated not only by the half-lives of the drugs in circulation and site specificity in the brain, but also by BBB permeability. Furthermore, many age-related pharmacokinetic changes occur in all older people (Hilmer et al., 2007) and alterations in blood–brain permeability in people with dementia means that they may be more sensitive to neurological and cognitive effects of medications than their peers (Farrall and Wardlaw, 2009). These pharmacokinetic changes are additional to drug–disease interactions that occur in dementia (Lindblad et al., 2006). The safety profile and efficacy of many medications in people with dementia are undetermined due to their active exclusion from 85% of published clinical trials (Van Spall et al., 2007).

Most commonly used therapeutic agents are inhibitors of AChE and memantine. These inhibitors do not treat the cause of dementia but provide relief from cognitive symptoms. Magnetic resonance imaging (MRI) and PET brain scanning studies have indicated that early signs of AD and PD-linked dementia pathology in patients appear ~4 to 17 years before the onset of dementia (Villemagne et al., 2013). In addition to the AChE inhibitors, other therapeutic agents that are commonly used at the present time are antioxidant and anti-inflammatory agents, statin therapy, memantine and nitro-memantine therapy, gene therapy, immunization with vaccines, insulin therapy, and stem cell therapy.

CHOLINERGIC STRATEGIES FOR THE TREATMENT OF DEMENTIA

According to the cholinergic hypothesis, AD type of dementia is caused by a reduction in the synthesis of acetylcholine in cholinergic neurons of the basal forebrain and the loss of cholinergic neurotransmission in the cerebral cortex. The decrease in acetylcholine significantly contributes to the deterioration in cognitive function seen in AD-linked dementia patients (Birks, 2006). Acetylcholinesterase (EC 3.1.1.7) belongs to a family of serine hydrolases. This enzyme breaks down acetylcholine, a neurotransmitter that plays an important role in learning, remembering, thinking, and cognition. The inhibition of acetylcholinesterase by AChE inhibitors reduces the breakdown of acetylcholine and increases the availability of acetylcholine at the cholinergic synapse, enhancing cholinergic transmission and restoring cognition and memory function. These drugs provide symptomatic short-term benefits, without clearly counteracting the progression of moderate AD and AD-linked dementia. Many AChE inhibitors have poor oral bioavailability, brain penetration ability, and pharmacokinetic parameters. To overcome these disadvantages, a new generation of AChE inhibitors, such as donepezil, galantamine, and rivastigmine, has been synthesized (Fig. 7.1). These drugs have been approved by the FDA for the treatment of mild-to-moderate AD in humans and animal models (Birks, 2006). Among these AChE inhibitors, rivastigmine is available as a transdermal patch. Although donepezil, galantamine, and rivastigmine share the same basic mode of action, they differ in terms of

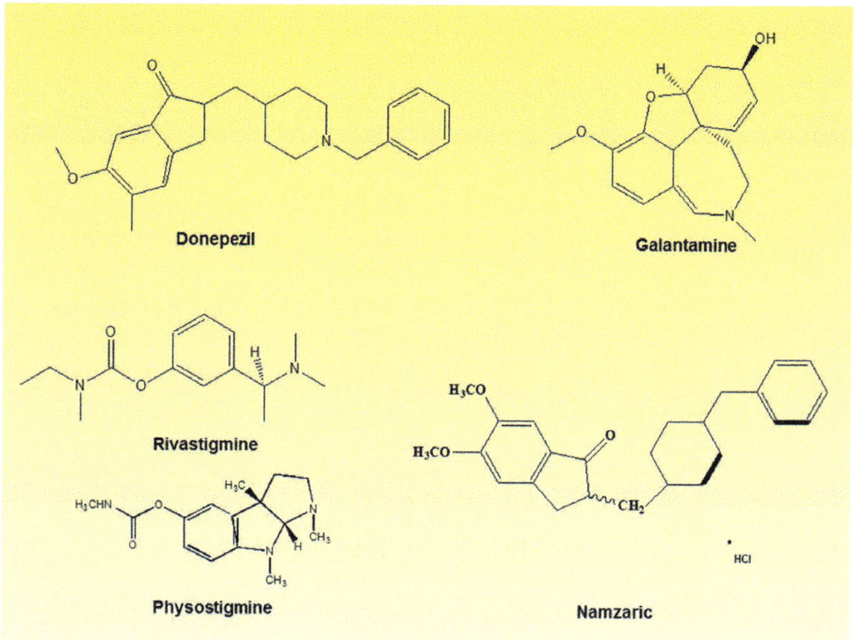

FIGURE 7.1 Chemical structures of acetylcholine receptor antagonists.

their pharmacologic characteristics and route of administration, which can impact their tolerability and safety profile. For instance, both donepezil and galantamine are selective reversible inhibitors of acetylcholinesterase, available in oral forms, and metabolized by the hepatic CYP-450 isoenzymes, mostly CYP 2D6/3A4. Rivastigmine, available in both oral and transdermal patch formulations, is a pseudo-irreversible (slowly reversible) dual inhibitor of acetyl and butyryl cholinesterase, selective for the gastrointestinal tract isoform of acetylcholinesterase, without hepatic metabolism by the CYP-450 system, leading to fewer drug–drug interactions (Table 7.1) (Grossberg, 2003). A combined donepezil–memantine drug with the brand name Namzaric was approved by the FDA in 2014 for the treatment of moderate-to-severe AD in people who are taking donepezil hydrochloride at the recommended clinically efficient dose of 10 mg/day (http://www.alz.org AD report). However, this combinative medicine may cause various side effects, including muscle problems, slow heartbeat and fainting, increased stomach acid levels, nausea, vomiting, and seizures.

Inhibitors of AChE not only reduce AChE activity, but also retard processing and deposition of Aβ (Muñoz-Torrero, 2008). In addition, these inhibitors also increase the cerebral blood flow in AD patients

TABLE 7.1 Cholinesterase Inhibitors Used for the Treatment of Dementia

Name of drug	Dose (mg/day)	Side effects	Reference
Donepezil	5–10	Nausea, vomiting, diarrhea, insomnia, muscle cramps, weight loss, and bradycardia	Grossberg (2003), Mohammad et al. (2017), Khoury et al. (2018)
Rivastigmine	3–12	Nausea, vomiting, diarrhea, insomnia, muscle cramps, weight loss, cardiorespiratory symptoms, and more gastrointestinal side effects	Grossberg (2003), Mohammad et al. (2017), Khoury et al. (2018)
Galantamine	8–24	Nausea, vomiting, diarrhea, insomnia, muscle cramps, weight loss, musculoskeletal symptoms, and more gastrointestinal side effects	Grossberg (2003), Mohammad et al. (2017), Khoury et al. (2018)
Physostigmine	7–18	Nausea, vomiting, diarrhea, insomnia, muscle cramps, weight loss, musculoskeletal symptoms, and more gastrointestinal side effects	Grossberg (2003), Mohammad et al. (2017), Khoury et al. (2018)

both after acute and fairly short periods of treatment (Nordberg, 1999). Numerous clinical trials and postmarketing studies have evaluated the safety of these medications (Mohammad et al., 2017; Khoury et al., 2018). Most AChE inhibitors produce only modest effects in delaying the progression of mild-to-moderate AD-linked dementia. The common adverse effects associated with cholinesterase inhibitors include gastrointestinal, cardiorespiratory, extrapyramidal, genitourinary, and musculoskeletal symptoms, as well as sleep disturbances. Few clinically significant drug–drug interactions with cholinesterase inhibitors have been identified. Three head-to-head trials of cholinesterase inhibitors in the treatment of AD have been published. These trials have limitations due to their open-label design, rates of titration, and the drug dosage levels utilized (Thompson et al., 2004; Alva and Cummings, 2008).

Some anticholinergic drugs are known to cause acute cognitive impairment, which is typically transient and reversible (Chuang et al., 2017). However, the continuous use of anticholinergic drugs (for 2–5 years) not only doubles the prevalence of both amyloid plaque and neurofibrillary tangle (NFT) densities in AD and PD patients, but also promotes brain atrophy (Perry et al., 2003; Chuang et al., 2017). This observation is further supported by recent animal studies (Caccamo et al., 2006; Haring et al., 1998). Caccamo and colleagues studied the effect of anticholinergic drugs on the development of Aβ peptides in transgenic mice that express several features similar to the human AD brain and found that a long-term blockade of the M_1 receptor with the use of anticholinergic increases the presence of Aβ peptides in the cortex, hippocampus, and amygdala (Caccamo et al., 2006). At the molecular level, anticholinergic drugs act by retarding the binding of acetylcholine with muscarinic and nicotinic receptors. This process results in numerous adverse drug events, especially in older adults. Thus prolonged exposure to anticholinergic drugs has been linked to long-term cognitive decline or dementia incidence among community-living cohorts and nursing home residents (Gray et al., 2015; Cai et al., 2013; Fox et al., 2011; Richardson et al., 2018).

MEMANTINE FOR THE TREATMENT OF ALZHEIMER'S DISEASE AND ALZHEIMER'S DISEASE TYPE OF DEMENTIA

Memantine (*Namenda* or *Ebixa*) is a derivative of amantadine and an uncompetitive antagonist of NMDA receptor (NMDAR) (Fig. 7.2). Memantine binds to NMDARs with a low-micromolar IC_{50} value. An important advantage of memantine is that it only interacts with the

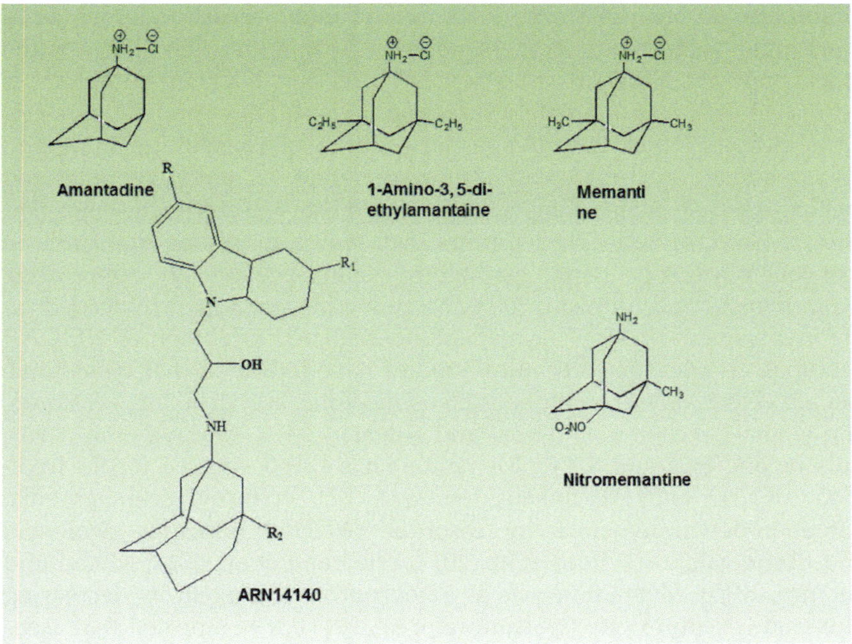

FIGURE 7.2 Chemical structures of memantine and memantine related compounds.

NMDAR channel when it is pathologically activated under an excessive glutamate concentration in the synaptic cleft. Memantine exhibits neuroprotective effects not only against Aβ toxicity (Hu et al., 2007) and tau phosphorylation (Song et al., 2008), but also against neuroinflammation and oxidative stress (Figueiredo et al., 2013; Liu et al., 2013). Since memantine is a low-affinity antagonist, it not only inhibits the NMDAR but is rapidly displaced from it, avoiding prolonged NMDAR blockade. It has no negative side effects on learning and memory, which are frequently observed with high affinity NMDAR antagonists (dissociative anesthetics, ketamine, and MK-801). Memantine also has suitable safety and tolerability limits that show a good therapeutic margin. Another advantage of Memantine interacts with the NMDAR channel when it is pathologically activated under an excessive glutamate concentration in the synaptic cleft, as is the case with AD type of dementia. It should be noted that researchers studying the mechanism of action of memantine have now focusing their attention on the extracellular Mg^{2+} (Mg_o^{2+}) instead of blockage of NMDAR subtypes (Cull-Candy and Leszkiewicz, 2004). Mg_o^{2+} is an endogenous NMDAR channel blocker that binds near memantine's binding site (Chen and Lipton, 2005). It is reported that a physiological concentration (1 mM) of Mg_o^{2+} decreases memantine inhibition of NR1/2A and NR1/2B receptors nearly 20-fold at a membrane voltage near rest. In contrast, memantine inhibition of the other principal NMDAR subtypes, NR1/2C and NR1/2D receptors, is decreased only approximately threefold. Quantitative modeling studies have indicated that the voltage dependence of memantine inhibition also is altered by 1 mM Mg_o^{2+} suggesting that currently hypothesized mechanisms of memantine action should be reconsidered, and that NR1/2C and/or NR1/2D receptors may play a more important role in cortical physiology and pathology than previously appreciated (Kotermanski and Johnson, 2009). More studies are urgently needed on the involvement of Mg_o^{2+} in memantine-mediated inhibition of NMDAR subtypes. In addition, preclinical studies have indicated that memantine can also block other receptors, such as nicotinic, acetylcholine, serotonin, and sigma-1 receptors (Allgaier and Allgaier, 2014; Aracava et al., 2005; Buisson and Bertrand, 1998). Memantine has also been used for the treatment of AD and AD-linked dementia, PD, epilepsy, schizophrenia, Attention-deficit/hyperactivity disorder (ADHD), vascular dementia, and fibromyalgia (Di Iorio et al., 2017; Friedman et al., 2012; Khalid and Soomro, 2015). Memantine acts as a neuroprotective agent by decreasing glutamate excitotoxicity (Nakamura et al., 2011). It is reported that accumulation of Aβ impairs brain-derived neurotrophic factor (BDNF) signaling due to truncation of BDNF receptor (TrkB-full length, TrkB-FL) (Fig. 7.3) (Tanqueiro et al., 2018). Such truncation is promoted by calpains. It results in the formation of an intracellular domain fragment leading to loss of BDNF function. Calpains are activated by Ca^{2+}, which enters into the cell due to the overstimulation of NMDAR. NMDAR antagonist, memantine, prevents excessive calpain activation and TrkB-FL truncation induced by $A\beta_{25\text{-}35}$. Inhibition of calpains by calpastatin results in a BDNF-mediated increase in the dendritic spine density of neurons after exposure to $A\beta_{25135}$. Moreover, memantine-mediated inhibition of NMDAR prevents the Aβ-driven deleterious impact of BDNF loss of function on structural (spine density) and functional outcomes (synaptic potentiation) supporting the view that these processes are driven by Aβ-mediated BDNF signaling disruption. Converging evidence suggests that memantine acts by increasing levels of BDNF, a growth factor, which modulates synaptic plasticity in rats (Fig. 7.4) (Picada et al., 2011). In addition, memantine not only blocks Kv1.3 potassium channels, inhibits CD3-antibody- and alloantigen-induced proliferation, but also suppresses chemokine-mediated migration of peripheral blood T cells of healthy donors. Furthermore, memantine may also normalize deviant immunopathology in AD and promote beneficial effects by inhibiting infection rate (Lowinus et al., 2016). In the Morris water maze, memantine improves acquisition performance, spatial accuracy, and increases durability of synaptic plasticity (Sahiner et al., 2011).

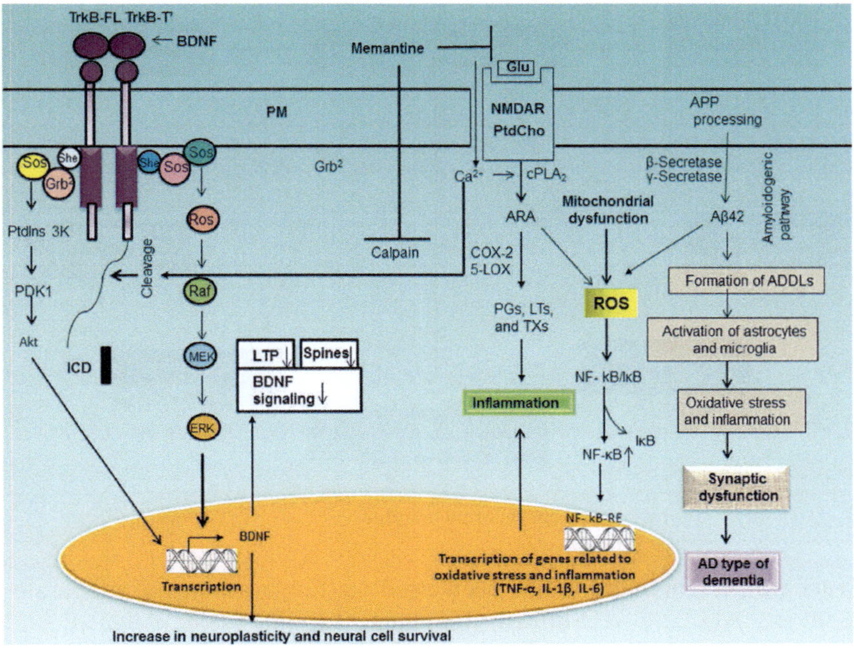

FIGURE 7.3 Hypothetical diagram showing the blocked of NMDA receptor and inhibition of calpain by memantine to promote BDNF-mediated signal transduction process. *Akt/PKB*, protein kinase B; *ARA*, arachidonic acid; *ARA-PtdCho*, arachidonic acid containing phosphatidylcholine; *BDNF-R*, brain-derived neurotrophic factor (BDNF) receptors; *COX-2*, cyclooxygenase-2; *cPLA*$_2$, cytosolic phospholipase A_2; *ERK*, extracellular signal regulated kinase; *IL-1β*, interleukin-1beta; *IL-6*, interleukin-6; *LOX*, lipoxygenase; *LTs*, leukotrienes; *MEK*, serine/tysosine/threonine/kinase; *NF-κB*, nuclear factor-kappa B; *NMDAR*, N-methyl-D-aspartate receptor; *PDK1*, pyruvate dehydrogenase lipoamide kinase isozyme 1; *PGs*, prostaglandins; *PtdIns 3K*, phosphatidylinositol 3-kinases; *PtdIns4,5-P*$_2$, phosphatidylinositol 4,5-bisphosphate; *raf*, serine/threonine protein kinase; *Ras*, small GTPase; *ROS*, reactive oxygen species; *Shc*, SH2-adaptor protein; *Sos*, son of sevenless; *TNF-α*, tumor necrosis factor-α; *TX*, thromboxane.

The clinically approved dose of memantine for humans starts with 5 mg/day, increasing progressively over a period of several weeks to 20 mg/day. This progressive dose adjustment may contribute to the drug's lack of side effects (Reisberg et al., 2003). At higher doses (7.5–20 mg/kg; s.c.), memantine attenuates morphine-induced tolerance, physical dependence, and drug-seeking effects in animals (Ribeiro Do Couto et al., 2004). Like the cholinesterase inhibitors, memantine provides symptomatic relief to some but has failed to provide universal benefit in AD. It produces side effects such as dizziness, anorexia, vomiting, and diarrhea (Alva and Cummings, 2008). In combination with acetylcholinesterase inhibitors (galantamine, donepezil, and

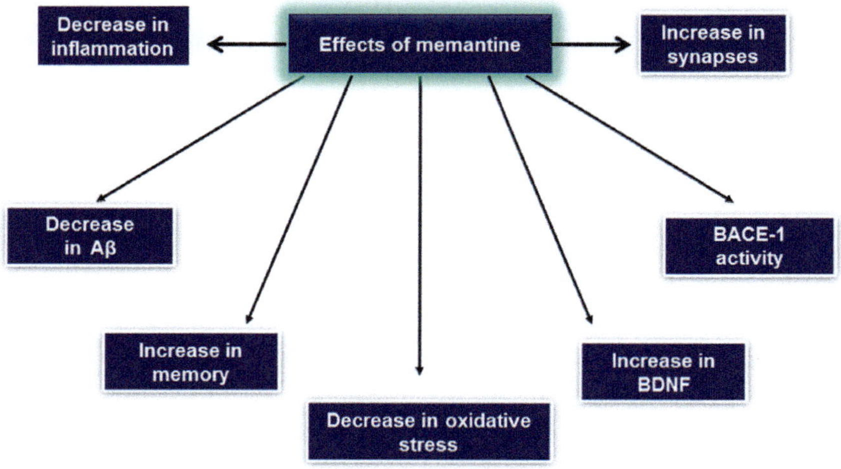

FIGURE 7.4 Neurochemical effects of memantine.

rivastigmine), memantine has been used for the treatment of moderate-to-severe AD and AD-linked dementia (Lipton, 2006; Allgaier and Allgaier, 2014; Nakamura et al., 2016). Thus investigators have used Namzaric (fixed-dose combination of memantine extended-release (ER)/donepezil 28/10 mg) for the treatment of patients with moderate-to-severe AD (Greig, 2015). This fixed-dose formulation is bioequivalent to coadministration of the individual drugs. In a 24-week phase III trial in patients with moderate-to-severe AD, addition of memantine ER 28 mg once daily to stable AChE inhibitor therapy produced a more effective effect than an add-on placebo on measures of cognition, global clinical status, dementia behavior, and semantic processing ability. Namzaric is generally well tolerated in the phase III trial, with diarrhea, dizziness, and influenza occurring at least twice (Greig, 2015). So far though, the biggest problem that is encountered with memantine for the treatment of AD type of dementia, is that it does not slow down the progression of the disease, but only produces a symptomatic effect (Areosa and Sherriff, 2003). That is why current research efforts on the therapy of AD type of dementia are focused on the development of new molecules that can modify the course of AD progression (Cummings et al., 2017; Piette et al., 2011).

Administration of memantine in various transgenic AD mice greatly improves cognitive deficits. However, these benefits of memantine in preclinical studies do not translate into clinical results of this drug, demonstrating only marginal and transient efficacy in moderate-to-severe AD. In AD patients, the administration of memantine results in statistically significant effects on cognition, behavior, and the ability to perform activities of daily living (McShane et al., 2006). A small reduction in agitation has been consistently observed. However, the trials examining memantine have been limited by high drop-out rates and the benefits identified, although results were statistically significant to a small magnitude. A recent 2-year trial has provided further evidence that memantine does not modify disease progression and is ineffective in mild AD (Dysken et al., 2014). Collective evidence suggests that like the cholinesterase inhibitors, memantine provides symptomatic relief to some but has failed to provide universal benefit in AD and AD-related dementias. It produces side effects such as dizziness, anorexia, vomiting, and diarrhea (Alva and Cummings, 2008). Recently, investigators have used Namzaric (fixed-dose combination of memantine ER/donepezil 28/10 mg) for the treatment of patients with moderate-to-severe AD (Greig, 2015). This fixed-dose formulation is bioequivalent to coadministration of the individual drugs. In a 24-week, phase III trial in patients with moderate-to-severe AD, addition of memantine ER 28 mg once daily to stable AChE inhibitor therapy produced a more effective effect than an add-on placebo on measures of cognition, global clinical status, dementia behavior, and semantic processing ability. Namzaric is generally well tolerated in the phase III trial, with diarrhea, dizziness, and influenza occurring at least twice (Greig, 2015). It is also stated that memantine and cholinesterase inhibitors have very limited value to improve agitation in AD patients (Matsunaga et al., 2015).

Recently, investigators have also developed a novel class of multi-target drugs, obtained by linking together two commercially available drugs for AD, galantamine, and memantine. This drug modulates the cholinergic and glutamatergic pathways, respectively (Simoni et al., 2012; Rosini et al., 2014). ARN14140 is a drug, which inhibits AChE and the NMDAR (Reggiani et al., 2016). It can penetrate BBB. The chronic infusion of ARN14140 in the lateral ventricles fully protects mice from the development of short-term memory deficits after i.c.v. injection of $A\beta_{25-35}$. Since the Y-maze performance mimics spatial learning and is driven at the hippocampal level, it is suggested that ARN14140 provides a functional neuroprotective effect for this type of memory. ARN14140 is also effective in the passive avoidance test, which is a fear-motivated test classically used to assess short- or long-term memory. Therefore the data also suggest ARN14140 has an effect on this type of short-term memory. Both memory types are typically affected in AD patients (Morris and Kopelman, 1986). The use of other drugs such as haloperidol, risperidone, apripiprazole, olanzapine, cholinesterase inhibitors, memantine, and benzodiazepines has also been described and it is reported that these drugs induce very little benefit, but produce numerous adverse effects, which are harmful for patients with dementia (Lonergan et al., 2002; Ballard and Howard, 2006; Tifratene et al., 2017; Hogan et al., 2008, Peisah et al., 2011).

Memantine has also been used for the treatment of frontotemporal dementia (FTD) (Lindquist et al., 2008; Vercelletto et al., 2011; Boxer et al., 2013). Lindquist et al have reported that treatment of FTD (R406W) with a dose of 10 mg twice a day stabilizes the progression of symptoms of disease (Lindquist et al., 2008). However, a multicenter, randomized, double-blind, placebo-controlled clinical trial (NCT00545974) evaluated the efficacy of MEM in mild-to-moderate FTD and the main conclusion of this study was that MEM treatment does not ameliorate these dementia symptoms (Vercelletto et al., 2011).

NONPHARMACOLOGICAL TREATMENT OF DEMENTIA

Nonpharmacological treatments to prevention or retard dementia involve a healthy lifestyle (healthy diet, regular exercise, optimal sleep, mental challenges, and socialization, as well as caloric restriction). In recent years investigators have used using aromatherapy, acupuncture, music therapy, cognitive behavioral therapy, animal-assisted therapy, electroconvulsive therapy, transcranial magnetic stimulation (TMS), and physical exercises, along with Yoga, meditation, and Tai Chi. Signal transduction mechanisms associated with the beneficial effects of the above therapies are not known. However, it is becoming increasingly evident that these therapies may activate specific pathways in several brain areas associated with emotional behaviors, such as the insular and cingulate cortex, hypothalamus, hippocampus, amygdala, and prefrontal cortex. Nonpharmacological therapies may produce their beneficial effects by promoting the release of neurotransmitters, neuropeptides, and neurochemical mediators (endorphins, endocannabinoids, dopamine and nitric oxide), which not only increase neuroplasticity, neurogenesis, and regeneration and repair mechanisms, but also stimulate neuroendocrine and neuropsychiatric mechanisms (Fang et al., 2017).

TREATMENT OF DEMENTIA WITH AROMATHERAPY

Aromatherapy has received considerable attention in recent years for the treatment of dementia (Forrester et al., 2014; Yang et al., 2015). Herbs, flowers, and essential oils are used as a source of aroma (Lee et al., 2012). In patients, aromatherapy can be introduced through inhaling or application to the skin. The mechanisms of aromatherapy are not known. However, it is proposed that aromatic molecules bind to olfactory epithelium acceptors that are specific for each different smell. The olfactory nerve system is responsible for the transmission of this stimulus to the hippocampus, limbic system, and amygdala and then to the hypothalamus with a consequent release of neuromediators (Jimbo et al., 2009). The involvement of the hippocampus and amygdala in the cognitive impairment characterizing dementia and the presence of NFTs in the entorhinal cortex, already in the early stages of AD (Jimbo et al., 2009; Braak and Braak, 1991; Gold et al., 2000), suggest an interesting link between olfaction and AD, further confirmed by the dysfunctional olfaction by which demented patients are often affected (Jimbo et al., 2009). It is also suggested that aromatherapy may promote neurogenesis in dentate gyrus of hippocampus (Jimbo et al., 2009; Eriksson et al., 1998). Accordingly, systemic absorption and distribution of pharmacologically active components of the phytocomplex are needed for aromatherapy to control Behavioral and Psychological Symptoms of Dementias (Fung et al., 2012). Aromatherapy reduces agitation in dementia patients. Aromatherapy with essential oils (*Melissa officinalis* (sage) and Lavender essential oils) has been reported to produce a calming and sedative effect in patients with dementia. Lavender essential oil is composed of over 100 constituents. Among them the principals are linalool (51%), linalyl acetate (35%), α-pinene, limonene, 1,8-cineole, *cis*- and *trans*-ocimene, 3-octanone, camphor, caryophyllene, terpinen-4-olandlavendulyl acetate, and cineole (Cavanagh and Wilkinson, 2002). Lavender essential oil constituents inhibit glutamate and GABA receptor binding (Huang et al., 2008; O'Connor et al., 2013). Furthermore, lavender has been shown to lower plasma cortisol levels (Shiina et al., 2008; Field et al., 2008) and reduces the need for analgesia during the postoperative period in humans (O'Connor et al., 2013). Another possible action of lavender essential oil is through tryptophan (Zeilmann et al., 2003; Fu et al., 2013). It is hypothesized that tryptophan promotes the relaxation response leading to a decrease in agitation and an increase in the sleeping time (Zeilmann et al., 2003). In addition, Lavender oil and lemon balm aroma also produce hypnotic, sedative, muscle-relaxant, antibacterial, and antispasmodic effects (Lee et al., 2012; Abuhamdah and Chazot, 2008). Aromatherapy with lavender oil has been used to reduce pain and anxiety during labor, and mothers have generally evaluated this approach as an appropriate method (Pollard, 2008). The mechanism of action of aroma on the brain is not known. However, it has been proposed that transdermal administration or inhalation of aroma from essential oil activates the autonomic nervous system and induces the reaction of the limbic system and hypothalamus (Cook and Lynch, 2008). Lemon balm hydroalcoholic extract contains anticholinesterase activity. The most active constituents of lemon balm are *cis*- and *trans*-rosmarinic acids and other rosmarinic acid derivatives. These derivatives have high anticholinesterase activity and free radical scavenger properties.

TREATMENT OF DEMENTIA WITH ACUPUNCTURE

Acupuncture is an ancient Chinese healing technique that treats disorders by inserting thin needles into the skin at specific locations (acupoints) of the body. This technique may be manipulated manually, electrically, or by heat (Li and Wang, 2013). According to this ancient Chinese practice, the mechanistic system of therapeutic needling is adjustment of the Qi (vital energy) flow that is believed to circulate in a network of 12 primary channels, also called meridians, which connect 360 principal acupuncture points (Kavoussi and Ross, 2007; Zhang et al., 2015). Stimulation of the needles is believed to elicit profound psychophysical responses by harmonizing or balancing the Qi energy, as well as blood flow throughout the body (Kim and Bae, 2010; Zhang et al., 2015, 2016). Acupuncture at particular acupoints activates afferent fibers that send signals to the spinal cord (Zhao, 2008) leading to anti-inflammatory effects (Fig. 7.5). Acupuncture has been used for the treatment of many neurological disorders including cardiovascular and psychiatric diseases, acute, and chronic pain, AD, and behavioral disturbances in vascular dementia (Abraha et al., 2017; Liu et al., 2016; Shi et al., 2014). Advances in acupuncture technology have been made and these days investigators are using electroacupuncture, lesser acupuncture, and acupoint injection to investigate the effects of acupuncture in animal models of dementia and stroke (Xu et al., 2013; Zhang et al., 2017; Yun et al., 2017). Clinical trials and meta-analysis have indicated the efficacy of acupuncture in improving balance function, reducing spasticity, and increasing muscle strength and general well-being poststroke (Chavez et al., 2017). The molecular mechanisms associated with the beneficial effects of acupuncture is not fully known. However, it is proposed that acupuncture may provide beneficial effects by regulating the expression of Bcl-2/Bax, caspase family, Fas/FasL, c-Fos, tumor necrosis factor-α (TNF-α), and NFκB, which suppress neural cell apoptosis and autophagy to reduce cell death in different pathological states especially ischemic stroke and vascular dementia (Luo et al., 2017). Apoptosis is characterized by cell rounding, membrane blebbing, cytoskeletal collapse, cytoplasmic condensation, and fragmentation, nuclear pyknosis, chromatin condensation/fragmentation, and formation of membrane-enveloped apoptotic bodies, that are

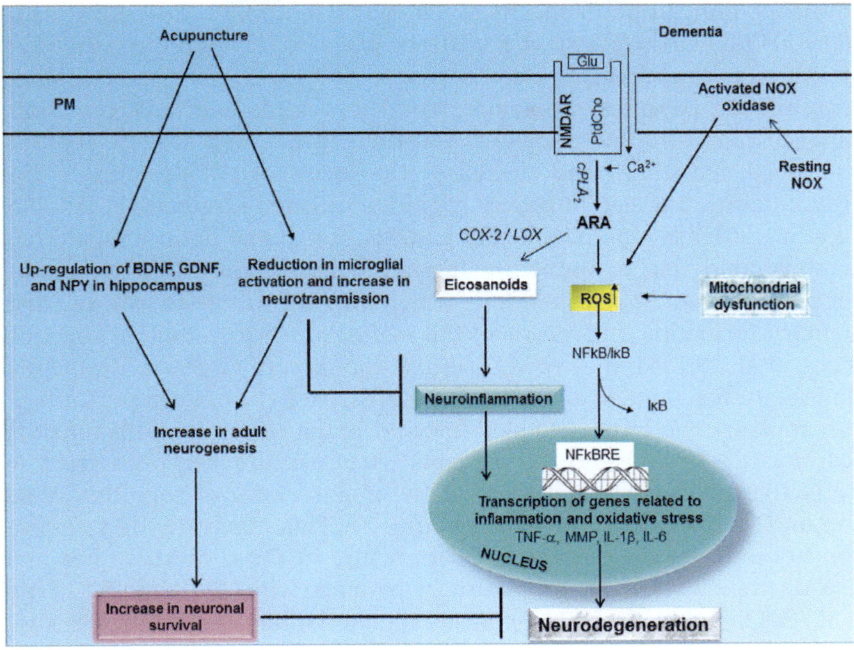

FIGURE 7.5 Neurochemical mechanisms contributing to anti-inflammatory effects of acupuncture.
ARA, arachidonic acid; *COX*, cyclooxygenase; *cPLA*$_2$, cytosolic phospholipase A_2; *IL-1β*, interleukin-1beta; *IL-6*, interleukin-6; *LOX*, lipoxygenase; *lyso-PtdCho*, lysophosphatidylcholine; *NF-κB*, nuclear factor-kappa B; *NF-κB-RE*, nuclear factor-kappa B response element; *NMDAR*, *N*-methyl-D-aspartate receptors; *NOX*, nicotinamide adenine dinucleotide phosphate (NADPH)-oxidase; *PAF*, platelet activating factor; *PM*, plasma membrane; *PtdCho*, phosphatidylcholine; *ROS*, reactive oxygen species; *TNF-α*, tumor necrosis factor-alpha.

rapidly phagocytosed by macrophages or neighboring cells. In contrast, autophagy is accompanied by pathways that target long-lived cytosolic proteins and damaged organelles. It involves a sequential set of events including double membrane formation, elongation, vesicle maturation and finally delivery of the targeted materials to the lysosome. Acupuncture has been reported to upregulate vascular endothelial growth factor (VEGF) expression through direct H3K9 acetylation at the VEGF promoter inducing angiogenesis in rat MI models (Fu et al., 2014). Electroacupuncture alleviates symptoms of chest pain related to myocardial ischemia (stable angina pectoris) through the modification of epigenetic markers including H3K4me1, H3K4me2, and H3K27ac (Wang et al., 2015a). These findings suggest a link to epigenetic regulation of regeneration and cellular apoptosis during and after MI via acupuncture (Wang et al., 2015b). Acupuncture exerts its therapeutic effect not only through the involvement of the micro-RNA-339/Sirt2/NFκB/FOXO1 axis (Wang et al., 2015b), but also through the regulation of glucose metabolism and increase in levels of acetylcholine. Acupuncture suppresses oxidative damage, retards neuroinflammation, enhances neurotrophin signaling, improves synaptic plasticity, reduces microglial activation, and decreases the levels of Aβ proteins in the hippocampus. but also improves cognitive function (Zeng et al., 2013a,b; Ye et al., 2017a,b). In AD and PD patients, functional brain imaging has demonstrated that acupuncture increases the neuronal activity in the temporal lobe and prefrontal lobe which may contribute to learning and memory formation and promote the maintenance of cognitive function (Zeng et al., 2013a,b; Lee et al., 2009). Although only a few acupuncture clinical studies on a small number of participants have been performed, they represent an important step forward in the research on the effect of acupuncture in AD and PD patients. Acupuncture has been used to study the molecular mechanism of the maintenance of cognitive function and its application on human subjects. This may provide information not only on the efficacy of acupuncture in the clinical studies, but also on the safety of acupuncture in patients with AD and PD (Zeng et al., 2013a,b; Lee et al., 2008). Based on animal model studies, it is suggested that acupuncture improves mitochondrial bioenergy parameters such as mitochondrial respiratory control rate and membrane potential. In addition, in animal models of AD, acupuncture improves cognitive function by regulating glucose metabolism and enhancing neurotransmission as well as reducing oxidative stress, Aβ protein deposition, and neuronal apoptosis. However, it is not known which specific signaling pathway contributes to the acupuncture effect. In animal models of ischemia, acupuncture has been reported to reverse bilateral common carotid arteries occlusion-mediated hippocampal mitochondrial dysfunction, which may contribute to its prevention of cognitive deficits. Acupuncture may also increase synaptic plasticity and blood vessel function. It is likely that no single factor can explain the protection provided by acupuncture (Zeng et al., 2013a,b; Lee et al., 2008).

TREATMENT OF DEMENTIA WITH MUSIC

Music, a universal art form that exists in every culture around the world, is an integral part of a number of social and courtship activities (Boso et al., 2006). Music is closely associated with other creative behaviors such as dancing and prayers (Qawali and Keertan). Music therapy includes both active forms of musical engagement such as song writing, singing, and playing musical instruments, as well as receptive forms of musical engagement such as listening to live or prerecorded music (Garrido and Schubert, 2015; Petrovsky et al., 2015). Recently, neuroimaging studies have provided information on the neural correlates of music processing and perception in the brain. Notably, musical stimuli have been shown to activate specific pathways in several brain areas (insular and cingulate cortex, hypothalamus, hippocampus, amygdala, and prefrontal cortex) associated with emotional behaviors. Music therapy has been used for the treatment of neuropsychiatric and behavioral symptoms of dementia, depression, autism, and schizophrenia (Särkämö et al., 2012; Vasionytė and Madison, 2013; Lin et al., 2011; Zhang et al., 2017). Many studies have demonstrated that music can improve multiple domains of cognitions in AD patients, including attention, psychomotor speed, memory, orientation, and executive functions (Satoh et al., 2015; Särkämö et al., 2014; Bruer et al., 2007; Ozdemir and Akdemir 2009). In AD, listening to the music increases the global cognition (Bruer et al., 2007). It is also reported that music therapy for 6 weeks not only improves the memory and orientation, but also decreases depression and anxiety in AD patients (Gómez Gallego and Gómez García, 2016). Collective evidence suggests that music is an important resource for achieving psychological, cognitive, and social goals. It promotes the calming of the emotional expression behavioral symptoms of dementia and also facilitates the motivation of rehabilitation activities supporting the view that music can be used for the rehabilitation of patients with age-related neurological diseases including dementia (Särkämö, 2017). In music therapy, a trained and registered music therapist treats dementia patients with a program of musical engagement based on established therapeutic practice (Garrido and Schubert, 2015). It is reported that singing or playing a musical instrument cannot only help in improving motor skills, but also enhances one's self-esteem (Fig. 7.6) (Baker et al., 2008). In addition, random control studies have indicated that music therapy significantly decreases anxiety and depression, significantly improves quality-of-life, and may decrease length of hospital stay (Baker et al., 2008). The mechanism of beneficial effects of music in dementia is not known. However, studies on the effect of music in AD type of dementia have indicated that music therapy produces beneficial effects by not only by increasing neuroplasticity, neurogenesis, and regeneration and repair mechanisms, but also stimulating the neuroendocrine and neuropsychiatric mechanisms (Fang et al., 2017).

A recent pilot study, where AD patients sang familiar songs with a karaoke device, has indicated that singing training improves the neural efficacy of cognition in AD patients (Raven, 1995; Guétin et al., 2009; Satoh et al., 2015). Another study has indicated that music therapy increases levels of melatonin leading to a relaxed and calm mood in AD

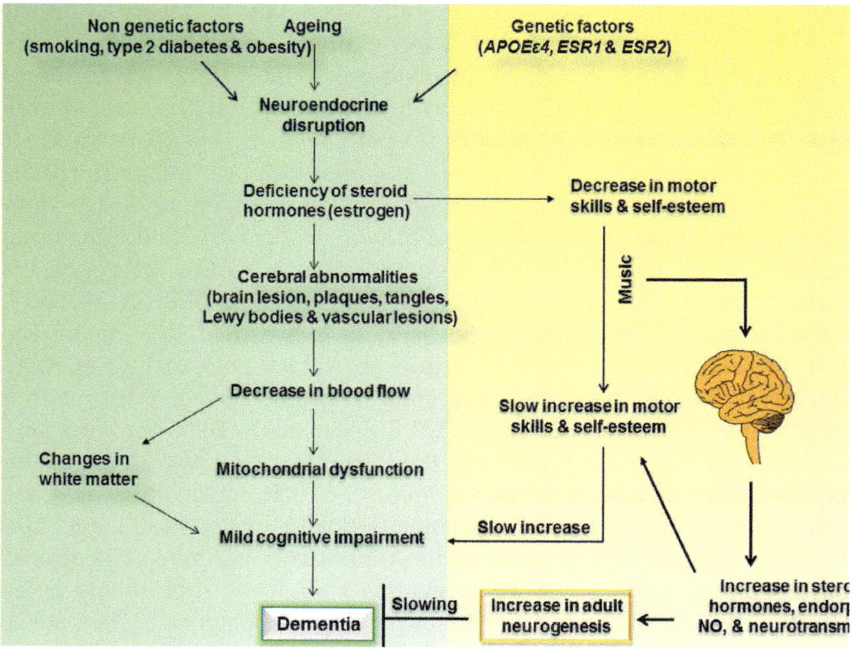

FIGURE 7.6 Neurochemical effects of music in the brain.
APOE-ε4, Apolipoprotein E-ε4; *ESR1*, estrogen receptor 1 gene; *ESR2*, estrogen receptor 2 gene.

patients (Kumar et al., 1999). It has also been shown that music influences cranial nerves from fetus to adult in humans. Listening to music facilitates the neurogenesis, the regeneration, and repair of cerebral nerves by adjusting and optimizing the secretion of steroid hormones, leading to an increase in neuroplasticity (Fig. 7.6) (Fukui and Toyoshima, 2008). Music may not only modulate levels of such steroids as cortisol, testosterone, and estrogen, but may also affect expression of genes related to steroid hormone receptors (Fukui and Toyoshima, 2008; Fukui et al., 2012). Earlier studies indicated that hormone replacement therapy can be useful in the treatment of AD and some types of dementia (Shumaker et al., 1998; Gouras et al., 2000). However, recently hormone replacement therapy has faced serious challenges due to its side effects (Rocca et al., 2012, 2014; Li et al., 2017). Some studies have indicated that music also facilitates the release of several neurotransmitters, neuropeptides, and other biochemical mediators such as endorphins, endocannabinoids, dopamine, and nitric oxide (Boso et al., 2006). These mediators may play a role in the musical experience. Furthermore, music therapy also affects the reward, stress and arousal, immunity, and social affiliation systems of humans (Chanda and Levitin, 2013). It is tempting to speculate that music therapy, a noninvasive and nonpharmacological intervention, can be used in the medical care of dementia patients as a safe and inexpensive option.

EFFECTS OF EXERCISE ON DEMENTIA

Aerobic exercise is known to produce beneficial effects on cognitive function in healthy seniors not only due to the increase in the mitochondrial biogenesis and upregulation of the mitophagy (Bori et al., 2012; Lanza and Sreekumaran, 2010; Guo et al., 2012; Erickson and Kramer, 2009; Erickson et al., 2009; Farooqui, 2013; Jedrziewski et al., 2014), but also because of the increase in neuroplasticity (Stranahan et al., 2009) and cognitive function (Hillman et al., 2008). Aerobic exercise also induces cardiorespiratory and muscular fitness by increasing energy consumption, improving insulin sensitivity, increasing blood flow, strengthening the immune system, reducing neuroinflammation, downregulating oxidative stress, promoting sleep, controlling weight, increasing gray matter volume, and inducing neurogenesis in the dentate gyrus along with the release of BDNF and insulin-like growth factor-1 (Fig. 7.7) (Farooqui, 2013). Exercise also reduces the brain's exposure to neurotoxic substances, including Aβ toxicity and excessive

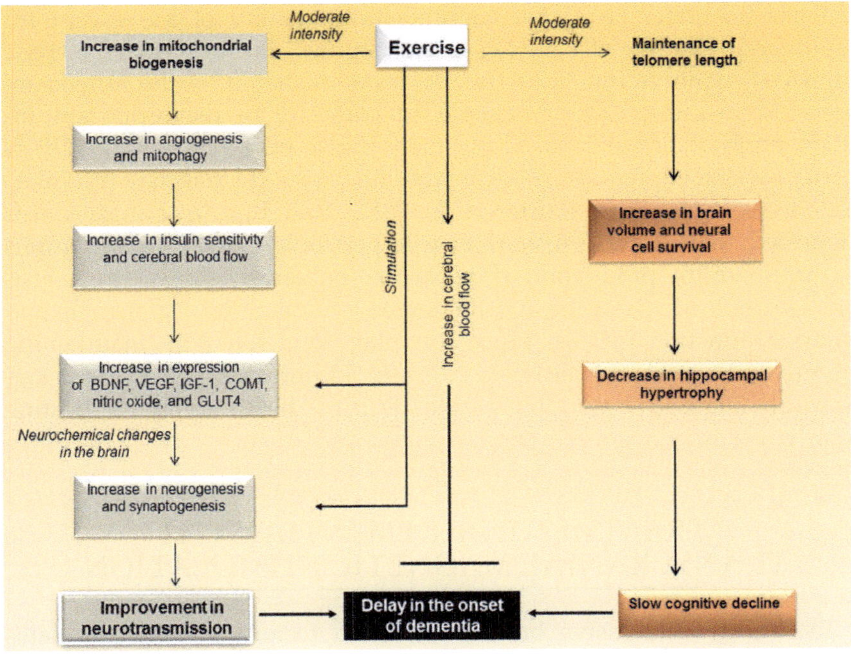

FIGURE 7.7 Effects of exercise on neurochemical processes that delay the onset of dementia.

glucose levels (Brown et al., 2013; Mehlig et al., 2014). At the same time exercise has mental stimulatory properties such as those that require eye-hand coordination and visuospatial memory, thus further augmenting their effects on cognitive functioning.

Studies on the effects of exercise in animal models of AD indicate that subjecting transgenic mice (mice expressing the skeletal muscle-specific mutant PS2 gene) to treadmill exercise for 3 months not only results in a reduction of Aβ-42 deposits, but also produces improvement in behavioral function, thereby restoring normal concentrations of total cholesterol, high-density lipoprotein cholesterol, low-density lipoprotein cholesterol, and triglyceride (Cho et al., 2003). Furthermore, 16 weeks of exercise on a treadmill by the NSE/APPsw Tg mice, indicated that exercise not only decreases levels of Aβ-42 peptides and produces antiapoptotic effects, but also inhibits the induction of glucose transporter-1 (GLUT-1) and BDNF (Um et al., 2008). Balance dysfunction, gait disturbances, and falls are common problems in later stages of AD and dementia compared with older people without these conditions (Manckoundia et al., 2006). Although in normal older people exercise reduces falls and improves their mood, in AD patients exercise induces marginal effects on fall reduction and mood improvement (Cho et al., 2003). Studies on the effect of aerobic exercise in mild-to-moderate AD patients have been controversial. Some studies have indicated that exercise has no effect on global cognition and quality of life except for depression (Yu et al., 2014). Other studies indicate that exercise improves cognitive functions such as tasks mediated by the hippocampus, and results in major changes in plasticity in the hippocampus. Interestingly, exercise-induced plasticity is also pronounced in *APOE ε4* carriers which express a risk factor for late-onset AD that may modulate the effect of treatments (Foster et al., 2011). Based on MRI studies, it is suggested that exercise may reduce hypertrophy in the hippocampus and promote the production of BDNF that enhances adult neurogenesis and neuroplasticity, including an increase in neuritic outgrowth and synaptic function. These processes play key roles in maintaining cognitive functions (Foster et al., 2011). In addition, aerobic exercise significantly increases hippocampal dentate gyrus blood volume indicating an increase in angiogenesis (Pereira et al., 2007).

TREATMENT OF DEMENTIA WITH TRANSCRANIAL MAGNETIC STIMULATION

TMS is a noninvasive brain stimulation technique that stimulates neurons via generation of brief pulses of high-intensity magnetic field. Application of pulses in a repetitive fashion, called repetitive transcranial magnetic stimulation (rTMS), results in persistent neural excitability (Elder and Taylor, 2014; Gervits et al., 2016), allowing noninvasive investigation and manipulation of brain circuit function and connectivity. The clinical applications of rTMS involves low intensity magnetic stimulation (pulse amplitude <100 mT) to treat dementia, depression, and pain (Di Lazzaro et al., 2013; Martiny et al., 2010; Shupak et al., 2004). Very little is known about cellular and molecular mechanisms underlying rTMS-induced neural plasticity. It is well known that local electrical stimulation activates a specific input to a neuron by depolarizing axons that are close to the stimulation electrode. In contrast to local stimulation, rTMS not only acts strictly via the depolarization of a specific set of axons (Bosch and Hayashi, 2012), but also targets other neural structures within the electric field within the stimulated network (up to several cm^3; Opitz et al., 2011). Thus rTMS-induced depolarization causes activation patterns distinct from local electrical stimulation (Edgley et al., 1997). This makes it difficult to predict which structures will be activated in the stimulated area of the brain. rTMS-induces its effects by a process similar to long-term potentiation (LTP), a specific, long-lasting increase in the strength of synaptic transmission when the pre- and postsynaptic neurons are activated simultaneously (Ziemann et al., 2008; Hoogendam et al., 2010). The mechanisms of LTP can be pre- or postsynaptic, but postsynaptic mechanisms seem most affected in dementia. It is hypothesized that the induction of LTP at low stimulation frequencies can be explained by the highly efficient recruitment of Hebbian-type plasticity mechanisms (Hebb, 1949) via the repeated activation of pre- and postsynaptic structures during rTMS (Müller-Dahlhaus and Vlachos, 2013).

The hemicerebellectomy (HCb) is a reliable and effective model for examining remote damage mechanisms (Viscomi and Molinari, 2014; Viscomi et al., 2015). It provides a testing ground for novel neuroprotective approaches such as rTMS. In rats, rTMS significantly reduces HCb-induced cell death of precerebellar neurons by blocking cyt-c-associated apoptosis. These findings are consistent with earlier reports on the antiapoptotic effects of rTMS in the perilesional area after traumatic brain injury (Yoon et al., 2015) and after transient cerebral ischemia (Gao et al., 2010). Although remote mechanisms differ substantially from those in perilesional areas after traumatic or ischemic insults (Viscomi and Molinari, 2014; Viscomi et al., 2015), results for the HCb model indicate the effectiveness of rTMS in counteracting apoptotic cell death in areas that are distant from the site of damage. Although the efficacy of rTMS in reducing apoptotic cell death in the HCb model is quite specific, further mechanistic studies are required to identify the signaling pathways of rTMS effects on precerebellar neurons. In addition to the effects on neurons, it is also shown that glial cells, specifically astrocytes and microglia, responded to rTMS stimulation. In fact, in a HCb model rTMS significantly reduces HCb-induced inflammatory responses, which have been shown to contribute to remote degeneration (Viscomi et al., 2015).

Based on above information, it can be proposed that the therapeutic effect of rTMS is largely attributed to its ability to dampen neuronal hyperexcitability, decrease neuroinflammation, alter BBB permeability, and promote neuronal survival (Cullen and Young, 2016).

It must be mentioned that rTMS treatment of human patients results in a considerable degree of variability in excitability among patients in different sessions. The frequency of rTMS-mediated stimulation of brain is the main determinant of the direction of excitability. It is becoming increasingly evident that there are interactions between frequency and several other stimulation parameters that also control the degree of modulation. In addition, the spatial interaction of the transient electric field induced by the TMS pulse with the cortical neurons is another contributor to variability. Application of rTMS to demented patients should take into account all the abovementioned factors in order to improve the consistency of the conditioning effect and to better understand the outcomes of rTMS therapy (Müller-Dahlhaus and Vlachos, 2013).

TREATMENT OF DEMENTIA WITH MEDITATION

The term "meditation" refers to a broad variety of strategies, which not only control complex emotions (stress, depression, anxiety, and neuroticism), but also promote self-relaxation, well-being, and emotional balance in human life (Lutz et al., 2008). Acute stress is self-limited. The organism can resolve it by adaptation. In contrast, chronic stress causes deleterious effects (Schneiderman, 2005). Thus chronic high-intensity stress not only leads to blunting of the hypothalamic–pituitary–adrenal axis, prolonged glucocorticoid secretion, alterations in synaptic plasticity, changes in corticotrophin releasing factor receptor signaling, but also produces a reduction in gray matter in several brain regions including the hippocampus (Gianaros et al., 2007). These processes lead to impaired learning and memory and dysregulated neuroendocrine activity (McEwen and Gianaros, 2010; Chen, 2010). Chronic stress also leads to loss of sleep (McCurry et al., 2007), alterations in mood (Schulz and Martire, 2004), and immunological dysfunction (Schulz and Martire, 2004) along with elevated risk for metabolic syndrome, cardiovascular disease (CVD), and mortality (Dimsdale, 2008). Chronic psychological stress also induces behavioral changes. It is also linked with an increased risk for mild cognitive impairment and dementia in older adults, and accelerated cognitive decline (Wilson et al., 2006, 2007).

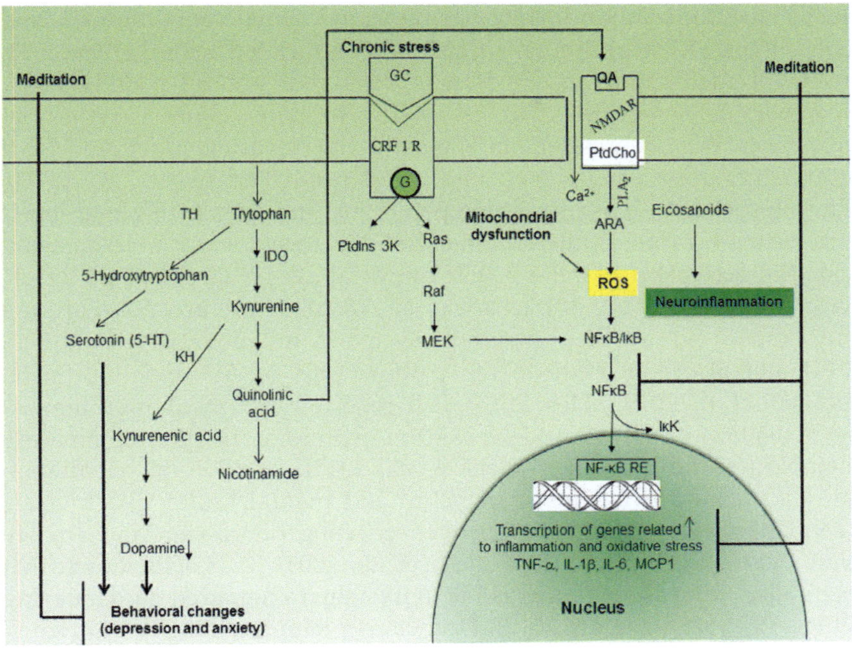

FIGURE 7.8 Hypothetical diagram showing the effect of meditation on behavioral changes in depression and dementia.
ARA, arachidonic acid; *cPLA*$_2$, cytosolic phospholipase A_2; *CRF1*, corticotrophin releasing factor receptor 1; *GC*, glucocorticoid; *IDO*, indolamine 2,3-dioxygenase; *IL-1*β, interleukin-1beta; *KMO*, kynurenine 3-monooxygenase; *NMDAR*, N-methyl-D-aspartate receptor; *PM*, plasma membrane; *PtdCho*, phosphatidylcholine; *PtdIns 3 kinase*, phosphoinositide 3 kinase; *ROS*, reactive oxygen species; *TH*, tryptophan hydroxylase; *TNF-*α, tumor necrosis factor-α. Quinolinic acid (QA) not only generates reactive oxygen species, but also stimulates NMDA receptor and promotes the generation of eicosanoids and expression of TNF-α, IL-1β, and IL-6. These products induce oxidative stress and facilitate neuroinflammation.

Furthermore, chronic stress produces deleterious neuroendocrine alterations, and the induction of inflammation, impaired synaptic plasticity, suppression of neurogenesis, and the reduction in neuronal survival in the hippocampus, prefrontal cortex, and other brain structures, leading to mood alterations, loss of sleep, and decline in memory and learning (Swaab et al., 2005; Lucassen et al., 2010). These processes not only increase the risk for the development and progression of AD (Martins et al., 2006), but also accelerate the onset of neuropsychiatric disorders, including Major Depressive Disorder and Generalized Anxiety Disorder as well as worsening other chronic diseases, such as artherosclerotic and CVD (Salvagioni et al., 2017). Importantly, it has also been shown that chronic stress can increase AD risk (Machado, 2014).

Meditation not only delays the onset of AD, but also decreases stress, depression, and anxiety (Fig. 7.8). In addition, meditation also produces beneficial effects in cardiovascular, cerebrovascular, autoimmune, and renal diseases. The molecular mechanisms associated with the beneficial effects of meditation on the human brain are not fully understood. However, it is reported that in healthy humans, acute psychosocial stress increases depression and anxiety by elevating blood levels of proinflammatory chemokines and cytokines (Bierhaus et al., 2003). Chronic social stress is also accompanied by an increase in blood levels of C-reactive protein, interleukin-6, soluble receptor for TNF-α, along with the activation of NF-κB (Chiang et al., 2012; Gruenewald et al., 2009; Bergamini et al., 2018; Pace et al., 2006). These processes are not only known to induce low-grade systemic inflammation (Rohleder, 2014), but also contribute to the pathogenesis of AD and depression (Dowlati et al., 2010; Maes, 2010). It is reported that meditation inhibits the activation of NF-κB and blocks the expression of proinflammatory chemokines and cytokines (Black et al., 2013). In addition, meditation can lead to improvements in physical and mental health (Black et al., 2009; Chiesa and Serretti, 2009) by decreasing depression and anxiety (Van Puymbroeck et al., 2007; Pace et al., 2009). It is also shown that meditation increases telomerase activity when compared to a relaxing activity (Lavretsky et al., 2012). Decrease in telomere length and reduction in telomerase activity are associated with premature mortality and predict a host of health risks for diseases (Lin et al., 2009), which may be regulated in part by psychological stress (Epel et al., 2009; Ornish et al., 2008). Over the long term, high telomerase activity likely promotes improvement in telomere maintenance and immune cell longevity (Jacobs et al., 2010).

The practice of regular meditation produces a positive effect on cognition (especially on attention and memory) and on brain structure and function, especially in frontal and limbic structures and insula (Berk et al., 2018; Chételat et al., 2018). Furthermore, neuroimaging studies have indicated that meditation increases gray matter volume and/or glucose metabolism in elderly subjects compared to age-matched controls. This increase in gray matter volume occurs in brain regions related to emotion regulation, learning, memory, and self-referential processes (Lazar et al., 2005; Vestergaard-Poulsen et al., 2009). These preliminary findings are important in the context of reserve and brain maintenance as they suggest that long-term meditation practice may help in preserving brain structure and function from progressive age-related decline (Gard et al., 2014; Chételat et al., 2018). Further studies are needed to confirm these results with larger samples and in randomized controlled trials and to investigate the mechanisms underlying these meditation-related effects. The European Commission has funded a project called Silver Santé Study. This study will address many aspects of meditation in a large elderly population. Two randomized controlled trials will be conducted to assess the effects of 2- and 18-month meditation, English learning, or health education training programs (vs a passive control) on behavioral, sleep, blood sampling, and neuroimaging measures. This study may provide the important information needed to delay or prevent dementia through meditation.

CONCLUSION

Dementia is one of the most important neurological disorders in the elderly. It is characterized by a constant decline in the function of multiple cognitive domains comprising memory impairment, behavioral problems, loss of initiative, loss of independence in daily activities, and loss of participation in social activities. Dementia is either caused by neurochemical changes associated with signal transduction processes in the brain or linked with the persisting chronic neurodegenerative diseases such as AD, PD, AIDS, and multiple sclerosis. The decrease in cognitive function not only reduces the quality of life in dementia patients and their caregivers, but also puts pressure on family relationships and friendships. Several trails have been performed in humans using acetylcholinesterase inhibitors and NMDA antagonist (memantine). These inhibitors provide symptomatic treatment and are not the cause of dementia. Thus there is no FDA-approved treatment for dementia. Many nonpharmacological treatments (aromatherapy, music therapy, acupuncture, lifestyle changes, and exercise) of dementia have been used to treat dementia. Among nonpharmacological treatments, aromatherapy, acupuncture, and rTMS improve cognitive function through the modulation of signaling pathways involved in neuronal survival and function, specifically, through promoting cholinergic dopaminergic and glutamatergic neural transmissions, enhancing neurotrophin signaling, suppressing oxidative stress, attenuating apoptosis, and reducing microglial activation. The nonpharmacological treatments have limited beneficial effects in dementia and more studies are required on pharmacological and nonpharmacological treatments of dementia.

References

Abraha, I., Rimland, J.M., Trotta, F.M., et al., 2017. Systematic review of systematic reviews of non-pharmacological interventions to treat behavioural disturbances in older patients with dementia. The SENATOR-OnTop series. BMJ Open 7, e012759.

Abuhamdah, S., Chazot, P.L., 2008. Lemon balm and lavender herbal essential oils: old and new ways to treat emotional disorders? Curr. Anaesth. Crit. Care 19, 221–226.

Allgaier, M., Allgaier, C., 2014. An update on drug treatment options of Alzheimer's disease. Front. Biosci. (Landmark Ed.) 19, 1345–1354.

Alva, G., Cummings, J.L., 2008. Relative tolerability of Alzheimer's disease treatments. Psychiatry (Edgmont) 5, 27–36.

Aracava, Y., Pereira, E.F., Maelicke, A., Albuquerque, E.X., 2005. Memantine blocks alpha7 nicotinic acetylcholine receptors more potently than n-methyl-D-aspartate receptors in rat hippocampal neurons. J. Pharmacol. Exp. Ther. 312, 1195–1205.

Areosa, S.A., Sherriff, F., 2003. Memantine for dementia. Cochrane Database Syst. Rev. (3), CD003154.

Baker, F., Wigram, T., Stott, D., McFerran, K., 2008. Therapeutic song writing in music therapy. Part I: who are the therapists, who are the clients, and why is song writing used. Nord. J. Music Ther. 17, 105–123.

Ballard, C., Howard, R., 2006. Neuroleptic drugs in dementia: benefits and harm. Nat. Rev. Neurosci. 7, 492–500.

Bergamini, G., Mechtersheimer, J., Azzinnari, D., Sigrist, H., Buerge, M., Dallmann, R., et al., 2018. Chronic social stress induces peripheral and central immune activation, blunted mesolimbic dopamine function, and reduced reward-directed behaviour in mice. Neurobiol. Stress 8, 42–56.

Berk, L., Warmenhoven, F., van Os, J., van Boxtel, M., 2018. Mindfulness training for people with dementia and their caregivers: rationale, current research, and future directions. Front. Psychol. 9, 982.

Berry, S.D., Placide, S.G., Mostofsky, E., et al., 2016. Antipsychotic and benzodiazepine drug changes affect acute falls risk differently in the nursing home. J. Gerontol. A Biol. Sci. Med. Sci. 71, 273–278.

Bierhaus, A., Wolf, J., Andrassy, M., Rohleder, N., Humpert, P.M., Petrov, D., et al., 2003. A mechanism converting psychosocial stress into mononuclear cell activation. Proc. Natl Acad. Sci. U.S.A. 100, 1920–1925.

Bierman, E.J., Comijs, H.C., Gundy, C.M., et al., 2007. The effect of chronic benzodiazepine use on cognitive functioning in older persons: good, bad or indifferent? Int. J. Geriatr. Psychiatry 22, 1194–1200.

Birks, J., 2006. Cholinesterase inhibitors for Alzheimer's disease. Cochrane Database Syst. Rev. CD005593.

Black, D.S., Cole, S.W., Irwin, M.R., Breen, E., St Cyr, N.M., Nazarian, N., et al., 2013. Yogic meditation reverses NF-κB and IRF-related transcriptome dynamics in leukocytes of family dementia caregivers in a randomized controlled trial. Psychoneuroendocrinol. 38, 348–355.

Black, D.S., Milam, J., Sussman, S., 2009. Sitting-meditation interventions among youth: a review of treatment efficacy. Pediatrics 124, 532–541.

Bori, Z., Zhao, Z., Koltai, E., Fatouros, I.G., Jamurtas, A.Z., Douroudos, I.I., et al., 2012. The effects of aging, physical training, and a single bout of exercise on mitochondrial protein expression in human skeletal muscle. Exp. Gerontol. 47, 417–424.

Bosch, M., Hayashi, Y., 2012. Structural plasticity of dendritic spines. Curr. Opin. Neurobiol. 2012 (22), 383–388.

Boso, M., Politi, P., Barale, F., Enzo, E., 2006. Neurophysiology and neurobiology of the musical experience. Funct. Neurol. 21, 187–191.

Boxer, A.L., Knopman, D.S., Kaufer, D.I., Grossman, M., Onyike, C., et al., 2013. Memantine in patients with frontotemporal lobar degeneration: a multicentre, randomised, double-blind, placebo-controlled trial. Lancet Neurol. 12, 149–156.

Braak, H., Braak, E., 1991. Neuropathological stageing of Alzheimer-related changes. Acta Neuropathol. 82, 239–259.

Brown, B.M., Peiffer, J.J., Taddei, K., Lui, J.K., Laws, S.M., et al., 2013. Physical activity and amyloid-β plasma and brain levels: results from the Australian imaging, biomarkers and lifestyle study of ageing. Mol. Psychiatry. 18, 875–881.

Bruer, R.A., Spitznagel, E., Cloninger, C.R., 2007. The temporal limits of cognitive change from music therapy in elderly persons with dementia or dementia-like cognitive impairment: a randomized controlled trial. J. Music Ther. 44, 308–328.

Buisson, B., Bertrand, D., 1998. Open-channel blockers at the human alpha4beta2 neuronal nicotinic acetylcholine receptor. Mol. Pharmacol. 53, 555–563.

Caccamo, A., Oddo, S., Billings, L.M., et al., 2006. M1 receptors play a central role in modulating AD-like pathology in transgenic mice. Neuron 49, 671–682.

Cai, X., Campbell, N., Khan, B., et al., 2013. Long-term anticholinergic use and the aging brain. Alzheimers Dement. 9, 377–385.

Cavanagh, H.M.A., Wilkinson, J.M., 2002. Biological activities of lavender essential oil. Phytother. Res. 16, 301–308.

Chanda, M.L., Levitin, D.J., 2013. The neurochemistry of music. Trends Cogn. Sci. 17, 179–193.

Chavez, L.M., Huang, S.S., MacDonald, I., Lin, J.G., Lee, Y.C., et al., 2017. Mechanisms of acupuncture therapy in ischemic stroke rehabilitation: a literature review of basic studies. Int. J. Mol. Sci. 18, pii: E2270.

Chen, Y., 2010. Correlated memory defects and hippocampal dendritic spine loss after acute stress involve corticotropin-releasing hormone signaling. Proc. Natl. Acad. Sci. U.S.A. 107, 13123–13128.

Chen, H.S.V., Lipton, S.A., 2005. Pharmacological implications of two distinct mechanisms of interaction of memantine with N-methyl-D-aspartate-gated channels. J. Pharmacol. Exp. Ther. 314, 961–971.

Chételat, G., Lutz, A., Arenaza-Urquijo, E., Collette, F., Klimecki, O., Marchant, N., 2018. Why could meditation practice help promote mental health and well-being in aging? Alzheimers Res. Ther. 10, 57.

Chiang, J.J., Eisenberger, N.I., Seeman, T.E., Taylor, S.E., 2012. Negative and competitive social interactions are related to heightened proinflammatory cytokine activity. Proc. Natl. Acad. Sci. U.S.A. 109, 1878–1882.

Chiesa, A., Serretti, A., 2009. Mindfulness-based stress reduction for stress management in healthy people: a review and meta-analysis. J. Altern. Complement. Med. 15, 593–600.

Cho, J.Y., Hwang, D.Y., Kang, T.S., Shin, D.H., Hwang, J.H., et al., 2003. Use of NSE/PS2m-transgenic mice in the study of the protective effect of exercise on Alzheimer's disease. J. Sports Sci. 21, 943–951.

Chuang, Y.F., Elango, P., Gonzalez, C.E., Thambisetty, M., 2017. Midlife anticholinergic drug use, risk of Alzheimer's disease, and brain atrophy in community-dwelling older adults. Alzheimers Dement. (NY) 3, 471–479.

Cook, N., Lynch, J., 2008. Aromatherapy: reviewing evidence for its mechanisms of action and CNS effects. Br. J. Neurosci. Nurs. 4, 595–601.

Cull-Candy, S.G., Leszkiewicz, D.N., 2004. Role of distinct NMDA receptor subtypes at central synapses. Sci STKE 2004, re16.

Cullen, C.L., Young, K.M., 2016. How does transcranial magnetic stimulation influence glial cells in the central nervous system? Front. Neural Circuits 10, 26.

Cummings, J., Lee, G., Mortsdorf, T., Ritter, A., Zhong, K., 2017. Alzheimer's disease drug development pipeline: 2017. Alzheimers Dement. (NY) 3, 367–384.

Di Iorio, G., Baroni, G., Lorusso, M., Montemitro, C., Spano, M.C., di Giannantonio, M., 2017. Efficacy of memantine in schizophrenic patients: a systematic review. J. Amino Acids 2017, 7021071.

Di Lazzaro, V., Capone, F., Apollonio, F., Borea, P.A., Cadossi, R., et al., 2013. A consensus panel review of central nervous system effects of the exposure to low-intensity extremely low-frequency magnetic fields. Brain Stimulat. 6, 469–476.

Dimsdale, J.E., 2008. Psychological stress and cardiovascular disease. J. Am. Coll. Cardiol. 51, 1237–1246.

Dowlati, Y., Herrmann, N., Swardfager, W., Liu, H., Sham, L., Reim, E.K., et al., 2010. A meta-analysis of cytokines in major depression. Biol. Psychiatr. 67, 446–457.

Dysken, M.W., Sano, M., Asthana, S., Vertrees, J.E., Pallaki, M., et al., 2014. Effect of vitamin E and memantine on functional decline in Alzheimer disease: the TEAM-AD VA cooperative randomized trial. JAMA 311, 33–44.

Edgley, S.A., Eyre, J.A., Lemon, R.N., Miller, S., 1997. Comparison of activation of corticospinal neurons and spinal motor neurons by magnetic and electrical transcranial stimulation in the lumbosacral cord of the anaesthetized monkey. Brain 120, 839–853.

Elder, G.J., Taylor, J.P., 2014. Transcranial magnetic stimulation and transcranial direct current stimulation: treatments for cognitive and neuropsychiatric symptoms in the neurodegenerative dementias? Alzheimers Res. Ther. 6, 74.

Epel, E., Daubenmier, J., Moskowitz, J.T., Folkman, S., Blackburn, E., 2009. Can meditation slow rate of cellular aging? Cognitive stress, mindfulness, and telomeres. Ann. N.Y. Acad. Sci. 1172, 34–53.

Erickson, K.I., Kramer, A.F., 2009. Aerobic exercise effects on cognitive and neural plasticity in older adults. Br. J. Sports Med. 43, 22–24.

Erickson, K.I., Prakash, R.S., Voss, M.W., Chaddock, L., Hu, L., Morris, K.S., et al., 2009. Aerobic fitness is associated with hippocampal volume in elderly humans. Hippocampus. 19, 1030–1039.

Eriksson, P.S., Perfilieva, E., Björk-Eriksson, T., et al., 1998. Neurogenesis in the adult human hippocampus. Nat. Med. 4, 1313–1317.

Fang, R., Ye, S., Huangfu, J., Calimag, D.P., 2017. Music therapy is a potential intervention for cognition of Alzheimer's Disease: a mini-review. Transl. Neurodegener. 6, 2.

Farooqui, A.A., 2013. Metabolic Syndrome: An Important Risk Factor for Stroke, Alzheimer Disease, and Depression. Springer, New York.

Farrall, A.J., Wardlaw, J.M., 2009. Blood-brain barrier: ageing and microvascular disease-systematic review and meta-analysis. Neurobiol. Aging 30, 337–352.

Field, T., Field, T., Cullen, C., et al., 2008. Lavender bath oil reduces stress and crying and enhances sleep in very young infants. Early Human Dev. 84, 399–401.

Figueiredo, C.P., Clarke, J.R., Ledo, J.H., Ribeiro, F.C., Costa, C.V., et al., 2013. Memantine rescues transient cognitive impairment caused by high-molecular-weight aβ oligomers but not the persistent impairment induced by low-molecular-weight oligomers. J. Neurosci. 33, 9626–9634.

Forrester, L.T., Maayan, N., Orrell, M., Spector, A.E., Buchan, L.D., et al., 2014. Aromatherapy for dementia. Cochrane Database Syst. Rev. CD003150.

Foster, P.P., Rosenblatt, K.P., Kuljiš, R.O., 2011. Exercise-Induced cognitive plasticity, implications for mild cognitive impairment and Alzheimer's disease. Front Neurol. 2, 28.

Fox, C., Richardson, K., Maidment, I.D., et al., 2011. Anticholinergic medication use and cognitive impairment in the older population: the medical research council cognitive function and ageing study. J. Am. Geriatr. Soc. 59, 1477–1483.

Friedman, D., Honig, L.S., Scarmeas, N., 2012. Seizures and epilepsy in Alzheimer's Disease. CNS Neurosci. Ther. 18, 285–294.

Fu, C.-Y., Moyle, W., Cooke, M., 2013. A randomised controlled trial of the use of aromatherapy and hand massage to reduce disruptive behaviour in people with dementia. BMC Complement. Altern. Med. 13, 165.

Fu, S.-P., He, S.-Y., Xu, B., Hu, C.J., Lu, S.F., et al., 2014. Acupuncture promotes angiogenesis after myocardial ischemia through H3K9 acetylation regulation at VEGF gene. PLoS One 9, e94604.

Fukui, H., Toyoshima, K., 2008. Music facilitate the neurogenesis, regeneration and repair of neurons. Med. Hypotheses 71, 765–769.

Fukui, H., Arai, A., Toyoshima, K., 2012. Efficacy of music therapy in treatment for the patients with Alzheimer's disease. Int. J. Alzheimers Dis. 2012, 531646.

Fung, J.K.K.M., Tsang, H.W.H., Chung, R.C.K., 2012. A systematic review of the use of aromatherapy in treatment of behavioral problems in dementia. Geriatr. Gerontol. Int. 12, 372–382.

Gao, F., Wang, S., Guo, Y., Wang, J., Lou, M., et al., 2010. Protective effects of repetitive transcranial magnetic stimulation in a rat model of transient cerebral ischaemia: a microPET study. Eur. J. Nucl. Med. Mol. Imaging 37, 954–961.

Gard, T., Hölzel, B.K., Lazar, S.W., 2014. The potential effects of meditation on age-related cognitive decline: a systematic review. Ann. N.Y. Acad. Sci. 1307, 89–103.

Garrido, S., Schubert, E., 2015. Moody melodies: do they cheer us up? A study of the effect of sad music on mood. Psychol. Music 43, 244–261.

Gervits, F., Ash, S., Coslett, H.B., Rascovsky, K., Grossman, M., et al., 2016. Transcranial direct current stimulation for the treatment of primary progressive aphasia: an open-label pilot study. Brain Lang. 162, 35–41.

Gianaros, P.J., Jennings, J.R., Sheu, L.K., Greer, P.J., Kuller, L.H., Matthews, K.A., 2007. Prospective reports of chronic life stress predict decreased grey matter volume in the hippocampus. Neuroimage 35, 795–803.

Godyń, J., Jończyk, J., Panek, D., Malawska, B., 2016. Therapeutic strategies for Alzheimer's disease in clinical trials. Pharmacol. Rep. 68, 127–138.

Gold, G., Bouras, C., Kövari, E., et al., 2000. Clinical validity of Braak neuropathological staging in the oldest-old. Acta Neuropathol. 99, 579–584.

Gómez Gallego, M., Gómez García, J., 2016. Music therapy and Alzheimer's disease: cognitive, psychological, and behavioural effects. Neurologia 32, 300–308.

Gouras, G.K., Xu, H., Gross, R.S., et al., 2000. Testosterone reduces neuronal secretion of Alzheimer's β-amyloid peptides. Proc. Natl. Acad. Sci. U.S.A. 97, 1202–1205.

Gray, S.L., Anderson, M.L., Dublin, S., et al., 2015. Cumulative use of strong anticholinergics and incident dementia: a prospective cohort study. JAMA Intern. Med. 175, 401–407.

Greig, S.L., 2015. Memantine ER/donepezil: a review in Alzheimer's Disease. CNS Drugs 29, 963–970.

Grossberg, G.T., 2003. Cholinesterase inhibitors for the treatment of Alzheimer's disease: getting on and staying on. Curr. Ther. Res. Clin. Exp. 64, 216–235.

Gruenewald, T.L., Cohen, S., Matthews, K.A., Tracy, R., Seeman, T.E., 2009. Association of socioeconomic status with inflammation markers in black and white men and women in the Coronary Artery Risk Development in Young Adults (CARDIA) study. Soc. Sci. Med. 69, 451–459.

Guétin, S., Portet, F., Picot, M.C., Pommié, C., Messaoudi, M., et al., 2009. Effect of music therapy on anxiety and depression in patients with Alzheimer's type dementia: randomised, controlled study. Dement. Geriatr. Cogn. Disord. 28, 36–46.

Guo, W., Wong, S., Li, M., Liang, W., Liesa, M., Serra, C., et al., 2012. Testosterone plus low-intensity physical training in late life improves functional performance, skeletal muscle mitochondrial biogenesis, and mitochondrial quality control in male mice. PLoS One. 7, e51180.

Haring, R., Fisher, A., Marciano, D., et al., 1998. Mitogen-activated protein kinase-dependent and protein kinase C-dependent pathways link the m1 muscarinic receptor to β-amyloid protein secretion. J. Neurochem. 71, 2094–2103.

Hebb, D.O., 1949. The Organization of Behavior: A Neuropsychological Theory. Wiley, New York.

Hillman, C.H., Erickson, K.I., Kramer, A.F., 2008. Be smart, exercise your heart: exercise effects on brain and cognition. Nat. Rev. Neurosci. 9, 58–65.

Hilmer, S.N., McLachlan, A.J., Le Couteur, D.G., 2007. Clinical pharmacology in the geriatric patient. Fundam. Clin. Pharmacol. 21, 217–230.

Hogan, D.B., Bailey, P., Black, S., et al., 2008. Diagnosis and treatment of dementia: 5. Nonpharmacologic and pharmacologic therapy for mild to moderate dementia. CMAJ 179, 1019–1026.

Hoogendam, J.M., Ramakers, G.M., Di Lazzarro, V., 2010. Physiology of repetitive transcranial magnetic stimulation of the human brain. Brain Stimul. 3, 95–118.

Hu, M., Schurdak, M.E., Puttfarcken, P.S., El Kouhen, R., Gopalakrishnan, M., et al., 2007. High content screen microscopy analysis of amyloid beta 1-42-induced neurite outgrowth reduction in rat primary cortical neurons: neuroprotective effects of alpha 7 neuronal nicotinic acetylcholine receptor ligands. Brain Res. 1151, 227–235.

Huang, L., Abuhamdah, S., Howes, M.-J.R., et al., 2008. Pharmacological profile of essential oils derived from *Lavandula angustifolia* and *Melissa officinalis* with anti-agitation properties: focus on ligand-gated channels. J. Pharm. Pharmacol. 60, 1515–1522.

Ijaopo, E.O., 2017. Dementia-related agitation: a review of non-pharmacological interventions and analysis of risks and benefits of pharmacotherapy. Transl. Psychiat. 7, e1250.

Jacobs, T.L., Epel, E.S., Lin, J., Blackburn, E.H., Wolkowitz, O.M., Bridwell, D.A., et al., 2010. Intensive meditation training, immune cell telomerase activity, and psychological mediators. Psychoneuroendocrinology 36, 664–681.

Jedrziewski, M.K., Ewbank, D.C., Wang, H., Trojanowski, J.Q., 2014. The impact of exercise, cognitive activities, and socialization on cognitive function: results from the national long-term care survey. Am. J. Alzheimers Dis. Other Demen. 29, 372–378.

Jimbo, D., Kimura, Y., Taniguchi, M., Inoue, M., Urakami, K., 2009. Effect of aromatherapy on patients with Alzheimer's disease. Psychogeriatrics 9, 173–179.

Kavoussi, B., Ross, B., 2007. The neuroimmune basis of anti-inflammatory acupuncture. Integr. Cancer Ther. 6, 251–257.

Khalid, S., Soomro, B.A., 2015. Efficacy of memantine in treating patients with dementia and fibromyalgia. J. Neurol. Stroke 3, 34–39.

Khoury, R., Rajamanickam, J., Grossberg, G.T., 2018. An update on the safety of current therapies for Alzheimer's disease: focus on rivastigmine. Ther. Adv. Drug Saf. 9, 171–178.

Kim, S.K., Bae, H., 2010. Acupuncture and immune modulation. Auton. Neurosci. 157, 38–41.

Kumar, A.M., Tims, F., Cruess, D.G., Mintzer, M.J., Ironson, G., et al., 1999. Music therapy increases serum melatonin levels in patients with Alzheimer's disease. Altern. Ther. Health Med. 5, 49–57.

Kotermanski, S.E., Johnson, J.W., 2009. Mg^{2+} imparts NMDA receptor subtype selectivity to the Alzheimer's drug memantine. J. Neurosci. 29, 2774–2779.

Lanza, I.R., Sreekumaran Nair, K., 2010. Regulation of skeletal muscle mitochondrial function: genes to proteins. Acta Physiol. (Oxf). 199, 529–547.

Lavretsky, H., Epel, E.S., Siddarth, P., Nazarian, N., Cyr, N.S., Khalsa, D.S., et al., 2012. A pilot study of yogic meditation for family dementia caregivers with depressive symptoms: effects on mental health, cognition, and telomerase activity. Int. J. Geriatr. Psychiatry 28, 57–65.

Lazar, S.W., Kerr, C.E., Wasserman, R.H., Gray, J.R., Greve, D.N., Treadway, M.T., et al., 2005. Meditation experience is associated with increased cortical thickness. Neuroreport 16, 1893–1897.

Lee, M.S., Shin, B.C., Kong, J.C., Ernst, E., 2008. Effectiveness of acupuncture for Parkinson's disease: a systematic review. Mov. Disord. 23, 1505–1515.

Lee, M.S., Shin, B.C., Ernst, E., 2009. Acupuncture for Alzheimer's disease: a systematic review. Int. J. Clin. Pract. 63, 874–879.

Lee, M.S., Choi, J., Posadzki, P., Ernst, E., 2012. Aromatherapy for health care: an overview of systematic reviews. Maturitas 71, 257–260.

Li, X., Wang, Q., 2013. Acupuncture therapy for stroke patients. Int. Rev. Neurobiol. 111, 159–179.

Li, W., Gan, C., Lv, Y., Wang, S., Cheng, H., 2017. Chemotherapy-induced prospective memory impairment in breast cancer patients with different hormone receptor expression. Medicine (Baltimore) 96, e6514.

Lin, J., Epel, E.S., Blackburn, E.H., 2009. Telomeres, telomerase stress and aging. In: Bernston, G.G., Cacioppo, J.T. (Eds.), Handbook of Neuroscience for the Behavioral Sciences. Wiley, Hoboken, NJ, (Chapter 65).

Lin, S.T., Yang, P., Lai, C.Y., Su, Y.Y., Yeh, Y.C., et al., 2011. Mental health implications of music: insight from neuroscientific and clinical studies. Harv. Rev. Psychiatry 19, 34–46.

Lindblad, C.I., Hanlon, J.T., Gross, C.R., Sloane, R.J., Pieper, C.F., et al., 2006. Clinically important drug-disease interactions and their prevalence in older adults. Clin. Ther. 28, 1133–1143.

Lindquist, S.G., Holm, I.E., Schwartz, M., Law, I., Stokholm, J., et al., 2008. Alzheimer disease-like clinical phenotype in a family with FTDP-17 caused by a MAPT R406W mutation. Eur. J. Neurol. 15, 377–385.

Lipton, S.A., 2006. Paradigm shift in neuroprotection by NMDA receptor blockade: memantine and beyond. Nat. Rev. Drug Discov. 5, 160–170.

Liu, W., Xu, Z., Deng, Y., Xu, B., Wei, Y., et al., 2013. Protective effects of memantine against methylmercury-induced glutamate dyshomeostasis and oxidative stress in rat cerebral cortex. Neurotox. Res. 24, 320–337.

Liu, Q., Wang, X.J., Zhang, Z.C., et al., 2016. Neuroprotection against vascular dementia after acupuncture combined with donepezil hydrochloride: P300 event related potential. Neural Regen. Res. 11, 460–464.

Lonergan, E., Luxenberg, J., Colford, J., 2002. Haloperidol for agitation in dementia. Cochrane Database Syst. Rev. CD002852.

Lowinus, T., Bose, T., Busse, S., Busse, M., Reinhold, D., Schraven, B., et al., 2016. Immunomodulation by memantine in therapy of Alzheimer's disease is mediated through inhibition of Kv1.3 channels and T cell responsiveness. Oncotarget 7, 53797–53807.

Lucassen, P.J., Meerlo, P., Naylor, A.S., van Dam, A.M., Dayer, A.G., Fuchs, E., et al., 2010. Regulation of adult neurogenesis by stress, sleep disruption, exercise and inflammation: implications for depression and antidepressant action. Eur. Neuropsychopharmacol. 20, 1–17.

Luo, D., Chen, R., Liang, F.X., 2017. Modulation of acupuncture on cell apoptosis and autophagy. Evid. Based Complement. Altern. Med. 2017, 8268736.

Lutz, A., Slagter, H.A., Dunne, J.D., Davidson, R.J., 2008. Attention regulation and monitoring in meditation. Trends Cogn. Sci. 12, 163–169.

Machado, A., 2014. Chronic stress as a risk factor for Alzheimer's disease. Rev. Neurosci. 25, 785–804.

Maes, M., 2010. Depression is an inflammatory disease, but cell-mediated immune activation is the key component of depression. Prog. Neuro-Psychopharmacol. Biol. Psychiatry 35, 664–675.

Manckoundia, P., Pfitzenmeyer, P., d'Athis, P., Dubost, V., Mourey, F., 2006. Impact of cognitive task on the posture of elderly subjects with Alzheimer's disease compared to healthy elderly subjects. Mov. Disord. 21, 236–241.

Martins, I.J., Hone, E., Foster, J.K., Sunram-Lea, S.I., Gnjec, A., Fuller, S.J., et al., 2006. Apolipoprotein E, cholesterol metabolism, diabetes, and the convergence of risk factors for Alzheimer's disease and cardiovascular disease. Mol. Psychiatry 11, 721–736.

Martiny, K., Lunde, M., Bech, P., 2010. Transcranial low voltage pulsed electromagnetic fields in patients with treatment-resistant depression. Biol. Psychiatry 68, 163–169.

Matsunaga, S., Kishi, T., Iwata, N., 2015. Memantine monotherapy for Alzheimer's disease: a systematic review and meta-analysis. PLoS One 10, e0123289.

McCurry, S.M., Logsdon, R.G., Teri, L., Vitiello, M.V., 2007. Sleep disturbances in caregivers of persons with dementia: contributing factors and treatment implications. Sleep Med. Rev. 11, 143–153.

McEwen, B.S., Gianaros, P.J., 2010. Central role of the brain in stress and adaptation: links to socioeconomic status, health, and disease. Ann. N.Y. Acad. Sci. 1186, 190.

McShane, R., Areosa Sastre, A., Minakaran, N., 2006. Memantine for dementia. Cochrane Database Syst. Rev. CD003154.

Mehlig, K., Skoog, I., Waern, M., Jonasson, J.M., Lapidus, L., et al., 2014. Physical activity, weight status, diabetes and dementia: a 34-year follow-up of the population study of women in Gothenburg. Neuroepidemiology 42, 252–259.

Mohammad, D., Chan, P., Bradley, J., Lanctôt, K., Herrmann, N., 2017. Acetylcholinesterase inhibitors for treating dementia symptoms – a safety evaluation. Expert Opin. Drug Saf. 16, 1009–1019.

Mondragon-Rodriguez, S., Basurto-Islas, G., Lee, H.G., Perry, G., Zhu, X., et al., 2010. Causes versus effects: the increasing complexities of Alzheimer's disease pathogenesis. Expert Rev. Neurother. 10, 683–10691.

Morris, R.G., Kopelman, M.D., 1986. The memory deficits in Alzheimer-type dementia: a review. Q. J. Exp. Psychol. A 38, 575–602.

Müller-Dahlhaus, F., Vlachos, A., 2013. Unraveling the cellular and molecular mechanisms of repetitive magnetic stimulation. Front. Mol. Neurosci. 6, 50.

Muñoz-Torrero, D., 2008. Acetylcholinesterase inhibitors as disease-modifying therapies for Alzheimer's disease. Curr. Med. Chem. 15, 2433–2455.

Nakamura, Y., Homma, A., Kitamura, S., Yoshimura, I., 2011. Phase III study of memantine hydrochloride, a new NMDA receptor antagonist, in patients with moderate to severe Alzheimer's disease – efficacy and safety. Jpn. J. Geriatr. Psychiatry 22, 464–473.

Nakamura, Y., Kitamura, S., Nagakubo, T., Kobayashi, M., Homma, A., 2016. Study of memantine hydrochloride in combination with donepezil hydrochloride in patients with moderate to severe Alzheimer's disease – efficacy and safety. Jpn. J. Geriatr. Med. 54, 1147–1158.

Nordberg, A., 1999. PET studies and cholinergic therapy in Alzheimer's disease. Rev. Neurol. (Paris) 155 (Suppl. 4), S53–S63.

O'Connor, D.W., Eppingstall, B., Taffe, J., Van Der Ploeg, E.S., 2013. A randomized, controlled cross-over trial of dermally-applied lavender (*Lavandula angustifolia*) oil as a treatment of agitated behaviour in dementia. BMC Complement. Altern. Med. 13, 315.

Opitz, A., Windhoff, M., Heidemann, R.M., Turner, R., Thielscher, A., 2011. How the brain tissue shapes the electric field induced by transcranial magnetic stimulation. Neuroimage 58, 849–859.

Ornish, D., Lin, J., Daubenmier, J., Weidner, G., Epel, E., Kemp, C., et al., 2008. Increased telomerase activity and comprehensive lifestyle changes: a pilot study. Lancet Oncol. 9, 1048–1057.

Ozdemir, L., Akdemir, N., 2009. Effects of multisensory stimulation on cognition, depression and anxiety levels of mildly-affected Alzheimer's patients. J. Neurol. Sci. 283, 211–213.

Pace, T.W., Mletzko, T.C., Alagbe, O., Musselman, D.L., Nemeroff, C.B., Miller, A.H., et al., 2006. Increased stress-induced inflammatory responses in male patients with major depression and increased early life stress. Am. J. Psychiatry 163, 1630–1633.

Pace, T.W., Negi, L.T., Adame, D.D., Cole, S.P., Sivilli, T.I., Brown, T.D., et al., 2009. Effect of compassion meditation on neuroendocrine, innate immune and behavioral responses to psychosocial stress. Psychoneuroendocrinology 34, 87–98.

Pagano, G., Yousaf, T., Politis, M., 2017. PET molecular imaging research of levodopa-induced dyskinesias in Parkinson's disease. Curr. Neurol. Neurosci. Rep. 17, 90.

Peisah, C., Chan, D.K., McKay, R., Kurrle, S.E., Reutens, S.G., 2011. Practical guidelines for the acute emergency sedation of the severely agitated older patient. Intern. Med. J. 41, 651–657.

Pereira, A.C., Huddleston, D.E., Brickman, A.M., et al., 2007. An in vivo correlate of exercise-induced neurogenesis in the adult dentate gyrus. Proc. Natl Acad. Sci. U.S.A. 104, 5638–5643.

Perry, E.K., Kilford, L., Lees, A.J., Burn, D.J., Perry, R.H., 2003. Increased Alzheimer pathology in Parkinson's disease related to antimuscarinic drugs. Ann. Neurol. 54, 235–238.

Petrovsky, D., Cacchione, P.Z., George, M., 2015. Review of the effect of music interventions on symptoms of anxiety and depression in older adults with dementia. Int. Psychogeriatr. 2015 (27), 1661–1670.

Picada, J., Flores, E., Cappelari, S., Pereira, P., 2011. Effects of memantine, a noncompetitive N-methyl-D-aspartate receptor antagonist, on genomic stability. Basic Clin. Pharmacol. Toxicol. 109, 413–417.

Piette, F., Belmin, J., Vincent, H., Schmidt, N., Pariel, S., et al., 2011. Masitinib as an adjunct therapy for mild-to-moderate Alzheimer's disease: a randomised, placebo-controlled phase 2 trial. Alzheimers Res. Ther. 3, 16.

Pollard, K., 2008. Introducing aromatherapy as a form of pain management into a delivery suite. J. Assoc. Chart Physiotherapists Womens Health 103, 12–16.

Porsteinsson, A.P., Drye, L.T., Pollock, B.G., Devanand, D.P., Frangakis, C., et al., 2014. Effect of citalopram on agitation in Alzheimer's Disease – The CitAD Randomized Controlled Trial. JAMA 311, 682–691.

Raven, J.C., 1995. Coloured Progressive Matrices Sets A, Ab, B. Oxford Psychologists Press, New York.

Reggiani, A.M., Simoni, E., Caporaso, R., Meunier, J., Keller, E., et al., 2016. In vivo characterization of ARN14140, a memantine/galantamine-based multi-target compound for Alzheimer's Disease. Sci. Rep. 6, 33172.

Reisberg, B., Doody, R., Stöffler, A., Schmitt, F., Ferris, S., et al., 2003. Memantine in moderate-to-severe Alzheimer's disease. N. Engl. J. Med. 348, 1333–1341.

Ribeiro Do Couto, B., Aguilar, M.A., Manzanedo, C., Rodriguez-Arias, M., Minarro, J., 2004. Effects of NMDA receptor antagonists (MK-801 and memantine) on the acquisition of morphine-induced conditioned place preference in mice. Prog. Neuropsychopharmacol. Biol. Psychiatry 28, 1035–1043.

Richardson, K., Fox, C., Maidment, I., Steel, N., Loke, Y.K., et al., 2018. Anticholinergic drugs and risk of dementia: case-control study. BMJ 361. Available from: https://doi.org/10.1136/bmj.k1315.

Rocca, W.A., Grossardt, B.R., Shuster, L.T., Stewart, E.A., 2012. Hysterectomy, oophorectomy, estrogen, and the risk of dementia. Neurodegener. Dis. 10, 175–178.

Rocca, W.A., Grossardt, B.R., Shuster, L.T., 2014. Oophorectomy, estrogen, and dementia: a 2014 update. Mol. Cell Endocrinol. 389, 7–12.

Rohleder, N., 2014. Stimulation of systemic low-grade inflammation by psychosocial stress. Psychosom. Med. 76, 181–189.

Rosini, M., Simoni, E., Minarini, A., Melchiorre, C., 2014. Multi-target design strategies in the context of Alzheimer's disease: acetylcholinesterase inhibition and NMDA receptor antagonism as the driving forces. Neurochem. Res. 39, 1914–1923.

Roy, R., Niccolini, F., Pagano, G., Politis, M., 2016. Cholinergic imaging in dementia spectrum disorders. Eur. J. Nucl. Med. Mol. Imaging 43, 1376–1386.

Sahiner, M., Erken, G., Kursunluoglu, R., Genc, O., Sahiner, T., 2011. Memantine improves learning in kindled rats. J. Neurol. Sci. 28, 322–329.

Salvagioni, D.A.J., Melanda, F.N., Mesas, A.E., González, A.D., Gabani, F.L., Andrade, S. M., 2017. Physical, psychological and occupational consequences of job burnout: a systematic review of prospective studies. PLoS One. 12, e0185781.

Särkämö, T., 2017. Music for the ageing brain: cognitive, emotional, social, and neural benefits of musical leisure activities in stroke and dementia. Dementia (London) 1471301217729237.

Särkämö, T., Laitinen, S., Tervaniemi, M., Numminen, A., Kurki, M., et al., 2012. Music, emotion, and dementia: insight from neuroscientific and clinical research. Music Med. 4, 153–162.

Särkämö, T., Tervaniemi, M., Laitinen, S., Numminen, A., Kurki, M., et al., 2014. Cognitive, emotional, and social benefits of regular musical activities in early dementia: randomized controlled study. Gerontologist 54, 634–650.

Satoh, M., Yuba, T., Tabei, K., Okubo, Y., Kida, H., et al., 2015. Music therapy using singing training improves psychomotor speed in patients with Alzheimer's disease: a neuropsychological and fMRI study. Dement. Geriatr. Cogn. Dis. Extra 5, 296–308.

Schneiderman, N., 2005. Stress and health: psychological, behavioral, and biological determinants. Annu. Rev. Clin. Psychol. 1, 607.

Schulz, R., Martire, L.M., 2004. Family caregiving of persons with dementia: prevalence, health effects, and support strategies. Am. J. Geriatr. Psychiatry 12, 240–249.

Shi, G.X., Liu, C.Z., Guan, W., et al., 2014. Effects of acupuncture on Chinese medicine syndromes of vascular dementia. Chin. J. Integr. Med. 20, 661–666.

Shiina, Y., Funabashi, N., Lee, K., et al., 2008. Relaxation effects of lavender aromatherapy improve coronary flow velocity reserve in healthy men evaluated by transthoracic Doppler echocardiography. Int. J. Cardiol. 129, 193–197.

Shumaker, S.A., Reboussin, B.A., Espeland, M.A., et al., 1998. The Women's Health Initiative Memory Study (WHIMS): a trial of the effect of estrogen therapy in preventing and slowing the progression of dementia. Control. Clin. Trials 19, 604–621.

Shupak, N.M., Prato, F.S., Thomas, A.W., 2004. Human exposure to a specific pulsed magnetic field: effects on thermal sensory and pain thresholds. Neurosci. Lett. 363, 157–162.

Simoni, E., Daniele, S., Bottegoni, G., Pizzirani, D., Trincavelli, M.L., et al., 2012. Combining galantamine and memantine in multitargeted, new chemical entities potentially useful in Alzheimer's disease. J. Med. Chem. 55, 9708–9721.

Song, M.S., Rauw, G., Baker, G.B., Kar, S., 2008. Memantine protects rat cortical cultured neurons against beta-amyloid-induced toxicity by attenuating tau phosphorylation. Eur. J. Neurosci. 28, 1989–2002.

Stranahan, A.M., Zhou, Y., Martin, B., Maudsley, S., 2009. Pharmacomimetics of exercise: novel approaches for hippocampally-targeted neuroprotective agents. Curr. Med. Chem. 16, 4668–4678.

Swaab, D.F., Bao, A.M., Lucassen, P.J., 2005. The stress system in the human brain in depression and neurodegeneration. Ageing Res. Rev. 4, 141–194.

Tanqueiro, S.R., Ramalho, R.M., Rodrigues, T.M., Lopes, L.V., Sebastião, A.M., Diógenes, M.J., 2018. Inhibition of NMDA receptors prevents the loss of BDNF function induced by amyloid β. Front. Pharmacol. 9, 237.

Thompson, S., Lanctôt, K.L., Herrmann, N., 2004. Combination therapy for Alzheimer's disease. Expert Opin. Drug Saf. 3, 425–440.

Tifratene, K., Manera, V., Fabre, R., et al., 2017. Antipsychotic prescribing for Alzheimer's disease and related disorders in specialized settings from 2010 to 2014 in France: a repeated cross-sectional study. Alzheimers Res. Ther. 34.

Um, H.S., Kang, E.B., Leem, Y.H., Cho, I.H., Yang, C.H., et al., 2008. Exercise training acts as a therapeutic strategy for reduction of the pathogenic phenotypes for Alzheimer's disease in an NSE/APPsw-transgenic model. J. Mol. Med. 22, 529–539.

Van Puymbroeck, M., Payne, L.L., Hsieh, P.C., 2007. A phase I feasibility study of yoga on the physical health and coping of informal caregivers. Evid. Based Complement. Altern. Med. 4, 519–529.

Van Spall, H.G.C., Toren, A., Kiss, A., Fowler, R.A., 2007. Eligibility criteria of randomized controlled trials published in high-impact general medical journals: a systematic sampling review. JAMA 297, 1233–1240.

Vasionytė, I., Madison, G., 2013. Musical intervention for patients with dementia: a meta-analysis. J. Clin. Nurs. 22, 1203–1216.

Vercelletto, M., Boutoleau-Bretonniere, C., Volteau, C., Puel, M., Auriacombe, S., et al., 2011. Memantine in behavioral variant frontotemporal dementia: negative results. J. Alzheimer Dis. 23, 749–759.

Vestergaard-Poulsen, P., van Beek, M., Skewes, J., Bjarkam, C.R., Stubberup, M., Bertelsen, J., et al., 2009. Long-term meditation is associated with increased gray matter density in the brain stem. Neuroreport 20, 170–174.

Villemagne, V.L., Burnham, S., Bourgeat, P., Brown, B., Ellis, K.A., et al., 2013. Amyloid beta deposition, neurodegeneration, and cognitive decline in sporadic Alzheimer's disease: a prospective cohort study. Lancet Neurol. 12, 357–367.

Viscomi, M.T., Molinari, M., 2014. Remote neurodegeneration: multiple actors for one play. Mol. Neurobiol. 50, 368–389.

Viscomi, M.T., Latini, L., Bisicchia, E., Sasso, V., Molinari, M., 2015. Remote degeneration: insights from the hemicerebellectomy model. Cerebellum 14, 15–18.

Wang, N., Lu, S.-F., Chen, H., Wang, J.F., Fu, S.P., et al., 2015a. A protocol of histone modification-based mechanistic study of acupuncture in patients with stable angina pectoris. BMC Complement. Altern. Med. 15, 139.

Wang, J.-Y., Li, H., Ma, C.-M., Wang, J.-L., Lai, X.-S., Zhou, S.-F., 2015b. Acupuncture may exert its therapeutic effect through MicroRNA-339/Sirt2/NFκB/FOXO1 axis. BioMed Res. Int. 2015, 249013.

Wilson, R.S., Arnold, S.E., Schneider, J.A., Kelly, J.F., Tang, Y., Bennett, D.A., 2006. Chronic psychological distress and risk of Alzheimer's disease in old age. Neuroepidemiology 27, 143–1453.

Wilson, R.S., Arnold, S.E., Schneider, J.A., Li, Y., Bennett, D.A., 2007. Chronic distress, age-related neuropathology, and late-life dementia. Psychosom. Med. 69, 47–53.

Xu, L., Xu, H., Gao, W., Wang, W., Zhang, H., Lu, D.P., 2013. Treating angina pectoris by acupuncture therapy. Acupunct. Electro-Ther. Res. 38, 17–35.

Yang, M.H., Lin, L.C., Wu, S.C., Chiu, J.H., Wang, P.N., et al., 2015. Comparison of the efficacy of aroma-acupressure and aromatherapy for the treatment of dementia-associated agitation. BMC Complement. Altern. Med. 15, 93.

Ye, Y., Zhu, W., Wang, X.R., Yang, J.W., Xiao, L.Y., et al., 2017a. Mechanisms of acupuncture on vascular dementia—a review of animal studies. Neurochem. Int. 107, 204–210.

Ye, Y., Li, H., Yang, J.W., et al., 2017b. Acupuncture attenuated vascular dementia-induced hippocampal long-term potentiation impairments via activation of D1/D5 receptors. Stroke 48, 1044–1051.

Yoon, K.J., Lee, Y.T., Chung, P.W., Lee, Y.K., Kim, D.Y., et al., 2015. Effects of repetitive transcranial magnetic stimulation on behavioral recovery during early stage of traumatic brain injury in rats. J. Korean Med. Sci. 30, 1496–1502.

Yu, F., Bronas, U.G., Konety, S., Nelson, N.W., Dysken, M., et al., 2014. Effects of aerobic exercise on cognition and hippocampal volume in Alzheimer's disease: study protocol of a randomized controlled trial (The FIT-AD trial). Trials 15, 394.

Yun, Y.C., Jang, D., Yoon, S.B., Kim, D., Choi, D.H., et al., 2017. Laser acupuncture exerts neuroprotective effects via regulation of Creb, Bdnf, Bcl-2, and Bax gene expressions in the hippocampus. Evid. Based Complement. Altern. Med. 2017, 7181637.

Zeilmann, C.A., Dole, E.J., Skipper, B.J., McCabe, M., Low Dog, T., Thyne, R.L., 2003. Use of herbal medicine by elderly Hispanic and non-His-panic white patients. Pharmacotherapy 23, 526–532.

Zeng, B.Y., Salvage, S., Jenner, P., 2013a. Effect and mechanism of acupuncture on Alzheimer's disease. Int. Rev. Neurobiol. 111, 181–195.

Zeng, B.Y., Salvage, S., Jenner, P., 2013b. Current development of acupuncture research in Parkinson's disease. Int. Rev. Neurobiol. 111, 141–158.

Zhao, Z.Q., 2008. Neural mechanism underlying acupuncture analgesia. Prog. Neurobiol. 85, 355–375.

Zhang, X.-C., Xu, X.-P., Xu, W.-T., et al., 2015. Acupuncture therapy for sudden sensorineural hearing loss: a systematic review and meta-analysis of randomized controlled trials. PLoS One 10, e0125240.

Zhang, Y., Lin, R., Tao, J., Wu, Y., Chen, B., et al., 2016. Electroacupuncture improves cognitive ability following cerebral ischemia reperfusion injury via CaM-CaMKIV-CREB signaling in the rat hippocampus. Exp. Ther. Med. 12, 777–782.

Zhang, Y., Cai, J., An, L., Hui, F., Ren, T., et al., 2017. Does music therapy enhance behavioral and cognitive function in elderly dementia patients? A systematic review and meta-analysis. Ageing Res. Rev. 35, 1–11.

Ziemann, U., Paulus, W., Nitsche, M.A., Pascual-Leone, A., Byblow, W.D., et al., 2008. Consensus: motor cortex plasticity protocols. Brain Stimul. 1, 164–182.

Further Reading

Dastmalchi, K., Ollilainen, V., Lackman, P., et al., 2009. Acetylcholinesterase inhibitory guided fractionation of *Melissa officinalis* L. Bioorg. Med. Chem. 17, 867–871.

Fayed, N., Olivan-Blázquez, B., Herrera-Mercadal, P., Puebla-Guedea, M., Pérez-Yus, M.C., et al., 2014. Changes in metabolites after treatment with memantine in fibromyalgia. A double-blind randomized controlled trial with magnetic resonance spectroscopy with a 6-month follow-up. CNS Neurosci. Ther. 20, 999–1007.

Irwin, M.R., Cole, S.W., 2011. Reciprocal regulation of the neural and innate immune systems. Nat. Rev. Immunol. 11, 625–632.

Petersen, M., Simmonds, M.S.J., 2003. Rosmarinic acid. Phytochemistry 62, 121–125.

CHAPTER

8

Potential Treatment Strategies for the Treatment of Dementia With Chinese Medicinal Plants

INTRODUCTION

In the absence of satisfactory knowledge about the molecular mechanisms of various types of dementias, little is known about Western pharmacological therapies for dementia (Schwarz et al., 2012; Ijaopo, 2017). The available pharmacological therapies with antidementia drugs have been largely symptomatic, with no permanent clinical benefits on functional, behavioral, and cognitive manifestations of dementia. According to the World Health Organization, about 70%–80% of the world's population relies on nonconventional medicines, mainly of herbal sources, in their healthcare (Jacqui, 2013). Public interest for the treatment with complementary and alternative medicine is mainly due to increased side effects in synthetic drugs, lack of curative treatment for several chronic diseases, high cost of new drugs, microbial resistance, and emerging diseases, etc. (Humber, 2002). People in Asian countries have turned to Traditional Chinese complementary medicine (China) and Ayurvedic medicine in India. Along with increased global interest in traditional medicines, efforts are underway to monitor and regulate herbal drugs and traditional medicines. China has been successful in promoting its therapies with more research and science-based approaches. In contrast, Ayurvedic medicine still needs more extensive scientific research and solid evidence on the usefulness of Ayurvedic medicines. China and India are the most populated countries of the world, where common people are very poor and cannot afford expensive allopathic antidementia medications. Ancient Chinese and Indian cultures have used common complementary medicine interventions using plants products (Zhou et al., 2016; Farooqui et al., 2018). Chinese herbal medicine is a traditional

Molecular Mechanisms of Dementia
DOI: https://doi.org/10.1016/B978-0-12-816347-4.00008-8

health practice that originated from Chinese philosophy and religion, holding the belief of holism and balance in the body. The use of plant products for the treatment of age-related disorders was documented in the literature more than 2000–3000 years ago in ancient China where Traditional Chinese medicines (TCM) were used to boost memory function and increase longevity (Liu and Chang, 2006). According to Chinese medicine theory, dementia is caused by (1) deficiency of vital energy of the kidney (*Shen*), marrow (*Sui*), heart (*Xin*), and spleen (*Pi*); and (2) stagnation of blood (*Xie*) and/or phlegm (*Tan*). Thus, herbs used for dementia are not specific for the nervous system but tend to be multifunctional (Ho et al., 2010). Collective evidence suggests that TCM is a complex system composed of multiple components and targets, and the function of each component is synergistic (Huang et al., 2014a). The composition of TCM is very complex. So, it is difficult to demonstrate the effect of the mixture of components on cognitive dysfunction and to elaborate it to the mechanism of action similar to modern medicine (Huang et al., 2014a). Early preclinical and clinical studies have indicated that TCM products can be used either as single preparation or as complex herbal formulations to treat dementia in experimental and clinical studies. A recent meta-analysis study has indicated that the use of TCM provides comparable efficacy and safety as Western medicine for improving the cognitive and behavior functions of patients with vascular cognitive impairment with no dementia (Feng et al., 2016). Therefore, it is proposed that TCM has great potential uses as preventive strategies against dementia and could have positive impacts on global public health. However, safety and clear pharmacological action mechanisms of TCM are still uncertain. It has been reported that at least two-thirds of the US population will be using one or more of the alternative therapeutic approaches to treat neurological disorders. Use of indigenous drugs of natural origin forms a major part of such therapies; more than 1500 herbals are sold as dietary supplements or ethnic traditional medicines (Legal Status of Traditional Medicine, Complementary/Alternative Medicine, 2001). A literature survey on the use of TCM for human subjects indicates that the top 10 TCM herb ingredients, including huperzine A, *Ginkgo biloba*, ginseng, *Anemarrhena* rhizome, green tea, danshen, and Radix puerariae, have been prioritized for highest potential benefit to dementia intervention. In TCM, at least, 236 formulae have been prepared from various herbs by renowned TCM doctors, over the past 10 centuries (Chinese Pharmacopoeia Commission, 2005). Pharmacological investigations have indicated that many TCM ingredients of these formulae can elicit memory-improving effects in vivo and in vitro via multiple mechanisms of action, covering estrogen-like, cholinergic, antioxidant, anti-inflammatory, antiapoptotic, neurogenetic, and anti-Aβ activities (Chinese Pharmacopoeia Commission, 2005).

HUPERZINE A AND DEMENTIA

Huperzine A (HupA) is a plant-based alkaloid and potent, selective, and well-tolerated reversible inhibitor of acetylcholinesterase (AChE) (Fig. 8.1) (Damar et al., 2016). It is derived from a Chinese herb called *Huperzia serrata*. HupA can cross the blood–brain barrier (BBB), has higher oral bioavailability, and longer duration of AChE inhibitory action. Hup A targets different sites on AChE, and its ability to inhibit AChE is eight- and twofold more effective than donepezil and rivastigmine, respectively (Wang and Tang, 2005). HupA protects not only against hydrogen peroxide, beta-amyloid (Aβ), and glutamate, but also against staurosporine-induced cytotoxicity and apoptosis (Fig. 8.2). These protective effects of HupA are related to its ability to attenuate oxidative stress, regulate the expression of apoptotic proteins Bcl-2, Bax, P53, and caspase-3, protect mitochondria, upregulate nerve growth factor and its receptors, and interfere with amyloid precursor protein (APP) metabolism (Zhang et al., 2008). Antagonizing effects of HupA on *N*-methyl-D-aspartate receptors and potassium currents may also contribute to its neuroprotection as well. HupA improves cognitive deficits in a broad range of animal models (Fig. 8.3) (Zhang et al., 2008). ZT-1 (*N*-[2-hydroxy-3-methoxy-5-chlorobenzylidene]) and methanesulfonyl fluoride (SNX-001) are pro-HupA drugs, which have been used for the

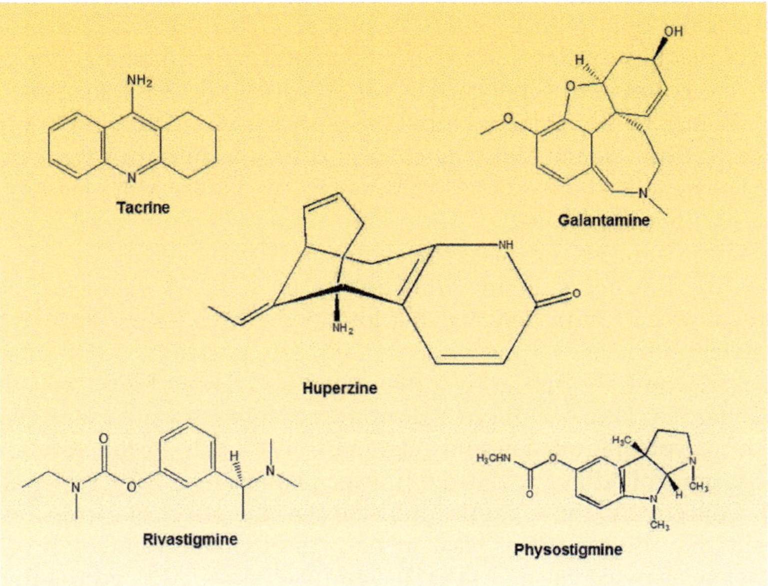

FIGURE 8.1 Chemical structures of huperzine and other acetyltransferase inhibitors.

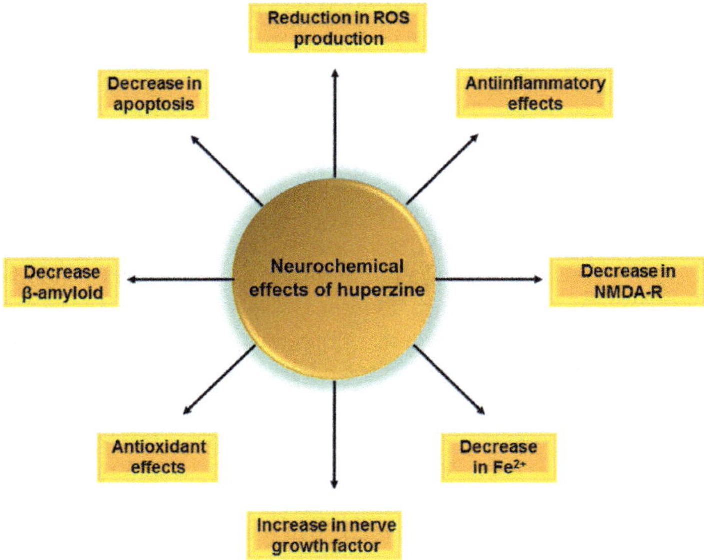

FIGURE 8.2 Neurochemical effects of huperzine.

treatment of Alzheimer's disease (AD) type of dementia in Phase I trial studies (Moss et al., 1999; Jia et al., 2013).

In addition to antiacetylcholinesterase activity, HupA produces neuroprotective effects through brain iron regulation (Fig. 8.3). HupA treatment not only reduces insoluble and soluble Aβ levels and ameliorates Aβ plaques formation, but also modulates hyperphosphorylation of tau in the cortex and hippocampus of APPswe/PS1dE9 transgenic AD mice (Wang et al., 2012). HupA also increases Disintegrin A and Metalloprotease Domain 10 (ADAM10) expression in HupA treated AD mice. The beneficial effects of HupA are largely abolished by feeding the animals with a high iron diet. Thus, HupA not only reduces iron content in the brain, but also decreases the expression of transferrin-receptor 1 as well as the transferrin-bound iron uptake in cultured neurons. Recent studies have also indicated that the loss of dendritic spine density and synaptotagmin levels in the brain of APPswe/presenilin-1 (PS1) transgenic mice can be significantly ameliorated by chronic HupA treatment suggesting that HupA-induced neuroprotection is associated with reductions in Aβ plaque burden and oligomeric Aβ levels in the cortex and hippocampus of APPswe/PS1dE9 transgenic mice (Fig. 8.3) (Wang et al., 2012). Collectively, these studies indicate that the effect of HupA on Aβ deposits may be caused at least in part, through the regulation of the expression of α-secretase (ADAM10) and excessive APP processing by β-secretase (BACE1) in these transgenic mice (Huang et al., 2014b).

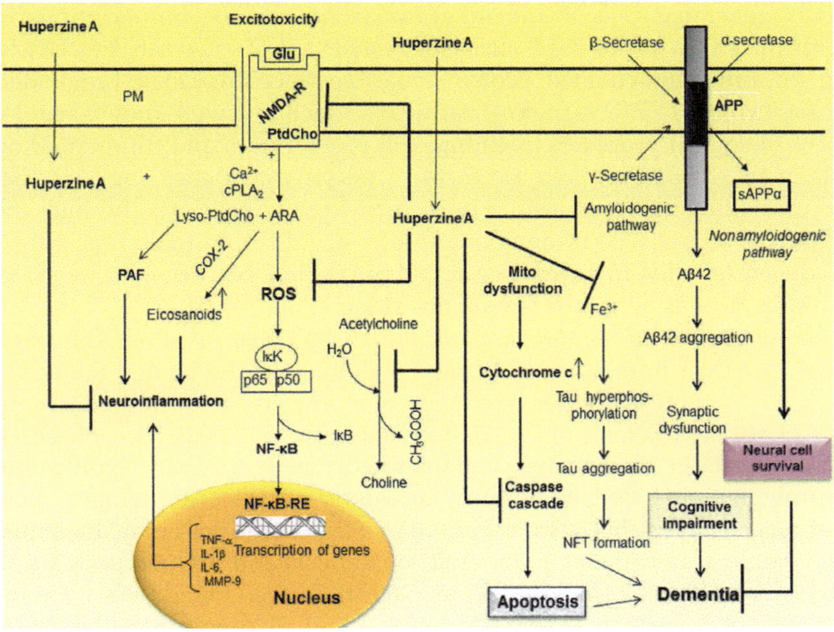

FIGURE 8.3 Hypothetical diagram showing effect of huperzine on oxidative stress, neuroinflammation, and APP processing in the brain.
Aβ, beta-amyloid; *APP*, amyloid precursor protein; *ARA*, arachidonic acid; *COX-2*, cyclooxygenase-2; *cPLA*$_2$, cytosolic phospholipase A_2; *Glu*, glutamate; *IL-1β*, interleukin-1β; *IL-6*, interleukin-6; *lyso-PtdCho*, lysophosphatidylcholine; *MMP*, matrix metalloproteinase; *NF-κB*, nuclear factor-κB; *NF-κB-RE*, nuclear factor-κB response element; *NMDA-R*, *N*-methyl-D-aspartate receptor; *PAF*, platelet activating factor; *PtdCho*, phosphatidylcholine; *ROS*, reactive oxygen species; *TNF-α*, tumor necrosis factor-α.

HupA also reduces the glutamate neurotoxicity via antagonizing the NMDA receptor and minimizing the level of synaptic loss along with neuronal cell death (Fig. 8.3) (Peters et al., 2016). It is well known that the brain-derived neurotropic factor (BDNF) is crucially important in learning and memory formation, because it regulates synaptic plasticity, neuronal differentiation, axonal sprouting, as well as long-term potentiation (LTP) (Shao, 2015). Levels of BDNF are diminished in AD patients as well as in demented subjects with mild cognitive impairment (Shao, 2015). HupA potentially exerts neuroprotective effects by upregulating the production of BDNF and minimizing the cognitive deficits and learning impairment induced by a reduced level of BDNF (Fig. 8.2) (Budni et al., 2016). In addition to the abovementioned pharmacological effects, HupA also induces hippocampal neurogenesis. It is reported that HupA not only promotes the proliferation of cultured mouse embryonic hippocampal neural stem cells (NSCs), but also increases the newly generated cells in the subgranular zone of the hippocampus in adult mice (Ma et al., 2013). It is suggested that Hup A acts by activating the mitogen-activated protein kinase/extracellular signal-regulated kinase (MAPK/ERK) signaling pathway, which is a well-known regulator of biological processes including cell proliferation and differentiation (Ma et al., 2013; Qian and Ke, 2014). Collective evidence suggests that by promoting neurogenesis, HupA provides a new insight into the treatment of AD type of dementia (Ma et al., 2013). In male rats, HupA produces lethally toxic effects at >4 mg/kg of bodyweight, whereas 1 mg/kg dose is lethal for female rats (Mao et al., 2014).

Based on the above information, HupA has been used for the treatment of AD in human clinical trials in China and the United States. So far, Chinese clinical studies have shown an improvement in the memory of AD patients (Wang et al., 2009a; Zhang et al., 2002). The phase IV clinical trials in China have demonstrated that HupA can significantly improve memory deficits not only in elderly people with benign senescent forgetfulness, but also in patients with AD and vascular dementia, with minimal peripheral cholinergic side effects and no unexpected toxicity (Zhang et al., 2008; Wang et al., 2009a). In the United States, a multicenter (29 centers in 17 states), double-blind placebo-controlled phase 2 clinical trial has shown that HupA treatment shows cognitive improvement in patients with mild to moderate AD and AD type of dementia (ClinicalTrials.gov; NCT00083590; Rafii et al., 2011).

GINKGO BILOBA AND DEMENTIA

G. biloba (Family Ginkgoaceae) is an important herb from the Chinese traditional system of medicine (Huh and Staba, 1992). Extract prepared from its leaves has been used in traditional medicine for several hundred years. *G. biloba* contains many phytochemicals, which are categorized into two main classes: terpene lactones (ginkgolides and bilobalide), and flavonoids (flavonols and flavone glycosides) (Fig. 8.4) (Solfrizzi and Panza, 2015; IARC Working Group, 2016). The triterpene ginkgolides A, B, and C are unique to *G. biloba* (Solfrizzi and Panza, 2015). Ginkgotoxin, which induces the epileptic seizures, is found in *G. biloba* seeds; and the phenolic type lipid bilobol, which has cytotoxic and antibacterial activities, is a component of *G. biloba* fruits that possess some specific biological effects (IARC Working Group, 2016; Tanaka et al., 2011).

The major constituents of *G. biloba* produce potent neurochemical effects in the brain (DiRenzo, 2000). These effects include modulation of neurotransmission, memory boosting effects, inhibition of apoptosis antioxidant and antiinflammatory effects, enhancement in neurogenesis, increase in cerebral blood flow, and improvement in cognitive function

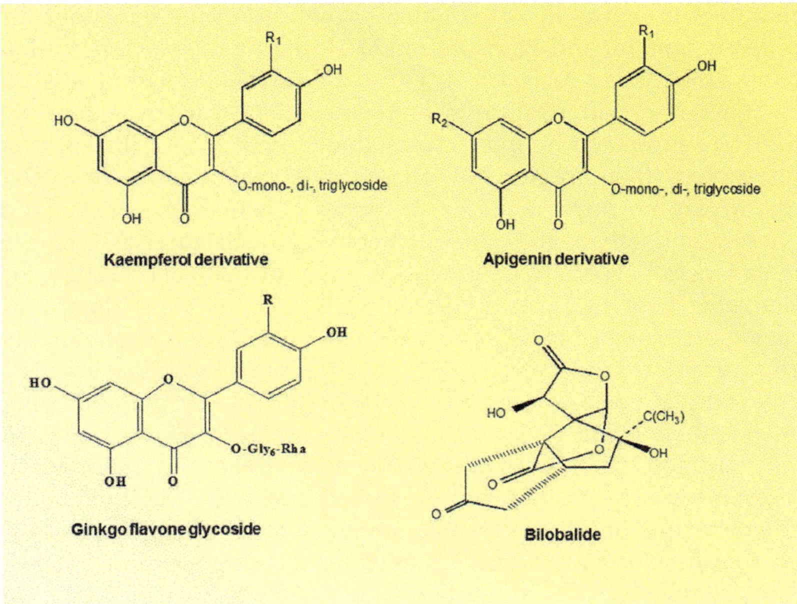

FIGURE 8.4 Chemical structures of kaempferol, apigenin, ginkgo flavone glycoside, and bilobalide.

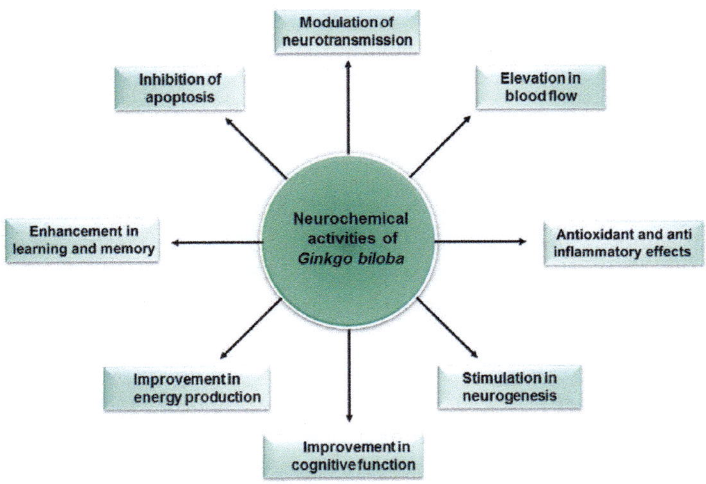

FIGURE 8.5 Neurochemical activities of *Ginkgo biloba*.

(Fig. 8.5) (Yoo et al., 2011; Zuo et al., 2017). *G. biloba* leaves extract is called EGb 761. A patent on this preparation was obtained by Beaufour-Ipsen Pharma (Paris, France) and Dr. Willmar Schwabe Pharmaceuticals

(Karlsruhe, Germany). EG761 contains glycosides of the flavonols quercetin, isorhamnetin, and kaempferol (24%), the terpene-lactones bilobalide and ginkgolides A, B, C, M, J, and bilobalide (6%), and less than 5 ppm ginkgolic acid (DeFeudis and Drieu, 2000). Many activities of EGb 761 are promoted by interactions among EGb 761, GABA, and glycine receptors that are located on neuronal cell membranes. These receptors play an important role in memory formation, consolidation, and cognition (Nathan, 2000; Ahlemeyer and Krieglstein, 2003). EGb 761 also enhances cholinergic processes in various cortical regions. Collectively, these studies support the view that the psychological and physiological benefits of EGb 761 are not only due to modulation of neurotransmitters and neurotransmitter receptors and scavenging of free radicals, but also associated with EGb 761-mediated improvement in small vessels blood flow and in the prevention of blood clot formation. In addition, EGb 761 also exerts its antioxidant and antiinflammatory effects via activation of the HO-1/Nrf2 pathway, VEGF regulation, and downregulation of various inflammatory mediators. The antioxidative action of EGb 761 is suggested to work in concert with its antiapoptotic mechanism. The antiinflammatory effects of the *G. biloba* polysaccharide are shown by its suppression of NO production (Yin et al., 2013).

EGb 761 has been effectively used for the symptomatic treatment of dementia. Daily oral treatment with EGb 761 reduces cognitive dysfunction in an animal model of stroke and vascular dementia in gerbils (Rocher et al., 2011). The molecular mechanism of neuroprotective effects of EGb 761 is very complex. However, based on animal model studies it is suggested that EGb 761 acts by inhibiting Aβ oligomer generation and its toxicity (Ramassamy et al., 2007). By inducing antioxidant and antiinflammatory effects, EGb 761may not only improve mitochondrial dysfunction (Eckert et al., 2003), reduce capillary fragility, decrease blood viscosity, and enhance microperfusion (Fig. 8.6) (Költringer et al., 1995; Pincemail et al., 1989; Clostre, 1999), but also antagonize the effects of platelet activating factor and modify energy metabolism particularly during hypoxia (Chung et al., 1987; Abdel-Wahab and Abd El-Aziz, 2012). Two constituents of Ginkgo-specific acylated flavonol glycosides of *G. biloba* (Q-ag and K-ag) have been reported to increase dopamine and acetylcholine levels in rat medial prefrontal cortex supporting the view that *G. biloba* improves the cognitive function (Kehr et al., 2012). Another constituent of *G. biloba*, cardanol (ginkgol), enhances the growth of NSC-34 immortalized motor neuron-like cells and increases working memory-related learning ability in young rats on chronic administration (Tobinaga et al., 2012). Treatment of 2-month-old APP/PS1 transgenic mouse (an animal model of AD) with EGb 761 daily for 6 months results in a marked reduction

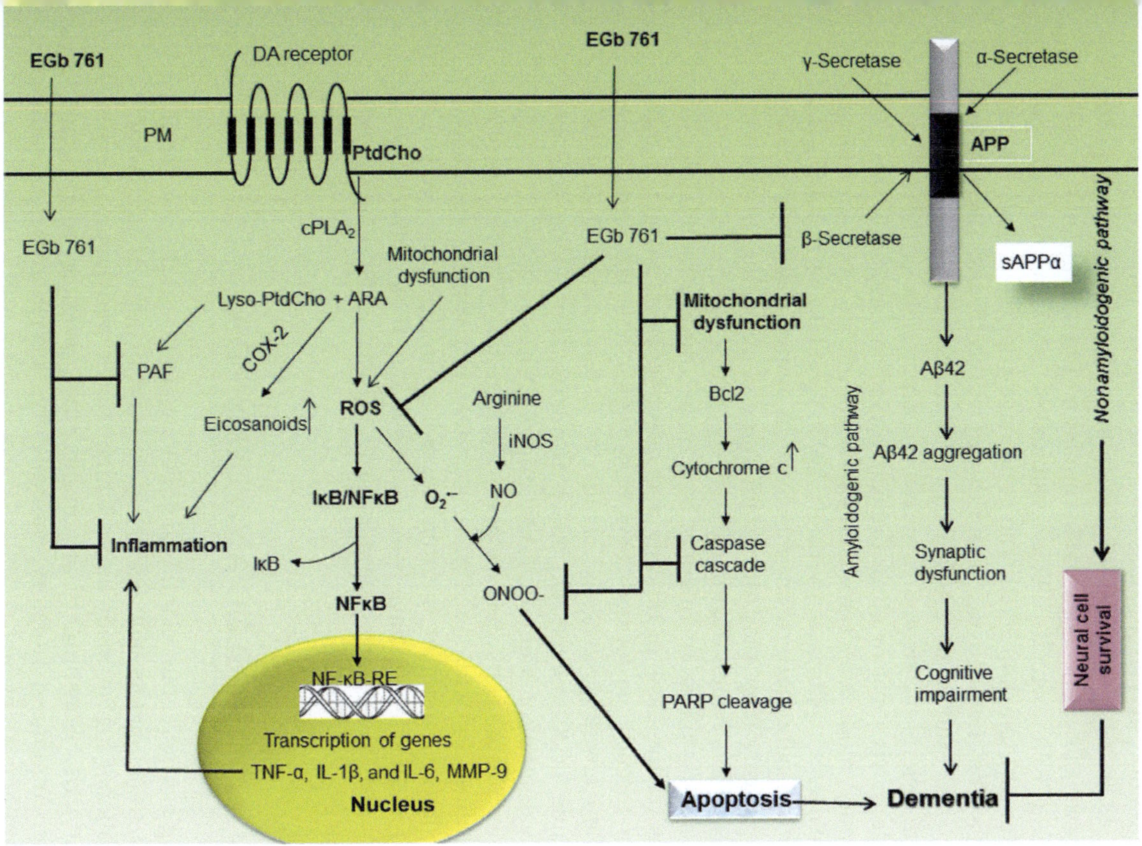

FIGURE 8.6 Hypothetical diagram showing effect of EGb 761 on oxidative stress, neuroinflammation, and APP processing in the brain. *Aβ*, beta-amyloid; *APP*, amyloid precursor protein; *ARA*, arachidonic acid; *Bcl2*, B-cell lymphoma 2; *COX-2*, cyclooxygenase-2; *$cPLA_2$*, cytosolic phospholipase A_2; *Glu*, glutamate; *IL-1β*, interleukin-1β; *IL-6*, interleukin-6; *iNOS*, inducible nitric oxide synthase; *lyso-PtdCho*, lysophosphatidylcholine; *MMP*, matrix metalloproteinase; *NF-κB*, nuclear factor-κB; *NF-κB-RE*, nuclear factor-κB response element; *NMDA-R*, N-Methyl-D-aspartate receptor; *$ONOO^-$*, peroxynitrite; *PAF*, platelet activating factor; *PtdCho*, phosphatidylcholine; *ROS*, reactive oxygen species; *TNF-α*, tumor necrosis factor-α.

in the levels of insoluble Aβ and proinflammatory inducible nitric oxide synthase, while the activity of arginase-1 is increased (Wan et al., 2016) indicating that EGb 761 plays a neuroprotective role in APP/PS1 mice by regulating the expression of Aβ and inflammatory cytokines, inhibiting inflammation, and stimulating nonamyloidogenic processing of APP (Fig. 8.6). EGb 761 also retards the toxic effects of Aβ peptides in the hippocampal cells of the aging rat by inhibiting the activation of protein kinase C (PKC), which blocks stimulated sodium nitroprusside (Bastianetto and Quirion, 2002). Yao et al. have suggested that free cholesterol may be involved in the production of APP and the amyloid β-peptide (Aβ) (Yao et al., 2004). It is reported that EGb 761 decreases the free cholesterol in rats and suppress the expressions of the APP and Aβ-peptide in vivo *and* in vitro. Collective evidence suggests that EGb 761 not only modulates serotonergic, dopaminergic, and cholinergic neurotransmission (Ramassamy et al., 1992; Yoshitake et al., 2010), but also improves neuronal insulin sensitivity in animal models of dementia. These effects may play important roles in retarding or delaying dementia and behavioral disorders.

Most earlier studies on humans have failed due to methodological limitations on the basis of efficacy and effectiveness. Two major trials have been recently performed. One trial, which had 176 participants, did not show any effectiveness (120 mg/day *G. biloba*) in mild to moderate dementia (McCarney et al., 2008). Similarly, the second trial (the Ginkgo Evaluation of Memory, GEM study) (DeKosky et al., 2008) did not show neuroprotective effects with 240 mg/day EGb 761 in older people without or with only mild cognitive impairment (Schneider, 2008). Another trial called the randomized controlled trial (RCT) has also failed and its results have been criticized due to its methodological problems and insufficient sample size (Ernst, 2009; Gaus, 2009; Weinmann et al., 2010).

GINSENG AND DEMENTIA

Ginseng is a perennial plant, which belongs to the Araliaceae family. Ginseng's roots, shoots, and leaves have been a popular and widely used traditional herbal medicine in China, Korea, and Japan for thousands of years. Constituents of ginseng root produce adaptogenic, restorative, immune-stimulatory, vasodilatory, antiinflammatory, antioxidant, antiaging, anticancer, antifatigue, antistress, and antidepressive effects in rodents and humans (Fig. 8.7) (Choo et al., 2003; Cheng et al., 2005; Wang et al., 2009b). Ginseng root contains more than 60 bioactive ginsenosides (Rg), such as Rb1, Rb2, Rb3, Rc, Rd, Re, Rg1, Rg2, and Rg3 (Fig. 8.8), as well as polysaccharides, oligopeptides, polyacetylenic

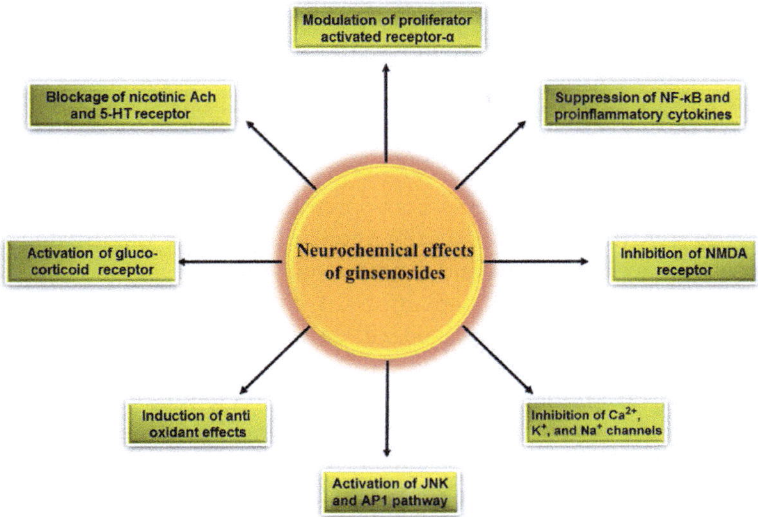

FIGURE 8.7 Neurochemical effects of ginsenoside in the brain.

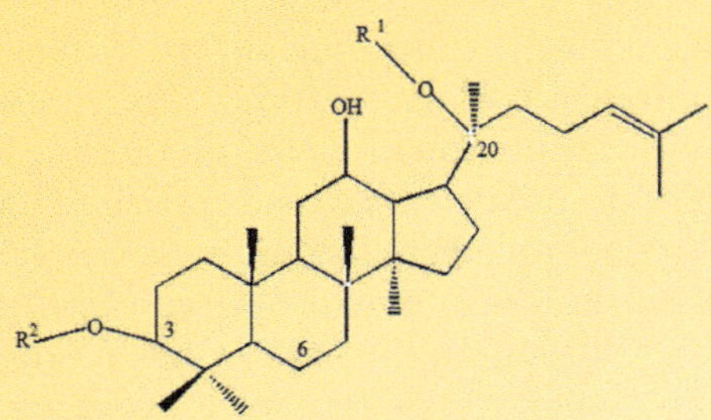

Ginsenoside	R^2	R^1
Rb1	Glc^{1}–^{2}Glc-	Glc^{1}–^{6}Glc-
Rb2	Glc^{1}–^{2}Glc-	Ara(p)^{1}–^{6}Glc-
Rb3	Glc^{1}–^{2}Glc-	Xyl^{1}–^{6}Glc-
Rc	Glc^{1}–^{2}Glc-	Ara(f)^{1}–^{6}Glc-
Rd	Glc^{1}–^{2}Glc-	Glc-
Rg3	Glc^{1}–^{2}Glc-	H
Rh2	Glc-	H
PPD	H	H

FIGURE 8.8 Chemical structure of ginsenosides.

alcohols, and fatty acids (Qi et al., 2010). Ginseng produces its neuroprotective effect in many neurological disorders including various types of dementia, stroke, PD, depression, and schizophrenia (Cho, 2012;

Ong et al., 2015; Rokot et al., 2016). Ginsenosides can cross the BBB and produce many neurochemical effects by modulating ion channels and neurotransmitter receptors (NMDA receptor, nicotinic acetylcholine, and 5-hydroxytryptamine type 3 receptors), decreasing oxidative stress and reducing neuroinflammation, and retarding memory deficit (Liu et al., 2010; Chen et al., 2010).

Neurochemical events in the abovementioned neurological disorders involve the release of glutamate, the overstimulation of glutamate receptor, nicotinic acetylcholine, and 5-hydroxytryptamine type 3 receptors, rapid calcium influx, and activation of calcium-dependent enzymes (phospholipase A_2, cyclooxygenase, and nitric oxide synthases) (Farooqui, 2010), and induction of oxidative stress and neuroinflammation (Fig. 8.9). In animal models of AD type dementia, ginseng acts by modulating the production of Aβ oligomers. Thus, the treatment of aged transgenic AD mice (TgmAPP mice) with ginsenoside Rg1 shows a marked decrease in cerebral Aβ levels, reverses neuropathological changes, and protects the ability to retain spatial learning and memory. At the molecular level, Rg1 as well as ginsenosides CK, F1, Rh1, and Rh2 also suppress γ-secretase (BACE1) in both TgmAPP mice and B103-APP cells, demonstrating the role of Rg1 in APP regulation (Karpagam et al., 2013; Huang et al., 2014c). Oral administration of ginsenosides increases the expression levels of enzymes involved in acetylcholine synthesis in the brain and alleviates Aβ-induced cholinergic deficits in AD models. In addition, administration of Rg1 promotes the activation of the PKA/CREB pathway in mAPP mice and cultured cortical neurons exposed to Aβ or glutamate-mediated synaptic stress (Fang et al., 2012). Similarly, some ginsenosides also enhance α-secretase activity and promote nonamyloidogenic processing of APP. Ginsenoside Rh2 treatment improves learning and memory performance at 14 months of age In Tg2576 model mice of AD (14 months); treatment with Ginsenoside Rh2 not only reduces senile plaques in the brains, but also improves learning and memory (Qiu et al., 2014). In mouse model of AD, ginseng (*Panax ginseng*) reduces tau hyperphosphorylation by enhancing the phosphatase activity of purified calcineurin (Tu et al., 2009). In addition, ginsenoside Rb1 reverses an aluminum-induced increase in p-GSK, and decreases PP2A level and tau phosphorylation in the cortex and hippocampus (Zhao et al., 2013). Similarly, ginsenoside Rg1 (20 mg/kg) also reverses memory impairments induced by okadaic acid by decreasing levels of phospho-tau, increasing phospho-GSK3β and suppressing the formation of Aβ in the brains of rats (Song et al., 2013). Furthermore, ginseng total saponins increased hippocampal glycogen synthase kinase-3β (GSK-3β) inhibitory phosphorylation (Chen et al., 2014). Together, results suggest an inhibitory effect of ginsenosides on tau hyperphosphorylation that may have beneficial effects on microtubule function in neurons.

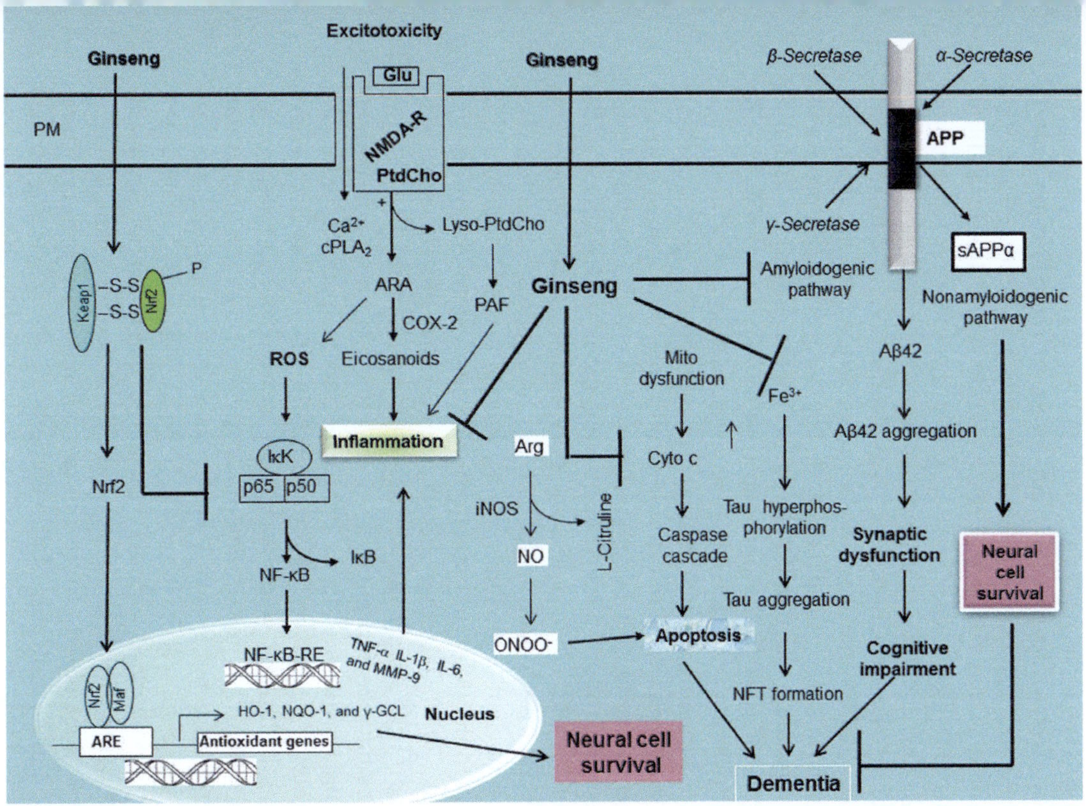

FIGURE 8.9 Hypothetical diagram showing effect of ginseng on oxidative stress, neuroinflammation, and APP processing in the brain. *Aβ*, beta-amyloid; *APP*, amyloid precursor protein; *ARA*, arachidonic acid; *ARE*, antioxidant response element; *COX-2* cyclooxygenase-2; *cPLA₂*, cytosolic phospholipase A_2; *γ-GCL*, γ-glutamate cystein ligase; *Glu*, glutamate; *HO-1*, hemoxygenase; *IL-1β*, interleukin-1β; *IL-6*, interleukin-6; *iNOS*, inducible nitric oxide synthase; *Keap1*, Kelch-like ECH-associated protein 1; *Keap1*, kelch-like erythroid Cap'n'Collar homologue-associated protein 1; *lyso-PtdCho*, lysophosphatidylcholine; *MMP*, matrix metalloproteinase; *NF-κB*, nuclear factor-κB; *NF-κB-RE*, nuclear factor-κB response element; *NMDA-R*, *N*-methyl-D-aspartate receptor; *NQO-1*, NADPH quinine oxidoreductase; *Nrf2*, nuclear factor-erythroid-2-related factor 2; *$ONOO^-$*, peroxynitrite; *PAF*, platelet activating factor; *PtdCho*, phosphatidylcholine; *ROS*, reactive oxygen species; *TNF-α*, tumor necrosis factor-α.

In animal models of PD type of dementia, ginsenosides produce neuroprotective effects (Cho, 2012). Thus, Rg1 ginsenoside increases dopamine and its metabolites levels in the striatum and upregulates the expression of TH in the substantia nigra (SN) of MPTP-treated C57BL/6 mice by attenuating elevated iron levels, decreasing divalent metal transport 1 expression, and increasing ferroportin1 expression in the SN (Wang et al., 2009c). Elevation of iron levels in the SN contribute to neuronal death in PD type of dementia by enhancing the generation of free radicals and oxidative stress (Lee and Andersen, 2010). In addition, pretreatment with Rg1 markedly reduces the generation of dopamine-induced ROS and the release of mitochondrial cytochrome C into the cytosol inhibiting the activation of caspase-3, induction of inducible nitric oxide synthase (iNOS) and generation of nitric oxide (NO) production in dopamine-induced PC12 cells (Chen et al., 2003). Rg1 also can protect SN neurons by regulating the insulin-like growth factor-I receptor signaling pathway (Shi et al., 2009), the phospho (p)-ERK1/2, and p-p38 MAPKs signaling pathways (Wang et al., 2008; Xu et al., 2009).

Panaxatriol saponins, which are major constituents of ginseng (*Panax notoginseng*) have been used to provide neuroprotection against a loss of dopaminergic neurons and behavioral impairment in 1-methyl-4-phenyl-1,2,3,6-tetrahydropyridine (MPTP) mouse model of Parkinson's disease (Luo et al., 2011). In addition, oral administration of ginseng extract G115 reduces dopaminergic cell loss, microgliosis, and accumulation of α-synuclein aggregates in a chronic mouse model of PD, induced by chronic dietary administration of phytosterolglucoside (Van Kampen et al., 2014). Together, results suggest that ginseng may have potential for prevention or treatment of PD-linked dementia.

Few studies have been performed on the effects of ginseng in patients with dementia. Treatment of AD type of dementia patients with total powder extract of *P. ginseng* for 3 or 6 months has indicated cognitive improvements in healthy volunteers (Reay et al., 2006; Kennedy et al., 2003), but not in AD type of dementia patients. However, these trials have limitations. It is not known whether the effects of ginseng are transient or not (Lee et al., 2008; Heo et al., 2008). A RCT with *P. notoginseng* and duxil was performed with 41 vascular dementia patients. Results indicate that memory function is significantly improved (Tian, 2003). In another trial with 64 older adults with lacunar infarction (cerebrovascular disease), it was indicated that injections of *P. notoginseng* extract (Xueshuantong) for 4 weeks resulted not only in significantly increased cerebral blood flow, but also in improvement in activities of daily living (ADL) scores, although MMSE scores showed no marked changes (Gui et al., 2013). Large-scale, long-term, double-blind studies using standardized extracts are required to confirm the clinical efficacy of ginseng therapy in patients with dementia.

ANEMARRHENA RHIZOME (RHIZOMA ANEMARRHENAE) AND DEMENTIA

Anemarrhena rhizome belongs to the family Liliaceae. It is a dry tuber commonly used in Chinese medicine for nourishment. The active constituents of *Anemarrhena* rhizome include sarsasapogenin, smilagenin, neogitogenin, and markosapogenin (Fig. 8.10). *Anemarrhena* rhizome and its products produce multiple pharmacological activities including antipyretic, antiinflammatory, and antidiabetic effects. Among the active constituents, sarsasapogenin and its glycosylated products have been reported to improve dementia symptoms not only through modulating the function of cholinergic system and suppressing neurofibrillary tangles, but also by inhibiting neuroinflammation (Huang et al., 2017). Sarsasapogenin-AA13 (AA13), a novel synthetic derivative of sarsasapogenin, has been reported to improve the spatial memory of scopolamine-treated mice in Morris water maze performance (Dong et al., 2017). Similar cognitive improvement efficacy is also observed in the mice model of Aβ-induced memory impairment (Dong et al., 2017). It is also reported that neuroprotective effects of AA13 in rat primary astrocytes are due to the upregulation of the BDNF expression, which

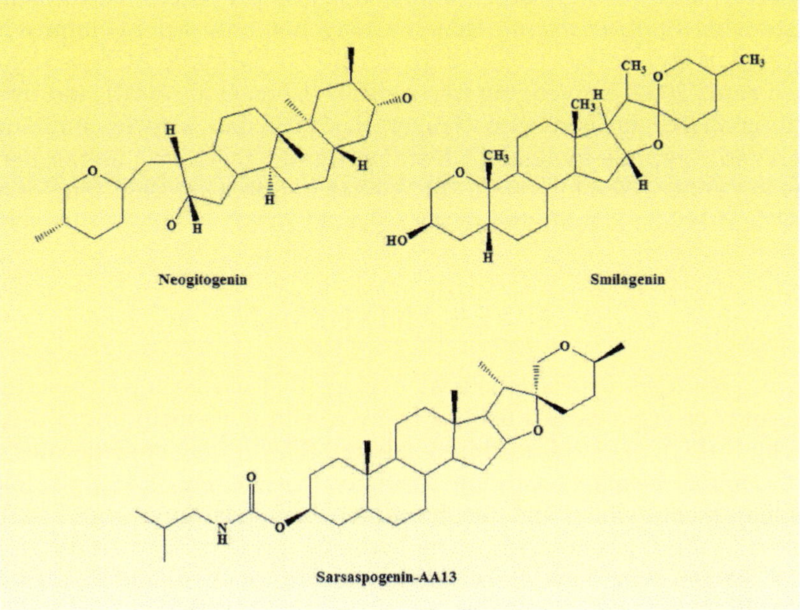

FIGURE 8.10 Chemical structures of neogitogenin, smilagenin, and sarsaspogenin-AA13.

may also contribute to the proliferation of astrocytes in rats and mice tissues leading to improvement in neuronal function.

Similarly, smilagenin (SMI), a steroidal sapogenin from *Anemarrhena* rhizome is known to improve memory in animal models of AD type of dementia. It acts by significantly elevating the declined muscarinic receptor (M receptor) density (Zhang et al., 2012). In cultured rat cortical neurons, pretreatment with SMI significantly attenuates the neurodegeneration caused by beta amyloid 25-35 (Aβ(25-35)). In addition, SMI restores levels of BDNF protein levels in the culture medium, which are decreased by Aβ(25-35) (Zhang et al., 2012). The glycosylated products of SMI are called Timosaponin-AI, -AII, -AIII, -AIV, -BI and -BII (Ji and Feng, 2010). Among them BII has been used as a neuroprotective agent in Chinese medicine. BII acts by suppressing the production of proinflammatory factors IL-1, IL-6, and TNF-α (Li et al., 2007; Lu et al., 2009). The dementia-palliative effect of Timosaponin-BII may involve multiple mechanisms and one of them is its potential antioxidative property. In addition, Timosaponin-BII diminishes the Aβ-induced oxidative impairment by promoting scavenging of superoxide radicals (Ouyang et al., 2005). Furthermore, Timosaponin-BII remarkably inhibits the upregulation of BACE1 and reduces the overproduction of β-CTF and Aβ in rat retina, which is induced by $FeCl_3$. The mechanism of Timosaponin-BII on BACE1 expression may be related to its antioxidant property. Based on these observations, it is also proposed that the active constituents of *Anemarrhena* rhizome may not only act by improving levels of acetylcholine (ACh) and density of M-type ACh receptors (Chen et al., 2004), scavenging free radicals (Chen et al., 2000), and upregulating BDNF in astrocytes (Hu et al., 2003), but also by inhibiting β-amyloid peptide-mediated learning and memory impairments (Ouyang et al., 2005; Liu et al., 2012), retarding ibotenic acid (Sun et al., 2004), and ischemic brain injury (Deng et al., 2005).

GREEN TEA AND DEMENTIA

Green tea (*Camellia sinensis*) is a beverage that has been consumed for thousands of years. Green tea has many constituents including: (−)-epicatechin (EC), (−)-epicatechin-3-gallate (ECG), (−)-epigallocatechin (EGC), and (−)-epigallocatechin-3-gallate (EGCG) (Fig. 8.11); alkaloids (caffeine, theophylline, and theobromine); flavonols (quercetin, kaempferol and rutin); amino acids; carbohydrates; proteins; and chlorophyll. Green tea catechins have three heterocyclic rings, A, B, and C, and the free radical scavenging property of green tea is attributed to the presence of trihydroxyl group on the B ring and the gallate moiety at the 3′ position in the C ring. Green tea catechin also chelates transition metal

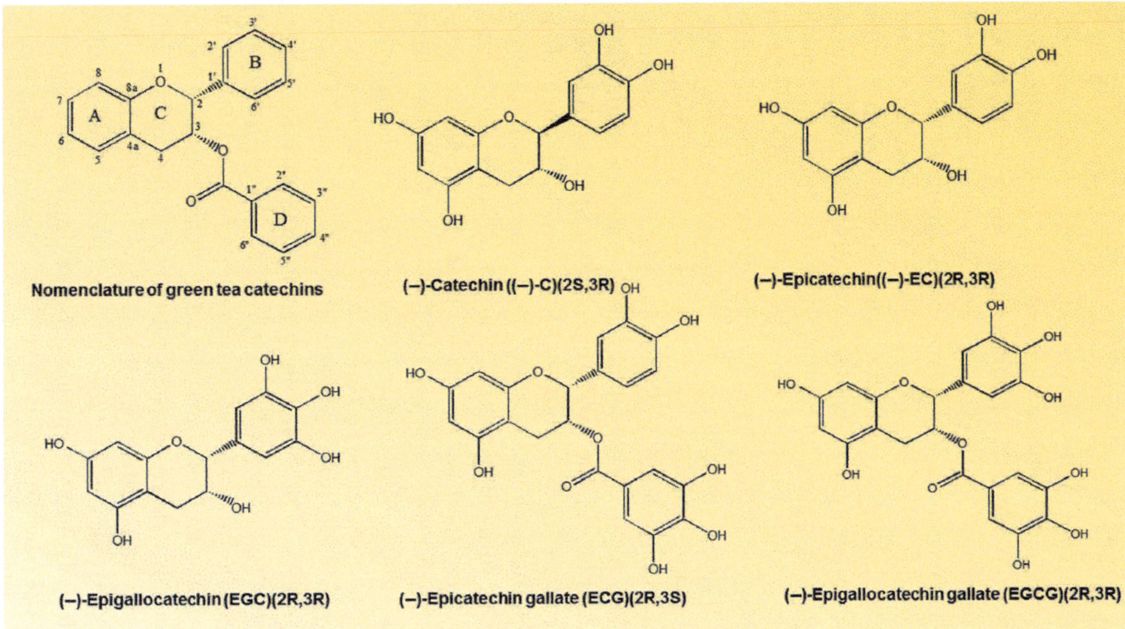

FIGURE 8.11 Chemical structures of various catechins in the green tea.

ions like iron and copper. There are two sites where metal ions bind to the catchin molecule: (1) *o*-diphenolic group in the 3′,4′-dihydroxy positions in the B ring; and (2) keto structure 4-keto, 3-hydroxy in the C ring of flavonols. Green tea catechins are brain permeable. Catechins are strong antioxidants that can quench reactive oxygen species (ROS) such as superoxide radical, singlet oxygen, hydroxyl radical, peroxyl radical, nitric oxide, nitrogen dioxide, and peroxynitrite (Feng, 2006). In addition, green tea also produces antihypertensive effects by suppressing angiotensin I converting enzyme. It not only suppresses appetite, hyperglycemia, and dyslipidemia, but also reduces blood pressure and improves insulin resistance and blood sugar (Wolfram et al., 2006; Thielecke and Boschmann, 2009; Liu et al., 2014). The most important bioactive component of green tea is ECCG. This catechin produces its biological effects (antiinflammatory, antiallergic, and antiproliferative effects) by interacting with laminin receptors (mol mass 67 kDa), which are found on neurons (Murakami and Ohnishi, 2012). EGCG produces its neuroprotective effects in AD type of dementia through a wide range of mechanisms including downregulation of proapoptotic genes, elevation of α-secretase activity, inhibition of β-secretase activity, inhibition of neuroinflammation, scavenging of ROS, and stabilization of mitochondrial function (Fig. 8.12) (Wang et al., 2010; Mandel et al., 2008; Sharangi, 2009). Signal transduction mechanisms associated with beneficial effects of green tea involve activation of PKC, iron chelation, and an increase in neurotrophins such as BDNF. This neurotrophin supports and maintains cognitive function (Spencer, 2008). Catechins also inhibit NADPH oxidase, xanthine oxidase, cyclooxygenase, lipoxygenase, suppress the activation of NF-κB, and activate adaptive cellular stress responses (Woo et al., 2005; Kim et al., 2010a). Catchins also interact with Nrf2, a transcription factor, which is located in the cytoplasm as a complex with Keap1. Catechins promote the release of free Nrf2 from Keap1–Nrf2 complex. Free Nrf2 migrate into the nucleus, where it, along with other transcription factors, e.g., sMaf, ATF4, JunD, and PMF-1, transactivates the antioxidant response elements (AREs) of many cytoprotective genes and enzymes (HO-1, catalase, SOD, epoxide hydrolase, UDP-glucuronosyltransferases, glutathione reductase, and thioredoxin). These enzymes induce neuroprotection by decreasing the oxidative stress (Wakabayashi et al., 2010). Upon recovery of cellular redox status, Keap1 travels into the nucleus and facilitates the dissociation of Nrf2 from ARE. Subsequently, the Nrf2–Keap1 complex is exported out of the nucleus by the nuclear export sequence in Keap1 (Wakabayashi et al., 2010). Once in the cytoplasm, the Nrf2–Keap1 complex associates with the Cul3-Rbx1 core ubiquitin machinery, leading to degradation of Nrf2 and termination of the Nrf2/ARE signaling pathway (Wakabayashi et al., 2010).

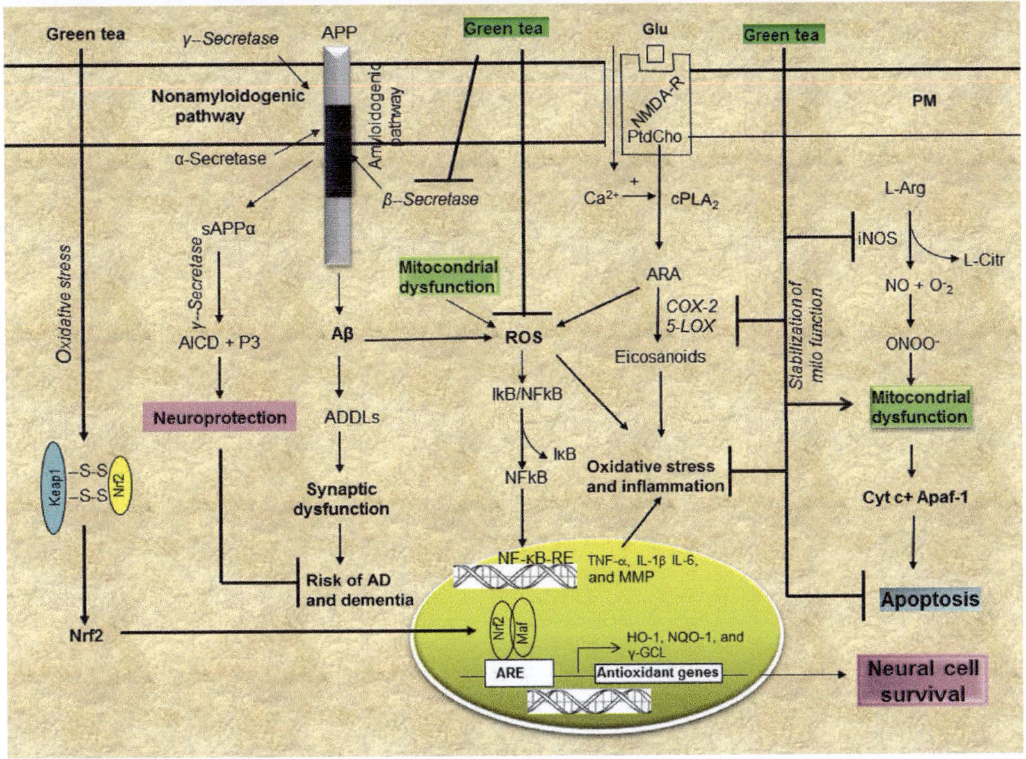

FIGURE 8.12 Hypothetical diagram showing effect of green tea catechins on oxidative stress, neuroinflammation, and APP processing in the brain. *Aβ*, beta-amyloid; *ADDL*, Aβ-derived diffusible ligand; *AICD*, APP intracellular domain; *APP*, amyloid precursor protein; *ARA*, arachidonic acid; *ARE*, antioxidant response element; *COX-2*, cyclooxygenase-2; *cPLA*$_2$, cytosolic phospholipase A_2; *γ-GCL*, γ-glutamate cystein ligase; *Glu*, glutamate; *HO-1*, hemoxygenase; *IL-1β*, interleukin-1β; *IL-6*, interleukin-6; *iNOS*, inducible nitric oxide synthase; *Keap1*, Kelch-like ECH-associated protein 1; *Keap1*, kelch-like erythroid Cap'n'Collar homologue-associated protein 1; *L-Arg*, L-arginine; *L-Citr*, L-citrulline; *lyso-PtdCho*, lysophosphatidylcholine; *MMP*, matrix metalloproteinase; *NF-κB*, nuclear factor-κB; *NF-κB-RE*, nuclear factor-κB response element; *NMDA-R*, *N*-methyl-D-aspartate receptor; *NQO-1*, NADPH quinine oxidoreductase; *Nrf2*, nuclear factor-erythroid-2-related factor 2; *ONOO*$^-$, peroxynitrite; *PAF*, platelet activating factor; *PtdCho*, phosphatidylcholine; *ROS*, reactive oxygen species; *TNF-α*, tumor necrosis factor-α.

Consumption of green tea has no effect on symptoms of AD. However, several in vitro studies in cell culture and animal models of AD have indicated that green tea extract protects neurons from the Aβ-induced toxicity (Ramassamy, 2006; Zhao, 2009). It is well known that APP is processed by two pathways: (1) a nonamyloidogenic pathway which involves cleavage of APP to soluble APP (sAPP) by the α-secretase activity; and (2) APP processing by an amyloidogenic β peptides pathway by the β- and γ-secretases. In neuronal cell cultures, green tea (EGCG) enhances the nonamyloidogenic α-secretase pathway via PKC dependent activation of α-secretase (Singh et al., 2008; Mandel et al., 2008), while EC reduces the formation of Aβ-fibrils. Similarly, in mouse model of AD, EGCG stimulates non-amyloidogenic processing of amyloid precursor protein (APP) by upregulating alpha-secretase through the oral bioavailability (Smith et al., 2010). This process prevents brain β-amyloid plaque formation and deposition, which is a hallmark of AD pathology. This is accompanied by a significant reduction in cerebral Aβ levels and β-amyloid plaques. Since sAPPα and Aβ are formed by two mutually exclusive mechanisms, stimulation of the secretory processing of sAPPα may retard the formation of the amyloidogenic Aβ. Recently a dual-inhibitor system containing EGCG and negatively charged polymeric nanoparticles (NP10) has been developed (Liu et al., 2017). It has been demonstrated that this dual-inhibitor system effectively inhibits Aβ ($A\beta_{42}$ and $A\beta_{40}$) aggregation and fibrillation at low concentrations (Liu et al., 2017). Converging evidence suggests that EGCG influences Aβ levels not only by translational inhibition of APP and by stimulating sAPPα secretion, but also by inhibiting the aggregation and fibrillation. It is tempting to speculate that green tea constituents may produce neuroprotective effects in AD type of dementia by enhancing the nonamyloidogenic pathway, but more research is needed on this important topic.

Studies have revealed that EGCG may produce benefits in PD-linked dementia patients by reducing dopaminergic degeneration (Renaud et al., 2015). In a rat model of PD-linked dementia, EGCG reverses pathological and behavioral modifications, demonstrating neuroprotection by decreasing rotational and increased locomotor activities. Additionally, EGCG improves cognitive dysfunction by inhibiting oxidative stress and neuroinflammation (Bitu Pinto et al., 2015). The molecular mechanisms underlying beneficial effects of EGCG have been investigated in animal models of PD-linked dementia. Thus, pretreatment of mice with either green tea extract (0.5 and 1 mg/kg) or EGCG (2 and 10 mg/kg) prevents dopaminergic neuronal death produced by *N*-methyl-4-phenyl-1,2,3,6-tetrahydropyridine (MPTP) (Levites et al., 2001; Kim et al., 2010b). It is proposed that the catechol-like structural in catechins may competitively inhibit the uptake by the presynaptic or vesicular transporters of the metabolite product of MPTP, 1-methyl-4-phenylpyridinium ion (MPP^+) (Pan et al., 2003), which has a structure similar to catechol. This competition may protect dopaminergic neuronal degeneration against the MPTP/MPP^+-mediated injury (Pan et al., 2003). In addition to its antioxidant effects, EGCG also acts as a chelating agent. It chelates iron and copper and reduces the production of ROS. A comparison of the beneficial effects of various catechins against iron-induced lipid peroxidation in synaptosomes indicates that the inhibitory effects of catechins decrease in the order of EGCG > ECG > EGC > EC (Guo et al., 1996). EGCG attenuates paraquat-mediated lipid oxidation in mice, a strong redox herbicide that contributes to the formation of ROS and to the toxicity of the nigrostriatal dopaminergic system (Liou et al., 2001). In mice, EGCG reduces oxidative stress and controls neurochemical deficits produced by MPTP treatment by regulating the iron-export protein ferroportin in SN (Xu et al., 2017). Collectively, these studies indicate that EGCG acts through scavenging free radicals and metal ion chelating properties in MPTP-, paraquat-, and 6-OHDA-induced animal models of PD. In addition, EGCG efficiently inhibits the fibrillogenesis of α-synuclein by directly binding to the natively unfolded polypeptides and preventing their conversion into toxic aggregated intermediates (Wanker, 2008). These observations support the view that green tea catechins produce a generic effect on aggregation pathways in neurodegenerative diseases (Wanker, 2008). Animal and epidemiological studies have suggested that drinking green tea confers protection to the brain against the aging process. An inverse correlation between tea consumption and the incidence of AD and PD has been suggested, although longitudinal and cross-sectional studies investigating the effect of green tea on cognitive function have produced mixed findings.

Green and black tea are known to contain theanine (*n*-ethylglutamic acid), a nonproteinaceous amino acid. Theanine readily crosses the BBB to produce a variety of neurophysiological and pharmacological effects (Lardner, 2014). Thus, theanine not only induces anxiolytic and calming effects due to the upregulation of inhibitory neurotransmitters (serotonin and dopamine) in selected areas of the brain, but also increases levels of BDNF. Theanine also improves cognitive function, including learning and memory, in human and animals via a decrease in NMDA-dependent CA1 LTP and an increase in NMDA-independent CA1-LTP (Lardner, 2014), supporting the view that this green tea component can produce beneficial effects in attention deficit hyperactivity disorder and neuropsychiatric disorders such as anxiety disorders, panic disorder, obsessive compulsive disorder, and bipolar disorder (Lardner, 2014)

Green tea induces stomach upset and constipation in some elderly. Two cups of green tea a day provides about 200 mg of caffeine. Five cups of green tea per day, which is known to decrease the risk for dementia may induce many side effects because of the increased caffeine content. These side effects can range from mild to serious and include headache, nervousness, sleep problems, vomiting, diarrhea, irritability, irregular heartbeat, tremor, heartburn, dizziness, ringing in the ears, convulsions, and confusion (Lavretsky, 2016).

INTEGRIPETAL RHODIOLA HERB AND DEMENTIA

Integripetal rhodiola herb is a perennial plant. It belongs to the Rhodiola family. The roots of this plant are a succulent rhizome. Studies on the effects of Integripetal rhodiola herb in rats have indicated that this herb not only improves learning and memory impairment in D-galactose, scopolamine, and β-amyloid peptide-induced neurotoxicity (Xie et al., 2003; Wu et al., 2004a; Xie et al., 2004), but also protects from hypoxia (Liu et al., 2003) and cerebral ischemia-reperfusion injury-mediated learning and memory loss (Liu et al., 2003; Song et al., 2005). This plant also contains Rhodosin, a component of Integripetal rhodiola herb, which contributes to memory enhancement in normal-aged rats (Jiang et al., 2001) due to its ability to increase ACh content and reduce cholinesterase activity in the brain (Wu et al., 2004b). Rhodosin also produces antioxidant effects in rats (Jiang et al., 2001) contributing to the retardation of neurodegenerative changes.

DANSHEN ROOT AND DEMENTIA

The dried roots of Danshen (*Salvia miltiorrhiza*) Bunge (SM) (Lamiaceae) are a very popular medicine in TCM. Danshen dried roots contain both lipophilic and hydrophilic constituents such as tanshinone I, tanshinone IIA, acetyltanshinone IIA, salvianolic acid A, and salvianolic acid B (Fig. 8.13) (Hung et al., 2016). These constituents are used in Korea, China, and Japan for the treatment of various diseases, including coronary heart disease (Su et al., 2015), cerebrovascular disease (Yu et al., 2007), AD (Zhang et al., 2016), Parkinson's disease (Zhang et al., 2016; Ren et al., 2015), and renal deficiency (Hu et al., 1996). Effects of Danshen roots in a rat model of stroke has indicated that Danshen root's active constituents act by retarding apoptosis and neuroinflammation (Lv et al., 2015). The beneficial effects of Danshen root may also be due to the induction of HO-1 expression through the PtdIns 3K/Akt-MEK1-Nrf2 pathway. Enhancement in the activity of this pathway results in a

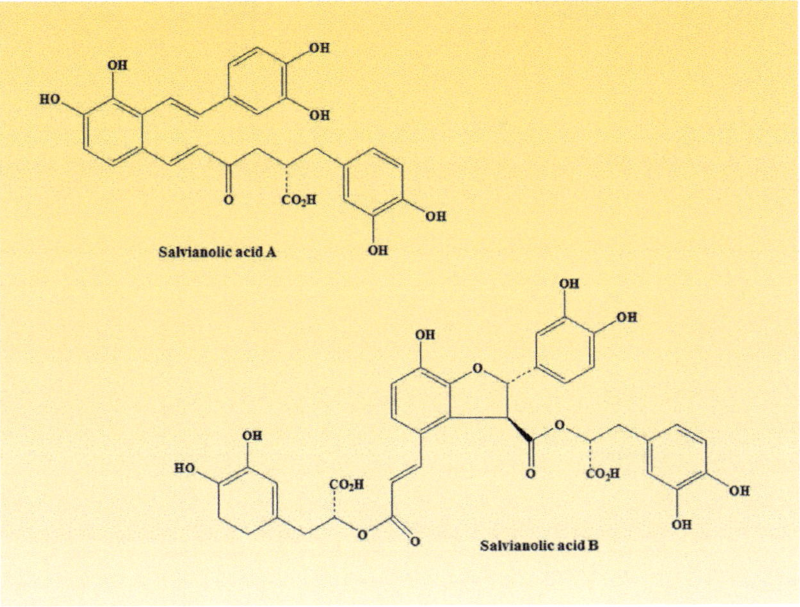

FIGURE 8.13 Chemical structures of salvianolic acid A and salvianolic acid B.

reduction in intracellular production of ROS via induction of heme oxygenase-1(HO-1) expression supporting the view that Danshen may produce a cytoprotective effect through the increased expression of HO-1 (Lee et al., 2012).

Induction of HO-1 protects against the cytotoxicity of oxidative stress and apoptotic cell death. More recently, HO-1 has been recognized to have major immunomodulatory and antiinflammatory properties, which have been demonstrated in HO-1 knockout mice and a human case of genetic HO-1 deficiency. Beneficial protective effects of HO-1 in inflammation are not only mediated via enzymatic degradation of proinflammatory free heme, but also via production of the antiinflammatory compounds bilirubin and carbon monoxide (Paine et al., 2010).

RADIX PUERARIAE (KUDZU ROOT) AND DEMENTIA

The active component of Radix puerariae root and leaves is a flavone called puerarin (Fig. 8.14). Radix puerariae is widely used in China for the treatment of cardiovascular and neurodegenerative diseases. Radix puerariae is also used for lowering blood pressure. In animal models of AD, puerarin produces neuroprotective effects by ameliorating learning

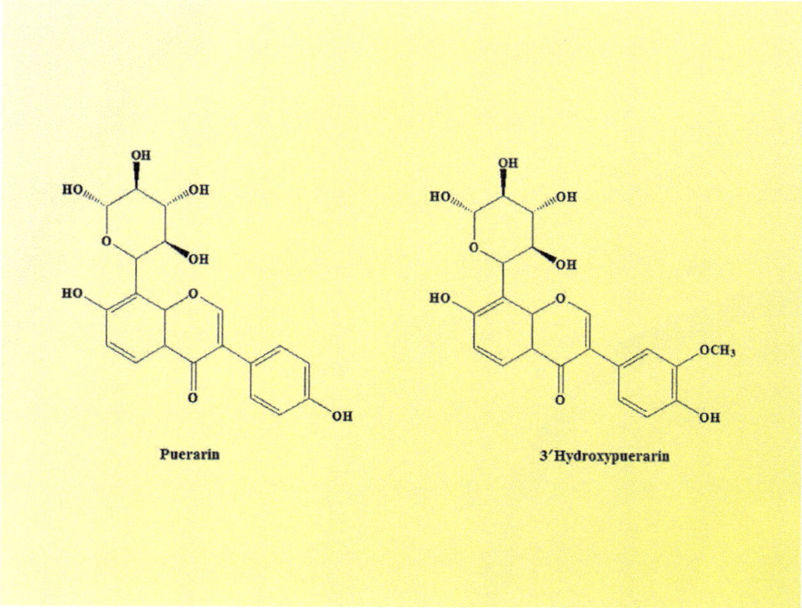

FIGURE 8.14 Chemical structures of puerarin and 3-hydroxypuerarin.

and memory deficits through the modulation of the glutamatergic/GABAergic system in the hippocampus (Xu and Zhao, 2002; Xu et al., 2004). In 6-hydroxydopamine-induced animal models of PD, puerarin produces neuroprotective effects not only by inhibiting apoptotic signaling pathways, but also by upregulating glial cell line-derived neurotrophic factor expression in the striatum (Li et al., 2003; Zhu et al., 2010). Similarly, in animal models of ischemia, puerarin produces neuroprotective effects. Thus, acute treatment with puerarin enhances the metabolism of both dopamine and serotonin (5-HT), and lowers the extracellular level of Glu, without changing GABA concentrations, This process may alleviate excitotoxicity under ischemic conditions leading to neuroprotective effects and neural cell survival (Xu et al., 2007; Chang et al., 2009; Wu et al., 2009). In vitro studies on primary cultured neurons have indicated that puerarin also alleviates mitochondrial oxidative stress (Xu and Zhao, 2002; Xiao et al., 2017).

CHINESE FORMULATIONS AND DEMENTIA

TCM is a holistic medicinal system that considers the human body as a whole. It emphasizes the importance of functions and emotions and considers patients as part of a system interacting with its environmental factors, such as diet, climate, and lifestyle. TCM has been used in China for thousands of years for maintaining the health of Chinese people. It has a high legal status in China, which parallels conventional medicine in the Western medical system. According to Chinese medical theory, the brain and bone marrow are the outgrowths of the kidneys. In Lingshu Meridians, it is stated that "at conception, essence is formed." After essence is formed, the brain and bone marrow are formed. The kidneys contain the essence, the essence sustains the marrow, and the marrow glorifies the brain. According to "Lingshu Discussion on Seas," humans have a marrow sea, a blood sea, a qi sea, and a water/grain sea (stomach). This is the meaning of the four seas. Among the four seas in the human body, the marrow sea refers to the brain. According to the "Category Text" Volume 9: where there is bone, there is marrow, and the brain has the most. Thus, all marrow belongs to the brain, and the brain is the sea of marrow (Ong et al., 2018).

TCM states that dementia is caused by (1) deficiency of vital energy of the kidney (*Shen*), marrow (*Sui*), heart (*Xin*), and spleen (*Pi*); and (2) stagnation of blood (*Xie*) and/or phlegm (*Tan*). Thus, herbs used for dementia are not specific for the nervous system but tend to be multifunctional (Fu, 1991; Ho et al., 2010). In TCM, the use of multiherb formulas rather than single herbs is common. Each herbal component in a formula has a specific role—sovereign, minister, assistant, and courier. Sovereign and minister herbs treat the main symptoms and have a major role in the formula. Assistant herbs assist the sovereign and minister herbs to treat the accompanying symptoms and reduce the side effects of the major herbs. Courier herbs help to lead the other components to the affected tissue or area (Ong et al., 2018). Interactions among herbs, such as mutual reinforcement, antagonism, or detoxification, help to determine the formula's therapeutic efficacy (Effertha et al., 2016; Ong et al., 2018). Chinese formulations for the treatment of dementia are prepared by mixing roots and leaves powders from Chinese medicinal plants (Table 8.1). During the past decades, a number of clinical trials have been conducted in China to investigate a series of Chinese formulations. Many of these formulations consist of a combination of 5–10 herbs obtained from Chinese medicinal plants. They produce beneficial effects in patients with dementia by reducing oxidative stress, retarding neuroinflammation, inhibiting apoptotic cell death, and increasing the expression of neurotropic factors (BDNF and GDNF) (Steiner et al., 2016; Qian and Ke, 2014; Lee et al., 2016; Ong et al., 2018).

TABLE 8.1 Names and Active Components of Chinese Formulations Used for the Treatment of Dementia

Name of formulation	Active components	Reference
Bushen-Yizhi prescription	*Cnidium* fruit, tree peony bark, ginseng root, Radix Polygoni Multiflori Preparata, barbary wolfberry fruit, and Fructus Ligustri Lucidi	Hou et al. (2014), Cai et al. (2018)
Yokukansan (Yi gan San) formula	*Angelica acutiloba* L., *Atractylodes lancea* DC, *Bupleurum falcatum* L., *Poria cocos* Wolf, *Glycyrrhiza uralensis*, *Cnidium officinale* Makino, and *Uncaria rhynchophylla* Schreb in a ratio of 3:4:2:4:1.5:3:3	Yu et al. (2014)
SaiLuo Tong formula	*Panax ginseng*, *Ginkgo biloba*, and *Crocus sativus*	Steiner et al. (2016)
Ming Dynasty prescription	Powder of Rehmannia with another seven plants mixed with honey	Iwasaki et al. (2004)
Gagamjungjuhwan formula	Ginseng, Acori Graminei Rhizoma, Uncariae Ramulus et Uncus, Polygalae Radic, and Frustus Euodiae	Lee et al. (2016)
Yi-Gan San formula	A mixture of seven different rootstock and branches, lyophilized dry extract	Iwasaki et al. (2005)
Kangen-karyu formula	Six herbs formula	Zhao et al. (2010)
Ba Wei Di Hunag wan		Ho et al. (2011)
Fumanjian formula	Radix Rehmanniae Recens; Radix Ophiopogoni; Radix Paeoniae Alba; Rhizoma acori tatarinowii; Herba Dendrobii; Cortex Moutan Radicis; Poria; Indian bread (*fuling*); the dried Sclerotia of *Poria cocos* Pericarpium Citri Reticulatae; Caulis Akebiae; Rhizoma Anemarrhenae in the ratio of 2 : 2 : 2 : 2 : 2 : 2 : 2 : 1 : 1.5 : 1.5 on a dry weight basis	Hu et al. (2014)
Tian-Ma-Gou-Teng-Yin formula	Neuroprotective and antineuroinflammatory effects against the progression of AD type of dementia	Wang et al. (2018)

CONCLUSION

Dementia is a syndrome associated with progressive impairments in memory and learning ability, cognitive skills, behavior, ADL, and quality of life. There are more than 47.5 million people with dementia worldwide and 7.7 million new cases are added to the dementia pool each year. TCM has long been used for the treatment of age-related memory disorders. A number of Chinese herbal formulas (huperzine, *G. biloba*, green tea, and ginseng), have been reported to produce anticholinergic, antioxidant, and antiinflammatory effects in demented human patients. Investigators are making attempts to study the molecular mechanisms of Chinese herbal formulas and explain these mechanisms on the basis of signal transduction processes and pathways.

References

Abdel-Wahab, B.A., Abd El-Aziz, S.M., 2012. *Ginkgo biloba* protects against intermittent hypoxia-induced memory deficits and hippocampal DNA damage in rats. Phytomedicine 19, 444–450.

Ahlemeyer, B., Krieglstein, J., 2003. Neuroprotective effects of *Ginkgo biloba* extract. Cell Mol. Life Sci. 60, 1779–1792.

Bastianetto, S., Quirion, R., 2002. Natural extracts as possible protective agents of brain aging. Neurobiol. Aging 23, 891–897.

Bitu Pinto, N., da Silva Alexandre, B., Neves, K.R., Silva, A.H., Leal, L.K., et al., 2015. Neuroprotective properties of the standardized extract from *Camellia sinensis* (green tea) and its main bioactive components, epicatechin and epigallocatechin gallate, in the 6-OHDA model of Parkinson's Disease. Evid. Based Complement. Altern. Med. 2015, 161092.

Budni, J., Bellettini-santos, T., Mina, F., Garcez, M.L., 2016. The involvement of BDNF, NGF and GDNF in aging and Alzheimer's disease. Aging Dis. 6, 331–341.

Cai, H., Luo, Y., Yan, X., Ding, P., Huang, Y., et al., 2018. The mechanisms of Bushen-Yizhi Formula as a therapeutic agent against Alzheimer's Disease. Sci. Rep. 8, 3104.

Chang, Y., Hsieh, C.Y., Peng, Z.A., Yen, T.L., Hsiao, G., Chou, D.S., et al., 2009. Neuroprotective mechanisms of puerarin in middle cerebral artery occlusion-induced brain infarction in rats. J. Biomed. Sci. 16, 9–21.

Chen, L., Dai, J., Wang, Z., Zhang, H., Huang, Y., Zhao, Y., 2014. Ginseng total saponins reverse corticosterone-induced changes in depression-like behavior and hippocampal plasticity-related proteins by interfering with gsk-3β-creb signaling pathway. Evid. Based Complement. Alternat. Med. 2014, 506735.

Chen, Q., Hu, Y.E., Xia, Z.Q., 2000. Action of Sapogenin from Zhimu on learning and memory ability and free-radical metabolism in mouse D-galactose dementia model. Pharmacol. Clin. Chin. Mater. Med. 16, 14–16.

Chen, Q., Cao, Y.G., Lin, Y.M., et al., 2004. Effects of sapogenin from zhimu (ZMS) and its isomer on learning and memory ability and muscarinic subtype M1 receptor density in aged rats. Chin. Pharmacol. Bull. 20, 561–564.

Chen, X.C., Zhu, Y.G., Zhu, L.A., Huang, C., Chen, Y., et al., 2003. Ginsenoside Rg1 attenuates dopamine-induced apoptosis in PC12 cells by suppressing oxidative stress. Eur. J. Pharmacol. 473, 1–7.

Chen, Z., Lu, T., Yue, X., Wei, N., Jiang, Y., et al., 2010. Neuroprotective effect of ginsenoside Rb1 on glutamate-induced neurotoxicity: with emphasis on autophagy. Neurosci. Lett. 482, 264–268.

Cheng, Y., Shen, L.H., Zhang, J.T., 2005. Anti-amnestic and anti-aging effects of ginsenoside Rg1 and Rb1 and its mechanism of action. Acta Pharmacol. Sin. 26, 143–149.

Chinese Pharmacopoeia Commission, 2005. Pharmacopoeia of the People's Republic of China, vol. 1. People's Medical Publishing House, Beijing.

Cho, I.H., 2012. Effects of *Panax ginseng* in neurodegenerative diseases. J. Ginseng Res. 36, 342–353.

Choo, M.K., Park, E.K., Han, M.J., Kim, D.H., 2003. Antiallergic activity of ginseng and its ginsenosides. Planta Med. 69, 518–522.

Chung, K.F., Dent, G., McCusker, M., Guinot, P., Page, C.P., Barnes, P.J., 1987. Effect of a ginkgolide mixture (BN 52063) in antagonising skin and platelet responses to platelet activating factor in man. Lancet 1 (8527), 248–251.

Clostre, F., 1999. *Ginkgo biloba* extract (EGb 761). State of knowledge in the dawn of the year 2000. Ann. Pharm. Fr. 57 (Suppl. 1), 1S8–1S88.

Damar, U., Gersner, R., Johnstone, J.T., Schachter, S., Rotenberg, A., 2016. Huperzine A as a neuroprotective and antiepileptic drug: a review of preclinical research. Expert Rev. Neurother. 16, 671–680.

DeFeudis, F.V., Drieu, K., 2000. *Ginkgo biloba* extract (EGb761) and CNS functions: basic studies and clinical applications. Curr. Drug Targets 1, 25–58.

DeKosky, S.T., Williamson, J.D., Fitzpatrick, A.L., Kronmal, R.A., Ives, D.G., et al., 2008. *Ginkgo biloba* for prevention of dementia: a randomized controlled trial. JAMA 300, 2253–2262.

Deng, Y., Ma, B.P., Xu, Q.P., et al., 2005. Effect and mechanism of effective component in Zhimu on ability of learning and memory in vascular dementia rats. Chin. Pharmacol. Bull. 21, 830–833.

Di Renzo, G., 2000. *Ginkgo biloba* and the central nervous system. Fitoterapia 71 (Suppl. 1), S43–S47.

Dong, D., Mao, Y., Huang, C., Jiao, Q., Pan, H., et al., 2017. Astrocytes mediated the nootropic and neurotrophic effects of Sarsasapogenin-AA13 via upregulating brain-derived neurotrophic factor. Am. J. Transl. Res. 9, 4015–4025.

Eckert, A., Keil, U., Kressmann, S., Schindowski, K., Leutner, S., et al., 2003. Effects of EGb 761 *Ginkgo biloba* extract on mitochondrial function and oxidative stress. Pharmacopsychiatry 36 (Suppl. 1), S15–S23.

Efferth, T., Shan, L., Zhang, Z.-W., 2016. Tonic herbs and herbal mixtures in Chinese Medicine. World J. Tradit. Chin. Med. 2, 10–25.

Ernst, E., 2009. A comment on *Ginkgo biloba* for mild to moderate dementia in a community setting by McCarney et al. Int. J. Geriatr. Psychiatry 24, 215–216.

Fang, F., Chen, X., Huang, T., Lue, L.F., Luddy, J.S., et al., 2012. Multi-faced neuroprotective effects of Ginsenoside Rg1 in an Alzheimer mouse model. Biochim. Biophys. Acta 1822, 286–292.

Farooqui, A.A., 2010. Neurochemical Aspects of Neurotraumatic and Neurodegenerative Diseases. Springer, New York.

Farooqui, A.A., Farooqui, T., Madan, A., Ong, J.H.-J., Ong, W.Y., 2018. Ayurvedic medicine for the treatment of dementia: mechanistic aspects. Evid. Based Complement. Altern. Med. 2018, 2481076.

Feng, M., Lu, J., May, B.H., Liy, S., Guo, X., et al., 2016. Chinese herbal medicine for patients with vascular cognitive impairment no dementia: protocol for a systematic review. BMJ Open 6, e010295.

Feng, W.Y., 2006. Metabolism of green tea catechin catechins: an overview. Curr. Drug Metab. 7, 755–809.

Fu, R.J., 1991. Guidelines for TCM diagnosis, classification and clinical research of senile dementia. J. Tradit. Chin. Med. 2, 56–57.

Gaus, W., 2009. An example for an underpowered study: a comment on *Ginkgo biloba* for mild to moderate dementia in a community setting by McCarney et al. Int. J. Geriatr. Psychiatry 24, 216–217.

Gui, Q.F., Yang, Y.M., Ying, S.H., Zhang, M.M., 2013. Xueshuantong improves cerebral blood perfusion in elderly patients with lacunar infarction. Neural Regen. Res. 8, 792–801.

Guo, Q., Zhao, B.L., Li, M.F., Shen, S.R., Xin, W.J., 1996. Studies on protective mechanisms of four components of green tea polyphenols against lipid peroxidation in synaptosomes. Biochem. Biophys. Acta 1304, 210–222.

Heo, J.H., Lee, S.T., Chu, K., Oh, M.J., Park, H.J., et al., 2008. An open-label trial of Korean red ginseng as an adjuvant treatment for cognitive impairment in patients with Alzheimer's disease. Eur. J. Neurol. 15, 865–868.

Ho, Y.S., So, K.F., Chang, R.C.C., 2010. Anti-aging herbal medicine--how and why can they be used in aging-associated neurodegenerative diseases? Ageing Res. Rev. 9, 354–362.

Ho, Y.S., So, K.F., Chang, R.C., 2011. Drug discovery from Chinese medicine against neurodegeneration in Alzheimer's and vascular dementia. Chin. Med. 6, 15.

Hou, X.Q., Wu, D.W., Zhang, C.X., Yan, R., Yang, C., et al., 2014. Bushen-Yizhi formula ameliorates cognition deficits and attenuates oxidative stress-related neuronal apoptosis in scopolamine-induced senescence in mice. Int. J. Mol. Med. 34, 429–439.

Hu, H.Y., Cui, Z.H., Li, H.Q., Wang, Y.R., Chen, X., et al., 2014. Fumanjian, a Classic Chinese Herbal Formula, can ameliorate the impairment of spatial learning and memory through apoptotic signaling pathway in the hippocampus of rats with $A\beta_{1-40}$-induced Alzheimer's Disease. Evid. Based Complement. Altern. Med. 2014, 942917.

Hu, L., Yu, T., Jia, Z., 1996. Experimental study of the protective effects of Astragalus and *Salvia miltiorrhiza* bunge on glycerol induced acute renal failure in rabbits. Zhonghua Wai Ke Za Zhi 34, 311–314.

Hu, Y.E., Sun, Q.X., Xia, Z.Q., 2003. The effect of ZMS, an active component of Zhimu on NGF and BDNF in brains of aged rats. Chin. Pharmacol. Bull. 19, 149–151.

Huang, C., Zheng, C., Li, Y., Wang, Y., Lu, A., Yang, L., 2014a. Systems pharmacology in drug discovery and therapeutic insight for herbal medicines. Brief. Bioinform. 15, 710–733.

Huang, X.T., Qian, Z.M., He, X., Gong, Q., Wu, K.C., et al., 2014b. Reducing iron in the brain: a novel pharmacologic mechanism of huperzine A in the treatment of Alzheimer's disease. Neurobiol. Aging 35, 1045–1054.

Huang, J., Wu, D., Wang, J., Li, F., Lu, L., et al., 2014c. Effects of *Panax notoginseng* saponin on alpha, beta, and gamma secretase involved in Abeta deposition in SAMP8 mice. Neuroreport 25, 89–93.

Huang, C., Dong, D., Jiao, Q., Pan, H., Ma, L., Wang, R., 2017. Sarsasapogenin-AA13 ameliorates Aβ-induced cognitive deficits via improving neuroglial capacity on Aβ clearance and antiinflammation. CNS Neurosci. Ther. 23, 498–509.

Huh, H., Staba, E.J., 1992. Botany and chemistry of *Ginkgo biloba* L. J. Herbs Spices Med. Plants 1, 92–124.

Humber, J.M., 2002. The role of complementary and alternative medicine: accommodating pluralism. J. Am. Med. Assoc. 288, 1655–1656.

Hung, Y.C., Pan, T.L., Hu, W.L., 2016. Roles of reactive oxygen species in anticancer therapy with *salvia miltiorrhiza bunge*. Oxid. Med. Cell Longev. 2016, 5293284.

IARC Working Group, 2016. Some Drugs and Herbal Products. Lyon: *Ginkgo biloba*. International Agency for Research and Cancer (IARC) Monographs, WHO.

Ijaopo, E.O., 2017. Dementia-related agitation: a review of non-pharmacological interventions and analysis of risks and benefits of pharmacotherapy. Transl. Psychiat. 7, e1250.

Iwasaki, K., Kobayashi, S., Chimura, Y., Taguchi, M., Inoue, K., Cho, S., et al., 2004. A randomized, double-blind, placebo-controlled clinical trial of the Chinese herbal medicine 'Ba wei di huang wan' in the treatment of dementia. J. Am. Geriatr. Soc. 52, 1518–1521.

Iwasaki, K., Satoh-Nakagawa, T., Maruyama, M., Monma, Y., Nemoto, M., Tomita, N., et al., 2005. A randomized, observer blind, controlled trial of the traditional Chinese medicine yi-gan san for improvement of behavioral and psychological symptoms and activities of daily living in dementia patients. J. Clin. Psychiatry 66, 248–252.

Jacqui, W., 2013. Herbal products are often contaminated, study finds. BMJ 347, f6138.

Ji, X., Feng, Y.F., 2010. Advances in studies on saponins in *Anemarrhena* asphodeloides. Chin. Tradit. Herbal Drugs 4, S12–S15.

Jia, J.Y., Zhao, Q.H., Liu, Y., Gui, Y.Z., Liu, G.Y., et al., 2013. Phase I study on the pharmacokinetics and tolerance of ZT-1, a prodrug of huperzine A, for the treatment of Alzheimer's disease. Acta Pharmacol. Sin. 34, 976–982.

Jiang, W.H., Meng, X.T., Hao, L.M., et al., 2001. Study of anti-aging and anti-dementia effects of Rhodosin on aging rats and experimental dementia rats. J. N Bethune Univ. Med. Sci. 27, 127–129.

Karpagam, V., Sathishkumar, N., Sathiyamoorthy, S., Rasappan, P., Shila, S., et al., 2013. Identification of BACE1 inhibitors from *Panax ginseng* saponins-an insilco approach. Comput. Biol. Med. 43, 1037–1044.

Kehr, J., Yoshitake, S., Ijiri, S., Koch, E., Noldner, M., et al., 2012. *Ginkgo biloba* leaf extract (EG b 761®) and its specific acylated flavonol constituents increase dopamine and acetylcholine levels in the rat medial prefrontal cortex: possible implications for the cognitive enhancing properties of EGb 761®. Int. Psychogeriatr. 24 (S1), S25–S34.

Kennedy, D.O., Scholey, A.B., Drewery, L., Marsh, V.R., Moore, B., et al., 2003. Electroencephalograph effects of single doses of *Ginkgo biloba* and *Panax ginseng* in healthy young volunteers. Pharmacol. Biochem. Behav. 75, 701–709.

Kim, J., Lee, H.J., Lee, K.W., 2010a. Naturally occurring phytochemicals for the prevention of Alzheimer's disease. J. Neurochem. 112, 1415–1430.

Kim, J.S., Kim, J.M., O, J.J., Jeon, B.S., 2010b. Inhibition of inducible nitric oxide synthase expression and cell death by (-)-epigallocatechin-3-gallate, a green tea catechin, in the 1-methyl-4-phenyl-1,2,3,6-tetrahydropyridine mouse model of Parkinson's disease. J. Clin. Neurosci. 17, 1165–1168.

Költringer, P., Langsteger, W., Eber, O., 1995. Dose-dependent hemorheological effects and microcirculatory modifications following intravenous administration of *Ginkgo biloba* special extract EGb 761. Clin. Hemorheol. 15, 649–656.

Lardner, A.L., 2014. Neurobiological effects of the green tea constituent theanine and its potential role in the treatment of psychiatric and neurodegenerative disorders. Nutr. Neurosci. 17, 145–155.

Lavretsky, H., 2016. Lifestyle medicine for prevention of cognitive decline: focus on green tea. Am. J. Geriatr. Psychiatry 24, 890–892.

Lee, B., Sur, B., Cho, S.-G., Yeom, M., Shim, I., Lee, H., et al., 2016. Wogonin attenuates hippocampal neuronal loss and cognitive dysfunction in Trimethyltin-intoxicated rats. Biomol. Ther. 24, 328.

Lee, D.W., Andersen, J.K., 2010. Iron elevations in the aging Parkinsonian brain: a consequence of impaired iron homeostasis? J. Neurochem. 112, 332–339.

Lee, S.E., Jeong, S.I., Yang, H., Jeong, S.H., Jang, Y.P., et al., 2012. Extract of *Salvia miltiorrhiza* (Danshen) induces Nrf2-mediated heme oxygenase-1 expression as a cytoprotective action in RAW 264.7 macrophages. J. Ethnopharmacol. 139, 541–548.

Lee, S.T., Chu, K., Sim, J.Y., Heo, J.H., Kim, M., 2008. *Panax ginseng* enhances cognitive performance in Alzheimer disease. Alzheimer Dis. Assoc. Disord. 22, 222–226.

Legal Status of Traditional Medicine, Complementary/Alternative Medicine: A Worldwide Review, 2001. World Health Organization, Geneva.

Levites, Y., Weinreb, O., Maor, G., Youdim, M.B., Mandel, S., 2001. Green tea polyphenol (-)-epigallocatechin-3-gallate prevents N-methyl-4-phenyl-1,2,3,6-tetrahydropyridine-induced dopaminergic neurodegeneration. J. Neurochem. 78, 1073–1082.

Li, T.J., Qiu, Y., Yang, P.Y., Rui, Y.C., Chen, W.S., 2007. Timosaponin B-II improves memory and learning dysfunction induced by cerebral ischemia in rats. Neurosci. Lett. 421, 147–151.

Li, X., Sun, S., Tong, E., 2003. Experimental study on the protective effect of puerarin to Parkinson disease. J. Huazhong Univ. Sci. Technol. Med. Sci. 23, 148–150.

Liou, H.H., Chen, R.C., Chen, T.H., Tsai, Y.F., Tsai, M.C., 2001. Attenuation of paraquat-induced dopaminergic toxicity on the substantia nigra by (-)-deprenyl in vivo. Toxicol. Appl. Pharmacol. 172, 37–43.

Liu, C.Y., Huang, C.J., Huang, L.H., Chen, I.J., Chiu, J.P., et al., 2014. Effects of green tea extract on insulin resistance and glucagon-like peptide 1 in patients with type 2 diabetes and lipid abnormalities: a randomized, double-blinded, and placebo-controlled trial. PLoS One 93, e91163.

Liu, H., Yu, L., Dong, X., Sun, Y., 2017. Synergistic effects of negatively charged hydrophobic nanoparticles and (-)-epigallocatechin-3-gallate on inhibiting amyloid β-protein aggregation. J. Colloid Interface Sci. 491, 305–312.

Liu, J., Chang, D., 2006. Vascular dementia. J. Complement. Med. 5, 14–20.

Liu, Y.W., Zhu, X., Lu, Q., Wang, J.Y., Li, W., et al., 2012. Total saponins from Rhizoma Anemarrhenae ameliorate diabetes-associated cognitive decline in rats: involvement of amyloid-beta decrease in brain. J. Ethnopharmacol. 139, 194–200.

Liu, Z.J., Zhao, M., Zhang, Y., Xue, J.F., Chen, N.H., 2010. Ginsenoside Rg1 promotes glutamate release via a calcium/calmodulin-dependent protein kinase II-dependent signaling pathway. Brain Res. 1333, 1–8.

Liu, Z.W., Wu, M.C., Chen, P., 2003. Effects of Rhodiola henryi extract on learning, memory and antihypoxia in mice. Acta Nutr. Sin. 25, 101–104.

Lu, W.Q., Qiu, Y., Li, T.J., Tao, X., Sun, L.N., Chen, W.S., 2009. Timosaponin B-II inhibits pro-inflammatory cytokine induction by lipopolysaccharide in BV2 cells. Arch. Pharm. Res. 32, 1301–1308.

Luo, F.C., et al., 2011. Protective effect of panaxatriol saponins extracted from *Panax notoginseng* against MPTP-induced neurotoxicity in vivo. J. Ethnopharmacol. 133, 448–453.

Lv, H., Wang, L., Shen, J., Hao, S., Ming, A., Wang, X., et al., 2015. Salvianolic acid B attenuates apoptosis and inflammation via SIRT1 activation in experimental stroke rats. Brain Res. Bull. 115, 30–36.

Ma, T., Gong, K., Yan, Y., Zhang, L., Tang, P., Zhang, X., et al., 2013. Huperzine A promotes hippocampal neurogenesis in vitro and in vivo. Brain Res. 1506, 35–43.

Mandel, S.A., Amit, T., Kalfon, L., Reznichenko, L., Youdim, M.B.H., 2008. Targeting multiple neurodegenerative diseases etiologies with multimodal-acting green tea catechins. J. Nutr. 138, 1578S–1583S.

Mao, X.Y., Cao, D.F., Li, X., Yin, J.Y., Wang, Z.B., Zhang, Y., et al., 2014. Huperzine a ameliorates cognitive deficits in streptozotocin-induced diabetic rats. Int. J. Mol. Sci. 15, 7667–7683.

McCarney, R., Fisher, P., Iliffe, S., van Haselen, R., Griffin, M., et al., 2008. *Ginkgo biloba* for mild to moderate dementia in a community setting: a pragmatic, randomised, parallel-group, double-blind, placebo-controlled trial. Int. J. Geriatr. Psychiatry 23, 1222–1230.

Moss, D.E., Berlanga, P., Hagan, M.M., Sandoval, H., Ishida, C., 1999. Methanesulfonyl fluoride (MSF): a double-blind, placebo-controlled study of safety and efficacy in the treatment of senile dementia of the Alzheimer type. Alzheimer Dis. Assoc. Disord. 13, 20–25.

Murakami, A., Ohnishi, K., 2012. Target molecules of food phytochemicals: food science bound for the next dimension. Food Funct. 3, 462–476.

Nathan, P., 2000. Can the cognitive enhancing effects of *Ginkgo biloba* be explained by its pharmacology? Med. Hypotheses 55, 491–493.

Ong, W.Y., Farooqui, T., Koh, H.L., Farooqui, A.A., Ling, E.A., 2015. Protective effects of ginseng on neurological disorders. Front. Aging Neurosci. 7, 129.

Ong, W.Y., Wu, Y.J., Farooqui, T., Farooqui, A.A., 2018. Qi Fu Yin-a Ming Dynasty prescription for the treatment of dementia. Mol. Neurobiol. 55, 7389–7400.

Ouyang, S., Sun, L.S., Guo, S.L., Liu, X., Xu, J.P., 2005. Effects of timosaponins on learning and memory abilities of rats with dementia induced by lateral cerebral ventricular injection of amyloid beta- peptide. Di Yi Jun Yi Da Xue Xue Bao 25, 121–126.

Paine, A., Eiz-Vesper, B., Blasczyk, R., Immenschuh, S., 2010. Signaling to heme oxygenase-1 and its anti-inflammatory therapeutic potential. Biochem. Pharmacol. 80, 1895–1903.

Pan, T., Fei, J., Zhou, X., Jankovic, J., Le, W., 2003. Effects of green tea polyphenols on dopamine uptake and on MPP^+-induced dopamine neuron injury. Life Sci. 72, 1073–1083.

Peters, C., Sepúlveda, F.J., Fernández-Pérez, E.J., Peoples, R.W., Aguayo, L.G., 2016. The level of NMDA receptor in the membrane modulates amyloid-β association and perforation. J. Alzheimer's Dis. 53, 197–207.

Pincemail, J., Dupuis, M., Nasr, C., et al., 1989. Superoxide anion scavenging effect and superoxide dismutase activity of *Ginkgo biloba* extract. Experientia 45, 708–712.

Qi, X., Ignatova, S., Luo, G., Liang, Q., Jun, F.W., Wang, Y., et al., 2010. Preparative isolation and purification of ginsenosides Rf, Re, Rd and Rb1 from the roots of *Panax ginseng* with a salt/containing solvent system and flow step-gradient by high performance counter-current chromatography coupled with an evaporative light scattering detector. J. Chromatogr. A 1217, 1995–2001.

Qian, Z.M., Ke, Y., 2014. Huperzine A: is it an effective disease-modifying drug for Alzheimer's Disease? Front. Aging Neurosci. 6, 216.

Qiu, J., Li, W., Feng, S.H., Wang, M., He, Z.Y., 2014. Ginsenoside Rh2 promotes nonamyloidgenic cleavage of amyloid precursor protein via a cholesterol-dependent pathway. Genet. Mol. Res. 13, 3586–3598.

Rafii, M.S., Walsh, S., Little, J.T., Behan, K., Reynolds, B., et al., 2011. A phase II trial of huperzine A in mild to moderate Alzheimer disease. Neurology 76, 1389–1394.

Ramassamy, C., 2006. Emerging role of polyphenolic compounds in the treatment of neurodegenerative diseases: a review of their intracellular targets. Eur. J. Pharmacol. 545, 51–64.

Ramassamy, C., Christen, Y., Clostre, F., Costentin, J., 1992. The *Ginkgo biloba* extract, EGb761, increases synaptosomal uptake of 5-hydroxytptamine: in-vitro and ex-vitro studies. J. Pharm. Pharmacol. 44, 943–945.

Ramassamy, C., Longpre, F., Christen, Y., 2007. *Ginkgo biloba* extract (EGb 761) in Alzheimer's disease: is there any evidence? Curr. Alzheimer Res. 4, 253–262.

Reay, J.L., Kennedy, D.O., Scholey, A.B., 2006. Effects of *Panax ginseng*, consumed with and without glucose, on blood glucose levels and cognitive performance during sustained 'mentally demanding' tasks. J. Psychopharmacol. 20, 771–781.

Ren, B., Zhang, Y.X., Zhou, H.X., Sun, F.W., Zhang, Z.F., Wei, Z., et al., 2015. Tanshinone IIA prevents the loss of nigrostriatal dopaminergic neurons by inhibiting NADPH oxidase and iNOS in the MPTP model of Parkinson's disease. J. Neurol. Sci. 348, 142–152.

Renaud, J., Nabavi, S.F., Daglia, M., Nabavi, S.M., Martinoli, M.G., 2015. Epigallocatechin-3-gallate, a promising molecule for Parkinson's Disease? Rejuvenation Res. 18, 257–269.

Rocher, M.N., Carré, D., Spinnewyn, B., Schulz, J., Delaflotte, S., et al., 2011. Long-term treatment with standardized *Ginkgo biloba* extract (EGb 761) attenuates cognitive deficits and hippocampal neuron loss in a gerbil model of vascular dementia. Fitoterapia 82, 1075–1080.

Rokot, N.T., Kairupan, T.S., Cheng, K.C., Runtuwene, J., Kapantow, N.H., et al., 2016. A role of ginseng and its constituents in the treatment of central nervous system disorders. Evid. Based Complement. Altern. Med. 2016, 2614742.

Schneider, L.S., 2008. *Ginkgo biloba* extract and preventing Alzheimer disease. JAMA 300, 2306–2308.

Schwarz, S., Froelich, L., Burns, A., 2012. Pharmacological treatment of dementia. Curr. Opin. Psychiatry 25, 542–550.

Shao, Z.-Q., 2015. Comparison of the efficacy of four cholinesterase inhibitors in combination with memantine for the treatment of Alzheimer's disease. Int. J. Clin. Exp. Med. 8, 2944–2948.

Sharangi, A.B., 2009. Medicinal and therapeutic potentialities of tea (*Camellia sinensis* L.)—a review. Food Res. Int. 42, 529–535.

Shi, C., Zhang, Y.X., Zhang, Z.F., 2009. Effect of phosphorylated-ERK1/2 on inducible nitric oxide synthase expression in the substantia nigra of mice with MPTP-induced Parkinson disease. Nan Fang Yi Ke Da Xue Xue Bao 29, 60–63.

Singh, M., Arseneault, M., Sanderson, T., Murthy, V., Ramassamy, C., 2008. Challenges for research on polyphenols from foods in Alzheimer's disease: bioavailability, metabolism, and cellular and molecular mechanisms. J. Agric. Food Chem. 56, 4855–4873.

Smith, A., Giunta, B., Bickford, P.C., Fountain, M., Tan, J., Shytle, R.D., 2010. Nanolipidic particles improve the bioavailability and alpha-secretase inducing ability of epigallocatechin-3-gallate (EGCG) for the treatment of Alzheimer's disease. Int. J. Pharm. 389, 207–212.

Solfrizzi, V., Panza, F., 2015. Plant-based nutraceutical interventions against cognitive impairment and dementia: meta-analytic evidence of efficacy of a standardized *Gingko biloba* extract. J. Alzheimers Dis. 43, 605–611.

Song, X.Y., et al., 2013. Ginsenoside Rg1 attenuates okadaic acid induced spatial memory impairment by the GSK3beta/tau signaling pathway and the Abeta formation prevention in rats. Eur. J. Pharmacol. 710, 29–38.

Song, Y.Y., Qi, G., Han, J.T., 2005. Protective effect of hongjingtian on hippocampal area and dentate gyrus of complete cerebral ischemia-reperfusion in rats. Chin. J. Clin. Rehabil. 9, 232–233.

Spencer, J.P., 2008. Flavonoids: modulators of brain function? Br. J. Nutr. 99 E (Suppl. 1), ES60–ES77.

Steiner, G.Z., Yeung, A., Liu, J.-X., Camfield, D.A., De Blasio, F.M., Pipingas, A., et al., 2016. The effect of Sailuotong (SLT) on neurocognitive and cardiovascular function in healthy adults: a randomised, double-blind, placebo controlled crossover pilot trial. BMC Complement. Altern. Med. 16, 15.

Su, C.Y., Ming, Q.L., Rahman, K., Han, T., Qin, L.P., 2015. *Salvia miltiorrhiza*: traditional medicinal uses, chemistry, and pharmacology. Chin. J. Nat. Med. 13, 163–182.

Sun, Q.X., Hu, Y.E., Guan, H., et al., 2004. Effects of ZDY101, an active component from Zhimu, on rat dementia model produced by intracranial injection of ibotenic acid. Nucl. Tech. 27, 297–300.

Tanaka, A., Arai, Y., Kim, S.-N., Ham, J., Usuki, T., 2011. Synthesis and biological evaluation of bilobol and adipostatin A. J. Asian Nat. Prod. Res. 13, 290–296.

Thielecke, F., Boschmann, M., 2009. The potential role of green tea catechins in the prevention of the metabolic syndrome—a review. Phytochemistry 701, 11–24.

Tian, J., 2003. Ginseng may improve memory in stroke dementia patients. In: Proceedings of the American Stroke Association Meeting, Augusta, Canada.

Tobinaga, S., Hashimoto, M., Utsunomiya, I., Taguchi, K., Nakamura, M., et al., 2012. Chronic administration of cardanol (ginkgol) extracted from *Ginkgo biloba* leaves and cashew nutshell liquid improves working memory-related learning in rats. Biol. Pharm. Bull. 35, 127–129.

Tu, L.H., Ma, J., Liu, H.P., Wang, R.R., Luo, J., 2009. The neuroprotective effects of ginsenosides on calcineurin activity and tau phosphorylation in SY5Y cells. Cell Mol. Neurobiol. 29, 1257–1264.

Van Kampen, J.M., Baranowski, D.B., Shaw, C.A., Kay, D.G., 2014. Panax ginseng is neuroprotective in a novel progressive model of Parkinson's disease. Exp. Gerontol. 50, 95–105.

Wakabayashi, N., Slocum, S.L., Skoko, J.J., Shin, S., Kensler, T.W., 2010. When NRF2 talks, who's listening? Antioxid. Redox Signal. 13, 1649–1663.

Wan, W., Zhang, C., Danielsen, M., Li, Q., Chen, W., et al., 2016. EGb761 improves cognitive function and regulates inflammatory responses in the APP/PS1 mouse. Exp. Gerontol. 81, 92–100.

Wang, B.S., Wang, H., Wei, Z.H., Song, Y.Y., Zhang, L., et al., 2009a. Efficacy and safety of natural acetylcholinesterase inhibitor huperzine A in the treatment of Alzheimer's disease: an updated meta-analysis. J. Neural Transm. 116, 457–465.

Wang, H., Peng, D., Xie, J., 2009b. Ginseng leaf-stem: bioactive constituents and pharmacological functions. Chin. Med. 4, 20.

Wang, J., Xu, H.M., Yang, H.D., Du, X.X., Jiang, H., Xie, J.X., 2009c. Rg1 reduces nigral iron levels of MPTP-treated C57BL6 mice by regulating certain iron transport proteins. Neurochem. Int. 54, 43–48.

Wang, Q., Zheng, H., Zhang, Z.F., Zhang, Y.X., 2008. Ginsenoside Rg1 modulates COX-2 expression in the substantia nigra of mice with MPTP-induced Parkinson disease through the P38 signaling pathway. Nan Fang Yi Ke Da Xue Xue Bao 28, 1594–1598.

Wang, R., Tang, X.C., 2005. Neuroprotective effects of huperzine A. A natural cholinesterase inhibitor for the treatment of Alzheimer's disease. Neurosignals 14, 71–82.

Wang, T., Wu, Z., Sun, L., Li, W., Liu, G., Tang, Y., 2018. A computational systems pharmacology approach to investigate molecular mechanisms of herbal formula Tian-Ma-Gou-Teng-Yin for treatment of Alzheimer's Disease. Front. Pharmacol. 9, 668.

Wang, Y., Huang, L.-Q., Tang, X.-C., Zhang, H.-Y., 2010. Retrospect and prospect of active principles from Chinese herbs in the treatment of dementia. Acta Pharmacol. Sin. 31, 649–664.

Wang, Y., Tang, X.C., Zhang, H.Y., 2012. Huperzine A alleviates synaptic deficits and modulates amyloidogenic and nonamyloidogenic pathways in APPswe/PS1dE9 transgenic mice. J. Neurosci. Res. 90, 508–517.

Wanker, E.E., 2008. EGCG redirects amyloidogenic polypeptides into unstructured, off-pathway oligomers. Nat. Struct. Mol. Biol. 15, 558–566.

Weinmann, S., Roll, S., Schwarzbach, C., Vauth, C., Willich, S.N., 2010. Effects of *Ginkgo biloba* in dementia: systematic review and meta-analysis. BMC Geriatr. 10, 14.

Wolfram, S., Wang, Y., Thielecke, F., 2006. Anti-obesity effects of green tea: from bedside to bench. Mol. Nutr. Food Res. 502, 176–187.

Woo, E.R., Lee, J.Y., Cho, I.J., Kim, S.G., Kang, K.W., 2005. Amentoflavone inhibits the induction of nitric oxide synthase by inhibiting NF-kappaB activation in macrophages. Pharmacol. Res. 51, 539–546.

Wu, H.Q., Chang, M.Z., Zhang, G.L., et al., 2004a. The mechanism of protective effects of puerarin on learning-memory disorder after global cerebral ischemic reperfusive injury in rats. J. Apoplexy Nerv. Dis. 21, 350–353.

Wu, Y.Q., Yao, W.B., Gao, X.D., et al., 2004b. Effects of the extracts of *Rhodiola rosea* L. on improving the ability of learning and memory in mice. J. China Pharm. Univ. 35, 69–72.

Wu, H.Q., Guo, H.N., Wang, H.Q., Chang, M.Z., Zhang, G.L., et al., 2009. Protective effects and mechanism of puerarin on learning-memory disorder after global cerebral ischemia-reperfusion injury in rats. Chin. J. Integr. Med. 15, 54–59.

Xiao, B., Sun, Z., Cao, F., Wang, L., Liao, Y., et al., 2017. Brain pharmacokinetics and the pharmacological effects on striatal neurotransmitter levels of *Pueraria lobata* isoflavonoids in rat. Front. Pharmacol. 8, 599.

Xie, G.Q., Sun, X.L., Tian, S.P., et al., 2003. Studies on the preventive and therapeutic effects of rhodosin on rats with Alzheimer's disease. Chin. J. Behav. Med. Sci. 12, 18–20.

Xie, G.Q., Sun, X.L., Tian, S.P., et al., 2004. Preventive effects of rhodosin and melatonin from damage induced by β-amyloid 1–40 in senile rats. J. Nanjing Med. Univ. 18, 203–206.

Xu, L., Chen, W.F., Wong, M.S., 2009. Ginsenoside Rg1 protects dopaminergic neurons in a rat model of Parkinson's disease through the IGF-I receptor signalling pathway. Br. J. Pharmacol. 158, 738–748.

Xu, Q., Langley, M., Kanthasamy, A.G., Reddy, M.B., 2017. Epigallocatechin gallate has a neurorescue effect in a mouse model of Parkinson Disease. J. Nutr. 147, 1926–1931.

Xu, X., Hu, Y., Ruan, Q., 2004. Effects of puerarin on learning – memory and amino acid transmitters of brain in ovariectomized mice. Planta Med. 70, 627–631.

Xu, X.H., Zhao, T.Q., 2002. Effects of puerarin on D-galactose-induced memory deficits in mice. Acta Pharmacol. Sin. 23, 587–590.

Xu, X.H., Zheng, X.X., Zhou, Q., Li, H., 2007. Inhibition of excitatory amino acid efflux contributes to protective effects of puerarin against cerebral ischemia in rats. Biomed. Environ. Sci. 20, 336–342.

Yao, Z.X., Han, Z., Drieu, K., Papadopoulos, V., 2004. *Ginkgo biloba* extract (Egb 761) inhibits beta-amyloid production by lowering free cholesterol levels. J. Nutr. Biochem. 15, 749–756.

Yin, Y., Ren, Y., Wu, W., Wang, Y., Cao, M., et al., 2013. Protective effects of bilobalide on Abeta(25-35) induced learning and memory impairments in male rats. Pharmacol. Biochem. Behav. 106, 77–84.

Yoo, D.Y., Nam, Y., Kim, W., Yoo, K.Y., Park, J., et al., 2011. Effects of *Ginkgo biloba* extract on promotion of neurogenesis in the hippocampal dentate gyrus in C57BL/6 mice. J. Vet. Med. Sci. 73, 71–76.

Yoshitake, T., Yoshitake, S., Kehr, J., 2010. The *Ginkgo biloba* extract EGb 761® and its main constituent flavonoids and ginkgolides increase extracellular dopamine levels in the rat prefrontal cortex. Br. J. Pharmacol 159, 659–668.

Yu, C.H., Ishii, R., Yu, S.C., Takeda, M., 2014. Yokukansan and its ingredients as possible treatment options for schizophrenia. Neuropsychiatr. Dis. Treat. 10, 1629–1634.

Yu, X.Y., Lin, S.G., Chen, X., Zhou, Z.W., Liang, J., et al., 2007. Transport of cryptotanshinone, a major active triterpenoid in *Salvia miltiorrhiza* Bunge widely used in the treatment of stroke and Alzheimer's disease, across the blood-brain barrier. Curr. Drug Metab. 8, 365–378.

Zhang, H.Y., Yan, H., Tang, X.C., 2008. Non-cholinergic effects of huperzine A: beyond inhibition of acetylcholinesterase. Cell Mol. Neurobiol. 28, 173–183.

Zhang, R., Wang, Z., Howson, P.A., Xia, Z., Zhou, S., et al., 2012. Smilagenin attenuates beta amyloid (25-35)-induced degeneration of neuronal cells via stimulating the gene expression of brain-derived neurotrophic factor. Neuroscience 210, 275–285.

Zhang, X.Z., Qian, S.S., Zhang, Y.J., Wang, R.Q., 2016. *Salvia miltiorrhiza*: a source for anti-Alzheimer's disease drugs. Pharm. Biol. 54, 18–24.

Zhang, Z., Wang, X., Chen, Q., Shu, L., Wang, J., et al., 2002. Clinical efficacy and safety of huperzine Alpha in treatment of mild to moderate Alzheimer disease, a placebo-controlled, double-blind, randomized trial. Zhonghua Yi Xue Za Zhi 82, 941–944.

Zhao, B., 2009. Natural antioxidants protect neurons in Alzheimer's disease and Parkinson's disease. Neurochem. Res. 34, 630–638.

Zhao, H.H., Di, J., Liu, W.S., Liu, H.L., Lai, H., Lü, Y.L., 2013. Involvement of GSK3 and PP2A in ginsenoside Rb1's attenuation of aluminum-induced tau hyperphosphorylation. Behav. Brain Res. 241, 228–234.

Zhao, Q., Yokozawa, T., Yamabe, N., Tsuneyama, K., Li, X., Matsumoto, K., 2010. Kangenkaryu improves memory deficit caused by aging through normalization of neuroplasticity-related signaling system and VEGF system in the brain. J. Ethnopharmacol. 131, 377–385.

Zhou, X., Cui, G., Tseng, H.H., Lee, S.M., Leung, G.P., Chan, S.W., et al., 2016. Vascular contributions to cognitive impairment and treatments with traditional Chinese medicine. Evid. Based Complement. Altern. Med. 2016, 9627258.

Zhu, G., Wang, X., Chen, Y., Yang, S., Cheng, H., et al., 2010. Puerarin protects dopaminergic neurons against 6-hydroxydopamine neurotoxicity via inhibiting apoptosis and upregulating glial cell line-derived neurotrophic factor in a rat model of Parkinson's disease. Planta Med. 6, 1820–1826.

Zuo, W., Yan, F., Zhang, B., Li, J., Mei, D., 2017. Advances in the studies of *Ginkgo biloba* leaves extract on aging-related diseases. Aging Dis. 8, 812–826.

Further Reading

Hofferberth, B., 1989. The effect of *Ginkgo biloba* extract on neurophysiological and psychometric measurement results in patients with psychotic organic brain syndrome. A double-blind study against placebo. Arzneimittelforschung. 39, 918–922.

Lim, S.L., Rodriguez-Ortiz, C.J., Kitazawa, M., 2015. Infection, systemic inflammation, and Alzheimer's disease. Microbes Infect. 17, 549–556.

Qiang, G., Wenzhai, C., Huan, Z., Yuxia, Z., Dongdong, Y., Sen, Z., et al., 2015. Effect of Sancaijiangtang on plasma nitric oxide and endothelin-1 levels in patients with type 2 diabetes mellitus and vascular dementia: a single-blind randomized controlled trial. J. Tradit. Chin. Med. 35, 375–380.

CHAPTER

9

Potential Treatment Strategies of Dementia With Ayurvedic Medicines

INTRODUCTION

Ayurvedic medicine is a personalized system of traditional medicine native to India and the Indian subcontinent. The holistic concepts of Ayurveda give emphasis to health promotion, disease prevention, early diagnosis, and personalized treatment. The term "Ayurveda" is derived by a combination of two Sanskrit words, ayur (life) and veda (science or knowledge). Three ancient books known as the Great Trilogy were written in Sanskrit more than 2000 years ago and are considered the main texts on Ayurvedic medicine—Caraka Samhita, Sushruta Samhita, and Astanga Hridaya. Ayurvedic medicine is based on a holistic view of treatment which promotes and support equilibrium in the different elements of human life, the body, the mind, the intellect, and the soul (Mishra et al., 2013). This is important because whenever equilibrium is lost, it results in vulnerability to disease processes. Ayurveda is also called the "science of longevity" because it offers a complete system to live a long healthy life. It offers programs to rejuvenate the body through diet and nutrition. Ayurveda offers treatment methods to cure many common diseases such as food allergies, which have few modern treatments (Mishra et al., 2013). However, it should be remembered that Ayurvedic nutrition is not a "magic bullet" system but requires the full participation of the patient to succeed. It is an interactive system that is user-friendly and educational. It teaches the patient to become responsible and self-empowered. In Ayurvedic medicinal system, each cell is considered to be inherently an essential expression of pure intelligence, hence it is called self-healing science (Lad, 1987). In addition, to the self-healing concept, the use of herbal treatment is equally important in this

Molecular Mechanisms of Dementia
DOI: https://doi.org/10.1016/B978-0-12-816347-4.00009-X

Indian traditional system of medicine. Thus, in Ayurvedic medicinal system, attempts are made to correct the equilibrium between tissues and cells so that our body can become a formidable fortress against diseases (Lad, 1987). Ayurvedic medicine dates back to the period of the Indus Valley civilization (about 3000 BCE) and has been passed on through generations of oral tradition, like the other four sacred texts (Rigveda, Yajurveda, Samaveda, and Atharvanaveda) which were composed between the 12th and 7th century BCE. Ayurvedic medicinal system has descriptions of over 5000 signs and symptoms of various diseases and 700 herbs and 6000 formulations to treat diseases. Historically, Ayurvedic medicinal system has been a holistic, inclusive, progressive, and continuously evolving knowledge system with universal attributes. The integrative approach to health care and cure has been the basic matrix of Ayurvedic medicine practice (Patwardhan and Vaidya, 2009). Treatment with Ayurvedic medicines depends on long-term usage of Ayurvedic medicinal plants. The need for scientific evaluation of Ayurvedic medicines has been recognized for a long time (Patwardhan, 2010). Attempts should be made not only to determine the molecular mechanisms of Ayurvedic drugs based on signal transduction processes, but also to collect information on efficacy, and safety of Ayurvedic medicine. Another challenge to Ayurvedic drugs is standardization and quality assurance and more research is needed on the development of new drugs and formulations. In recent years, pharmacological and toxicological studies have begun providing verification of pharmacological and therapeutic activities of Ayurvedic medicines based on the signal transduction process in the brain (Farooqui, 2016; Farooqui et al., 2018). In recent years, there has been a huge resurgence of the use of herbal products not only due to the side effects of modern drugs, but also due to the failure of modern drugs against chronic diseases, and microbial resistance. According to World Health Organization, the majority of the Indian subcontinent population uses Ayurvedic medicine exclusively or in combination with conventional Western medicine. In addition, Ayurvedic medicinal system has gained more popularity in the Europe and United States not only due to its easy availability, low cost, congeniality, and better accessibility, but also due to its higher safety than allopathic medicinal system (Meulenbeld, 2002; https://nccih.nih.gov/health/ayurveda/introduction.htm.). Despite its Asian origin, a recent study has indicated that nearly 59% of Asian Indians living in the United States used Ayurvedic medicines and almost all were aware of Ayurveda (Satow et al., 2008).

In the Ayurvedic tradition, the human brain is considered the sixth sense that coordinates the other five senses. All bodily functions are controlled through the brain, spinal cord, and nerves, which carry and deliver messages to different parts and tissues of the body. The brain not only processes cognition function and retains memories, but it also coordinates and controls other physiological functions such as body movements, hunger, thirst, and excretion. It is the site of mind and buddhi, the latter translated as rationality, discretion, and creativity (Vaidya and Raut, 2006). Although a direct reference to dementia in the ancient Ayurvedic text has not been made, the concept of forgetfulness, memory loss, and brain cell loss (vismriti) have been described (Mishra et al., 2013). Ayurvedic literature mentions and explains the use of several herbs and their qualities and energetics for nervous system disorders, including memory loss typically seen in older adults, but only recently have there been mechanistic studies on the role of these herbs in nervous system disorders and dementias, including dementia associated with AD (Manyam, 1999; Farooqui et al., 2018).

Studies on recent pharmacological treatment of dementia using drugs (haloperidol, risperidone, apripiprazole, olanzapine, cholinesterase inhibitors (tacrine, donepezil, rivastigmine, metrifonate, and galantamine), memantine, and benzodiazepines) have failed. These drugs produce numerous adverse effects such as dizziness, anorexia, vomiting, and diarrhea (Sink et al., 2005; Lonergan et al., 2002; Ballard and Howard, 2006; Alva and Cummings, 2008; Tifratene et al., 2017; McShane et al., 2006; Hogan et al., 2008; Peisah et al., 2011). AChE inhibitors and memantine have only modest effects in delaying the progression of AD (Hansen et al., 2007; Raina et al., 2008). Furthermore, the tolerability of these drugs is compromised by their side effects, for instance dizziness, anorexia, vomiting, and diarrhea (Alva and Cummings, 2008). Over the past two decades, considerable efforts have been made to develop safe and effective pharmacological treatment of AD (Hansen et al., 2007; Raina et al., 2008).

INDIAN MEDICINAL PLANTS FOR THE TREATMENT OF DEMENTIA

As stated above, Ayurvedic medicine is a traditional healthcare system of India and the Indian subcontinent since ancient times. Ayurvedic medicine are used to modulate the neuroendocrinoimmune systems and have been found to be a rich source of antioxidants and antiinflammatory compounds (Brahma and Debnath, 2003; Rege et al., 1999). It is suggested that Ayurvedic medicinal herbs not only promote intellect and enhance memory, but also rejuvenate cognitive function (Schlebusch et al., 2000; Govindarajan et al., 2005). Several Ayurvedic medicines have been exploited for the treatment and management of various acute and chronic neurological and visceral diseases in human beings. Examples of popular Ayurvedic medications (formulation) are Brāhmī Ghṛita, Divya Medha Kwath, and Brento Forte. These formulations produce specific influences on brain functions such as increase in blood flow, maintenance of memory, and prevention of misfolded protein accumulation (Manyam, 1999).

WITHANIA SOMNIFERA AND DEMENTIA

Withania somnifera (Ashwagandha) belongs to the family Solanaceae. It is also called Indian Ginseng, a common Indian herb used in Ayurvedic medicine as an adaptogen or antistress agent. Chemically, powder from roots of *W. somnifera* contains a large variety of compounds including 12 alkaloids, 40 withanolides, and several sitoindosides and flavonoids isolated from different parts of the plant (Mishra et al., 2000; Dar et al., 2015; Mirjalili et al., 2009). Withanolide A and Withaferin A (WL-A) are two constituents to show similar pharmacokinetic profiles, except that the relative oral bioavailability for WL-A is 1.44 times greater than Withanolide A (Fig. 9.1) (Patil et al., 2013). These components produce antistress, antioxidant, and immunomodulatory effects in acute models of experimental stress (Bhattacharya et al., 1987; Agarwal et al., 1999; Oh et al., 2008). According to Ayurvedic medicine, *W. somnifera* constituents provide a number of healthful effects such as a youthful state of physical and mental health and expansion of

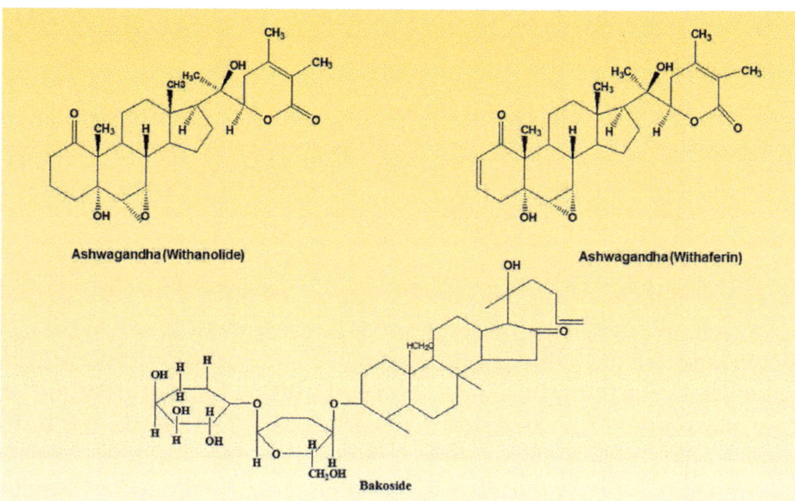

FIGURE 9.1 Chemical structures of active constituents of Ahswagandha and *Bacopa monnieri*.

happiness. It is given to small children as a tonic. Consumption of *W. somnifera* by the middle-aged and elderly men increases their longevity (Abbas and Singh, 2006; Ven Murthy et al., 2011). Recent studies have indicated that *W. somnifera* root powder improves the body's defense against chronic diseases not only by improving the cell-mediated immunity, but also by producing potent antioxidant and anti-inflammatory effects, which protect against cellular damage caused by free radicals and inflammatory mediators (Fig. 9.2) (Abbas and Singh, 2006; Ven Murthy et al., 2010). The treatment of human neuroblastoma SK-N-SH cells with methanolic extracts of *W. somnifera* roots retards the generation of Aβ and promotes production of neurite outgrowth. In addition, it also facilitates synaptic reconstruction (Tohda et al., 2000; Kuboyama et al., 2005). The treatment of cultured rat cortical neurons with Aβ (25–35) (10 μM) produces axonal and dendritic atrophy. Subsequent treatment with WL-A (1 μM) not only results in significant regeneration of both axons and dendrites, but also promotes the reconstruction of pre- and postsynapses in the neurons (Fig. 9.2) (Kuboyama et al., 2005; Kuboyama et al., 2014). The molecular mechanism of action of WL-A is not fully understood. However, it is proposed that WL-A interacts with Keap-Nrf2 complex and promotes the release of free Nrf2, leading to its translocation to the nucleus, where after its binding with antioxidant response-element it upregulates the expression of neuroprotective proteins, such as hemoxygenase-1 (HO-1), superoxide dismutase (SOD), NADPH quinine oxidoreductase (NQO-1), and; γ-glutamate cysteine ligase (γ-GCL) (Fig. 9.2) (Narayan et al., 2015; Sun et al., 2016). In addition, WL-A also attenuates expression of Semaphorin 3A (Sema3A) promoting neuronal regeneration. Based on the above information, it is proposed that WL-A can be used as a therapeutic agent for the treatment of dementia and neurodegenerative diseases (Alzheimer's disease (AD), Parkinson's disease (PD), and amyotrophic lateral sclerosis), as it is able to reconstruct neuronal networks (Singh et al., 2011; Sandhir and Sood, 2017). The beneficial effects of *W. somnifera* root constituents in dementia and neurodegenerative diseases may also be due to its GABA mimetic, dendrite promoting, antioxidant, antiinflammatory, antiapoptotic, and anxiolytic activities, which improve energy levels and mitochondrial dysfunction along with increases in levels of reduced glutathione (Singh et al., 2011; Sun et al., 2016; Sandhir and Sood, 2017). The proposed hypothetical mechanism for the action of *W. somnifera* root power is presented in Fig. 9.2. At the molecular level, *W. somnifera* root powder may produce beneficial effects in AD type of dementia by inhibiting the activation of NF-κB, blocking Aβ production, retarding apoptotic cell death, restoring synaptic function, and enhancing antioxidant effects through the migration of Nrf2 to the nucleus, where it increases the expression of antioxidant

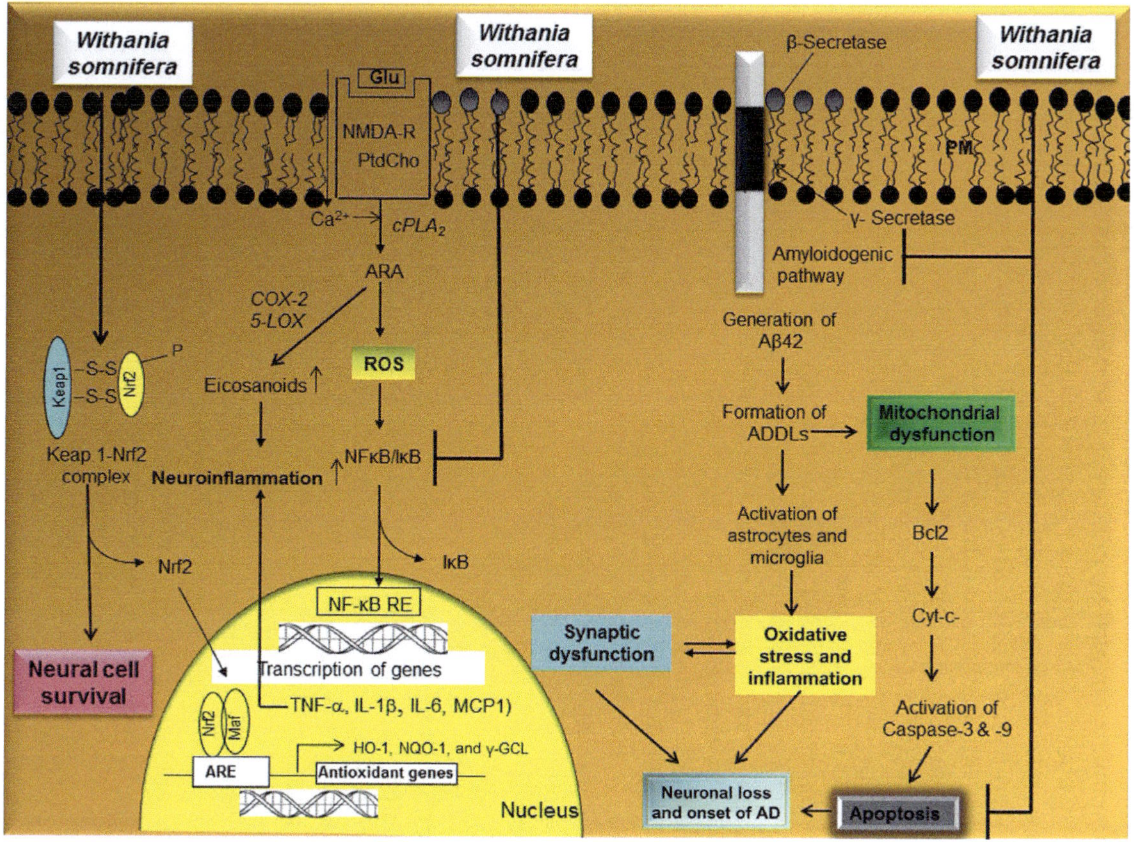

FIGURE 9.2 Hypothetical diagram showing target sites for the action of Ashwagandha (*Withania somnifera*).
Aβ, β-amyloid; *AD*, Alzheimer's disease; *ADDL*, Aβ-derived diffusible ligand; *APP*, amyloid precursor protein; *ARA*, arachidonic acid; *ARE*, antioxidant response element; *Bcl-2*, B-cell lymphoma 2; *COX-2*, cyclooxygenase-2; *cPLA$_2$*, cytosolic phospholipase A_2; *cyto-c*, cytochrome; *γ-GCL*, γ-glutamate cystein ligase; *Glu*, glutamate; *HO-1*, heme oxygenase; *I-κB*, inhibitory subunit of NF-κB; *IL-1β*, interleukin-1β; *IL-6*, interleukin-6; *Keap1*, kelch-like ECH-associated protein 1; *5-LOX*, 5-lipoxygenase; *lyso-PtdCho*, lyso-phosphatidylcholine; *Maf*, small leucine zipper proteins; *MCP-1*, monocyte chemoattractant protein-1; *NF-κB*, nuclear factor-κB; *NF-κB-RE*, nuclear factor-κB-response element; *NMDA-R*, NMDA receptor; *NQO-1*, NADPH quinine oxidoreductase; *Nrf2*, nuclear factor E2-related

enzymes (Sandhir and Sood, 2017). Administration of Withanone (WS-2), a compound isolated from root extract of *W. somnifera* to wistar rats for a duration of 21 days results in significant improvement in the cognitive skill through inhibition of amyloid β-42 formation and attenuation of the elevated levels of proinflammatory cytokines, like TNF-α, IL-1β, IL-6, MCP-1, nitric oxide, and lipid peroxidation. Administration of WS-2 also significantly reverses the decline in acetylcholine and Glutathione (GSH) activity. WS-2 shows promising results in the treatment of AD because of its cognitive benefits and, more importantly, the mechanisms of action with respect to the fundamental pathophysiology of the disease, which is not limited to the inhibition of AChE, but also includes the modification of Aβ processing, protection against oxidative stress, and antiinflammatory effects (Pandey et al., 2018). *W. somnifera* is a safe herb, which has no toxicity even when large amounts are consumed for 6 months. However, a few people experience diarrhea or nausea. *W. somnifera* can cross the blood–brain barrier (BBB) and lower inflammation in the brain (Vareed et al., 2014). The half-life of *W. somnifera* in circulation and the brain are not known. Large multicenter clinical trials of *W. somnifera* in demented patients have not been performed. *W. somnifera* should not be taken with barbiturate-type sedatives because the herb can increase the effectiveness of these drugs.

CURCUMIN AND DEMENTIA

Curcumin ($C_{21}H_{20}O_6$) or diferuloylmethane (bis-α,β-unsaturated β-diketone) (Fig. 9.3) is a hydrophobic polyphenolic compound (mol mass of 368.38) present in the Indian spice turmeric (curry powder). It is derived from the rhizomes of *Curcuma longa*, which belongs to the family Zingiberaceae. Curcumin root extract consists of three major curcuminoid analogs: diferuloylmethane (curcumin I), desmethoxycurcumin (curcumin II), and bisdesmethoxycurcumin (curcumin III); and two minor curcuminoids: curcumin IV and curcumin X (Sharma et al., 2005). In vivo and in vitro studies have indicated that curcumin produces antioxidant, antiinflammatory, antiallergy, antitumor, and antidementia effects (Khar et al., 1999; Ram et al., 2003; Lim et al., 2001; Goel et al., 2008). Curcumin has been declared safe not only by the FDA (http://www.accessdata.fda.gov/scripts/fcn/gras_notices/GRN000460.pdf) in the United States and the Natural Health Products Directorate of Canada, but also by the Joint Expert Committee of the Food and Agriculture Organization/World Health Organization (FAO/WHO) (National Cancer Institute, 1996). Curcumin reduces oxidative damage and improves cognitive functions related to the aging process. Both

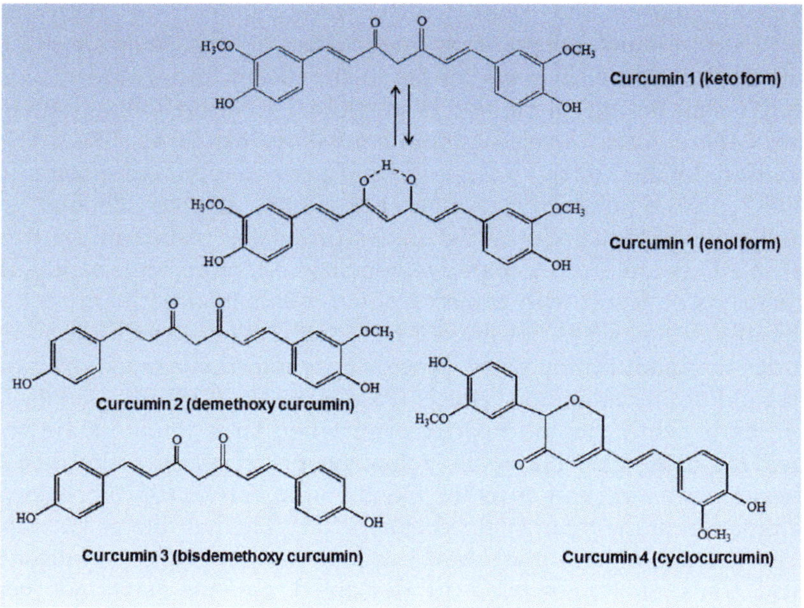

FIGURE 9.3 Chemical structures of various curcumins.

in vitro and in vivo studies have indicated that curcumin binds with Aβ and inhibits its aggregation (Yang et al., 2005; Hong et al., 2009), as well as fibril and oligomer formation (Yang et al., 2005). In vivo studies have shown that dietary curcumin not only crosses the BBB and significantly decreases Aβ deposition and plaque burden in AD transgenic mice (Yang et al., 2005), but markedly inhibits Tau phosphorylation (Fig. 9.4) (Ma et al., 2009). The absorption rate and bioavailability of curcumin can be increased by consuming it with black pepper (*Piper nigrum*). Several studies have indicated that piperine, the active ingredient of black paper increases the bioavailability and bioefficacy of curcumin by inhibiting glucuronidation (Shoba et al., 1998). In addition, administration of piperine and curcumin also protect against the chronic unpredictable stress-induced cognitive impairment and associated oxidative damage in mice (Rinwa and Kumar, 2012; Sehgal et al., 2012). The antioxidant activity of curcumin may be due to the presence of phenolic and the methoxy group on the phenyl ring and the 1,3-diketone systems. The degradation of curcumin in vivo produces smaller phenols like ferulic acid (trans-4-hydroxy-3-methoxycinnamic acid) (Ghosh et al., 2015), which is capable of producing neuroprotective effects. Recently, investigators have synthesized curcumin analogs (Priyadarsini, 2013). It is well known that an increase in levels of redox

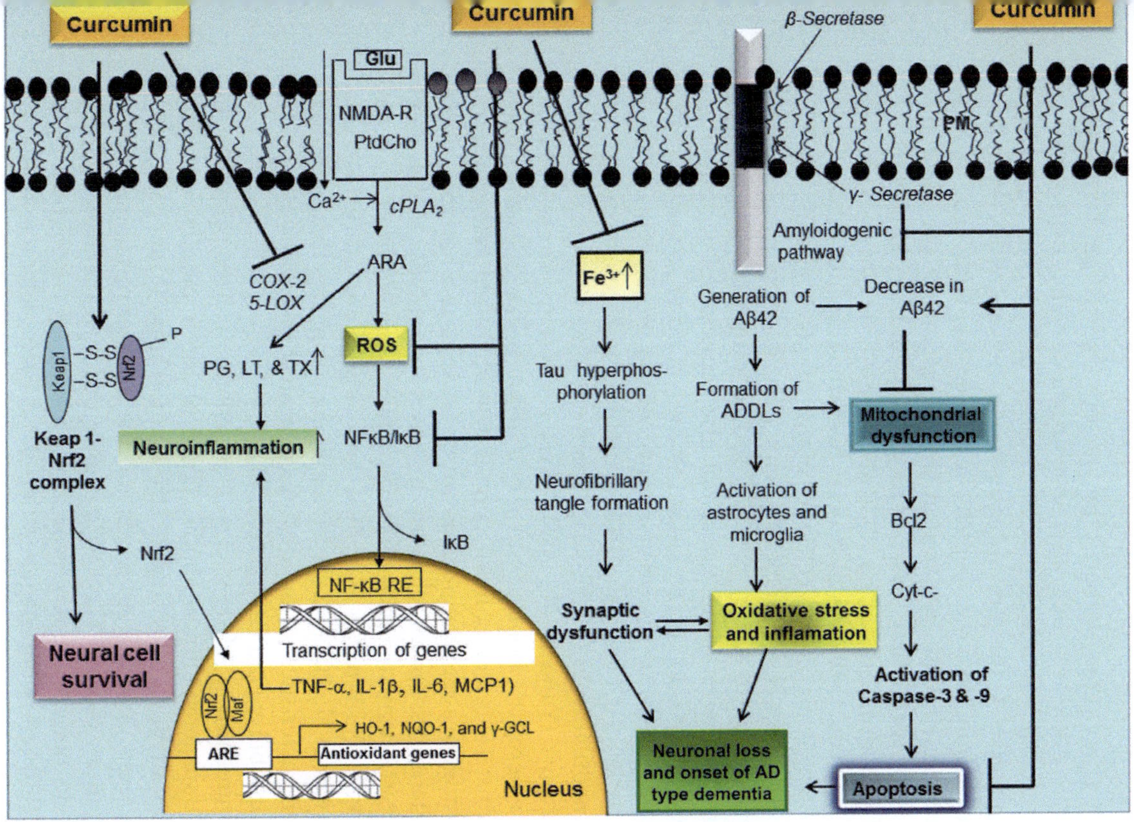

FIGURE 9.4 Hypothetical diagram showing target sites for the action of curcumin.
Aβ, β-amyloid; *AD*, Alzheimer's disease; *ADDL*, Aβ-derived diffusible ligand; *APP*, amyloid precursor protein; *ARA*, arachidonic acid; *ARE*, antioxidant response element; *Bcl-2*, B-cell lymphoma 2; *COX-2*, cyclooxygenase-2; *cPLA*$_2$, cytosolic phospholipase A$_2$; *cyto-c*, cytochrome; *γ-GCL*, γ-glutamate cystein ligase; *Glu*, glutamate; *HO-1*, heme oxygenase; *I-κB*, inhibitory subunit of NF-κB; *IL-1β*, interleukin-1β; *IL-6*, interleukin-6; *Keap1*, kelch-like ECH-associated protein 1; *5-LOX*, 5-lipoxygenase; *lyso-PtdCho*, lyso-phosphatidylcholine; *Maf*, small leucine zipper proteins; *MCP-1*, monocyte chemoattractant protein-1; *NF-κB*, nuclear factor-κB; *NF-κB-RE*, nuclear factor-κB-response element; *NMDA-R*, NMDA receptor; *NQO-1*, NADPH quinine oxidoreductase; *Nrf2*, nuclear factor E2-related factor 2; *PtdCho*, phosphatidylcholine; *ROS*, reactive oxygen species; *TNF-α*, tumor necrosis factor-α.

active Fe^{3+} and Cu^{2+} promote the generation of reactive oxygen species (ROS) in AD. This leads to DNA damage via the production of hydroxyl and superoxide radicals (Fenton reaction). Curcumin binds Cu^{2+} and Fe^{3+} and forms tight and inactive complexes leading to the protection of neural cell DNA against ROS and singlet oxygen (Kim et al., 2005). Curcumin not only inhibits cyclooxygenase-2 (COX-2), 5-lipoxygenase (5-LOX), phospholipases, transcription factors (NF-κB), and release of inflammatory cytokines, but also downregulates the expression of inducible nitric oxide synthase (iNOS) resulting in a marked decrease in neuroinflammation. Curcumin upregulates neprilysin (NEP), an important Aβ-degrading enzyme, which is decreased with age and is inversely correlated with Aβ accumulation supporting the view that there may be a correlation between its activity and the late-onset AD (Fig. 9.4) (Farooqui, 2016). The ability of these polyhydroxycurcuminoids to upregulate NEP has also been confirmed by mRNA and protein expression levels in the cell and mouse models. Finally, treating feeding monohydroxylated demethoxycurcumin or dihydroxylated bisdemethoxycurcumin to APP_{swe}/PS_1dE_9 double transgenic mice not only upregulates NEP levels in the brain, but also reduces Aβ accumulation in the hippocampus and cortex. Future studies on polyhydroxycurcuminoids may offer hope in the prevention of AD in more animal models and in AD patients (Chen et al., 2016). Curcumin induces its antioxidant effects by modulating the Nrf2-keap1 pathway and reducing genomic instability events (Yang et al., 2005). Nrf2 is present primarily in the cytoplasm, where it is bound with the BTB-Kelch-like ECH-associated protein 1 (Keap1). Interactions of curcumin with Keap1 releases Nrf2, which then migrates into the nucleus where it binds as a heterodimer to the antioxidant responsive element in DNA to initiate target gene expression. Nrf2-regulated genes include antioxidants enzymes, molecular chaperones, DNA repair enzymes, and antiinflammatory response proteins (Bryan et al., 2013). Collective evidence suggests that curcumin fulfills the characteristics for an ideal neuroprotective agent with its low toxicity, affordability, and easy accessibility.

Oral administration of low doses of curcumin (160 ppm) to an Alzheimer transgenic mouse model (Tg2576) for 6 months not only reduces inflammation and oxidative stress in the brain (Lim et al., 2001), but also decreases levels of Aβ levels and the number of plaques in different brain areas (Lim et al., 2001). Higher doses of dietary curcumin (5000 ppm) do not reduce Aβ levels. Infusion of Aβ contributes to its deposition leading to neurodegeneration in rat brain, which is similar to the deposition of Aβ in patients with AD. Treatment with dietary curcumin (2000 ppm) results in the reduction of oxidative damage and the increase in microglial reaction near Aβ deposit sites (Frautschy et al., 2001). Administration of low doses of curcumin (160 ppm) does not produce spatial memory impairment in rat treated with Aβ infusion. Subsequent studies in Tg2576 mice show that curcumin reduces Aβ oligomer and fibril formation (Yang et al., 2005). Another study performed in the same mouse model reported that oral administration of low doses of curcumin (500 ppm) for 4 months produced a reduction in plaque burden and Aβ levels (Begum et al., 2008). Using another Alzheimer mouse model (APP-swe/PS1dE9), it is shown that administration of curcumin (7.7 mg/kg/day) for 7 days enhances clearance of Aβ deposit in mouse brain (Garcia-Alloza et al., 2006). after the intravenous administration of curcumin (7.7 mg/kg/day) for 7 days. Collectively, these studies suggest that in AD curcumin produces beneficial effects not only through antiamyloid and metal iron chelating properties, but also due to its antioxidant and antiinflammatory activities.

Very little information is available on the pharmacokinetics of curcumin in humans. The first phase I and II clinical trials of curcumin have been performed in patients with advanced colorectal cancer for up to 4 months at several doses (500, 1000, 2000, 4000, 8000, and 12,000 mg/day) without any toxicity (Cheng et al., 2001; Sharma et al., 2001; Anand et al., 2007). The serum concentration of curcumin usually peaks 1–2 hours after oral intake of curcumin and gradually declines within 12 hours. The average peak serum concentrations after taking 4000, 6000, and 8000 mg of curcumin were 0.51 ± 0.11, 0.63 ± 0.06, and 1.77 ± 1.87 μM, respectively (Cheng et al., 2001; Sharma et al., 2001; Anand et al., 2007). So far, an upper level of toxicity has not been established for curcumin. Studies have shown that a dosage as high as 12 g/day is safe and tolerable to humans with a few reporting mild side-effects (Cheng et al., 2001; Sharma et al., 2001). Epidemiological studies have revealed that in India, where dietary curcumin is consumed daily in the form of curry, the morbidity rate attributed to AD for Indian elders (70–79 years old) is 4.4 times lower compared to the same age group of Americans (Jorm and Jolley, 1998; Ganguli et al., 2000). The consumption of curry containing food by healthy elderly individuals results in a better cognitive performance (Ng et al., 2006) than seniors who did not consume curry (Ng et al., 2006).

Little information is available on clinical trials of AD patients with curcumin. First randomized, double blind clinical trial was performed in 34 patients with AD (Baum et al., 2008) using 1 or 4 g/day of oral curcumin. In this trial patients were also treated with 120 mg/day ginkgo leaf extract. No significant differences have been observed on levels of Aβ40 and F2-isoprostane in serum, as well as MMSE scores after 6 months of curcumin supplementation (Baum and Ng, 2004). In this study, biomarkers were only measured in blood, which may not be necessarily correlated with the levels in the brain or CSF (Baum et al., 2008). The second trial a was randomized, double-blind, placebo-controlled study. In this trial 36 mild-to-moderate AD were treated with curcumin (2 or 4 g/day) for 24 weeks. Treatment with curcumin did not produce any neuroprotective effects in AD patients (Ringman et al., 2012). One small study with three AD patients was performed in Japan and it was reported that there was significant improvement in behavioral symptoms of AD patients after curcumin treatment for 12 weeks (Hishikawa et al., 2012). In one case, the mini-mental state examination (MMSE) score was up by 5 points, from 12/30 to 17/30. In the other two cases, no significant change was observed in the MMSE; however, patient came to recognize their family within 1 year of treatment. In all cases that have been consuming curcumin for more than 1 year, reexacerbation of behavioral and psychological symptoms of dementia (BPSD) have not been observed. Though it is a small sample size, three AD cases treated with curcumin suggested a significant improvement of the cognitive and behavioral symptoms, suggesting a probable benefit in the use of curcumin in individuals with AD for BPSD. Limitations to these trials include small number of AD patients and shorter trial duration. However, patients were able to tolerate high doses of curcumin without any harmful effects. At present, four clinical trials are underway to test the efficacy of curcumin for the treatment of AD. Two trials have been performed in China and the United States, but no significant differences have been observed in cognitive function between placebo and curcumin groups. No results have been reported on the two other remaining clinical trials. It is important to determine the effective dose and optimal bioavailability for the successful curcumin trials in normal human subjects and AD patients. It is tempting to speculate that multicenter randomized double-blind clinical trials should be performed in AD patients using an effective dose.

BRAHMI (*BACOPA MONNIERI*) AND DEMENTIA

Bacopa monnieri is a creeping perennial with small oblong leaves and purple flowers, found in warm wetlands, and native to Australia and India. Commonly found as a weed in rice fields, *B. monnieri* grows throughout East Asia and the United States. The entire plant is used medicinally. It belongs to the family Scrophulariaceae (Satyavati et al., 1976). *B. monnieri* leaves extracts exert antiamnesic (Prabhakar et al., 2008), antioxidant (Harsahay et al., 2012), antistress (Gohil and Patel, 2010), anxiolytic (Khan et al., 2008), memory enhancing (Sudharani et al., 2011), and antiulcerogenic effects (Gohil and Patel, 2010). In addition, *B. monnieri* also exerts antiinflammatory (Viji et al., 2010a) and antiarthritis activities (Viji et al., 2010b). Pretreatment with *B. monnieri* extracts protects the human neuroblastoma cell line SK-N-SH against H_2O_2 and acrolein by modulating the activity of several redox regulated proteins, that is, NF-κB, Sirt1, ERK1/2, and p66Shc, so as to favor cell survival in response to oxidative stress (Hosamani, 2010). Because of the abovementioned activities, *B. monnieri* extracts have been used for the treatment of insomnia, anxiety, epilepsy, and also for enhancing the memory and intellect (Roodenrys et al., 2002).

Detailed investigations have indicated that *B. monnieri* extracts contain 12 Bacosides (Rauf et al., 2013). In addition, *B. monnieri* also contains hersaponin, apigenin, D-mannitol, monnierasides I-III, plantainoside B, cucurbitacin, and the alkaloids brahmine, herpestine, and nicotine. Bacoside A is the most studied and potent constituent of *Bacopa*, which is composed of bacoside A3, bacopasaponin C, bacopaside II, and bacopaside X (Srivastava et al., 2012; Deepak and Amit, 2013; Singh et al., 2014). The beneficial effects of *B. monnieri* extract in dementia are due to the presence of triterpenoid saponins (bacosides) present in the plant extract. Bacosides have been shown to enhance nerve impulse transmission (Mathur et al., 2016). The bacosides promote the repair of damaged neurons by upregulating neuronal synthesis and kinase activity. The bacosides also aid in the restoration of synaptic activity, which ultimately leads to nerve impulse transmission (Singh and Dhawan, 2007). The nerve impulse transmission, plays a vital role in promoting healthy cognitive functions like attention span, focus, concentration, learning, and memory. There is evidence which suggests that *Bacopa*, by the virtue of containing active constituents like bacosides, influences the synthesis and availability of the neurotransmitter, serotonin; therefore, *Bacopa* helps to maintain neurotransmitter balance (Charles et al., 2011; Rauf et al., 2012). Collective evidence suggests that the memory enhancing properties of *B. monnieri* are due to the presence of bacoside A, assigned as 3-(a-L-arabinopyranosyl)-*O*-β-D-glucopyranoside-10, 20-dihydroxy-16-keto-dammar-24-ene (Chatterji et al., 1963), and bacoside B (Singh and Dhawan, 1992) (Fig. 9.1).

In addition, bacosides may not only act by inhibiting lipoxygenase activity and scavenging free radicals, but may also protect neural cells in the prefrontal cortex, hippocampus, and striatum against cytotoxicity and DNA damage implicated in AD type of dementia (Fig. 9.5). In PD and Lewy body dementia, bacosides promote their action by increasing glutathione peroxidase activity, chelating iron, and inhibiting aggregation of α-synuclein (Anand et al., 2010; Chaudhari et al., 2017) (Fig. 9.6). As stated above, Brahmi has been reported to increase the level of serotonin and trigger 5-HT3A receptors and CREB in hippocampus of postpartum rats, thereby facilitating its learning abilities (Rajan et al., 2011). Treatment of postnatal rats with *B. monniera* leaf extract elevates the 5-HT level by upregulating tryptophan hydroxylase-2 (TPH2) and

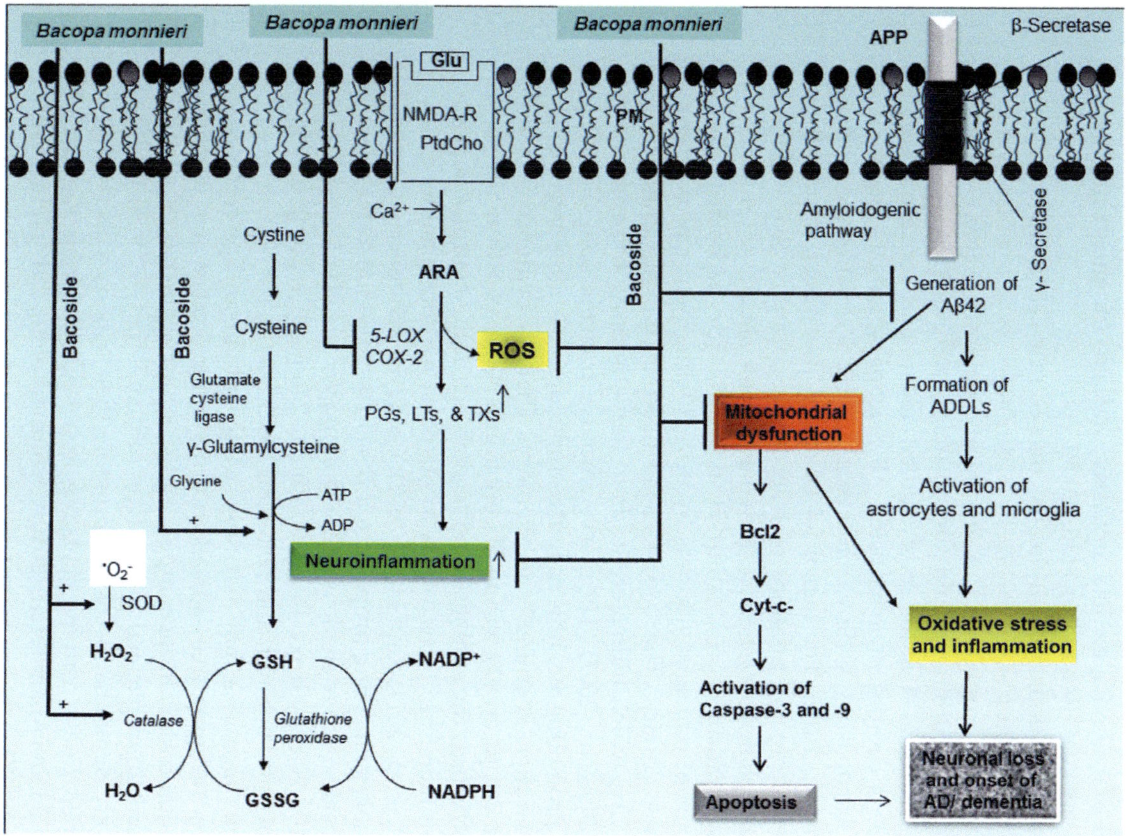

FIGURE 9.5 Hypothetical diagram showing target sites for the action of *Bacopa monnieri*.
Aβ, β-amyloid; *AD*, Alzheimer's disease; *ADDL*, Aβ-derived diffusible ligand; *APP*, amyloid precursor protein; *ARA*, arachidonic acid; *ARE*, antioxidant response element; *Bcl-2*, B-cell lymphoma 2; *COX-2*, cyclooxygenase-2; *cPLA*$_2$, cytosolic phospholipase A_2; *cyto-c*, cytochrome; *γ-GCL*, γ-glutamate cystein ligase; *Glu*, glutamate; *HO-1*, heme oxygenase; *I-κB*, inhibitory subunit of NF-κB; *IL-1β*, interleukin-1β; *IL-6*, interleukin-6; *Keap1*, kelch-like ECH-associated protein 1; *5-LOX*, 5-lipoxygenase; *lyso-PtdCho*, lyso-phosphatidylcholine; *Maf*, small leucine zipper proteins; *MCP-1*, monocyte chemoattractant protein-1; *NF-κB*, nuclear factor-κB; *NF-κB-RE*, nuclear factor-κB-response element; *NMDA-R*, NMDA receptor; *NQO-1*, NADPH quinine oxidoreductase; *Nrf2*, nuclear factor E2-related factor 2; *PtdCho*, phosphatidylcholine; *ROS*, reactive oxygen species; *TNF-α*, tumor necrosis factor-α.

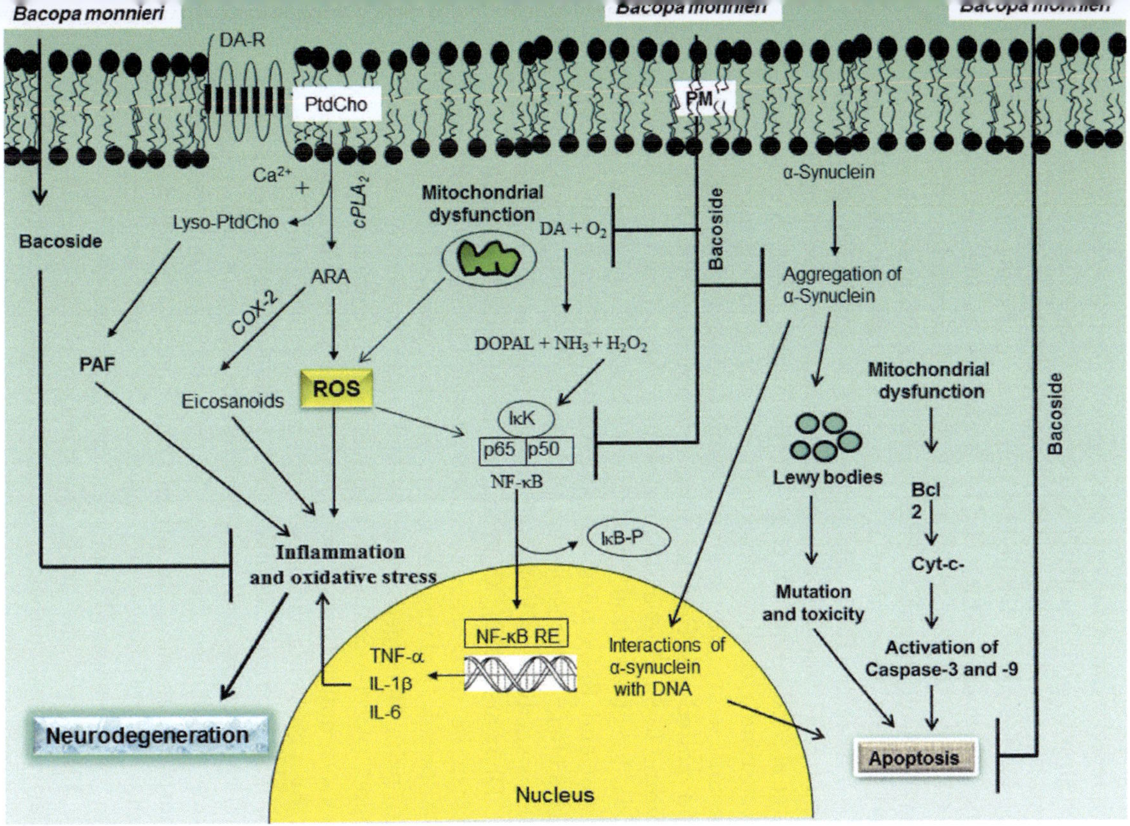

FIGURE 9.6 Hypothetical diagram showing target sites for the action of *Bacopa monnieri*.
Aβ, β-amyloid; *AD*, Alzheimer's disease; *ADDL*, Aβ-derived diffusible ligand; *APP*, amyloid precursor protein; *ARA*, arachidonic acid; *ARE*, antioxidant response element; *Bcl-2*, B-cell lymphoma 2; *COX-2*, cyclooxygenase-2; *cPLA*$_2$, cytosolic phospholipase A_2; *cyto-c*, cytochrome; *DA-R*, dopamine receptor; *HO-1*, heme oxygenase; *I-κB*, inhibitory subunit of NF-κB; *IL-1β*, interleukin-1β; *IL-6*, interleukin-6; *Keap1*, kelch-like ECH-associated protein 1; *lyso-PtdCho*, lyso-phosphatidylcholine; *Maf*, small leucine zipper proteins; *MCP-1*, monocyte chemoattractant protein-1; *NF-κB*, nuclear factor-κB; *NF-κB-RE*, nuclear factor-κB-response element; *NQO-1*, NADPH quinine oxidoreductase; *Nrf2*, nuclear factor E2-related factor 2; *PAF*, platelet activating factor; *PtdCho*, phosphatidylcholine; *ROS*, reactive oxygen species; *TNF-α*, tumor necrosis factor-α; *γ-GCL*, γ-glutamate cystein ligase.

serotonin transporter (SERT) expression (Rajathei et al., 2014; Prisila et al., 2011). Thus, the elevated 5-HT level influences learning and memory by modulating other ACh and Glu, acting through the 5-HT_3 receptor (Rajan et al., 2011). The mechanisms contributing to memory are not known. However, it is revealed that major active compounds bacoside A_3 and A interact with different residues of TPH through hydrogen bonds. Interestingly, Tyr235, Thr265, and Glu317 are the key residues among them, but none of them are either at the tryptophan or BH4 binding regions. However, detailed investigation has shown that Tyr235 is a catalytic sensitive residue, Thr265 is present in the flexible loop region and Glu317 is known to interact with Fe. Interactions with these residues may critically regulate TPH function and thus serotonin synthesis. Based on this observation, it is suggested that the interaction of bacosides (A_3/A) with TPH might upregulate its activity to elevate the biosynthesis of 5-HT, thereby enhancing learning and memory formation (Wang et al., 2002; Rajathei et al., 2014). Finally, bacosides enhance nitric oxide-mediated cerebral vasodilation leading to improvement in the total memory score, and maximum improvement was seen in logical memory and paired associate learning in humans (Chaudhari et al., 2017). According to Singh et al, bacosides may also act by triggering the membrane dephosphorylation (Singh et al., 1990). This leads to elevation in protein and RNA turnover observed in certain brain regions. The nootropics effect of bacosides is mediated through the enhancement of protein kinase activity and production of protein in the hippocampus (Singh and Dhawan, 1982). Collectively, these studies indicate that *B. monnieri* can be used as a memory enhancing, antiinflammatory, analgesic, antipyretic, sedative, and antiepileptic agent, which acts as a nootropic (repairs damaged neurons and improve brain function). A double-blind, placebo-controlled trial has been performed in humans and results have indicated that *B. monnieri* is a safe Ayurvedic drug, which acts by enhancing cognitive performance in aging (Calabrese et al., 2008). A randomized, double-blind, parallel design, single-center study on 300 children has indicated that the treatment of Indian school children with *B. monnieri* beverage for 60 and 121 days does not improve short-term memory or any of the secondary outcomes tested relative to the "control" beverage. However, the spatial working memory "strategy" score showed significant improvement on Day 60, but not on Day 121 due to the active intervention (Mitra-Ganguli et al., 2017). Similarly, in vivo studies in rat model of AD have indicated that *B. monnieri* not only improves the Morris water test results, but also protects neurons from neurodegeneration (Uabundit et al., 2010).

B. monnieri is used in the preparation of formulation called Brāhmī Ghṛita. This formulation contains *B. monnieri*, Vacā (*Acorus calamus*), Kuṣṭha (*Sassurea lappa*), Śaṅkhapuṣpī (*Convolvulus pluricaulis*), and

Purāṇa Ghṛita (old clarified butter). The roots and rhizomes of *A. calamus* are used in Ayurvedic medicine on a regular basis for the treatment of insomnia, melancholia, neurosis, loss of memory, and remittent fevers. *C. pluricaulis* is known to significantly improve learning and memory (Kumar et al., 2006). *S. lappa* has been reported to produce anti-inflammatory activity (Cho et al., 2000). Clarified butter is described in Ayurveda as a memory enhancer, anticonvulsant, and antiinflammatory agent (Yadav et al., 2012, 2014). This formulation can also be used for the treatment of a number of neurological disorders such as anxiety and dementia (Singh and Dhawan, 1982). The molecular mechanism associated with beneficial effects of Brāhmī Ghṛita is not fully understood. However, it is reported that in animal model of dementia, Brāhmī Ghṛita may act not only by reversing the depletion of acetylcholine, reducing the choline acetyltransferase activity, and decreasing muscarinic, cholinergic receptor binding in the frontal cortex and hippocampus (Russo and Borrelli, 2005), but also by alleviating cholinergic degeneration (Sara, 1989), lowering norepinephrine, and increasing 5-hydroxytryptamine levels in the hippocampus, hypothalamus, and cerebral cortex (Saraf et al., 2011). Rhizomes of *A. calamus*, another constituent of Brāhmī Ghṛita, are also used as brain tonic in a weak memory psychoneurosis and epilepsy. Methanolic extracts of the *A. calamus* roots, contain essential oil β-asarone, which inhibits acetylcholinesterase (Oh et al., 2008). Finally, old clarified butter is especially good for healing the mind (Chandre et al., 2004). The half-life of Brāhmī Ghṛita in the circulation and brain are not known. Large multicenter clinical trials of Brāhmī Ghṛita in demented patients have not been performed.

Brahmi has been used in the form of memory enhancer for many years. Brahmi improves motor skills, acquisition, and consolidation of memory in mice (Singh and Dhawan, 1982). To study the efficacy of Brahmi in producing the reversal of amnesia, several behavioral studies have been performed by inducing amnesic agents in animals. Some of the potential amnesic agents, including benzodiazepines, scopolamine, quinoline derivatives, and phenytoin, cause amnesia by interrupting long-term potentiation (LTP). The process of LTP is probably interfered with by the involvement of the gamma-aminobutyric acid-benzodiazepine pathway. Saraf et al. demonstrated that administration of diazepam (1.75 mg/kg) in mice induces amnesia, which can be significantly reversed by oral treatment with Brahmi (120 mg/kg) (Prabhakar et al., 2008). Studies on the effect of diazepam on the downstream signaling molecules related to LTP in amnesiac mice indicate that diazepam upregulates the gene expression of iNOS, mitogen activated protein kinase (MAP kinase), and phosphorylated CREB (pCREB) whereas it reduces the expression levels of cAMP response element binding protein (CREB), cyclic adenosine monophosphate (cAMP), and total nitrite and nitrate. The levels of calmodulin are not affected by diazepam (Prabhakar et al., 2008). On the contrary, administration of Brahmi inhibits the increased expression of iNOS, pCREB, and MAP kinase molecules and restores nitrite level to normal. The levels of cAMP, total CREB, total nitrite, nitrate, and PDE are not affected by Brahmi. These behavioral findings provide tempting conclusions that Brahmi reverses amnesia induced by diazepam and can be used in the treatment of AD and schizophrenia (Prabhakar et al., 2008).

Apart from alleviating memory, Brahmi is used for the treatment of PD and PD-linked dementia, pathological conditions characterized by the loss of neurons, which produces dopamine in substancia nigra, and α-synuclein protein accumulation in the Lewy bodies (Feany and Bender, 2000). Lewy bodies are not specific for PD but can occur in other clinical syndromes, such as AD-, PD-linked dementia, psychosis, and dysautonomia (Dickson et al., 2008). Lewy bodies are partly produced by protein misfolding (Braak and Del Tredici, 2008). In 1-methyl-4-phenyl-1,2,3,6-tetrahydropyridine (MPTP)-induced animal model of PD, ethanolic extract of *B. monnieri* rescues the motor behavior (Rotarod, Grip Strength and Foot Printing test). Furthermore, ethanolic extracts of *B. monnieri* produce neuroprotective effects not only by inhibiting lipid peroxidation and improving catalase, SOD, glutathione reductase, and glutathione peroxidase, but also by enhancing levels of dopamine, DOPAC, and HVA (Fig. 9.6). In addition, there is a significant reduction in tyrosine hydroxylase immunoreactivity in the substantia nigra in the MPTP treated group (Singh et al., 2017). This decrease in tyrosine hydroxylase can be restored by the use of *B. monnieri* extract. *B. monnieri* extracts also induce neuroprotection by inhibiting caspase-3 activity and increasing the expression of Bcl2 (Fig. 9.6). Collectively, these studies suggest that *B. monnieri* treatment provides nigrostriatal dopaminergic neuroprotection against MPTP-induced Parkinsonism by the modulation of oxidative stress and apoptotic machinery, possibly accounting for the behavioral effects (Singh et al., 2017). In a transgenic *Drosophila* fruit fly model of PD, Brahmi improves the climbing ability as well as activity pattern, reduces oxidative stress, and prevents apoptosis in a dose-dependent manner (Siddique et al., 2014). In addition, Brahmi treatment also attenuates behavioral deformities, and reduces the oxidative stress and neuronal cell death in the brains of PD model flies. Similar results were obtained by Jansen et al., who reported that Brahmi alleviates the climbing activity of fruit flies compared to non-treated fruit flies (Jansen et al., 2014). In zebra fish model of PD, pre-treatment with *B. monnieri* -containing nanoparticles significantly reverses the toxic effects of MPTP by increasing the levels of dopamine, its metabolites, GSH, and activities of GPx, catalase, SOD, and complex I, and reducing levels of MDA along with enhanced locomotor activity

(Nellore et al., 2013). In *Caenorhabditis elegans* model of PD, Brahmi exposures reduces α-synuclein accumulation, prevents dopaminergic cell death, and restores the lipid content. Collective evidence suggests that Brahmi can be considered as a possible anti-Parkinsonian medication and further research is needed on this important topic in PD and PD-linked dementia patients (Jadiya et al., 2011). The use of *B. monnieri* should be avoided by pregnant and breastfeeding women so as to avoid possible repercussions. While there are no studies that prove that *B. monnieri* causes side effects, people have observed that excessive intake of *B. monnieri* may lead to stomach upset, diarrhea, and nausea. To avoid the risk of suffering from these adverse effects, it would be a good idea to gauge your tolerance for this herb.

SHANKPUSHPI AND DEMENTIA

Shankhpushpi (*C. pluricaulis*) is a common plant in India. It belongs to the family Convolaceae. The whole plant of Shankhpushpi is used in various formulae as a nervine tonic for the improvement of memory and cognitive function (Bihaqi et al., 2011; Malik et al., 2011). The major bioactive components of *C. pluricaulis* include glycosides, flavonoids, coumarins, anthocyanins, alkaloids, and kampferol derivatives. Sitosterol glycoside, octacosanol tetracosane, triterpenes, xanthones, hydroxy cinnamic acid, and glucose have also been isolated from the plant (Fig. 9.7) (Jatwa et al., 2014). These metabolites contribute to nootropic and memory-enhancing properties along with its other pharmacological activities (Malik et al., 2011; Mukherjee et al., 2008; Sethiya et al., 2009). Shankhpushpi is known to calm the nerves by regulating the body's production of the stress hormones, adrenaline and cortisol (Sethiya et al., 2009). Shankhpushpi is recommended for nervous system disorders such as stress, anxiety, mental fatigue, and insomnia (Kumar, 2006; Singh et al., 2008; Malik et al., 2011). The ethanolic extract of Shankhpushpi also improves learning and memory (Nahata et al., 2008) and induces antioxidant effects in rats (Nahata et al., 2008). Furthermore, ethanolic extracts of the whole Shankhpushpi plant, when administered to cholesterol-fed gerbils, reduce serum cholesterol, low-density lipoprotein cholesterol, triglycerides, and phospholipids significantly (Malik et al., 2011). Ethanolic extracts of Shankhpushpi not only produce nootropic effects in rats, but also attenuate scopolamine-induced memory impairment in Wistar rats (Nahata et al., 2008). In addition, Shankhpushpi extracts also cause a decrease in the elevated acetylcholine esterase activity. These extracts also lower lipid peroxidation and protein carbonyl levels as well as restoration of the altered

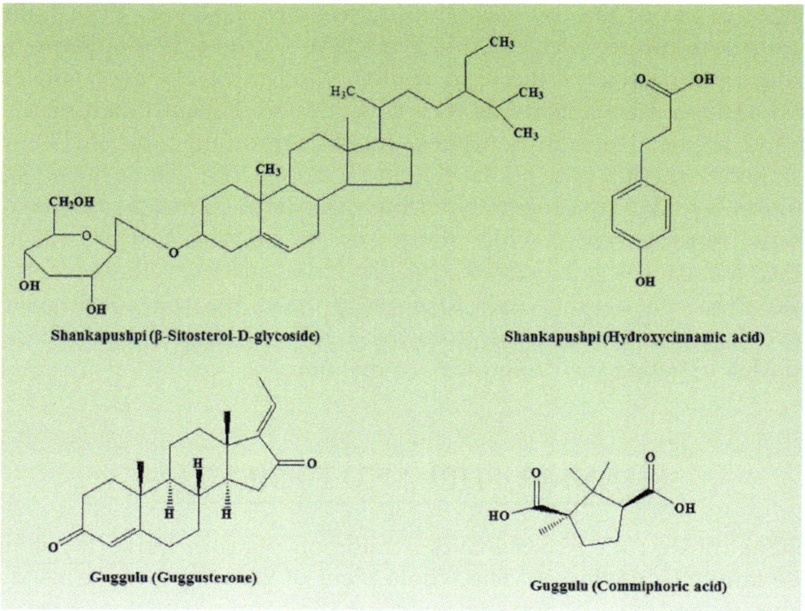

FIGURE 9.7 Chemical structures of active constituents of Shankhapushpi and Guggulu.

levels of antioxidants associated with scopolamine administration (Bihaqi et al., 2011). Aqueous extracts of Shankhpushpi also decrease scopolamine-mediated increase in protein and mRNA levels of AD-related biomarkers, namely AβPP and tau (Bihaqi et al., 2012).

Acetylcholine (ACh) is an important neurotransmitter in the brain. It binds to two major acetylcholine receptors (AChRs). They are called nicotinic acetylcholine receptors (nAChRs) and muscarinic receptors (mAChRs). In the brain, AChE inhibitors not only enhance the activity of cholinergic neurons, but also suppress inflammation. The administration of ethanolic extract of Shankhpushpi increases acetylcholine esterase, a serine hydrolase responsible for the termination of impulse signaling at cholinergic synapses. In addition, treatment with ethanolic extract of Shankhpushpi increases acetylcholine content in the CA1 and CA3 area of hippocampus in a dose-dependent manner (Sharma et al., 2010; Rai et al., 2002). The increase in acetylcholine content is accompanied by a significant increase in dendritic intersections, branching points, and dendritic processes arising from the soma of neurons in the amygdale region in comparison with age-matched saline controls, suggesting that ethanolic extract of Shankhpushpi enhances memory by increasing the functional growth of neurons (Rai et al., 2001, 2002). It is also reported that Shankhpushpi extracts reverse the scopolamine-induced amnesia. The drugs which reverse the scopolamine-induced amnesia produce their effects by regulating the cholinergic system, mainly by modulating acetylcholine levels in the brain. Moreover, it is well known that scopolamine is a muscarinic antagonist, which acts by increasing the phosphorylation of tau proteins leading to the formation of β-amyloid in the brain resulting in memory impairment (Francis et al., 1999). Thus, cholinergic or muscarinic agonists have been found to prevent the formation of β-amyloids through the GSK-3 enzyme pathway and improve memory (Forlenza et al., 2000). Therefore, it has been hypothesized that Shankhpushpi extracts enhance memory by two different pathways: (1) by modulating acetylcholine levels in brain either by increasing its synthesis or by inhibiting the acetylcholinesterase enzyme; and/or (2) by acting as muscarinic agonist which further reduce the β-amyloid formation. Collectively, these studies suggest that constituents of Shankhpushpi produce anxiolytic, tranquillizing, antidepressant, antistress, antiamnesic, antioxidant, hypolipidemic, immunomodulatory, analgesic, antifungal, antibacterial, and antidiabetic effects in animals and humans (Sethiya et al., 2009). The half-life of Shankhpushpi in the circulation and brain are not known. Large multicenter clinical trials of Shankhpushpi in demented patients have not been performed.

GOTU KOLA AND DEMENTIA

Gotu kola (*Centella asiatica*) is a perennial creeping herb with long thick stems and smooth fan leaves. It belongs to the family Umbellifere (Apiceae). The primary active ingredients of gotu kola are saponins (also called triterpenoids), which include asiaticosides, in which a trisaccharide moiety is linked to the aglycone asiatic acid, madecassoside and madasiatic acid (Singh and Rastogi, 1969). These triterpene saponins and their sapogenins contribute to the wound healing and vascular effects by inhibiting the production of collagen at the wound site. Studies on the neuroregenerative capacity of *C. asiatica* in the brain have indicated that active components of *C. asiatica* include asiatic acid, madecassic acid, asiaticoside, and madecassoside. These components promote the elongation of neurites (Soumyanath et al., 2005). This has been confirmed by Rao et al. who observed that fresh *C. asiatica* leaf extract significantly increases the dendritic arborization in hippocampal CA3 neurons in vivo, supporting the view that *C. asiatica* extract acts as a nerve tonic for promoting brain growth and improving memory (Appa Rao et al., 1973; Rao et al., 2006; Chivapat et al., 2011). The molecular mechanisms associated with the elongation of neurites and memory improvement are not understood. However, it is suggested that mechanisms may involve MEK/ERK and PtdIns 3K/Akt signaling pathways (Wanakhachornkrai et al., 2013). Other active components isolated from Gotu kola, such as brahmoside and brahminoside, may be responsible for CNS and utero-relaxant actions, but are yet to be confirmed by clinical studies. In Tg2576 mouse model of AD, water extracts of *C. asiatica* attenuates Aβ-induced cognitive impairments (Soumyanath et al., 2012). These mice express a mutant form of human amyloid precursor protein (APP) leading to age-dependent Aβ accumulation in the hippocampus and cortex, and concomitant learning and memory deficits (Hsiao et al., 1996). It is reported that 2 weeks of treatment with *C. asiatica* in the drinking water normalizes the behavioral deficits normally seen in aged Tg2576 animals (Soumyanath et al., 2012). It is also reported that *C. asiatica* improves cognitive performance in aged wild-type animals as well, and this behavioral improvement is accompanied by increased synaptic gene expression in the brains of treated animals (Gray et al., 2014). At the molecular level, asiaticoside derivatives from gotu kola (asiatic acid and asiaticoside) are capable of reducing hydrogen peroxide-induced cell death, decreasing free radical levels, and inhibiting beta-amyloid-mediated neural cell death in vitro, suggesting a possible role for gotu kola in the treatment and prevention of AD type of dementia and beta-amyloid toxicity (Fig. 9.8) (Dhanasakarann et al., 2009; Gray et al., 2014, 2015) and supporting the view that gotu kola decreases the oxidative stress by retarding Aβ-mediated neurotoxicity (Gray et al., 2017). In the Tg2576 mouse model of AD, water extract of gotu kola attenuates the cognitive impairments without altering plaque burden (Soumyanath et al., 2012; Gray et al., 2014). Although, the molecular mechanisms remain unknown, studies in other models of neurotoxicity show that gotu kola extract not only possesses antioxidant activity, but can also alter mitochondrial function (Fig. 9.8) (Prakash, 2013; Shinomol, 2008). Because mitochondrial dysfunction is a common process that contributes to neurodegeneration in many neurodegenerative diseases (Farooqui, 2010), there are potentially broad implications for the use of water extract of gotu kola. In another study, it is reported that oral treatment with 50 mg/kg/day of crude methanol extract of *C. asiatica* for 14 days significantly increases the antioxidant enzymes, like SOD, catalase and glutathione peroxidase (GSHPx) in lymphoma-bearing mice, and the antioxidants like glutathione (GSH) and ascorbic acid are decreased in these animals (Jayashree et al., 2003). In the Ayurvedic system of medicine, water extract of gotu kola is used not only for the rejuvenation and restoration of neural cells, but also the stimulation of healthy sleep. It has a powerful effect on quality of life in disorders like epilepsy. High doses of *C. asiatica* make consumers drowsy. It is widely used as a blood purifier as well as for treating high

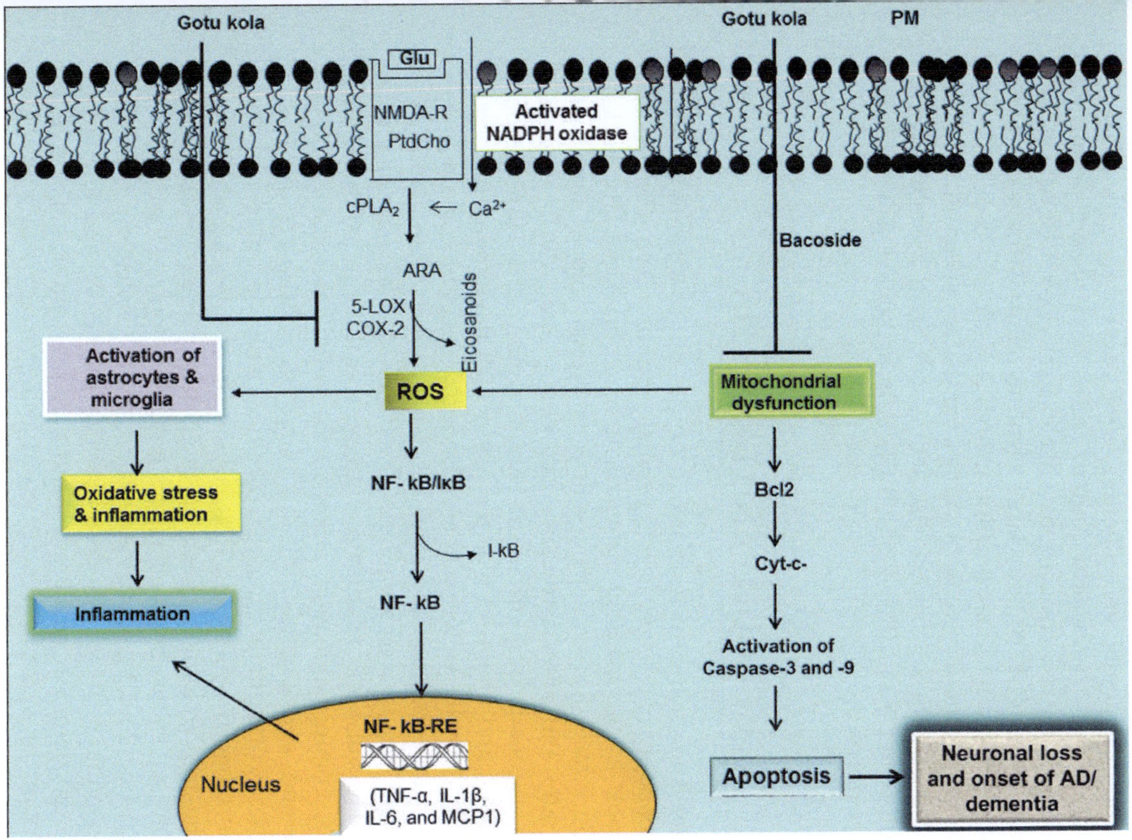

FIGURE 9.8 Hypothetical diagram showing target sites for the action of Gotu kola.
ARA, arachidonic acid; *Bcl-2*, B-cell lymphoma 2; *COX-2*, cyclooxygenase-2; *cPLA*$_2$, cytosolic phospholipase A_2; *cyto-c*, cytochrome; *Glu*, glutamate; *I-κB*, inhibitory subunit of NF-κB; *5-LOX*, 5-lipoxygenase; *lyso-PtdCho*, lyso-phosphatidylcholine; *NF-κB*, nuclear factor-κB; *NF-κB-RE*, nuclear factor-κB-response element; *NMDA-R*, NMDA receptor; *PtdCho*, phosphatidylcholine; *ROS*, reactive oxygen species.

blood pressure, for memory enhancement, and promoting longevity. Tea or powder of goto kola can be very helpful for relieving tension, relaxing the mind, and soothing anxiety. As a nervine adaptogen, constituents of gotu kola are capable of increasing intelligence, longevity, and memory (Shinomol and Bharath, 2011; Cervenka and Jahodar, 2006). Studies on a few healthy human adults have shown promising cognitive-enhancing effects of gotu kola extract (Dev RDOMS et al., 2009; Wattanathorn, 2008), but large rigorous clinical trials of water extract of gotu kola in either healthy or impaired elderly people is still needed to determine if gotu kola extract can be beneficial in humans. The half-life of gotu kola in the circulation and brain are not known. Large multicenter clinical trials of gotu kola in demented patients have not been performed.

GUGGULU AND DEMENTIA

Guggulu is an oleo-gum resin which exudes out as a result of injury from the bark of *Commiphora wightii* (Arnott) Bhandari [syn. *Commiphora mukul* (Hook. Ex Stocks) Engl; *Balsamodendron mukul* (Hook. Ex Stocks); Family, Burseraceae]. It is a pale yellow or brown colored mass with aromatic odor and bitter astringent taste (Sarup et al., 2015). Guggulu preparations contain 30%–60% water-soluble gum, 20%–40% alcohol-soluble resins, and about 8% volatile oils, which have many biological activities. Chemically water-soluble extracts of guggulu contain mucilage, sugars, and proteins, and alcohol-soluble extracts of guggulu have commiphoric acids, commiphorinic acid, and heerabomyrrhols (Fig. 9.7). The volatile constituents of guggulu include terpenes, sesquiterpenoids, cuminic aldehyde, eugenol, the ketone steroids Z-and E-guggulsterone, and guggulsterols I, II, and III (Ulbricht et al., 2005). Among these compounds guggulsterones is known to antagonize two nuclear hormone receptors and decrease cholesterol levels, which may explain the hypolipidemic effects of guggulu's extracts (Nohr et al., 2009). There is a link between cholesterol, APP processing, and AD (Vestergaard et al., 2010; Morley and Bank, 2010). Cholesterol is known to be an essential modulator of the physicochemical state and functional activity in physiological membranes, and thus plays an essential role in the regulation of synaptic function and cell plasticity. In vitro and in vivo modulation of membrane cholesterol levels affect different cholesterol pools within the plasma membrane bilayer that are differentially sensitive to Aβ's disrupting effects. Membrane acyl-chains in the hydrocarbon core are most susceptible to Aβ. In this neural membrane region, cholesterol attenuates the membrane disordering effects of Aβ (Eckert et al., 2003). It is likely that beneficial effects of guggulu on AD type of dementia may be due to cholesterol-lowering effects of guggulu (Vestergaard et al., 2010). Decreased neuronal cholesterol levels, in turn, inhibit the beta-amyloid-forming amyloidogenic pathway, possibly by removing APP from cholesterol and sphingolipid-enriched membrane microdomains (Fig. 9.9). These intriguing relationships raise the hopes that cholesterol-lowering strategies may influence the progression of dementia associated with AD (Eckert et al., 2003; Vestergaard et al., 2010). In addition, in animal models and in humans, administration of guggulipid (Z-guggulsterone) (Fig. 9.7) significantly lowers both serum LDL cholesterol and triglyceride levels, supporting the view that giggulipids may produce beneficial effects in the cardiovascular system (Saxena et al., 2007; Szapary et al., 2003). Guggulsterones have also been reported to regulate gene expression by exhibiting control over other molecular targets including transcription factors such as nuclear factor (NF)-κB, signal transducer, and activator of transcription (STAT) and steroid receptors (Shah et al., 2012). Z-guggulsterone can also attenuate the behavioral abnormalities induced by neuroinflammation in the forced swimming test and tail suspension test (Huang et al., 2016), and also prevents memory impairment in a scopolamine-induced memory impairment model through activation of the CREB-BDNF signal (Chen et al., 2016). Guggulu also contains ferulic acids, phenols, and other nonphenolic aromatic acids that are potent scavengers of superoxide radicals and can potentially be of importance for the treatment of dementia, AD, and other neurodegenerative diseases associated with oxidative stress (Perluigi et al., 2006; Sultana et al., 2005). A recent study indicated that in streptozotocin-induced memory deficit model of dementia, gugulipid produces beneficial effects, which can be attributed to its cholesterol-lowering, antioxidant, and antiacetylcholine esterase activity suggesting that gugulipid as a potential antidementia drug (Saxena et al., 2007). The half-life of guggulu compounds in circulation and brain are not known. Large multicenter clinical trials of guggulu compounds in demented patients have not been performed.

RHODIOLA ROSEA AND DEMENTIA

Rhodiola rosea is also known as golden root and arctic root. It is known to improve cognitive function (Spasov et al., 2000), enhance memory and learning (Lazarova et al., 1986), and protect the brain (Chen et al., 2008). *R. rosea* is found in India and China. It belongs to the Crassulaceae family. Approximately 140 compounds have been isolated

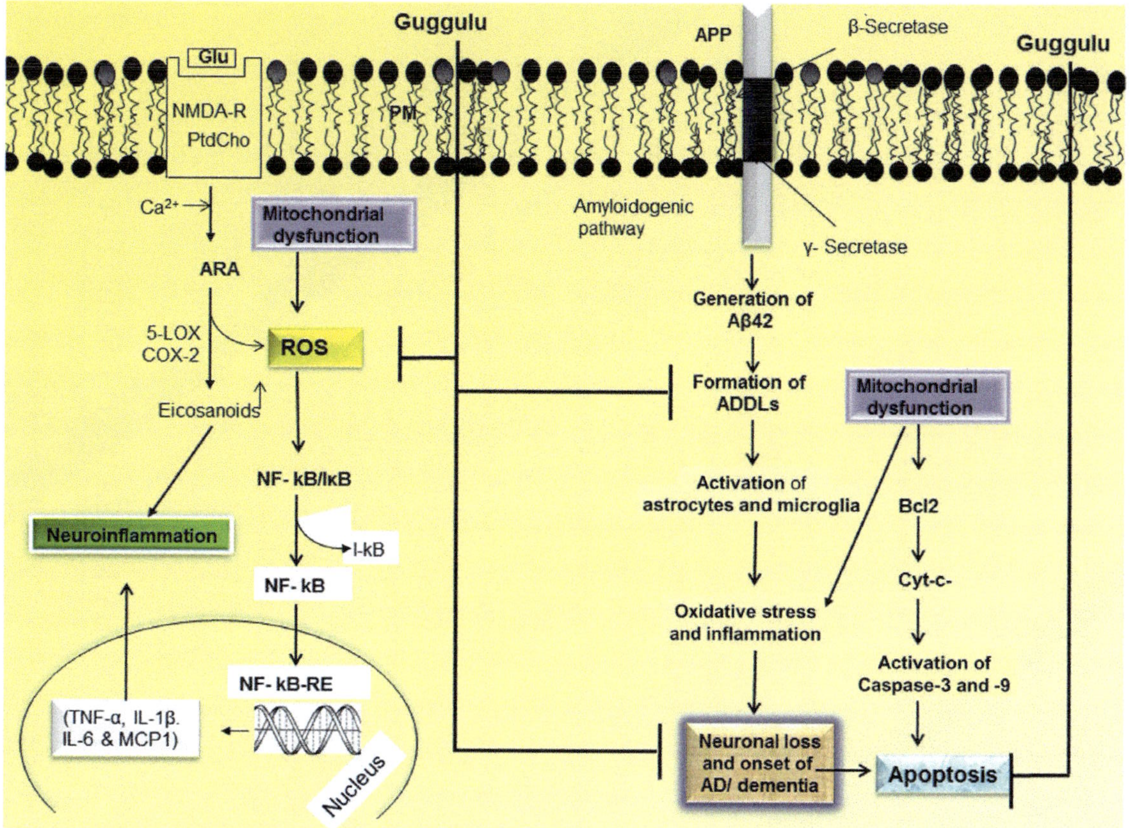

FIGURE 9.9 Hypothetical diagram showing target sites for the action of guggulu.
ADDL, Aβ-derived diffusible ligand; *APP*, amyloid precursor protein; *ARA*, arachidonic acid; *Aβ*, β-amyloid; *Bcl-2*, B-cell lymphoma 2; *COX-2*, cyclooxygenase-2; *cPLA*$_2$, cytosolic phospholipase A_2; *cyto-c*, cytochrome; *Glu*, glutamate; *5-LOX*, 5-lipoxygenase; *lyso-PtdCho*, lyso-phosphatidylcholine; *NMDA-R*, NMDA receptor; *PtdCho*, phosphatidylcholine; *ROS*, reactive oxygen species.

from roots and rhizome of *R. rosea* including monoterpene alcohols and their glycosides, cyanogenic glycosides, aryl glycosides, phenylethanoids, phenylpropanoids and their glycosides, flavonoids, flavonlignans, proanthocyanidins, and gallic acid derivatives (Panossian et al., 2010). *Rhodiola* compounds exhibit adaptogenic effects such as cardioprotective, antifatigue, anxiolytic, nootropic, life span increasing effects, and CNS stimulating activity. In addition, *R. rosea* produces antidepressant effects not only by interacting with HPA-system (cortisol-reducing), but also by modulating protein kinases p-JNK, nitric oxide, and defense mechanism proteins (e.g., heat shock proteins Hsp 70 and FoxO/DAF-16) (Panossian et al., 2010). *R. rosea* root also contains salidroside, a 2-[4-hydroxyphenyl]ethyl β-D-glucopyranoside, which has been reported to produce antiaging, anticancer, antiinflammatory, and antioxidative functions (Li et al., 2012). In addition, salidroside also restores the capacity of the dentate gyrus to generate new neurons and intercepts learning and memory decays in mice during aging (Jin et al., 2016; Wang et al., 2002). The molecular mechanism of action of salidroside is not known. However, it is proposed that salidroside targets CREB transcription factor for the development and survival of new neurons in the dentate gyrus of old mice (Jin et al., 2016). Thus, salidroside is therapeutically effective against learning and memory decays via stimulation of CREB-dependent functional neurogenesis in aging. Some of *R. rosea* constituents increase the level of 5-HT and NE in the cerebral, prefrontal, and frontal cortex (Lazarova et al., 1986). At the same time, treatment with *R. rosea* produces the upregulation of dopamine and acetylcholine in the limbic system. It is reported that intake of *R. rosea* may contribute to emotional calming (Furmanowa et al., 2014) due to the antioxidant properties of *R. rosea* contstituents. Several studies have indicated that salidroside not only produces antioxidant, antiinflammatory, and antiamyloidogenic effects in the brain, but also improves cognitive function, and prevents epilepsy across a wide therapeutic time window (Cao et al., 2006; Li et al., 2010; Xing et al., 2014; Lai et al., 2015; Wang et al., 2015). At 120 and 240 μM salidroside protected primary hippocampal neurons against glutamate-induced apoptosis via stimulating p-Akt in vitro (Xian et al., 2014). Another study reports that salidroside (1−100 μM) regulates MMP, suppresses mitochondrial cyt c release into cytosol, and attenuated caspase activation via an apoptosis pathway in H_2O_2-induced PC12 cells (Cai et al., 2008). Notably, salidroside (200 μM) effectively suppresses BACE1 expression, Aβ generation, and β-secretase activity, and triggers soluble APP secretion in hypoxia-induced SH-SY5Y cells, suggesting that salidroside may be useful in the prevention and treatment of AD type of dementia (Li et al., 2010). In MPP^+-induced model of PD-linked dementia, it is reported that salidroside (1−100 μM) inhibits apoptosis, attenuates MMP^+ induced oxidative stress, promotes condensing of chromatin, and promotes the release of lactate dehydrogenase in PC12 cells through the involvement of NO and the PtdIns 3K/Akt pathway (Li et al., 2011; Zhang et al., 2012). Another study indicates that the pretreatment with salidroside (15 and 45 mg/kg) improves behavior disorders in an MPTP-induced PD mouse model when administered once a day for 7 consecutive days (Wang et al., 2015; Zhang et al., 2016). In these studies, treatment with salidroside ameliorates tyrosine hydroxylase-positive neuron loss in substantia nigra by increasing monoamine substances levels. These processes produce neuroprotection not only via the involvement of PtdIns 3K/Akt/GSK3β, but also ROS-NO-related pathways (Wang et al., 2015; Zhang et al., 2016). More recently, salidroside (25–100 μM) is also reported to significantly reduce MPP^{+}-induced neuronal injury via DJ-1-Nrf2 antioxidant pathway in SH-SY5Y cells (Wu et al., 2017). Besides, this it has been demonstrated that salidroside (100 μg/mL) can induce the differentiation of rat mesenchymal stem cells to dopaminergic neurons with increased DA release (Zhao et al., 2014). Salidroside has also been reported to protect primary cortical neurons and SN4741 cells from endoplasmic reticulum stress (Tao et al., 2016).

Studies in animal models of stroke have indicated that salidroside-pretreatment produce neuroprotective effects on global cerebral ischemia/reperfusion rats. Thus, salidroside improves neurological severity scores and reduces the degree of cerebral edema at a dose of 12 mg/kg, whereas the underlying mechanism of action is a possible contribution to free radical scavenging. Recent studies have indicated that repeated administration of salidroside at a dose of 24 mg/kg for 7 days not only reduces brain edema and suppresses TNF-α release, but also ameliorates neurological scores and reduces the infarction volume in focal cerebral ischemia/reperfusion model in rats (Shi et al., 2012; Han, 2013). Collectively, the abovementioned studies indicate that due to its antioxidant and antiinflammatory properties, *R. rosea* be used against oxidative and inflammatory damage in AD, PD, and stroked-linked dementia. The treatment with *R. rosea*'s constituents can also enhance the learning and memory in AD, PD, and stroke-linked dementia (Qu et al., 2009).

TULSI AND DEMENTIA

Tulsi (*Ocimum tenuiflorum* L.) is an aromatic shrub, which belongs to the family Lamiaceae (tribe ocimeae) (Bast et al., 2014). In the Ayurvedic medicine tulsi is called "The Queen of Herbs," or "elixir of life" for its healing powers (Singh et al., 2010). It provides a vast array of health benefits that are just beginning to be confirmed by modern science. Daily consumption of tulsi not only prevents disease and promotes general health, well-being, and longevity, but also assists in dealing with the stresses of daily life. Tulsi produces potent pharmacological effects including adaptogenic (Jothie Richard et al., 2016), metabolic (Suanarunsawat et al., 2016), immunomodulatory (Raja et al., 2016), anticancer (Lin et al., 2014), antiinflammatory (Tanko et al., 2008), antioxidant (Kelm et al., 2000), hepatoprotective (Chattopadhyay et al., 1992), and antidiabetic effects (Rai et al., 1997). Tulsi is used for the treatment of anxiety, dementia, cough, asthma, diarrhea, fever, dysentery, arthritis, eye diseases, otalgia, indigestion, hiccups, vomiting, gastric, cardiac and genitourinary disorders, back pain, skin diseases, ringworm, insect, snake, scorpion bites, and malaria (Singh et al., 2010; Mahajan et al., 2013; Mohan et al., 2011; Pattanayak et al., 2010; Mondal et al., 2009). The tulsi plant leaves (dried or fresh) contain several bioactive compounds including eugenol, ursolic acid, β-caryophyllene, linalool, and 1,8-cineole (Bernhardt et al., 2015). Eugenol lowers blood glucose not only by preventing the binding of glucose to serum albumin, but also by inhibiting the conversion of complex carbohydrate to glucose (Singh et al., 2016). Other potentially active secondary metabolites of tulsi include phenylpropanoids (methyl eugenol, rosmarinic acid), monoterpenes (ocimene), and sesquiterpenes (germacrene). These metabolites may produce beneficial health effects among elderly and demented subjects alone or synergistically through their antioxidant, antiinflammatory, and antidiabetic effects (Singh et al., 2015).

JYOTISHMATI AND DEMENTIA

Jyotishmati (*Celastrus paniculatus*) seeds produce heat in the body and give a feeling of warmness after their consumption. Jyotishmati seeds are rich in oil content (around 30%), and contain several alkaloids and other phytochemicals including glycosides, cumarines, tannins, flavonoids, saponins, steroids, and triterpenoids. The seeds also contain chromium, copper, iodine, iron, manganese, selenium, and zinc. Powder from Jyotishmati seeds not only improves concentration and alertness, but also enhances cognitive functions. It is proposed that Jyotishmati seed powder produces its beneficial effects by increasing acetylcholine level in the brain, which enhances the cognition (Bhanumathy et al., 2010). Aqueous extracts of Jyotishmati seeds also produce antioxidant effects. Jyotishmati seed extracts protected neuronal cells against H_2O_2-induced toxicity in part by virtue of their antioxidant properties and their ability to induce antioxidant enzymes (Godkar et al., 2003). Jyotishmati seed extracts also protected neuronal cells against excitotoxicity by modulating glutamate receptor function. In addition, Jyotishmati seed extracts protect neuronal cells by inducing the antioxidant enzyme catalase (da Rocha et al., 2011; Godkar et al., 2003, 2004, 2006; Kumar and Gupta, 2002; Bhagya et al., 2016). In addition, aqueous extracts of CP seed have dose-dependent cholinergic activity, thereby improving memory performance (Bhanumathy et al., 2010).

NARDOSTACHYS JATAMANSI AND DEMENTIA

Nardostachys jatamansi is a flowering plant, which belongs to the Valerian family. It grows in north India, Nepal, Bhutan, Myanmar, and southwest China. It is rich in essential oil, and contains Jatamansone (valeranone), acacin, ursolic acid, octacosanol, nardosinone, nardosinone, oleanolic acid, calarene, aristolene, valerena-4/7(11)-diene, desoxo-narchinol-A, and beta-sitosterol. It increases levels of GABA, norepinephrine, dopamine, serotonin, and 5-hydroxyindoleatic acid in rat brain (Prabhu et al., 1994). *N. jatamansi* produces antifungal, antimicrobial, antioxidant, hepatoprotective, and cardioprotective effects (Sahu et al., 2016). *N. jatamansi* also produces antiischemic, antioxidant, anticonvulsant, and neuroprotective effects. It has also been used as a memory enhancer. Thus, administration of *N. jatamansi* (200 mg/kg) for 8 successive days in young and aged mice not only improves learning and memory in young mice, but also reverses the amnesia induced by diazepam (1 mg/kg, i.p.) and scopolamine (0.4 mg/kg, i.p.) (Joshi and Parle, 2006). In addition, *Nardostachys jatmansi* treatment also reverses aging-induced amnesia due to the natural aging of mice. As scopolamine-induced amnesia is reversed, it is possible that the memory improvement may be because of facilitation of cholinergic transmission in the brain. Hence, *N. jatamansi* might prove to be a useful memory restorative agent in the treatment of dementia seen in elderly persons. It is proposed that the underlying mechanism of action can be attributed to its antioxidant property (Joshi and Parle, 2006).

CONCLUSION

Ayurveda is one of the oldest healthcare systems that evolved in the Indian Subcontinent about 3000 BCE. Ayurvedic perspective is based on equilibrium and interplay among different elements of human life, the body, the mind, the intellect, and the soul. Ayurveda not only involves a healthy lifestyle, health promotion, and sustenance, but also disease prevention, diagnosis, and treatment. Popular Ayurvedic medicinal plants for the treatment of dementia are Ashwagandha, Brahmi, Gotu kola, Sankhapuspi, and guggul. Long-term consumption of these plants may not only retard brain aging and induce antistress and memory enhancing effects, which help in regeneration of neural tissues, but also induce antioxidant, antiinflammatory, antiamyloidogenic, nutritional, and immune-supportive effects in the human body. This may lead to retardation of dementia, neurological, and visceral disorders. Scientific validation and the documentation of Ayurvedic medicines are very essential for its quality evaluation and global acceptance. Therapeutic efficacy of Ayurvedic herbal formulation can be enhanced not only by achieving purity, but also by studying the physical and biological properties of Ayurvedic plant products using modern scientific methods. These days attempts are underway to achieve this goal. Once it is done, then large multicenter clinical trials of Ayurvedic medicine can be planned and performed in patients with dementia and other neurodegenerative disorders.

References

Abbas, S.S., Singh, N., 2006. Anti-STRESS AGENTS (Herbs) of Indian Origin – Herbal Drugs, A Twenty-First Century Perspective. D.R.a.D.O. Delhi: Institute of Nuclear Medicine and Allied Sciences Ed., New Delhi, pp. 578–591.

Agarwal, R., Diwanay, S., Patki, P., Patwardhan, B., 1999. Studies on immunomodulatory activity of *Withania somnifera* (Ashwagandha) extracts in experimental immune inflammation. J. Ethnopharmacol. 67, 27–35.

Alva, G., Cummings, J.L., 2008. Relative tolerability of Alzheimer's disease treatments. Psychiatry 5, 27–36.

Anand, P., Kunnumakkara, A.B., Newman, R.A., Aggarwal, B.B., 2007. Bioavailability of curcumin: problems and promises. Mol. Pharm. 4, 807–818.

Anand, P., Nair, H.B., Sung, B., et al., 2010. Design of curcumin-loaded PLGA nanoparticles formulation with enhanced cellular uptake, and increased bioactivity in vitro and superior bioavailability in vivo. Biochem. Pharmacol. 79, 330–338.

Appa Rao, M.V.R., Srinivasan, K., Rao, K.T., 1973. The effect of mandookaparni (*Centella asiatica*) on the general mental ability (medhya) of mentally retarded children. J. Res. Indian Med. 8, 9–16.

Ballard, C., Howard, R., 2006. Neuroleptic drugs in dementia: benefits and harm. Nat. Rev. Neurosci. 7, 492–500.

Bast, F., Rani, P., Meena, D., 2014. Chloroplast DNA phylogeography of holy basil (*Ocimum tenuiflorum*) in Indian subcontinent. Sci. World J. 2014, 847482.

Baum, L., Ng, A., 2004. Curcumin interaction with copper and iron suggests one possible mechanism of action in Alzheimer's disease animal models. J. Alzheimer Dis. 2004 (6), 367–377.

Baum, L., Lam, C.W., Cheung, S.K., Kwok, T., Lui, V., Tsoh, J., et al., 2008. Six-month randomized, placebo-controlled, double-blind, pilot clinical trial of curcumin in patients with Alzheimer disease. J Clin Psychopharmacol. 28, 110–113.

Begum, A.N., Jones, M.R., Lim, G.P., et al., 2008. Curcumin structure-function, bioavailability, and efficacy in models of neuroinflammation and Alzheimer's disease. J. Pharmacol. Exp. Ther. 326, 196–208.

Bernhardt, B., Szabó, K., Bernáth, J., 2015. Sources of variability in essential oil composition of *Ocimum americanum* and *Ocimum tenuiflorum*. Acta Alimentaria 44, 111–118.

Bhagya, V., Christofer, T., Shankaranarayana Rao, B.S., 2016. Neuroprotective effect of *Celastrus paniculatus* on chronic stress-induced cognitive impairment. Indian J. Pharmacol. 48, 687–693.

Bhanumathy, M., Harish, M.S., Shivaprasad, H.N., Sushma, G., 2010. Nootropic activity of *Celastrus paniculatus* seed. Pharm. Biol. 48, 324–327.

Bhattacharya, S.K., Goel, R.K., Kaur, R., Ghosal, S., 1987. Anti-stress activity of sitoindosides VII and VIII, new acylsterylglucosides from *Withania somnifera*. Phytother. Res. 1, 32–37.

Bihaqi, S.W., Singh, A.P., Tiwari, M., 2011. In vivo investigation of the neuroprotective property of *Convolvulus pluricaulis* in scopolamine-induced cognitive impairments in Wistar rats. Indian J. Pharmacol. 43, 520–525.

Bihaqi, S.W., Singh, A.P., Tiwari, M., 2012. Supplementation of *Convolvulus pluricaulis* attenuates scopolamine-induced increased tau and Amyloid precursor protein (AβPP) expression in rat brain. Indian J. Pharmacol. 44, 593–598.

Braak, H., Del Tredici, K., 2008. Nervous system pathology in sporadic Parkinson disease. Neurology 70, 1916–1925.

Brahma, S.K., Debnath, P.K., 2003. Therapeutic importance of Rasayana drugs with special reference to their multi-dimensional actions. Aryavaidyan 16, 160–163.

Bryan, H.K., Olayanju, A., Goldring, C.E., Park, B.K., 2013. The Nrf2 cell defence pathway: Keap1-dependent and -independent mechanisms of regulation. Biochem. Pharmacol. 85, 705–717.

Cai, L., Wang, H., Li, Q., Qian, Y., Yao, W., 2008. Salidroside inhibits H_2O_2-induced apoptosis in PC12 cells by preventing cytochrome c release and inactivating of caspase cascade. Acta Biochim. Biophys. Sin. 40, 796–802.

Calabrese, C., Gregory, W.L., Leo, M., Kraemer, D., Bone, K., et al., 2008. Effects of a standardized *Bacopa monnieri* extract on cognitive performance, anxiety, and depression in the elderly: a randomized, double-blind, placebo-controlled trial. J. Altern. Complement. Med. 14, 707–713.

Cao, L.L., Du, G.H., Wang, M.W., 2006. The effect of salidroside on cell damage induced by glutamate and intracellular free calcium in PC12 cells. J. Asian Nat. Prod. Res. 8, 159–165.

Cervenka, F., Jahodár, L., 2006. Plant metabolites as nootropics and cognitives. Ceska Slov Farm. 55, 219–229.

Chandre, R., Murthy, K.N., Singh, R.H., 2004. Evaluation of the efficacy of kushmandaghrta in the management of depressive illness. Aryavaidyan 18, 87–90.

Charles, P.D., Ambigapathy, G., Geraldine, P., Akbarsha, M.A., Rajan, K.E., 2011. *Bacopa monniera* leaf extract up-regulates tryptophan hydroxylase (tph2) and serotonin transporter (sert) expression: implications in memory formation. J. Ethnopharmacol. 134, 55–61.

Chaudhari, K.S., Tiwari, N.R., Tiwari, R.R., Sharma, R.S., 2017. Neurocognitive effect of nootropic drug Brahmi (*Bacopa monnieri*) in Alzheimer's Disease. Ann. Neurosci. 24, 111–122.

Chatterji, N., Rastogi, R., Dhar, M.L., 1963. Chemical examination of *Bacopa monniera* Wettst.: Part I-Isolation of chemical constituents. Indian J. Chem. 1, 212–215.

Chattopadhyay, R., Sarkar, S., Ganguly, S., Medda, C., Basu, T., 1992. Hepatoprotective activity of *Ocimum sanctum* leaf extract against paracetamol induced hepatic damage in rats. Ind. J. Pharmacol. 24, 163.

Chen, X., Liu, J., Gu, X., Ding, F., 2008. Salidroside attenuates glutamate-induced apoptotic cell death in primary cultured hippocampal neurons of rats. Brain Res. 1238, 189–198.

Chen, P.T., Chen, Z.T., Hou, W.C., Yu, L.C., Chen, R.P., 2016. Polyhydroxycurcuminoids but not curcumin upregulate neprilysin and can be applied to the prevention of Alzheimer's disease. Sci. Rep. 6, 29760.

Cheng, A.L., Hsu, C.H., Lin, J.K., Hsu, M.M., Ho, Y.F., et al., 2001. Phase I clinical trial of curcumin, a chemopreventive agent, in patients with high-risk or pre-malignant lesions. Anticancer Res. 21, 2895–2900.

Chivapat, S., Chavalittumrong, P., Tantisira, M.H., 2011. Acute and sub-chronic toxicity studies of a standardized extract of *Centella asiatica* ECa 233. Thail. J. Pharmacol. Sci. 35, 55–64.

Cho, J.Y., Baik, K.U., Jung, J.H., Park, M.H., 2000. In vitro anti-inflammatory effect of cynaropicrin, a serquiterpene lactone, from Saussurea lappa. Eur. J Pharmacol. 398, 399–407.

da Rocha, M.D., Viegas, F.P., Campos, H.C., Nicastro, P.C., Fossaluzza, P.C., Fraga, C.A., et al., 2011. The role of natural products in the discovery of new drug candidates for the treatment of neurodegenerative disorders II: Alzheimer's disease. CNS Neurol. Disord. Drug Targets 10, 251–270.

Dar, N.J., Hamid, A., Ahmad, M., 2015. Pharmacologic overview of *Withania somnifera*, the Indian Ginseng. Cell Mol. Life Sci. 72, 4445–4460.

Deepak, M., Amit, A., 2013. 'Bacoside B'—the need remains for establishing identity. Fitoterapia 87, 7–10.

Dev, R.D.O., 2009. Comparison on cognitive effects of Centella asiatica in healthy middle age female and male volunteers. Ann. Nutr. Metab. 55, 709.

Dhanasekaran, M., Holcomb, L.A., Hitt, A.R., Tharakan, B., Porter, J.W., Young, K.A., et al., 2009. *Centella asiatica* extract selectively decreases amyloid β levels in hippocampus of Alzheimer's disease animal model. Phytother Res. 23, 14–19.

Dickson, D., 2008. Neuropathology. In: Factor, S., Weiner, W. (Eds.), Parkinson's Disease: Diagnosis and Clinical Management, second ed. Demos, New York.

Eckert, G.P., Kirsch, C., Leutz, S., Wood, W.G., Muller, W.E., 2003. Cholesterol modulates amyloid beta-peptide's membrane interactions. 36, S136–S143.

Farooqui, A.A., 2010. Neurochemical aspects of neurotraumatic and neurodegenerative diseases, 1–401. Springer, New York.

Farooqui, A.A., 2016. Therapeutic Potentials of Curcumin for Alzheimer Disease. Springer, New York.

Farooqui, A.A., Farooqui, T., Madan, A., Ong, J.H.-J., Ong, W.Y., 2018. Ayurvedic medicine for the treatment of dementia: mechanistic aspects. Evid. Based Complement. Altern. Med. 2018 2481076.

Feany, M.B., Bender, W.W., 2000. A Drosophila model of Parkinson's disease. Nature 404, 394–398.

Forlenza, O.V., Spink, J.M., Dayanandan, R., Anderton, B.H., Olesen, O.F., et al., 2000. Muscarinic agonists reduce tau phosphorylation in non-neuronal cells via GSK-3beta inhibition and in neurons. J. Neural Transm. 107, 1201–1212.

Francis, P.T., Palmer, A.M., Snape, M., Wilcock, G.K., 1999. The cholinergic hypothesis of Alzheimer's disease: a review of progress. J. Neurol. Neurosurg. Psychiatr 66, 137–147.

Frautschy, S.A., Hu, W., Kim, P., et al., 2001. Phenolic anti-inflammatory antioxidant reversal of Aβ-induced cognitive deficits and neuropathology. Neurobiol. Aging 22, 993–1005.

Furmanowa, M., Skopińska-Rozewska, E., Rogala, E., Hartwich, M., 2014. *Rhodiola rosea* in vitro culture—phytochemical analysis and antioxidant action. Acta Soc. Bot. Pol. 67, 69–73.

Ganguli, M., Chandra, V., Kamboh, M.I., Johnston, J.M., Dodge, H.H., Thelma, B.K., et al., 2000. Apolipoprotein E polymorphism and Alzheimer disease: the Indo-US Cross-National Dementia Study. Arch. Neurol. 57, 824–830.

Garcia-Alloza, M., Robbins, E.M., Zhang-Nunes, S.X., et al., 2006. Characterization of amyloid deposition in the APPswe/PS1dE9 mouse model of Alzheimer disease. Neurobiol. Dis. 24, 516–524.

Ghosh, S., Banerjee, S., Sil, P.C., 2015. The beneficial role of curcumin on inflammation, diabetes and neurodegenerative disease: a recent update. Food Chem. Toxicol. 83, 111–124.

Godkar, P., Gordon, R.K., Ravindran, A., Doctor, B.P., 2003. *Celastrus paniculatus* seed water soluble extracts protect cultured rat forebrain neuronal cells from hydrogen peroxide-induced oxidative injury. Fitoterapia 74, 658–669.

Godkar, P.B., Gordon, R.K., Ravindran, A., Doctor, B.P., 2004. *Celastrus paniculatus* seed water soluble extracts protect against glutamate toxicity in neuronal cultures from rat forebrain. J. Ethnopharmacol. 93, 213–219.

Godkar, P.B., Gordon, R.K., Ravindran, A., Doctor, B.P., 2006. *Celastrus paniculatus* seed oil and organic extracts attenuate hydrogen peroxide- and glutamate-induced injury in embryonic rat forebrain neuronal cells. Phytomedicine 13, 29–36.

Goel, A., Kunnumakkara, A.B., Aggarwal, B.B., 2008. Curcumin as "Curecumin": from kitchen to clinic. Biochem. Pharmacol. 75, 787–809.

Gohil, K.J., Patel, J.J., 2010. A review on *Bacopa monniera*: current research and future prospects. Int. J. Green Pharm. 4, 1–9.

Govindarajan, R., Vijayakumar, M., Pushpangadan, P., 2005. Antioxidant approach to disease management and the role of Rasayana herbs of Ayurveda. J. Ethnopharmacol. 99, 165–178.

Gray, N.E., Morré, J., Kelley, J., Maier, C.S., Stevens, J.F., et al., 2014. Caffeoylquinic acids in *Centella asiatica* protect against amyloid-β toxicity. J. Alzheimers Dis. 40, 359–373.

Gray, N.E., Sampath, H., Zweig, J.A., Quinn, J.F., Soumyanath, A., 2015. *Centella asiatica* attenuates amyloid-β-induced oxidative stress and mitochondrial dysfunction. J. Alzheimers Dis. 45, 933–946.

Gray, N.E., Zweig, J.A., Matthews, D.G., Caruso, M., Quinn, J.F., et al., 2017. *Centella asiatica* attenuates mitochondrial dysfunction and oxidative stress in *Aβ*-exposed hippocampal neurons. Oxid. Med. Cell Longev. 2017, 7023091.

Han, T., 2013. Effects of salidroside pretreatment on expression of tumor necrosis factor-alpha and permeability of blood brain barrier in rat model of focal cerebralischemia-reperfusion injury. Asian Pac. J. Trop. Med. 6, 156–158.

Hansen, R.A., Gartlehner, G., Lohr, K.N., et al., 2007. Functional outcomes of drug treatment in Alzheimer's disease: a systematic review and meta-analysis. Drugs Aging 24, 155–167.

Harsahay, M., Hemant, K.P., Pankaj, P., Mahesh, C.A., Zakwan, A., 2012. Evaluation of antioxidant activity of two important memory enhancing medicinal plants *Baccopa monnieri* and *Centella asiatica*. Ind. J. Pharmacol. 44, 114–117.

Hishikawa, N., Takahashi, Y., Amakusa, Y., Tanno, Y., Tuji, Y., et al., 2012. Effects of turmeric on Alzheimer's disease with behavioral and psychological symptoms of dementia. Ayu 33, 499–504.

Hogan, D.B., Bailey, P., Black, S., et al., 2008. Diagnosis and treatment of dementia: 5. Nonpharmacologic and pharmacologic therapy for mild to moderate dementia. CMAJ 179, 1019–1026.

Hong, H.S., Rana, S., Barrigan, L., et al., 2009. Inhibition of Alzheimer's amyloid toxicity with a tricyclic pyrone molecule in vitro and in vivo. J. Neurochem. 108, 1097–1108.

Hosamani, R., 2010. Muralidhara Prophylactic treatment with *Bacopa monnieri* leaf powder mitigates paraquatinduced oxidative perturbations and lethality in drosophila melanogaster. Ind. J. Biochem. Biophys. 47, 75–82.

Hsiao, K., Chapman, P., Nilsen, S., Eckman, C., Harigaya, Y., Younkin, S., et al., 1996. Correlative memory deficits, Abeta elevation, and amyloid plaques in transgenic mice. Science 274, 99–102.

Huang, C., Wang, J., Lu, X., et al., 2016. Z-guggulsterone negatively controls microglia-mediated neuroinflammation via blocking IκB-α-NF-κB signals. Neurosci. Lett. 619, 34–42.

Jadiya, P., Khan, A., Sammi, S.R., Kaur, S., Mir, S.S., et al., 2011. Anti-parkinsonian effects of *Bacopa monnieri*: insights from transgenic and pharmacological *Caenorhabditis elegans* models of parkinson's disease. Biochem. Biophys. Res. Commun. 413, 605–610.

Jansen, R., Brogan, B., Whitworth, A., Okello, E., 2014. Effects of five ayurvedic herbs on locomotor behaviour in a *drosophila melanogaster* parkinson's disease model. Phytother. Res. 28, 1789–1795.

Jatwa, V., Khirwadkar, P., Dashora, K., 2014. Indian traditional memory enhancing herbs and their medicinal benefits. Ind. J. Res. Pharm. Biotechnol. 2, 1030.

Jayashree, G., Kurup Muraleedhara, G., Sudarslal, S., Jacob, V.B., 2003. Anti-oxidant activity of *Centella asiatica* on lymphoma-bearing mice. Fitoterapia 74, 431–434.

Jin, H., Pei, L., Shu, X., Yang, X., Yan, T., et al., 2016. Therapeutic intervention of learning and memory decays by salidroside stimulation of neurogenesis in aging. Mol. Neurobiol. 53, 851–866.

Jorm, A.F., Jolley, D., 1998. The incidence of dementia: a meta-analysis. Neurology 51, 728–733.

Joshi, H., Parle, M., 2006. *Nardostachys jatamansi* improves learning and memory in mice. J. Med. Food 9, 113–118.

Jothie Richard, E., Illuri, R., Bethapudi, B., et al., 2016. Anti-stress activity of *Ocimum sanctum*: possible effects on hypothalamic–pituitary–adrenal axis. Phytother. Res. 30, 805–814.

Kelm, M.A., Nair, M.G., Strasburg, G.M., DeWitt, D.L., 2000. Antioxidant and cyclooxygenase inhibitory phenolic compounds from *Ocimum sanctum* Linn. Phytomedicine 7, 7–13.

Khan, R., Krishnakumar, A., Paulose, C.S., 2008. Decreased glutamate receptor binding and NMDA R1 gene expression in hippocampus of pilocarpine-induced epileptic rats: neuroprotective role of *Bacopa monnieri* extract. Epilepsy Behav. 12, 54–60.

Khar, A., Ali, A.M., Pardhasaradhi, B.V., Begum, Z., Anjum, R., 1999. Antitumor activity of curcumin is mediated through the induction of apoptosis in AK-5 tumor cells. FEBS Lett. 19, 165–168.

Kim, G.Y., Kim, K.H., Lee, S.H., Yoon, M.S., Lee, H.J., et al., 2005. Curcumin inhibits immunostimulatory function of dendritic cells: MAPKs and translocation of NF-B as potential targets. J. Immunol. 174, 8116–8124.

Kuboyama, T., Tohda, C., Komatsu, L., 2005. Neuritic regeneration and synaptic reconstruction induced by withanolide A. Br. J. Pharmacol. 144, 961–971.

Kuboyama, T., Tohda, C., Komatsu, K., 2014. Effects of Ashwagandha (roots of *Withania somnifera*) on neurodegenerative diseases. Biol. Pharm. Bull. 37, 892–897.

Kumar, M.H., Gupta, Y.K., 2002. Antioxidant property of *Celastrus paniculatus* willd.: a possible mechanism in enhancing cognition. Phytomedicine 9, 302–311.

Kumar, V., 2006. Potential medicinal plants for CNS disorders: an overview. Phytother. Res. 20, 1023–1035.

Lad, V., 1987. Ayurveda, the Science of Self-Healing: A Practical Guide, second ed. Lotus Press, New Delhi.

Lai, W., Zheng, Z., Zhang, X., et al., 2015. Salidroside-mediated neuroprotection is associated with induction of early growth response genes (Egrs) across a wide therapeutic window. Neurotox. Res. 28, 108–121.

Lazarova, M.B., Petkov, V.D., Markovska, V.L., Mosharrof, A., 1986. Effects of meclofenoxate and Extr. *Rhodiolae roseae* L. on electroconvulsive shock-impaired learning and memory in rats. Methods Find. Exp. Clin. Pharmacol. 8, 547–552.

Li, Q.Y., Wang, H.M., Wang, Z.Q., Ma, J.F., Ding, J.Q., Chen, S.D., 2010. Salidroside attenuates hypoxia-induced abnormal processing of amyloid precursor protein by decreasing BACE1 expression in SH-SY5Y cells. Neurosci. Lett. 481, 154–158.

Li, X., Ye, X., Li, X., Sun, X., Liang, Q., Tao, L., et al., 2011. Salidroside protects against MPP(+)-induced apoptosis in PC12 cells by inhibiting the NO pathway. Brain Res. 1382, 9–18.

Li, X., Sipple, J., Pang, Q., Du, W., 2012. Salidroside stimulates DNA repair enzyme Parp-1 activity in mouse HSC maintenance. Blood 119, 4162–4173.

Lim, G.P., Chu, T., Yang, F., Beech, W., Frautschy, S.A., Cole, G.M., 2001. The curry spice curcumin reduces oxidative damage and amyloid pathology in an Alzheimer transgenic mouse. J. Neurosci. 21, 8370–8377.

Lin, C.-C., Chao, P.-Y., Shen, C.-Y., et al., 2014. Novel target genes responsive to apoptotic activity by ocimum gratissimum in human osteosarcoma cells. Am. J. Chin. Med. 42, 743–767.

Lonergan, E., Luxenberg, J., Colford, J., 2002. Haloperidol for agitation in dementia. Cochrane Database Syst. Rev. CD002852.

Ma, Q.L., Yang, F., Rosario, E.R., et al., 2009. Beta-amyloid oligomers induce phosphorylation of tau and inactivation of insulin receptor substrate via c-Jun N-terminal kinase signaling: suppression by omega-3 fatty acids and curcumin. J. Neurosci. 29, 9078–9089.

Mahajan, N., Rawal, S., Verma, M., Poddar, M., Alok, S., 2013. A phytopharmacological overview on *Ocimum*species with special emphasis on *Ocimum sanctum*. Biomed. Prev. Nutr. 3, 185–192.

Malik, J., Karan, M., Vasisht, K., 2011. Nootropic, anxiolytic and CNS-depressant studies on different plant sources of shankhpushpi. Pharmace Biol. 49, 1234–1242.

Manyam, B.V., 1999. Dementia in Ayurveda. J. Altern. Complement. Med. 5, 81–88.

Mathur, D., Goyal, K., Koul, V., Anand, A., 2016. Effects of five ayurvedic herbs on locomotor behaviour in a *Drosophila melanogaster* Parkinson's Disease Model. Front. Pharmacol. 7, 44.

McShane, R., Areosa Sastre, A., Minakaran, N., 2006. Memantine for dementia. Cochrane Database Syst. Rev. CD003154.

Meulenbeld, G.J., 2002. A History of Indian Medical Literature. Koninklijke Brill NV, Leiden, The Netherlands, p. 3872.

Mirjalili, M.H., Moyano, E., Bonfill, M., Cusido, R.M., Palazon, J., 2009. Steroidal lactones from *Withania somnifera*, an ancient plant for novel medicine. Molecules 14, 2373–2393.

Mishra, L.C., Singh, B.B., Dagenais, S., 2000. Scientific basis for the therapeutic use of *Withania somnifera* (ashwagandha): a review. Altern. Med. Rev. 5, 334–346.

Mishra, S., Trikamji, B., Singh, S., Singh, P., Nair, R., 2013. Historical perspective of Indian neurology. Ann. Indian Acad. Neurol. 16, 467–477.

Mitra-Ganguli, T., Kalita, S., Bhushan, S., Stough, C., Kean, J., et al., 2017. A randomized, double-blind study assessing changes in cognitive function in indian school children receiving a combination of *Bacopa monnieri* and micronutrient supplementation vs. placebo. Front. Pharmacol. 8, 678.

Mohan, L., Amberkar, M.V., Kumari, M., 2011. Ocimum sanctum linn. (TULSI)-an overview. Int. J. Pharm. Sci. Rev. Res. 7, 51–53.

Mondal, S., Mirdha, B.R., Mahapatra, S.C., 2009. The science behind sacredness of Tulsi (*Ocimum sanctum* Linn.). Indian J. Physiol. Pharmacol. 53, 291–306.

Morley, J.E., Banks, W.A., 2010. Lipids and cognition. 20, 737–747.

Morley, J.E., Banks, W.A., 2010. Lipids and cognition. J Alzheimer's Disease 20, 737–747.

Mukherjee, P.K., Kumar, V., Kumar, V.N.S., Heinrich, M., 2008. The Ayurvedic medicine Clitoria ternatea-From traditional use to scientific assessment. J Ethnopharmacol. 120, 291–301.

Narayan, M., Seeley, K.W., Jinwal, U.K., 2015. Identification and quantitative analysis of cellular proteins affected by treatment with withaferin a using a SILAC-based proteomics approach. J. Ethnopharmacol. 175, 86–92.

National Cancer Institute Dictionary, 1996.

Nahata, A., Patil, U.K., Dixit, V.K., 2008. Effect of *Convolvulus pluricaulis* choisy. on learning behaviour and memory enhancement activity in rodents. Nat. Prod. Res. 22, 1472–1482.

Nellore, J., Pauline, C., Amarnath, K., 2013. *Bacopa monnieri* phytochemicals mediated synthesis of platinum nanoparticles and its neurorescue effect on 1-methyl 4-phenyl 1,2,3,6 tetrahydropyridine-induced experimental parkinsonism in zebrafish. J. Neurodegener. Dis. 2013, 972391.

Ng, T.P., Chiam, P.C., Lee, T., Chua, H.C., Lim, L., Kua, E.H., 2006. Curry consumption and cognitive function in the elderly. Am. J. Epidemiol. 164, 898–906.

Nohr, L.A., Rasmussen, L.B., Straand, J., 2009. Resin from the mukul myrrh tree, guggul, can it be used for treating hypercholesterolemia? A randomized, controlled study. Complement Ther Med. 17, 16–22.

Oh, J.H., Lee, T.J., Park, W., Kwon, T.K., 2008. Withaferin A inhibits iNOS expression and nitric oxide production by Akt inactivation and down-regulating LPS-induced activity of NF-kappaB in RAW 264.7 cells. Eur. J. Pharmacol. 599, 11–17.

Pandey, A., Bani, S., Dutt, P., Kumar Satti, N., Avtar Suri, K., et al., 2018. Multifunctional neuroprotective effect of Withanone, a compound from *Withania somnifera* roots in alleviating cognitive dysfunction. Cytokine 102, 211–221.

Panossian, A., Wikman, G., Sarris, J., 2010. Rosenroot (*Rhodiola rosea*): traditional use, chemical composition, pharmacology and clinical efficacy. Phytomedicine 17, 481–493.

Patil, D., Gautam, M., Mishra, S., et al., 2013. Determination of withaferin A and withanolide A in mice plasma using high-performance liquid chromatography-tandem mass spectrometry: application to pharmacokinetics after oral administration of *Withania somnifera* aqueous extract. J. Pharm. Biomed. Anal. 80, 203–212.

Patwardhan, B., 2010. Ayurveda, evidence-base and scientific rigor. J. Ayurveda Integr. Med. 1, 169–170.

Patwardhan, B., Vaidya, A.D., 2009. Ayurveda: scientific research and publications. Curr. Sci. 97, 1117–1121.

Pattanayak, P., Behera, P., Das, D., Panda, S.K., 2010. *Ocimum sanctum* Linn. A reservoir plant for therapeutic applications: an overview. Pharmacogn. Rev. 4, 95–105.

Peisah, C., Chan, D.K., McKay, R., Kurrle, S.E., Reutens, S.G., 2011. Practical guidelines for the acute emergency sedation of the severely agitated older patient. Intern. Med. J. 41, 651–657.

Perluigi, M., Joshi, G., Sultana, R., et al., 2006. In vivo protective effects of ferulic acid ethyl ester against amyloid-beta peptide 1-42-induced oxidative stress. J. Neurosci. Res. 84, 418–426.

Prabhakar, S., Saraf, M.K., Pandhi, P., Anand, A., 2008. *Bacopa monniera* exerts antiamnesic effect on diazepam-induced anterograde amnesia in mice. Psychopharmacology (Berl.) 200, 27–37.

Prabhu, V., Karanth, K.S., Rao, A., 1994. Effects of *Nardostachys jatamansi* on biogenic amines and inhibitory amino acids in the rat brain. Planta Med. 60, 114–117.

Prakash, A., Kumar, A., 2013. Mitoprotective effect of centella asiatica against aluminum-induced neurotoxicity in rats: Possible relevance to its anti-oxidant and anti-apoptosis mechanism. Neuro Sci. 34, 1403–1409.

Prisila, D.C., Ambigapathy, G., Geraldine, P., Akbarsha, M.A., Rajan, K.E., 2011. Bacopa monniera leaf extract up-regulates tryptophan hydroxylase (TPH2) and serotonin transporter (SERT) expression: implications in memory formation, J. Ethnopharmacol. 134, 55–61.

Priyadarsini, K.I., 2013. Chemical and structural features influencing the biological activity of curcumin. Curr. Pharm. Des. 19, 2093–2100.

Qu, Z.-Q., Zhou, Y., Zeng, Y.-S., Li, Y., Chung, P., 2009. Pretreatment with *Rhodiola rosea* extract reduces cognitive impairment induced by intracerebroventricular streptozotocin in rats: implication of anti-oxidative and neuroprotective effects. Biomed. Environ. Sci. 22, 318–326.

Rai, V., Iyer, U., Mani, U.V., 1997. Effect of Tulasi (*Ocimum sanctum*) leaf powder supplementation on blood sugar levels, serum lipids and tissues lipids in diabetic rats. Plant Foods Hum. Nutr. 50, 9–16.

Rai, K.S., Murthy, K.D., Karanth, K.S., Rao, M.S., 2001. Clitoria ternatea (Linn) root extract treatment during growth spurt period enhances learning and memory in rats. Ind. J. Physiol. Pharmacol. 45, 305–313.

Rai, K., Murthy, K.D., Karanth, K.S., Nalini, K., Rao, M.S., Srinivasan, K.K., 2002. Clitoria ternatea root extract enhances acetylcholine content in rat hippocampus. Fitoterapia 73, 685–689.

Raina, P., Santaguida, P., Ismaila, A., Patterson, C., Cowan, D., et al., 2008. Effectiveness of cholinesterase inhibitors and memantine for treating dementia: evidence review for a clinical practice guideline. Ann. Intern. Med. 148, 379–397.

Raja, T., Reddy, R.R.N., Priyadharshini, M.B., 2016. An evaluation of anti-hyperglycemic activity of *Ocimum sanctum* Linn (leaves) in Wister rats. Pharma Innov. J. 5, 1–3.

Rajan, K.E., Singh, H.K., Parkavi, A., Charles, P.D., 2011. Attenuation of 1-(m-chlorophenyl)biguanide induced hippocampus-dependent memory impairment by a standardised extract of *Bacopa monniera* (beseb cdri-08). Neurochem. Res. 36, 2136–2144.

Rajathei, D.M., Preethi, J., Singh, H.K., Rajan, K.E., 2014. Molecular docking of bacosides with tryptophan hydroxylase: a model to understand the bacosides mechanism. Nat. Prod. Bioprospect. 4, 251–255.

Ram, A., Das, M., Ghosh, B., 2003. Curcumin attenuates allergen-induced airway hyperresponsiveness in sensitized guinea pigs. Biol. Pharm. Bull. 26, 1021–1024.

Rao, M.G.K., Rao, M.S., Rao, G.S., 2006. *Centella asiatica* (L.) leaf extract treatment during the growth spurt period enhances hippocampal CA3 neuronal dendritic arborization in rats. Evid. Based Complement Altern. Med. 3, 349–357.

Rauf, K., Subhan, F., Sewell, R.D., 2012. A bacoside containing *Bacopa monnieri* extract reduces both morphine hyperactivity plus the elevated striatal dopamine and turnover. Phytother. Res. 26, 758–763.

Rauf, K., Subhan, F., Al-Othman, A., Khan, I., Zarrelli, A., Shah, M., 2013. Preclinical profile of bacopasides from *Bacopa monnieri* (bm) as an emerging class of therapeutics for management of chronic pains. Curr. Med. Chem. 20, 1028–1037.

Rege, N.N., Thatte, U.M., Dahanukar, S.A., 1999. Adaptogenic properties of six rasayana herbs used in Ayurvedic medicine. Phytother. Res. 13, 275–291.

Ringman, J.M., Frautschy, S.A., Teng, E., Begum, A.N., Bardens, J., et al., 2012. Oral curcumin for Alzheimer's disease: tolerability and efficacy in a 24-week randomized, double blind, placebo-controlled study. Alzheimers Res. Ther. 4, 43.

Rinwa, P., Kumar, A., 2012. Piperine potentiates the protective effects of curcumin against chronic unpredictable stress-induced cognitive impairment and oxidative damage in mice. Brain Res. 1488, 38–50.

Roodenrys, S., Booth, D., Bulzomi, S., Phipps, A., Micallef, C., et al., 2002. Chronic effects of Brahmi (*Bacopa monnieri*) on human memory. Neuropsychopharmacology 27, 279–281.

Russo, A., Borrelli, F., 2005. Bacopa monniera, a reputed nootropic plant: an overview. Phytomedicine. 12, 305–317.

Sahu, R., Dhongade, H., Pandey, A., Sahu, P., Patel, D., et al., 2016. Medicinal Properties of *Nardostachys jatamansi* (a review). Orient. J. Chem. 32, 55–57.

Sandhir, R., Sood, A., 2017. Neuroprotective potential of *Withania somnifera* (Ashwagandha) in neurological conditions. In: Kaul, S., Wadhwa, R. (Eds.), Science of Ashwagandha: Preventive and Therapeutic Potentials. Springer, Cham, pp. 373–387.

Sara, S.J., 1989. Noradrenergic-cholinergic interaction: its possible role in memory dysfunction associated with senile dementia. Arch Gerontol. Geriat. 1, 99–108.

Saraf, M.K., Prabhakar, S., Khanduja, K.L., Anand, A., 2011. *Bacopa monniera* attenuates scopolamine-induced impairment of spatial memory in mice. Evid. Based Complement. Altern. Med. 236186.

Sarup, P., Bala, S., Kamboj, S., 2015. Pharmacology and Phytochemistry of Oleo-Gum Resin of Oleo-Gum Resin of Commiphora wightii (Guggulu). Scientifica (Cairo) 2015, 1–14.

Satow, Y.E., Kumar, P.D., Burke, A., Incardi, J.F., 2008. Exploring the prevalence of Ayurveda use among Asian Indians. J. Altern. Complement. Med. 14, 1249–1253.

Satyavati, G.V., Raina, M.K., Sharma, M., et al., 1976. Medicinal Plants of India. Indian Council of Medical Research, New Delhi.

Saxena, G., Singh, S.P., Pal, R., Singh, S., Pratap, R., Nath, C., 2007. Gugulipid, an extract of *Commiphora whighitii* with lipid lowering properties, has protective effects against streptozotocin-induced memory deficits in mice. Pharmacol. Biochem. Be. 86, 797–805.

Schlebusch, L., Bosch, B.A., Polglase, G., Kleinschmidt, I., Pillay, B.J., Cassimjee, M.H., 2000. A double-blind, placebo-controlled, double-centre study of the effects of an oral multivitamin-mineral combination on stress. S. Afr. Med. J. 90, 1216–1223.

Sehgal, A., Kumar, M., Jain, M., Dhawan, D.K., 2012. Piperine as an adjuvant increases the efficacy of curcumin in mitigating benzo(a)pyrene toxicity. Hum. Exp. Toxicol. 31, 473–482.

Sethiya, N.K., Nahata, A., Mishra, S.H., Dixit, V.K., 2009. An update on Shankhpushpi, a cognition-boosting Ayurvedic medicine. Zhong Xi Yi Jie He Xue Bao 7, 1001–1022.

Shah, R., Gulati, V., Palombo, E.A., Shah, R., Gulati, V., Palombo, E.A., 2012. Pharmacology and phytochemistry of oleo-gum resin of *Commiphora wightii* (Guggulu). Phytother. Res. 26, 1594–1605.

Sharma, R.A., McLelland, H.R., Hill, K.A., Ireson, C.R., Euden, S.A., et al., 2001. Pharmacodynamic and pharmacokinetic study of oral *Curcuma* extract in patients with colorectal cancer. Clin. Cancer Res. 7, 1894–1900.

Sharma, R.A., Gescher, A.J., Steward, W.P., 2005. Curcumin: the story so far. Eur. J. Cancer 41, 1955–1968.

Sharma, K., Bhatnagar, M., Kulkarni, S.K., 2010. Effect of Convolvulus pluricaulis Choisy. and Asparagus racemosus willd on learning and memory in young and old mice: A comparative evaluation. Ind. J Expert Biol. (IJEB) 48, 479–485.

Shi, T.Y., Feng, S.F., Xing, J.H., Wu, Y.M., Li, X.Q., Zhang, N., et al., 2012. Neuroprotective effects of Salidroside and its analogue tyrosol galactoside against focal cerebral ischemia in vivo and H_2O_2-induced neurotoxicity in vitro. Neurotox. Res. 21, 358–367.

Shinomol, G.K., Muralidhara, 2008. Prophylactic neuroprotective property of *Centella asiatica* against 3-nitropropionic acid induced oxidative stress and mitochondrial dysfunctions in brain regions of prepubertal mice. Neurotoxicol 29, 948–957.

Shinomol, G.K., Bharath, M.M., 2011. Exploring the role of 'Brahmi' (*Bocopa monnieri* and *Centella asiatica*) in brain function and therapy. Recent Pat Endocr Metab Immune Drug Discov. 5, 33–49.

Shoba, G., Joy, D., Joseph, T., Majeed, M., Rajendran, R., et al., 1998. Influence of piperine on the pharmacokinetics of curcumin in animals and human volunteers. Planta Med. 64, 353–356.

Siddique, Y.H., Mujtaba, S.F., Faisal, M., Jyoti, S., Naz, F., 2014. The effect of *Bacopa monnieri* leaf extract on dietary supplementation in transgenic *Drosophila* model of Parkinson's disease. Eur. J. Int. Med. 6, 571–580.

Singh, B., Rastogi, R.P., 1969. A reinvestigation of the triterpenes of Centella asiatica. Phytochem. 5, 917–921.

Singh, H., Dhawan, B., 1982. Effect of *Bacopa monniera* linn.(brâhmi) extract on avoidance responses in rat. J. Ethnopharmacol. 5, 205–214.

Singh, H.K., Srimal, R.C., Srivastava, A.K., Garg, N.K., Dhawan, B.N., 1990. Neuropsychopharmacological effects of bacosides A & B. Proc. Fourth Conf. Neurobiol. Learning Memory, Abst. No. 79, Irvine, California.

Singh, H.K., Dhawan, B.N., 1992. Drugs affecting learning and memory. Lect. Neurobiol. 1, 189–207.

Singh, H., Dhawan, B., 2007. Neuropsychopharmacological effects of the ayurvedic nootropic *Bacopa monniera* linn.(brahmi). Ind. J. Pharmacol. 29, 359.

Singh, A.S., Mulder, C., Twisk, J.W., van Mechelen, W., Chinapaw, M.J., 2008. Tracking of childhood overweight into adulthood: a systematic review of the literature. Obes Rev. 9, 474–488.

Singh, N., Hoette, Y., Miller, R., 2010. Tulsi: The Mother Medicine of Nature, second ed. International Institute of Herbal Medicine, Lucknow, pp. 28–47.

Singh, N., Bhalla, M., de Jager, P., Gilca, M., 2011. An overview on ashwagandha: a Rasayana (rejuvenator) of Ayurveda. Afr. J. Tradit. Complement. Altern. Med. 5 (Suppl.), 208–213.

Singh, R., Ramakrishna, R., Bhateria, M., Bhatta, R.S., 2014. *In vitro* evaluation of *Bacopa monniera* extract and individual constituents on human recombinant monoamine oxidase enzymes. Phytother. Res. 28, 1419–1422.

Singh, P., Kalunke, R.M., Giri, A.P., 2015. Towards comprehension of complex chemical evolution and diversification of terpene and phenylpropanoid pathways in *Ocimum* species. RSC Adv. 5, 106886–106904.

Singh, P., Jayaramaiah, R.H., Agawane, S.B., et al., 2016. Potential dual role of eugenol in inhibiting advanced glycation end products in diabetes: proteomic and mechanistic insights. Sci. Rep. 6.

Singh, B., Pandey, S., Yadav, S.K., Verma, R., Singh, S.P., et al., 2017. Role of ethanolic extract of *Bacopa monnieri* against 1-methyl-4-phenyl-1,2,3,6-tetrahydropyridine (MPTP) induced mice model via inhibition of apoptotic pathways of dopaminergic neurons. Brain Res. Bull. 135, 120–128.

Sink, K.M., Holden, K.F., Yaffe, K., 2005. Pharmacological treatment of neuropsychiatric symptoms of dementia: a review of the evidence. JAMA 293, 596–608.

Soumyanath, A., Zhong, Y.-P., Gold, S.A., et al., 2005. *Centella asiatica* accelerates nerve regeneration upon oral administration and contains multiple active fractions increasing neurite elongation in vitro. J. Pharm. Pharmacol. 57, 1221–1229.

Soumyanath, A., Zhang, Y., Henson, E., Wadsworth, T., Bishop, J., Gold, B.G., et al., 2012. *Centella asiatica* extract improves behavioral deficits in a mouse model of Alzheimer's Disease: investigation of a possible mechanism of action. Int. J. Alzheimers Dis. 381974.

Spasov, A.A., Wikman, G.K., Mandrikov, V.B., Mironova, I.A., Neumoin, V.V., 2000. A double-blind, placebo-controlled pilot study of the stimulating and adaptogenic effect of *Rhodiola rosea* SHR-5 extract on the fatigue of students caused by stress during an examination period with a repeated low-dose regimen. Phytomedicine 7, 85–89.

Srivastava, P., Raut, H.N., Puntambekar, H.M., Desai, A.C., 2012. Stability studies of crude plant material of *Bacopa monnieri* and quantitative determination of bacopaside i and bacoside a by hplc. Phytochem. Anal. 23, 502–507.

Suanarunsawat, T., Anantasomboon, G., Piewbang, C., 2016. Anti-diabetic and anti-oxidative activity of fixed oil extracted from *Ocimum sanctum* L. leaves in diabetic rats. Exp. Ther. Med. 11 (3), 832–840.

Sudharani, D., Krishna, K.L., Deval, K., Safia, A.K., 2011. Priya Pharmacological profiles of *Bacopa monnieri*: a review. Int. J. Pharm. 1, 15–23.

Sultana, R., Ravagna, A., Mohmmad-Abdul, H., Calabrese, V., Butterfield, D.A., 2005. Ferulic acid ethyl ester protects neurons against amyloid β-peptide(1–42)-induced oxidative stress and neurotoxicity: relationship to antioxidant activity. J. Neurochem. 92, 749–758.

Sun, G.Y., Li, R., Cui, J., et al., 2016. *Withania somnifera* and its withanolides attenuate oxidative and inflammatory responses and up-regulate antioxidant responses in BV-2 microglial cells. Neuromol. Med. 18, 241–252.

Szapary, P.O., Wolfe, M.L., Bloedon, L.T., Cucchiara, A.J., DerMarderosian, A.H., Cirigliano, M.D., Rader, D.J., 2003. Guggulipid for the treatment of hypercholesterolemia: a randomized controlled trial. JAMA. 290, 765–772.

Tanko, Y., Magaji, G.M., Yerima, M., Magaji, R.A., Mohammed, A., 2008. Anti-nociceptive and anti-inflammatory activities of aqueous leaves extract of *Ocimum gratissimum* (Labiate) in rodents. Afr. J. Tradit. Complement. Altern. Med. 5, 141–146.

Tao, K., Wang, B., Feng, D., Zhang, W., Lu, F., Lai, J., et al., 2016. Salidroside protects against 6-hydroxy-dopamine-induced cytotoxicity by attenuating ER stress. Neurosci. Bull. 32, 61–69.

Tifratene, K., Manera, V., Fabre, R., et al., 2017. Antipsychotic prescribing for Alzheimer's disease and related disorders in specialized settings from 2010 to 2014 in France: a repeated cross-sectional study. Alzheimers Res. Ther. 9, 34.

Tohda, C., Kuboyama, T., Komatsu, K., 2000. Dendrite extension by methanol extract of Ashwagandha (roots of *Withania somnifera*) in SK-N-SH cells. Neuroreport 11, 1981–1985.

Uabundit, N., Wattanathorn, J., Mucimapura, S., Ingkaninan, K., 2010. Cognitive enhancement and neuroprotective effects of *Bacopa monnieri* in Alzheimer's disease model. J. Ethnopharmacol. 127, 26–31.

Ulbricht, C., Basch, E., Szapary, P., Hammerness, P., Axentsev, S., Boon, H., et al., 2005. Guggul for hyperlipidemia: A review by the Natural Standard Research Collaboration. Complement Ther Med. 13, 279–290.

Vaidya, A.D.B., Raut, A., 2006. Evidence-Based Ayurveda–Sorting Fact from Fantasy, Ayurveda and Its Scientific Aspects. AYUSH–CSIR Publication, New Delhi, pp. 1–39.

Vareed, S.K., Bauer, A.K., Nair, K.M., Liu, Y., Jayaprakasam, B., et al., 2014. Blood-brain barrier permeability of bioactive withanamides present in *Withania somnifera* fruit extract. Phytother. Res. 28, 1260–1264.

Ven Murthy, M.R., Ranjekar, P.K., Ramassamy, C., Deshpande, M., 2010. Scientific basis for the use of Indian ayurvedic medicinal plants in the treatment of neurodegenerative disorders: ashwagandha. Cent Nerv Syst Agents Med Chem. 10, 238–246.

Vestergaard, M., Hamada, T., Morita, M., Takagi, M., 2010. Cholesterol, lipids, amyloid beta, and Alzheimer's. Curr Alzheimer Res. 7, 262–270.

Viji, V., Shobha, B., Kavitha, S.K., Ratheesh, M., Kripa, K., Helen, A., 2010a. Betulinic acid isolated from Bacopa monniera (L.) Wettst suppresses lipopolysaccharide stimulated interleukin-6 production through modulation of nuclear factor-κB in peripheral blood mononuclear cells. Int. Immunopharmacol. 10 (8), 843–849.

Viji, V., Kavitha, S.K., Helen, A., 2010b. Bacopa monniera (L.) Wettst inhibits type II collagen-induced arthritis in rats. Phytother. Res. 24, 1377–1383.

Wanakhachornkrai, O., Pongrakhananon, V., Chunhacha, P., et al., 2013. Neuritogenic effect of standardized extract of *Centella asiatica* ECa233 on human neuroblastoma cells. BMC Complement. Altern. Med. 13, 1–7.

Wang, L., Erlandsen, H., Haavik, J., Knappskog, P.M., 2002. Three-dimensional structure of human tryptophan hydroxylase and its implications for the biosynthesis of the neurotransmitters serotonin and melatonin. R.C. Biochem. 41, 12569–12574.

Wang, S., He, H., Chen, L., Zhang, W., Zhang, X., Chen, J., 2015. Protective effects of salidroside in the MPTP/MPP(+)-induced model of Parkinson's disease through ROS-NO-related mitochondrion pathway. Mol. Neurobiol. 51, 718–728.

Wattanathorn, J., Mator, L., Muchimapura, S., Tongun, T., Pasuriwong, O., Piyawatkul, N., et al., 2008. Positive modulation of cognition and mood in the healthy elderly volunteer following the administration of *Centella asiatica*. J. Ethonpharmacol. 116, 325–332.

Wu, L., Xu, H., Cao, L., Li, T., Li, R., Feng, Y., et al., 2017. Salidroside protects against MPP + -induced neuronal injury through DJ-1-Nrf2 antioxidant pathway. Evid. Based Complement. Altern. Med. 2017, 5398542.

Xian, H., Zhao, J., Zheng, Y., et al., 2014. MADP, a salidroside analog, protects hippocampal neurons from glutamate induced apoptosis. Life. Sci. 103, 34–40.
Xing, S., Yang, X., Li, W., et al., 2014. Salidroside stimulates mitochondrial biogenesis and protects against H_2O_2-induced endothelial dysfunction. Oxid. Med. Cell Longev. 2014, 904834.
Yadav, K.D., Reddy, K.R., Kumar, V., 2012. Study of Brāhmī Ghṛta and piracetam in amnesia. Anc Sci Life. 32, 11–15.
Yadav, K.D., Reddy, K.R., Kumar, V., 2014. Beneficial effect of *Brahmi Ghrita* on learning and memory in normal rat. Ayu. 35, 325–329.
Yang, F., Lim, G.P., Begum, A.N., et al., 2005. Curcumin inhibits formation of amyloid β oligomers and fibrils, binds plaques, and reduces amyloid in vivo. J. Biol. Chem. 280, 5892–5901.
Zhang, L., Ding, W., Sun, H., Zhou, Q., Huang, J., Li, X., et al., 2012. Salidroside protects PC12 cells from MPP(+)-induced apoptosis via activation of the PI3K/Akt pathway. Food Chem. Toxicol. 50, 2591–2597.
Zhang, W., He, H., Song, H., et al., 2016. Neuroprotective effects of salidroside in the MPTP mouse model of Parkinson's disease: involvement of the PI3K/Akt/GSK3β pathway. Parkinson's Dis. 2016, 9450137.
Zhao, H.B., Ma, H., Ha, X.Q., Zheng, P., Li, X.Y., Zhang, M., et al., 2014. Salidroside induces rat mesenchymal stem cells to differentiate into dopaminergic neurons. Cell Biol. Int. 38, 462–471.

Further Reading

Borovikova, L.V., Ivanova, S., Zhang, M., Yang, H., Botchkina, G.I., Watkins, L.R., et al., 2000. Vagus nerve stimulation attenuates the systemic inflammatory response to endotoxin. Nature 405, 458–462.
Das, U.N., 2007. Is metabolic syndrome X a disorder of the brain with the initiation of low-grade systemic inflammatory events during the perinatal period? J. Nutr. Biochem. 18, 701–713.

CHAPTER

10

Summary and Perspective for Future Research on Dementia

INTRODUCTION

Dementias are a group of irreversible and progressive syndromes, which afflict elderly persons (65 years or above). Dementia is accompanied by serious decline in cognitive performance. Symptoms of dementia include memory and learning problems, decrease in attention, orientation problems, abnormal executive function, decrease in sensory perception (vision, hearing, touch, smell, and taste), abnormal motor coordination, and increased in agitation and/or aggression (Alexander et al., 2012). Molecular mechanisms contributing to the pathogenesis of dementias are not fully understood. Mounting evidence indicates that dementias are accompanied by mutation, aggregation, and accumulation of misfolded proteins [Aβ in Alzheimer's disease (AD), α-synuclein in Parkinson's disease (PD), huntingtin in Huntington disease (HD), and TDP-43 in amyotrophic lateral sclerosis (ALS)] (Rodrigue et al., 2012; Farooqui, 2017); decrease in cerebral blood flow (CBF); and neurovascular unit dysfunction (Zlokovic, 2011; Girouard and Iadecola, 2006; Bangen et al., 2014). Onset of the abovementioned processes may have multiple consequences. Most recent evidence suggests that the abovementioned processes may result in mitochondrial dysfunction, alterations in calcium homeostasis, induction of oxidative stress, onset of neuroinflammation, abnormal gene transcription, early synaptic disconnection, autophagy, and/or endosomal transport, and late apoptotic cell death (Lashuel et al., 2013; Winner et al., 2011; Umeda et al., 2011; Meraz-Ríos et al., 2010). Dementias are associated with multiple cognitive deficits that include progressive impairment in memory and at least one of the following cognitive disturbances: aphasia, apraxia, agnosia, or a disturbance in executive functioning (Scott and Barrett, 2007). The other cognitive functions that can be affected in

DOI: https://doi.org/10.1016/B978-0-12-816347-4.00010-6

dementias include general intelligence, learning, language, problem solving, orientation, perception, attention and concentration, judgment, and social abilities. The personality is also affected. Agitation or withdrawal, hallucinations, delusions, and insomnia are also common in dementias. Collective evidence suggests that cognitive and neuropsychiatric symptoms are the key clinical features of dementia (Assal and Cummings, 2002). Dementias should not be confused with normal aging, which is characterized by minor memory problems but it does not become severe and does not interfere significantly with a person's social or occupational behavior. Aging is an important risk factor for the dementia. In addition to aging, dementias are also associated with acute and chronic and progressive neurological disorders (stroke, AD, and PD) (Ritchie and Lovestone, 2002; Blennow et al., 2006). Other neurological disorders, which predispose humans to dementia include multiple sclerosis (MS) and AIDS. These pathological conditions produce indirect damage to the brain through immune activated macrophages (Navia and Rostasy, 2005). These brain disorders not only affect memory and learning capacity, but also alter the ability to perform activities of daily living, language, and judgment. (Sosa-Ortiz et al., 2012). Several types of dementia have been reported to occur in the human population including Alzheimer's type of dementia, vascular dementia, mixed dementia, Lewy body dementia, frontotemporal dementia (FTD)/degeneration, and infective dementia. Among these dementias, AD is the major cause of dementia. Among the vascular risk factors for AD type of dementia, hypertension is the most important factor, as it doubles the risk for AD type of dementia in the elderly (Israeli-Korn et al., 2010; Marr and Hafez 2014; Joas et al., 2012). This observation has led to several modifications and expansions of the original vascular hypothesis of AD and AD type of dementia, invoking hypertension-induced microvascular injury in various pathological manifestations of AD type of dementia, from cerebral microhemorrhages (Ungvari et al., 2013) to blood–brain barrier (BBB) disruption and consequent neuroinflammation (Zlokovic, 2008, 2011). The brain is the most metabolically active organ of the body with limited intracellular energy storage. Its function critically depends on CBF. AD and AD type of dementia is accompanied by cerebrovascular and cardiovascular dysfunctions leading to reduction in CBF resulting in cerebral hypoperfusion, hypertension, and impairment in blood pressure (Farooqui, 2017). Long-term hypertension not only damages the blood vessels, but also results in increased expression of hypoxia-sensitive genes (*HIF-1α*, etc.) and molecular cascades during its hypoxic phase. Induction of neuroinflammation occurs due to the synthesis of proinflammatory eicosanoids (PGs, LTs, and TXs) and the release of proinflammatory cytokines (TNF-α, IL-1β, and IL-6). These processes in turn, disrupt the BBB resulting in the induction of the expression of adhesion molecules in endothelial cells and thereby contributing to leukocyte and platelet adhesion and microvascular occlusion (Rosenberg, 2017).

Studies on transgenic mouse models have confirmed the association between hypertension and AD type of dementia by showing that prolonged hypertension increases microvascular amyloid deposition in Tg2576 mice and enhances β-secretase mediated amyloid precursor protein cleavage (Diaz-Ruiz et al., 2009; Faraco et al., 2016). It is also reported that transverse aortic coarctation-mediated hypertension exacerbates Aβ deposition in the mouse brain, promoting cognitive decline (Fig. 10.1) (Carnevale and Lembo, 2011; Carnevale et al., 2012a,b). Furthermore, interaction between hypertension and aging promotes amyloidogenic gene expression in the mouse brain (Csiszar et al., 2013). Importantly, the effects of hypertension on Aβ deposition in the mouse

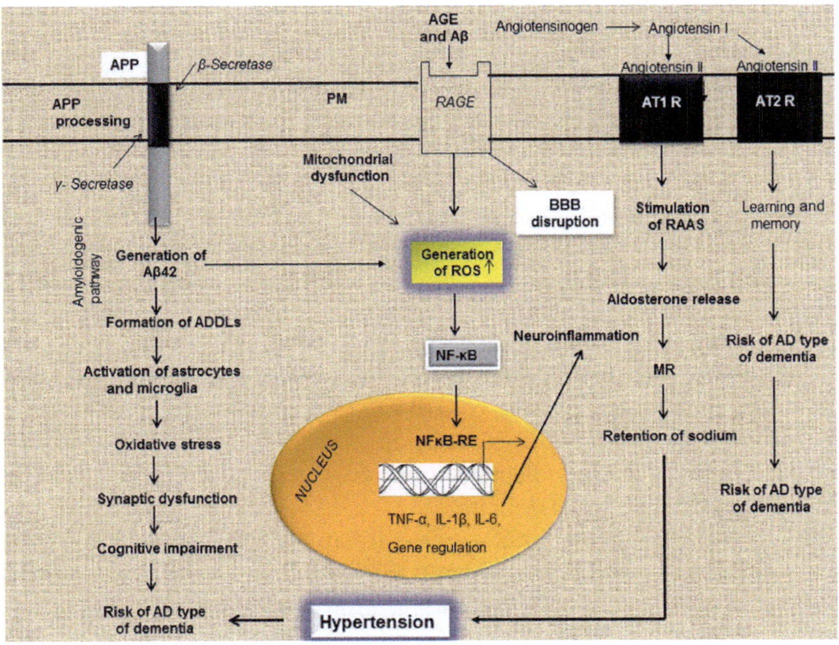

FIGURE 10.1 Relationship between pathogenesis of Alzheimer's disease and hypertension. *Aβ*, beta-amyloid; *AD*, Alzheimer's disease; *ADDLs*, Aβ-derived diffusible ligands; *AGE*, advanced glycation end products; *APP*, amyloid precursor protein; *AT1 receptor*, angiotensin II type-1 receptor; *BBB*, blood–brain barrier; *IκB*, inhibitory subunit of NF-κB; *IL-1β*, interleukin-1β; *MR*, mineralocorticoid receptors; *NF-κB*, nuclear factor-kappaB; *NF-κB-RE*, nuclear factor-kappaB response element; *NFTs*, neurofibrillary tangles; *PM*, plasma membrane; *RAAS*, renin-angiotensin-addosterone; *RAGE*, receptor for advanced glycation end products; *TNF-α*, tumor necrosis factor-α.

brain are manifested within 4 weeks after induction of hypertension (Carnevale and Lembo, 2011; Carnevale et al., 2012a,b), suggesting that early hypertension-induces cerebromicrovascular impairment sufficient to trigger molecular processes contributing to the pathogenesis of AD type of dementia. Collective evidence suggests that cerebrovascular disease in AD type of dementia is caused by cerebral hypoperfusion, which is present in patients several years before the onset of clinical symptoms. The diffusion pattern of cerebral hypoperfusion is stereotyped in AD type of dementia: the first affected area is the precuneus, where it appears 10 years before the onset of AD type of dementia, followed by the cingulate gyrus and the lateral part of the parietal lobe, then the frontal and temporal lobes, and the eventually the cerebrum (Love and Miners, 2016). The main mechanism of cerebral hypoperfusion in AD type of dementia may be nonstructural (Love and Miners, 2016). In vivo and in vitro studies have shown cerebral hypoperfusion increases the production of Aβ and tau hyperphosphorylation, reduces the clearance of Aβ, then aggravates the progress of AD type of dementia (Lee et al., 2011; Qiu et al., 2016; Shang et al., 2016; Zhai et al., 2016). Furthermore, there are studies that support the view that the receptor for advanced glycation end products (RAGEs) activation in the cerebral microvessels is a crucial mechanism by which hypertension promotes AD type of dementia pathologies (Carnevale et al., 2012b). RAGE is known to control the BBB transport of Aβ into the brain (Deane et al., 2003). RAGE activation is associated not only with the development of diabetes, but also with pathogenesis of AD type of dementia in murine models (Deane et al., 2003). However, there is no definitive evidence of whether blood pressure challenge can activate RAGE in brain vessels, triggering and sustaining Aβ precipitation in the brain. It is quite likely that several other mechanisms equally contribute to pathogenesis of AD type of dementia (Nicolakakis et al., 2008; Tong et al., 2012) and that inhibiting one or more of these molecular mechanisms can limit the onset of microvascular-related AD deficits.

Blood pressure abnormalities due to autonomic dysfunction also occur in the early stages of PD type of dementia. These abnormalities often precede the onset of the classic motor symptoms of PD-linked dementia (Asahina et al., 2013). In addition to orthostatic and postprandial hypotension, PD type of dementia patients also experience nocturnal and supine hypertension, suggesting that BP regulation is impaired in PD type of dementia patients (Asahina et al., 2013; Tsukamoto et al., 2013). Since supine hypertension may be a sign of premotor PD (Sharabi and Goldstein, 2011), it has been hypothesized that preexisting hypertension may promote faster progression of nigral dopaminergic neurodegeneration and related motor symptoms. In contrast, epidemiological studies have provided inconclusive results to date (Qiu et al., 2011; Cereda et al., 2012). The mechanisms for linking hypertension and motor stage of PD type of dementia are not known. However, it has been speculated that PD-linked dementia patients with hypertension may show a decrease in resting CBF. This may reduce the delivery of oxygen to ischemia-sensitive brain regions such as the substantia nigra, promoting neurodegeneration of dopaminergic neurons and subsequent motor deficits. In addition, chronic hypertension may promote neurovascular unit dysfunction in multiple brain regions, including the basal ganglia, resulting in dopaminergic neurodegeneration in the substantia nigra and producing a decrease in dopamine transmitters in the striatum (Qiu et al., 2011). Collectively, these studies suggest that hypertension in neurodegenerative diseases may be caused by cerebrovascular diseases through a number of mechanisms, including atherosclerotic changes (Sander et al., 2000), BBB dysfunction (Zumkeller et al., 1991), lipohyalinosis (Munoz, 2003), carotid stenosis (Crouse et al., 1996), and hemorrhage (Zia et al., 2007). Among these factors, BBB dysfunction precedes cognitive decline in AD type of dementia (Bell and Zlokovic, 2009; Zlokovic, 2011). During normal aging, cerebral neuronal function and signaling are protected from blood-borne potentially neurotoxic macromolecules as a consequence of the restricted transport and maintenance of BBB. Alterations in BBB not only result in cerebral extravasation of plasma molecules, but also in increased generation of mediators, which produce neuroinflammation and oxidative stress. These processes result in progressive loss of neuronal function and neuronal apoptosis (Ramirez et al., 2009). These features provide mechanistic insight as to how cerebrovascular disease may heighten the risk for AD type of dementia via a cerebrovascular axis. In addition, onset of hypertension can also contribute to cognitive decline among older adults.

Increase in blood pressure in mid- and late-life is also associated with stroke and white matter hyperintensity (WMH) volume (DeCarli et al., 1999). Studies on association between hypertension and cerebrovascular disease have indicated that blood pressure measurements have significant impact in elderly patients on the development and onset of cerebrovascular diseases. In clinical settings, absolute blood pressure level is used as a therapeutic target to prevent clinical stroke and heart disease, but blood pressure fluctuation over long periods and its impact on cerebrovascular disease are typically not considered.

Human brain requires a constant flow of blood through a network of cerebral arteries and veins not only to deliver oxygen, glucose, and other essential nutrients, but also to remove carbon dioxide, lactic acid, and other metabolic products. CBF in adults represents ~15%–20% of the total cardiac output, while the brain accounts for only 2% of total body weight. Regional blood flow, which is tightly regulated to meet the metabolic demands of the brain, varies significantly between gray and white matter, and among different gray matter regions (Bentourkia et al., 2000). After adolescence, CBF stays relatively stable for a long period, after which it steadily declines. In fact, in middle-aged and elderly adults, aging accounts for a decrease of ~0.45%–0.50% in global CBF per year (Parkes et al., 2004; Zhang et al., 2017). Thus perfusion through both cortical regions of the cerebral cortex decreases with age, especially in the frontal, temporal, and parietal lobes, and subcortical regions (Zhang et al., 2017). Regular aerobic exercise and healthy diet improves cerebrovascular and cardiovascular function not only by increasing CBF and decreasing blood pressure, but also by reducing the risk of dementias (Fig. 10.2). In AD type of dementia, chronic cerebral hypoperfusion and glucose hypometabolism precede decades before the cognitive decline (Farooqui, 2017). These conditions not only upregulate neuroinflammation through the expression of proinflammatory cytokines (TNF-α, IL-1β, IL-6), endothelin-1, and nitric oxide production, but also through promotion of long-term damage involving fatty acids, proteins, DNA, and mitochondria. These processes amplify and perpetuate several feedforward and feedback pathological loops

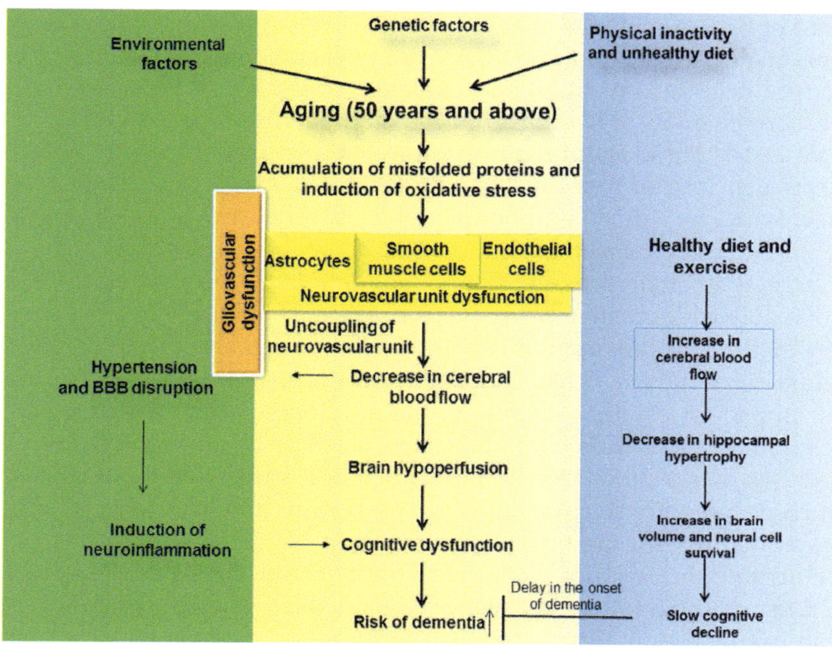

FIGURE 10.2 Effect of environmental factors, genetic factors, and lifestyle on cerebral blood flow and their effects on the pathogenesis of dementia.

(Farooqui, 2017). The latter includes compromised mitochondrial ATP production, β-amyloid generation, endothelial dysfunction, and BBB leading to neurodegeneration (Farooqui, 2017). Other factors, which may contribute to dementia include environmental factors and gene mutation. These factors predispose the carriers for different forms of dementia, and these include mutations in the MAPT gene, which encodes tau. These mutations can lead to FTD, corticobasal degeneration, and other forms of dementia (Jellinger, 2009). In addition, mutations in the PGRN gene encoding progranulin can cause FTD (Yu et al., 2010; van Swieten and Heutink, 2008). Secondary causes of dementia include vascular abnormalities, CNS infections, traumatic brain injury, metabolic derangements, and other reversible/treatable causes, such as type 2 diabetes, stroke, AIDS, or MS (Kabasakalian and Finney, 2009; Ironside and Bell, 2007; Bello and Schultz, 2011). Some dementia-induced changes are reversible (pseudodementia) while others are irreversible. Pseudodementia is caused by depression, malnourishment (vitamin deficiency), dehydration, medications, sleep deprivation, metabolic problems, excessive drinking, smoking, and infections. In contrast, irreversible dementias stem from progressive molecular and cellular changes that lead to irreversible neuronal destruction that occurs in some brain regions of demented patients (Kabasakalian and Finney, 2009; Gallucci Neto et al., 2005; Sosa-Ortiz et al., 2012).

EPIDEMIOLOGY OF DEMENTIA

As stated above, advancing age is the strongest risk factor for dementia and cognitive decline. After 65 years of age, both the prevalence and the incidence of dementia double approximately every 5–6 years until age 90, and ~30% of people aged ≥85 years may be affected by dementia (Winblad et al., 2016; Prince et al., 2013). In addition, ~80% of dementia cases occur in people aged ≥75 years (Winblad et al., 2016; Fratiglioni and Qiu, 2011). Increase in life expectancy and prevalence of dementia are critical issues for public health and health policy development because of the fact that the oldest old people (e.g., octogenarians, nonagenarians, and centenarians) are the fastest growing segment of the population, and that dementia has already posed a huge burden to our aging society (Xu et al., 2018). In 2015 the World Health Organization (WHO) estimated that 47.5 million people were living with dementia worldwide and the cost of managing dementia related illnesses was about US$818 billion in 2015 (Wimo et al., 2017). It is reported that 7.7 million new cases are added to the dementia pool each year.

This number is projected to reach 135.5 million by 2050 (Alzheimer's disease facts and figures, 2010, 2014; World Health Organization, 2015, 2016). East Asia has largest number of people with dementia (9.8 million), followed by Western Europe (7.4 million), South Asia (5.1 million), and North America (4.8 million). The top ten countries with the highest number of people with dementia in 2015 are China (9.5 million), United States (4.2 million), India (4.1 million), Japan (3.1 million), Brazil (1.6 million), Germany (1.6 million), Russia (1.3 million), Indonesia, France, and Italy (1.2 million each) (http://www.alz.co.uk/research/WorldAlzheimerReport2015.pdf). The overall prevalence of dementia among people aged 65 years and above is between 5% and 10%, varying among different global regions (Alzheimer's Disease International, 2015). Although, younger-onset dementia has been reported in humans, dementia is most commonly a disease that affects the seniors (65–85 years).

Maintenance of cognitive function is an important feature of successful aging. It not only promotes the quality of life and functional independence, but also decreases the risk of institutionalization (Fiocco and Yaffe, 2010). As stated above, diet plays an important role in improvement of health status and quality of life in the old age. Consumption of a diet rich in saturated fatty acids has negative effects on age-related cognitive decline and mild cognitive impairment (MCI) (Farooqui, 2015). In contrast, consumption of a diet rich in fish (omega 3-fatty acids) reduces the risk of cognitive decline and dementia (Farooqui, 2009; Solfrizzi et al., 2011). Light-to-moderate alcohol may reduce the risk of dementia and AD. Furthermore, consumption of a diet rich in fruits and vegetables protects against cognitive decline, dementia, and AD. Long-term consumption of a Mediterranean diet, which includes high consumption of olive oil, legumes, unrefined cereals, fruits and vegetables, fish, garlic, and red wine, not only maintains cognitive function and lowers the risk of risk of cardiovascular, cerebrovascular, and metabolic diseases, but also delays the onset of dementia and AD (Farooqui and Farooqui, 2018). According to the US Centers for Disease Control and Prevention, a healthy brain is one that performs all the mental processes and maintains all cognitive functions such as the ability to learn and judge, the use of language, and memory (Centers for Disease Control and Prevention. Healthy Aging, 2017). It has been proposed by The American Heart Association/American Stroke Association that maintenance of cardiovascular and cerebrovascular health during aging is a giant step in cognitive function and health (Gorelick et al., 2017).

Currently, no specific biomarkers have been proven to robustly discriminate vulnerable patients from those with a better prognosis or to discriminate AD type of dementia from poststroke dementia (PSD). Neuroimaging is an important diagnostic tool for the diagnosis of various types of dementia. Thus computerized tomography is used for demonstrating not only type and location of primary lesion, but also for locating atrophy and severe white matter changes. Magnetic resonance imaging (MRI) is a neuroimaging technique, which is used for detecting pathological changes such as small vessel disease (SVD). Advanced multimodal imaging is used for studying the fiber tracking and detecting changes in the neuronal network. Positron emission tomography (PET) is used to study CBF and interactions between vascular and metabolic changes. Current American Academy of Neurology guidelines for dementia diagnosis recommend neuroimaging to identify structural brain diseases that can cause cognitive impairment (de Leon et al., 1989). Thus the diagnosis of dementia is made by a broad range of neuroimaging techniques such as computed tomography, MRI, and PET (Rosen et al., 2002; McKhann et al., 2011; Crutch et al., 2017; Sacks et al., 2017). Advances in radionuclide tracers have allowed for more accurate imaging that reflects the actions of numerous neurotransmitters, energy metabolism utilization, inflammation, and pathological protein accumulation. All of these achievements in molecular brain imaging have broadened our understanding of brain function in neurodegenerative diseases and their related dementias. The implementation of molecular imaging has not only resulted in more accurate diagnosis, but also in assessment of therapeutic outcome. Early diagnosis can be performed by neuroimaging using curcumin analogs that can detect the earliest pathological and metabolic alterations that occur in AD (Yanagisawa et al., 2015; Yang et al., 2017; Chen et al., 2018). It should be mentioned that the diagnosis of dementia is a challenging issue and early and moderate stages of dementia cannot be detected easily. This limits the potential for early intervention in various types of dementia at an early stage. Discovering tests for early diagnosis of dementia is a very important issue and many investigators are trying to discover early specific biomarkers with the ability to identify the predementia before the onset of cognitive decline and neurodegeneration (Mckhann et al., 2011; Cairns et al., 2007).

Another challenge is the differential diagnoses of various types of dementia. This cannot be done using neuroimaging techniques, as there is an overlap in the clinical features across the different dementia types (Woodward et al., 2010; Kalaria and Ballard, 1999). There is currently no single specific biomarker available that can differentiate among AD- and PD-linked dementia, vascular dementia, and PSD. Hence, there is an urgent need for the discovery of specific biomarkers that can distinguish among various types of dementias.

INTERVENTIONS TO DELAY THE ONSET OF DEMENTIA

Many factors are known to modulate cognitive impairment and the onset of dementia (Fig. 10.3). These factors include age, educational period, gender (Li et al., 2016; Kim, 2010; Park and Song, 2016; Park et al., 2015), health life factors such as drinking and smoking (Kim, 2010), depression (Barnes and Yaffe, 2011), social factors such as social activity and occupation, history of disease, and body mass index (Oh and Lee, 2016). These factors can be modified through multimodal interventions, especially at critical time windows over a life course (Qiu, 2012). Multimodal interventions utilize a combination of components—such as physical activity, balanced diet, social engagement, and cognitive training—that target multiple dementia risk factors simultaneously (Fig. 10.4). In addition, the impact of nonpharmacological interventions for dementia using noninvasive brain stimulation should also be included under multimodal interventions. Multimodal interventions not only lead to a reduction in the cardiovascular risk burden (e.g., optimal control of hypertension, diabetes, and high cholesterol), and decrease stress reduction, but also increase cognitive reserve (e.g., education, social engagement, and mental activities). These interventions may delay or prevent the onset of dementia (Stern, 2012). The use of multimodal approaches may be more effective in delaying or retarding complex conditions such as age-related cognitive decline, MCI, and AD type of dementia (Etgen et al., 2011; Gottesman, 2016) by targeting several

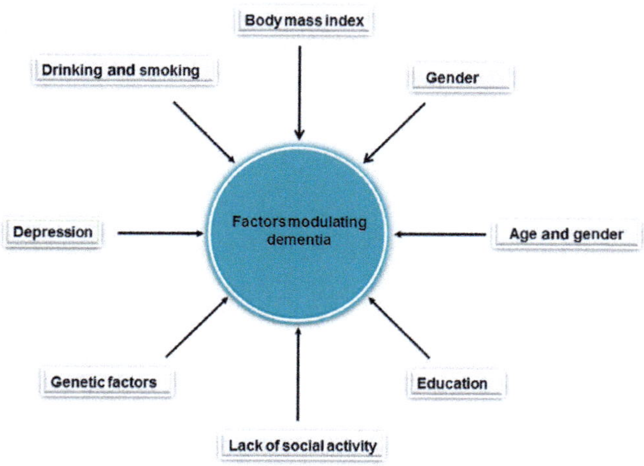

FIGURE 10.3 Factors modulating the onset of dementia.

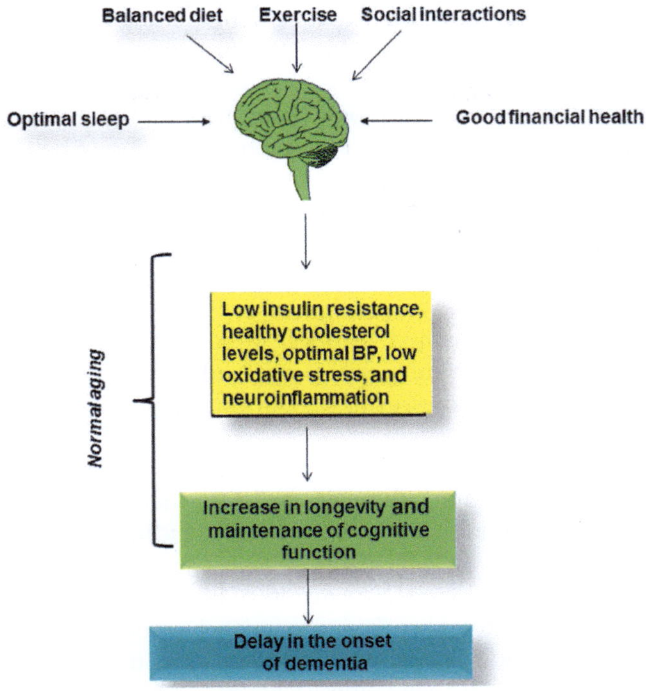

FIGURE 10.4 Effect of multimodal intervention on the onset of dementia.

putative pathways that contribute to AD type of dementia by acting synergistically (Downey, 2017; Rakesh et al., 2017). This approach may also modulate and promote multiple endogenous activities that can help in maintaining cognitive function and reducing dementia risk (Verghese, 2016). Keeping brain functions as normal as possible in old age is an important task for seniors and their clinicians. The effectiveness of a multimodal intervention will not only depend on baseline levels of risk factors, as well as other factors (e.g., dose, adherence, schedule), but also on multiple independent studies testing the same combination of component elements. The use of multimodal lifestyle interventions may also delay the onset of age-related neurological disorders, such as AD, PD, HD, and ALS. The causes of these diseases are not fully understood. However, it is proposed that accumulation of misfolded proteins, induction of oxidative stress, and neuroinflammation are closely associated with the pathogenesis of these pathological conditions. Among these processes, the brain is especially vulnerable to oxidative damage because of its high oxygen consumption rate, high content of lipids, and relative paucity of antioxidant enzymes compared with other organs (Farooqui, 2014).

Pharmacological approaches to treat dementia have failed (Lancet, 2016; Farooqui, 2017, 2018) raising at least two important issues. Current treatments and interventions are unlikely to be effective in individuals with overt disease symptoms. However, treatment can be effective if targeted very early in the pathogenesis of dementia, before the appearance of clinical signs. Hence the need for clear specific biomarkers that may allow timely diagnosis and accurate characterization of the underlying processes resulting in dementia. Second, there is a need for the discovery of accurate risk prediction models and identification of the full range of genetic risk variants, as well as environmental factors, which could be obtained through large epidemiological studies. This will also facilitate the categorization of subgroups within the population most suited for studies of new pharmacological and nonpharmacological interventions.

Herbal medicines have a long history of treating these conditions in Asian countries (China and India). It is believed that many of the medicinal herbs have antiaging properties. Recent studies have shown that some medicinal herbs are effective in intervention or prevention of aging-associated neurological disorders. In Chapter 8, Potential Treatment Strategies for the Treatment of Dementia With Chinese Medicinal Plants, and Chapter 9, Potential Treatment Strategies of Dementia With Ayurvedic Medicines, I have described the use of traditional Chinese medicinal plants and Ayurvedic medicinal plants to treat neurological disorders and their related dementias. Still more studies and multicenter double-blind human trials are required using traditional Chinese medicinal plants and Ayurvedic medicinal plants.

Healthy aging in general is critical for healthy brain aging. According to Vemuri (2018), the aging process produces a number of biological mechanisms at the cellular or tissue level that lead to loss of reserve and function (Fabbri et al., 2015). Prominent age-related changes occur in the brain during midlife, and more so in the sixth to seventh decades. Midlife also represents the time during which (neurodegenerative and cerebrovascular) pathologies are observed in brain autopsies (Nelson et al., 2011). Even in the absence of pathologies, individuals suffer from age-related structural and functional alterations not only in the brain, but also in the cardiovascular and cerebrovascular systems (Jagust, 2013; Fjell et al., 2014), along with alterations in gene expression (Berchtold et al., 2008) starting in midlife. The presence of neurodegenerative and cerebrovascular pathologies contributes to greater structural and functional deteriorations of the brain than in their absence. This accelerated decline in brain health due to neurodegenerative and cerebrovascular pathologies is the primary observed cause of dementia. By age 80, >60% of clinically unimpaired individuals have either onset of neurodegenerative disease-linked dementia or cerebrovascular disease.

The introduction of multimodal lifestyle interventional changes such as physical activity and ideal levels of cardiovascular health (Arenaza-Urquijo et al., 2017; Okonkwo et al., 2014) in midlife may delay the onset of dementias in human subjects. Thus optimal functioning of the brain depends upon CBF, which is regulated by cerebrovascular and cardiovascular systems (Farooqui, 2013). Engagement in regular aerobic exercise enhances systemic arterial endothelial function, reduces large elastic artery stiffness, and decreases the risk of arterial atherosclerosis in middle-aged and older adults (Taddei et al., 2000; Kramer et al., 1999). Regular aerobic exercise and meditation have been reported to increase not only CBF, but also stimulate angiogenesis, synaptogenesis, and neurogenesis (especially in dentate gyrus in the hippocampus) (Rakesh et al., 2017). In addition, animal studies have shown that aerobic exercise initiates the upregulation of several neurotrophins, such as brain-derived neurotrophic factor (BDNF), in the brain (Vaynman et al., 2006; Hillman et al., 2008). This neurotrophin is primarily synthesized there during exercise (Reichardt, 2006). BDNF can also enter the brain via freely diffusing across the BBB (Mousavi and Jasmin, 2006). Furthermore, during exercise, proteins and their metabolic derivatives secreted from peripheral muscles, such as cathepsin B and FNDC5/irisin, also cross the BBB to promote and mediate BDNF expression in the hippocampus and subsequent neurogenesis and memory improvement (Wrann et al., 2013; Moon et al., 2016). Indeed, mice injected with skeletal muscle endurance factors had elevated levels of hippocampal neurogenesis and increased spatial memory (Kobilo et al., 2010). Increased expression of BDNF improves the endothelial cell function, increases the levels of nitric oxide, and reduces the risk of various types of dementias. In AD type of dementia, chronic cerebral hypoperfusion and glucose hypometabolism precede decades before the cognitive decline (Farooqui, 2017). This not only increases oxidative stress and upregulates inflammation through the production of proinflammatory cytokines (TNF-α, IL-1β, IL-6), but also promotes long-term damage to fatty acids, proteins, and DNA through the process of oxidation. These neurochemical events amplify and perpetuate several feedforward and feedback pathological loops along with generation of β-amyloid production leading to BBB disruption and neurodegeneration (Farooqui, 2017). Long-term hypertension, diabetes, unbalanced diet, and obesity have been reported to promote the onset of dementia. Long-term consumption of a healthy diet, aerobic exercise (45–60 min/day), meditation, and optimal sleep may decrease the risk of dementia by reversing cognitive decline (Gorelick et al., 2017). Indeed, health promotion programs targeting the risk reduction of dementia have been developed in several countries (e.g., Australia, the United States, Canada, France, and the United Kingdom) and by professional organizations (e.g., WHO,

Alzheimer's Disease International, and Age UK). At present, there is no disease-modifying treatment for any type of dementia, worsening the burden of dementia in the aging population. It is time to accelerate innovation research in the diagnosis, prevention, treatment, and management of various types of dementias. Successful discovery of dementia-modifying drugs and the development of biomarkers to following the action of drugs will help in providing information on predementia stage of various types of dementias. This information will allow differential diagnosis of various subtypes of dementias. This would allow inclusion of the right patients in the clinical trials, monitoring of the treatment efficacy, and exclusion of patients that have already reached a point-of-no-return and would not see any beneficial effect of a given intervention (Cummings, 2011; Blennow, 2010). Future studies on dementia should include: (1) more information on understanding of the genetics behind neurodegeneration in various types of dementia; (2) more information on the initial symptoms and signs associated with neurodegeneration in dementia; (3) more studies towards the discovery of specific biomarkers for neurodegeneration in various types of dementia; and (4) the early detection and identification of subpopulations of dementia patients is required for the treatment of various types of dementias at an earlier stage.

CONCLUSION

Dementia is a debilitating syndrome of unknown pathology, which is characterized by cognitive and language deficits, impaired visuospatial skills, and a loss of executive function and attention. These deficits interfere with the activities of daily life. The exact symptoms of a person experiencing a dementia episode depend on the disease that is causing dementia. Symptoms of dementia depend on the parts of the brain that are damaged and the complexity of these conditions is such that even within common underlying conditions, the presentation of symptoms differs between individuals. Cerebral hypoperfusion is a major underlying pathophysiological mechanism which contributes to cognitive decline and degenerative processes leading to dementia. Sustained cerebral hypoperfusion is suggested to be the cause of white matter attenuation, a key feature common to various types of dementia associated with cerebral SVD. White matter changes also increase the risk for stroke, dementia, and disability. Dementias have enormous impact not only on individuals, but also on the society medically and economically. The most common form of dementia is AD type of dementia, which accounts for approximately two-thirds of all cases. The remaining cases result from other types of dementia with diverse etiologies. As stated above, dementias are accompanied by the deposition of misfolded proteins (β-amyloid and tau in AD, α-synuclein in PD, huntingtin in HD, and TDP-43 in ALS), decrease in CBF, and neurovascular unit dysfunction. These neurochemical and neuropathological changes are the primary underlying causes of neurodegeneration and cognitive dysfunction which ultimately leads to dementia. Importantly, the onset of these changes in the brain commences long before clinical manifestations and progress. This scenario creates the prospect for developing interventions that aim at early identification and treatment of the preclinical stages of dementias. Emerging evidence also suggests that the integrity of BBB is central to the onset and progression of neurodegeneration, cognitive impairment, and dementia. Multimodal interventions utilize a combination of components—such as physical activity, healthy diet, social engagement, and cognitive training—that target multiple dementia risk factors simultaneously and may prevent or delay the onset of cognitive decline and dementia.

References

Alexander, G.E., Ryan, L., Bowers, D., Foster, T.C., Bizon, J.L., Geldmacher, D.S., et al., 2012. Characterizing cognitive aging in humans with links to animal models. Front. Aging Neurosci. 4, 21.

Alzheimer's Disease International, 2015. World Alzheimer's Report 2015: Global Impact of Dementia.

Alzheimer's disease facts and figures, 2010. Alzheimers Dement. 6, 158–194.

Alzheimer's disease facts and figures, 2014. Alzheimers Dement: J. Alzheimer's Assoc. 10, e47–e92.

Arenaza-Urquijo, E.M., de Flores, R., Gonneaud, J., Wirth, M., Ourry, V., et al., 2017. Distinct effects of late adulthood cognitive and physical activities on gray matter volume. Brain Imaging Behav. 11, 346–356.

Asahina, M., Vichayanrat, E., Low, D.A., Iodice, V., Mathias, C.J., 2013. Autonomic dysfunction in parkinsonian disorders: assessment and pathophysiology. J. Neurol. Neurosurg. Psychiatry 84, 674–680.

Assal, F., Cummings, J.L., 2002. Neuropsychiatric symptoms in the dementias. Curr. Opin. Neurol. 15, 445–450.

Bangen, K.J., Nation, D.A., Clark, L.R., Harmell, A.L., Wierenga, C.E., et al., 2014. Interactive effects of vascular risk burden and advanced age on cerebral blood flow. Front. Aging Neurosci. 6, 159.

Barnes, D.E., Yaffe, K., 2011. The projected effect of risk factor reduction on Alzheimer's disease prevalence. Lancet Neurol. 10, 819–828.

Bell, R.D., Zlokovic, B.V., 2009. Neurovascular mechanisms and blood-brain barrier disorder in Alzheimer's disease. Acta Neuropathol. 118, 103–113.

Bello, V.M.E., Schultz, R.R., 2011. Prevalence of treatable and reversible dementias: a study in a dementia outpatient clinic. Dement. Neuropsychol. 5, 44–47.

Bentourkia, M., Bol, A., Ivanoiu, A., Labar, D., Sibomana, M., Coppens, A., et al., 2000. Comparison of regional cerebral blood flow and glucose metabolism in the normal brain: effect of aging. J. Neurol. Sci. 181, 19–28.

Berchtold, N.C., Cribbs, D.H., Coleman, P.D., Rogers, J., Head, E., Kim, R., et al., 2008. Gene expression changes in the course of normal brain aging are sexually dimorphic. Proc. Natl. Acad. Sci. U.S.A. 105, 15605–15610.

Blennow, K., 2010. Biomarkers in Alzheimer's disease drug development. Nat. Med. 16, 1218–1220.

Blennow, K., de Leon, M.J., Zetterberg, H., 2006. Alzheimer's disease. Lancet (Lond.). 368, 387–403.

Cairns, N.J., Bigio, E.H., Mackenzie, I.R.A., Neumann, M., Lee, V.M.Y., et al., 2007. Neuropathologic diagnostic and nosologic criteria for frontotemporal lobar degeneration: consensus of the Consortium for Frontotemporal Lobar Degeneration. Acta Neuropathol. 114, 5–122.

Carnevale, D., Lembo, G., 2011. 'Alzheimer-like' pathology in a murine model of arterial hypertension. Biochem. Soc. Trans. 39, 939–944.

Carnevale, D., Mascio, G., Ajmone-Cat, M.A., D'Andrea, I., Cifelli, G., Madonna, M., et al., 2012a. Role of neuroinflammation in hypertension-induced brain amyloid pathology. Neurobiol. Aging 33, 205e19–205e29.

Carnevale, D., Mascio, G., D'Andrea, I., Fardella, V., Bell, R.D., Branchi, I., et al., 2012b. Hypertension induces brain beta-amyloid accumulation, cognitive impairment, and memory deterioration through activation of receptor for advanced glycation end products in brain vasculature. Hypertension 60, 188–197.

Centers for Disease Control and Prevention. Healthy Aging. What is a healthy brain? New research explores perceptions of cognitive health among diverse older adults. <https://www.cdc.gov/aging/pdf/perceptions_of_cog_hlth_factsheet.pdf> (accessed 11.04.17.).

Cereda, E., Cassani, E., Barichella, M., Spadafranca, A., Caccialanza, R., et al., 2012. Low cardiometabolic risk in Parkinson's disease is independent of nutritional status, body composition and fat distribution. Clin. Nutr. 31, 699–704.

Chen, M., Du, Z.Y., Zheng, X., Li, D.L., Zhou, R.P., et al., 2018. Use of curcumin in diagnosis, prevention, and treatment of Alzheimer's disease. Neural Regen. Res. 13, 742–752.

Crouse, J.R., Goldbourt, U., Evans, G., Pinsky, J., Sharrett, A.R., et al., 1996. Risk factors and segment-specific carotid arterial enlargement in the Atherosclerosis Risk in Communities (ARIC) cohort. Stroke 27, 69–75.

Crutch, S.J., Schott, J.M., Rabinovici, G.D., et al., 2017. Alzheimer's Association ISTAART Atypical Alzheimer's Disease and Associated Syndromes Professional Interest Area. Consensus classification of posterior cortical atrophy. Alzheimers Dement. 13, 870–884.

Csiszar, A., Tucsek, Z., Toth, P., Sosnowska, D., Gautam, T., et al., 2013. Synergistic effects of hypertension and aging on cognitive function and hippocampal expression of genes involved in beta-amyloid generation and Alzheimer's disease. Am. J. Physiol. Heart Circ. Physiol. 305, H1120–H1130.

Cummings, J.L., 2011. Biomarkers in Alzheimer's disease drug development. Alzheimers Dement. 7, e13–e44.

Deane, R., Du Yan, S., Submamaryan, R.K., LaRue, B., Jovanovic, S., et al., 2003. RAGE mediates amyloid-beta peptide transport across the blood-brain barrier and accumulation in brain. Nat. Med. 9, 907–913.

DeCarli, C., Miller, B.L., Swan, G.E., Reed, T., Wolf, P.A., et al., 1999. Predictors of brain morphology for the men of the NHLBI twin study. Stroke 30, 529–536.

de Leon, M.J., George, A.E., Stylopoulos, L.A., Smith, G., Miller, D.C., 1989. Early marker for Alzheimer's disease: the atrophic hippocampus. Lancet 2, 672–673.

Diaz-Ruiz, C., Wang, J., Ksiezak-Reding, H., Ho, L., Qian, X., et al., 2009. Role of hypertension in aggravating Abeta neuropathology of AD type and tau-mediated motor impairment. Cardiovasc. Psychiatry Neurol. 2009, 107286.

Downey, A., 2017. Preventing dementia and cognitive impairment: a way forward. In: Leshner, A.I., Landis, S., Stroud, C., Downey, A. (Eds.), National Academies of Sciences, Engineering, and Medicine Division; Board on Health Sciences Policy. The National Academies Press.

Etgen, T., Sander, D., Bickel, H., Forstl, H., 2011. Mild cognitive impairment and dementia: the importance of modifiable risk factors. Deutsch. Ärzteblatt Int. 108, 743–750.

Fabbri, E., Zoli, M., Gonzalez-Freire, M., Salive, M.E., Studenski, S.A., et al., 2015. Aging and multimorbidity: new tasks, priorities, and frontiers for integrated gerontological and clinical research. J. Am. Med. Dir. Assoc. 16, 640–647.

Faraco, G., Park, L., Zhou, P., Luo, W., Paul, S.M., et al., 2016. Hypertension enhances Abeta-induced neurovascular dysfunction, promotes beta-secretase activity, and leads to amyloidogenic processing of APP. J. Cereb. Blood Flow Metab. 36, 241–252.

Farooqui, A.A., 2009. Beneficial Effects of Fish Oil on Human Brain. Springer, New York.

Farooqui, A.A., 2013. Metabolic Syndrome: An Important Risk Factor for Stroke, Alzheimer Disease, and Depression. Springer, New York.

Farooqui, A.A., 2014. Inflammation and Oxidative Stress in Neurological Disorders. Springer International Publishing, Switzerland.

Farooqui, A.A., 2015. High Calorie Diet and Human Brain. Springer International Publishing, Switzerland.

Farooqui, A.A., 2017. Neurochemical Aspects of Alzheimer's Disease. Academic Press, An Imprint of Elsevier, San Diego, CA.

Farooqui, A.A., 2018. Ischemic and Traumatic Brain and Spinal Cord Injuries. Academic Press, An Imprint of Elsevier, San Diego, CA.

Farooqui, A.A., Farooqui, T., 2018. Effects of Mediterranean diet components on neurodegenerative diseases. In: Farooqui, T., Farooqui, A.A. (Eds.), Role of the Mediterranean Diet in the Brain and Neurodegenerative Diseases. Academic Press, An Imprint of Elsevier, San Diego, CA.

Fiocco, A.J., Yaffe, K., 2010. Defining successful aging: the importance of including cognitive function over time. Arch. Neurol. 67, 876–880.

Fjell, A.M., McEvoy, L., Holland, D., Dale, A.M., Walhovd, K.B., et al., 2014. What is normal in normal aging? Effects of aging, amyloid and Alzheimer's disease on the cerebral cortex and the hippocampus. Prog. Neurobiol. 117, 20–40.

Fratiglioni, L., Qiu, C., 2011. Prevention of cognitive decline in ageing: dementia as the target, delayed onset as the goal. Lancet Neurol. 10, 778–779.

Gallucci Neto, J., Tamelini, M.G., Forleza, O.V., 2005. Diagnóstico diferencial das demências. Rev. Psiquiatr. Clin. 32, 119–130.

Girouard, H., Iadecola, C., 2006. Neurovascular coupling in the normal brain and in hypertension, stroke, and Alzheimer disease. J. Appl. Physiol. 100, 328–335.

Gorelick, P.B., Furie, K.L., Iadecola, C., Smith, E.E., Waddy, S.P., et al., 2017. Defining optimal brain health in adults: a presidential advisory from the American Heart Association/American Stroke Association. Stroke 48, e284–e303.

Gottesman, R.F., 2016. Preventing dementia and cognitive impairment—vascular factors, diabetes, and obesity: observational evidence. In: Paper Presented at Preventing Dementia and Cognitive Impairment: A Workshop, Washington, DC, October 25.

Hillman, C.H., Erickson, K.I., Kramer, A.F., 2008. Be smart, exercise your heart: exercise effects on brain and cognition. Nat. Rev. Neurosci. 9, 58–65.

Ironside, J.W., Bell, J.E., 2007. Infective dementias. Panminerva Med. 49, 209–225.

Israeli-Korn, S.D., Masarwa, M., Schechtman, E., Abuful, A., Strugatsky, R., et al., 2010. Hypertension increases the probability of Alzheimer's disease and of mild cognitive impairment in an Arab community in northern Israel. Neuroepidemiology 34, 99–105.

Jagust, W., 2013. Vulnerable neural systems and the borderland of brain aging and neurodegeneration. Neuron 77, 219–234.

Jellinger, J.A., 2009. Recent advances in our understanding of neurodegeneration. J. Neural Transmission 116, 1111–1162.

Joas, E., Backman, K., Gustafson, D., Ostling, S., Waern, M., et al., 2012. Blood pressure trajectories from midlife to late life in relation to dementia in women followed for 37 years. Hypertension 59, 796–801.

Kabasakalian, A., Finney, G.R., 2009. Reversible dementias. Int. Rev. Neurobiol. 84, 283–302.

Kalaria, R.N., Ballard, C., 1999. Overlap between pathology of Alzheimer disease and vascular dementia. Alzheimer Dis. Assoc. Disord. 13 (Suppl. 3), S115–S123.

Kim, E.J., 2010. Factors influencing cognitive impairment of the elderly residents. J. East West Nurs. Res. 16, 122–130.

Kobilo, T., Yuan, C., van Praag, H., 2010. Endurance factors improve hippocampal neurogenesis and spatial memory in mice. Learn. Mem. 18, 103–107.

Kramer, A.F., Hahn, S., Cohen, N.J., Banich, M.T., McAuley, E., et al., 1999. Ageing, fitness and neurocognitive function. Nature 400, 418–419.

Lancet, E.T., 2016. Alzheimer's disease: expedition into the unknown. Lancet 388, 2713.

Lashuel, H.A., Overk, C.R., Oueslati, A., Masliah, E., 2013. The many faces of α-synuclein: from structure and toxicity to therapeutic target. Nat. Rev. Neurosci. 14, 38–48.

Lee, J.S., Im, D.S., An, Y.S., Hong, J.M., Gwag, B.J., Joo, I.S., 2011. Chronic cerebral hypoperfusion in a mouse model of Alzheimer's disease: an additional contributing factor of cognitive impairment. Neurosci. Lett. 489, 84–88.

Li, W., Wang, T., Xiao, S., 2016. Type 2 diabetes mellitus might be a risk factor for mild cognitive impairment progressing to Alzheimer's disease. Neuropsychiatr. Dis. Treat. 12, 2489–2495.

Love, S., Miners, J.S., 2016. Cerebrovascular disease in ageing and Alzheimer's disease. Acta Neuropathol. 131, 645–658.

Marr, R.A., Hafez, D.M., 2014. Amyloid-beta and Alzheimer's disease: the role of neprilysin-2 in amyloid-beta clearance. Front. Aging Neurosci. 6, 187.

McKhann, G.M., Knopman, D.S., Chertkow, H., et al., 2011. The diagnosis of dementia due to Alzheimer's disease: recommendations from the National Institute on Aging-Alzheimer's Association workgroups on diagnostic guidelines for Alzheimer's disease. Alzheimers Dement. 7, 263–269.

Meraz-Ríos, M.A., Lira-De León, K.I., Campos-Peña, V., De Anda-Hernández, M.A., Mena-López, R., 2010. Tau oligomers and aggregation in Alzheimer's disease. J. Neurochem. 112, 1353–1367.

Moon, H.Y., Becke, A., Berron, D., Becker, B., Sah, N., Benoni, G., et al., 2016. Running-induced systemic cathepsin B secretion is associated with memory function. Cell Metab. 24, 332–340.

Mousavi, K., Jasmin, B.J., 2006. BDNF is expressed in skeletal muscle satellite cells and inhibits myogenic differentiation. J. Neurosci. 26, 5739–5749.

Munoz, D.G., 2003. Small vessel disease: neuropathology. Int. Psychogeriatr. 15 (Suppl. 1), 67–69.

Navia, B.A., Rostasy, K., 2005. The AIDS dementia complex: clinical and basic neuroscience with implications for novel molecular therapies. Neurotox. Res. 8, 3–24.

Nelson, P.T., Head, E., Schmitt, F.A., Davis, P.R., Neltner, J.H., et al., 2011. Alzheimer's disease is not "brain aging": neuropathological, genetic, and epidemiological human studies. Acta Neuropathol. 121, 571–587.

Nicolakakis, N., Aboulkassim, T., Ongali, B., Lecrux, C., Fernandes, P., et al., 2008. Complete rescue of cerebrovascular function in aged Alzheimer's disease transgenic mice by antioxidants and pioglitazone, a peroxisome proliferator-activated receptor gamma agonist. J. Neurosci. 28, 9287–9296.

Oh, E., Lee, A.Y., 2016. Mild cognitive impairment. J. Korean Neurol. Assoc. 34, 167–175.

Okonkwo, O.C., Schultz, S.A., Oh, J.M., Larson, J., Edwards, D., et al., 2014. Physical activity attenuates age-related biomarker alterations in preclinical AD. Neurology 83, 1753–1760.

Park, H.K., Song, H.J., 2016. Predictors of cognitive function decline of elderly: using living conditions and welfare needs of older Korean persons panel data. Korean J. Health Serv. Manag. 10, 147–158.

Park, J.S., Lee, Y.W., Kim, H.S., 2015. Cognitive impairment risk factors in elders. J. Korean Gerontol. Nurs. 17, 121–130.

Parkes, L.M., Rashid, W., Chard, D.T., Tofts, P.S., 2004. Normal cerebral perfusion measurements using arterial spin labeling: reproducibility, stability, and age and gender effects. Magn. Reson. Med. 51, 736–743.

Prince, M., Bryce, R., Albanese, E., Wimo, A., Ribeiro, W., et al., 2013. The global prevalence of dementia: a systematic review and metaanalysis. Alzheimers Dement. 9, 63. e2–75.e2.

Qiu, C., 2012. Preventing Alzheimer's disease by targeting vascular risk factors: hope and gap. J. Alzheimers Dis. 32, 721–731.

Qiu, C., Hu, G., Kivipelto, M., Laatikainen, T., Antikainen, R., et al., 2011. Association of blood pressure and hypertension with the risk of Parkinson disease: the National FINRISK Study. Hypertension 57, 1094–1100.

Qiu, L., Ng, G., Tan, E.K., Liao, P., Kandiah, N., Zeng, L., 2016. Chronic cerebral hypoperfusion enhances tau hyperphosphorylation and reduces autophagy in Alzheimer's disease mice. Sci. Rep. 6, 23964.

Rakesh, G., Szabo, S.T., Alexopoulos, G.S., Zannas, A.S., 2017. Strategies for dementia prevention: latest evidence and implications. Ther. Adv. Chronic Dis. 8, 121–136.

Ramirez, S.H., Potula, R., Fan, S., Eidem, T., Papugani, A., et al., 2009. Methamphetamine disrupts blood-brain barrier function by induction of oxidative stress in brain endothelial cells. J. Cereb. Blood Flow Metab. 29, 1933–1945.

Reichardt, L.F., 2006. Neurotrophin-regulated signaling pathways. Philos. Trans. R. Soc. B 361, 1545–1564.

Ritchie, K., Lovestone, S., 2002. The dementias. Lancet 360, 1759–1766.

Rodrigue, K.M., Kennedy, K.M., Devous Sr., M.D., Rieck, J.R., Hebrank, A.C., et al., 2012. beta-Amyloid burden in healthy aging: regional distribution and cognitive consequences. Neurology 78, 387–395.

Rosen, H.J., Gorno-Tempini, M.L., Goldman, W.P., Perry, R.J., Schuff, N., et al., 2002. Patterns of brain atrophy in frontotemporal dementia and semantic dementia. Neurology 58, 198–208.

Rosenberg, G.A., 2017. Extracellular matrix inflammation in vascular cognitive impairment and dementia. Clin. Sci. (Lond.) 131, 425–437.

Sacks, C.A., Avorn, J., Kesselheim, A.S., 2017. The failure of solanezumab—how the FDA saved taxpayers billions. N. Engl. J. Med. 376, 1706–1708.

Sander, D., Kukla, C., Klingelhofer, J., Winbeck, K., Conrad, B., 2000. Relationship between circadian blood pressure patterns and progression of early carotid atherosclerosis: a 3-year follow-up study. Circulation 102, 1536–1541.

Scott, K.R., Barrett, A.M., 2007. Dementia syndromes: evaluation and treatment. Expert Rev. Neurother. 7, 407–422.

Shang, J., Yamashita, T., Zhai, Y., Nakano, Y., Morihara, R., Fukui, Y., et al., 2016. Strong impact of chronic cerebral hypoperfusion on neurovascular unit, cerebrovascular remodeling, and neurovascular trophic coupling in Alzheimer's disease model mouse. J. Alzheimers Dis. 52, 113–126.

Sharabi, Y., Goldstein, D.S., 2011. Mechanisms of orthostatic hypotension and supine hypertension in Parkinson disease. J. Neurol. Sci. 310, 123–128.

Solfrizzi, V., Panza, F., Frisardi, V., Seripa, D., Logroscino, G., et al., 2011. Diet and Alzheimer's disease risk factors or prevention: the current evidence. Expert Rev. Neurother. 11, 677–708.

Sosa-Ortiz, A.L., Acosta-Castillo, I., Prince, M.J., 2012. Epidemiology of dementias and Alzheimer's disease. Arch. Med. Res. 43, 600–608.

Stern, Y., 2012. Cognitive reserve in ageing and Alzheimer's disease. Lancet Neurol. 11, 1006–1012.

Taddei, S., Galetta, F., Virdis, A., Ghiadoni, L., Salvetti, G., et al., 2000. Physical activity prevents age-related impairment in nitric oxide availability in elderly athletes. Circulation 101, 2896–2901.

Tong, X.K., Lecrux, C., Rosa-Neto, P., Hamel, E., 2012. Age-dependent rescue by simvastatin of Alzheimer's disease cerebrovascular and memory deficits. J. Neurosci. 32, 4705–4715.

Tsukamoto, T., Kitano, Y., Kuno, S., 2013. Blood pressure fluctuation and hypertension in patients with Parkinson's disease. Brain Behav. 3, 710–714.

Umeda, T., Tomiyama, T., Sakama, N., et al., 2011. Intraneuronal amyloid β oligomers cause cell death via endoplasmic reticulum stress, endosomal/lysosomal leakage, and mitochondrial dysfunction in vivo. J. Neurosci. Res. 89, 1031–1042.

Ungvari, Z., Tucsek, Z., Sosnowska, D., Toth, P., Gautam, T., et al., 2013. Aging-induced dysregulation of dicer1-dependent microRNA expression impairs angiogenic capacity of rat cerebromicrovascular endothelial cells. J. Gerontol. A Biol. Sci. Med. Sci. 68, 877–891.

van Swieten, J.C., Heutink, P., 2008. Mutations in progranulin (GRN) within the spectrum of clinical and pathological phenotypes of frontotemporal dementia. Lancet Neurol. 7, 965–974.

Vaynman, S., Ying, Z., Wu, A., Gomez-Pinilla, F., 2006. Coupling energy metabolism with a mechanism to support brain-derived neurotrophic factor-mediated synaptic plasticity. Neuroscience 139, 1221–1234.

Vemuri, P., 2018. "Exceptional brain aging" without Alzheimer's disease: triggers, accelerators, and the net sum game. Alzheimers Res. Ther. 10, 53.

Verghese, J., 2016. Physical activity and diet. In: Paper Presented at Preventing Dementia and Cognitive Impairment: A Workshop, Washington, DC, October 25.

Wimo, A., Guerchet, M., Ali, G.C., Wu, Y.T., Prina, A.M., et al., 2017. The worldwide costs of dementia 2015 and comparisons with 2010. Alzheimers Dement. 13, 1–7.

Winblad, B., Amouyel, P., Andrieu, S., Ballard, C., Brayne, C., et al., 2016. Defeating Alzheimer's disease and other dementias: a priority for European science and society. Lancet Neurol. 15, 455–532.

Winner, B., Jappelli, R., Maji, S.K., et al., 2011. In vivo demonstration that alpha-synuclein oligomers are toxic. Proc. Natl. Acad. Sci. U.S.A. 108, 4194–4199.

Woodward, M., Jacova, C., Black, S.E., Kertesz, A., Mackenzie, I.R., et al., 2010. Differentiating the frontal variant of Alzheimer's disease. Int. J. Geriatr. Psychiatry 25, 732–810.

World Health Organization (WHO) Fact Sheet. 362. WHO, 2015. Dementia.

World Health Organization (WHO) Fact Sheet. 362. WHO, 2016. Dementia.

Wrann, C.D., White, J.P., Salogiannis, J., Laznik-Bogoslavski, D., Wu, J., Ma, D., et al., 2013. Exercise induces hippocampal BDNF through a PGC-1α/FNDC5 pathway. Cell Metab. 18, 649–659.

Xu, X., Parker, D., Shi, Z., Byles, J., Hall, J., Hickman, L., 2018. Dietary pattern, hypertension and cognitive function in an older population: 10-year longitudinal survey. Front. Public Health 6, 201.

Yanagisawa, D., Taguchi, H., Morikawa, S., Kato, T., Hirao, K., et al., 2015. Novel curcumin derivatives as potent inhibitors of amyloid β aggregation. Biochem. Biophys. Rep. 4, 357–368.

Yang, J., Zhang, X., Yuan, P., Yang, J., Xu, Y., et al., 2017. Oxalate-curcumin-based probe for micro- and macroimaging of reactive oxygen species in Alzheimer's disease. Proc. Natl. Acad. Sci. U.S.A. 114, 12384–12389.

Yu, C.E., Bird, T.D., Bekris, L.M., et al., 2010. The spectrum of mutations in progranulin: a collaborative study screening 545 cases of neurodegeneration. Arch. Neurol 67, 161–170.

Zhai, Y., Yamashita, T., Nakano, Y., Sun, Z., Shang, J., et al., 2016. Chronic cerebral hypoperfusion accelerates Alzheimer's disease pathology with cerebrovascular remodeling in a novel mouse model. J. Alzheimers Dis. 53, 893–905.

Zhang, N., Gordon, M.L., Goldberg, T.E., 2017. Cerebral blood flow measured by arterial spin labeling MRI at resting state in normal aging and Alzheimer's disease. Neurosci. Biobehav. Rev. 72, 168–175.

Zia, E., Hedblad, B., Pessah-Rasmussen, H., Berglund, G., Janzon, L., et al., 2007. Blood pressure in relation to the incidence of cerebral infarction and intracerebral hemorrhage. Hypertensive hemorrhage: debated nomenclature is still relevant. Stroke 38, 2681–2685.

Zlokovic, B.V., 2008. The blood-brain barrier in health and chronic neurodegenerative disorders. Neuron 57, 178–201.

Zlokovic, B.V., 2011. Neurovascular pathways to neurodegeneration in Alzheimer's disease and other disorders. Nat. Rev. Neurosci. 12, 723–738.

Zumkeller, M., Hollerhage, H.G., Reale, E., Dietz, H., 1991. Ultrastructural changes in the blood-brain barrier after nimodipine treatment and induced hypertension. Exp. Neurol. 113, 315–321.

Further Reading

Karantzoulis, S., Galvin, J.E., 2011. Distinguishing Alzheimer's disease from other major forms of dementia. Expert Rev. Neurother. 11, 1579–1591.

World Health Organization, 2012. Dementia: A Public Health Priority. World Health Organization.

Index

Note: Page numbers followed by "*f*" and "*t*" refer to figures and tables, respectively.

CPI Antony Rowe
Eastbourne, UK
June 18, 2019